Atlas of Canine & Feline DENTAL RADIOGRAPHY

Thomas W. Mulligan, DVM • Mary Suzanne Aller, DVM
Charles A. Williams, DVM

Mary Suzanne Aller, Editor

VETERINARY LEARNING SYSTEMS

Yardley, Pennsylvania

Published by Veterinary Learning Systems
Yardley, Pennsylvania

Medical Illustrations by Chris Jouan

Printed in Canada by Printcrafters, Inc.

5 4 3 2

ISBN 1-884254-36-5

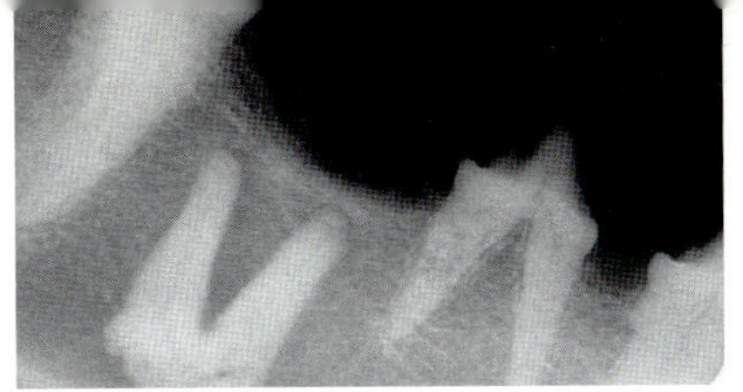

FOREWORD

When I first entered practice in the mid 1940s, small animal dentistry essentially consisted of two procedures: scrape and pull —with the emphasis on pull. Little to no thought was given to the deleterious effects that dental infections could have on the body as a whole. Some vague thoughts suggested that periodontal disease might be associated with bacterial endocarditis and perhaps contribute in some manner to the prevalence of mitral disease in older dogs. Veterinary patients were fed soft diets, however, and the function of their teeth was relegated to a minor role.

As ultrasonic dental scalers became a popular tool for prophylaxis, technologic adaptations of the equipment used in human dentistry to that used in veterinary dentistry could be credited with more specialized delivery of veterinary dental services. Practitioners also began to recognize that the bacteria-laden ultrasonic spray used during dental procedures generated a vast amount of environmental contamination. In an experiment I devised, we found that tagged bacteria in the spray were found as far as four feet from the patient.[1] This prompted AAHA to ban ultrasonic cleaning during surgery and to recommend the use of surgical masks by technicians during ultrasonic scaling.

A greater awareness of small animal dentistry became evident during the 1960s when Dr. Donald Ross, a small animal veterinarian in Texas, became interested in dentistry as a specialty and enrolled in several courses in dental schools. The papers he subsequently presented at veterinary meetings encouraged small animal practitioners to pay heed to the great potential of properly performed dental procedures. Small animal dentistry began to be perceived as a field that was lucrative as well as beneficial to the well-being of animals.

Even into the 1970s, there was nearly a complete absence of literature dealing with small animal dental radiography. During this time, I was prompted to undertake a resident project on the subject. I had returned to school to complete a residency in radiology after 25 years of small animal practice. The result was two papers[2,3] that helped lay the groundwork for dental radiography in veterinary medicine.

During the past decade, veterinary interest in small animal dentistry has literally exploded. In 1988, the American Veterinary Dental College was granted provisional recognition by the AVMA as the certifying body for specialists in animal dentistry, truly a milestone in the progress and sophistication of the veterinary profession. Drs. Don Ross and Peter Emily, a dentist, are among the professionals whose diligent study of this burgeoning specialty has helped to establish it as an accepted veterinary division.

Although a number of excellent small animal dentistry textbooks that touch on the techniques of dental radiography have been published, this particular book fills a much needed void. The *Atlas of Canine and Feline Dental Radiography* is an invaluable addition to the excellent array of reference material available to veterinary professionals. It is indeed a great honor for me to have been asked to write the foreword for a book that deals with a subject so dear to my heart. I congratulate Drs. Aller, Mulligan, and Williams for their persistence and hard work in producing this fine textbook.

—William J. Zontine, DVM, MS
Diplomate, American College of Veterinary Radiology

REFERENCES

1. Zontine WJ, Sims S, Donovan ML: Bacterial environmental contamination associated with ultrasonic dental procedures in dogs. *JAAHA* 5(3):150–154, 1969.
2. Zontine WJ: Dental radiographic technique and interpretation. *Vet Clin North Am* 4(4):741–762, 1974.
3. Zontine WJ: Canine dental radiology: Radiographic technique, development, and anatomy of the teeth. *J Am Vet Rad Soc* 16(3):75–82, 1975.

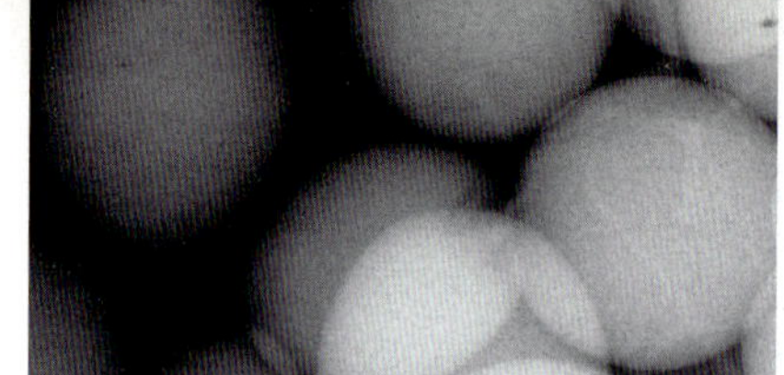

PREFACE

At the time this book was conceived, as now, a textbook or manual on dental radiography or its interpretation did not exist. Specifically, a vast collection of radiographs to demonstrate a wide spectrum of oral and dental pathology has been lacking. All three authors of this book realized that the intraoral radiographs we had collected over the years represented a wealth of knowledge that was so unique it seemed mandatory for us to share it with our colleagues.

As background, the use of intraoral radiography for small animal dentistry was almost unheard of prior to the 1970s, except for a few rare case reports or research projects. During the 1970s and 1980s, a small group of pioneers in the field of veterinary dentistry recognized the value and unique capabilities afforded by small intraoral films and dental x-ray machines.

Like small animal dentistry as a whole, the use of intraoral radiography burgeoned in the late 1980s and has become increasingly advanced as well as popular in the 1990s. Thus, we offer this atlas of dental radiography to provide our fellow veterinary practitioners with an easy-to-use reference manual. We hope that the book will become a template for recognizing, identifying, and diagnosing a variety of oral pathologic conditions and a valuable teaching tool for any student of small animal health care.

Although all three of us authored this project, we each contributed our specific expertise. Chuck Williams can be credited with the initial concept of the book and many of its radiographs. Tom Mulligan provided his share of the radiographs, scanned much of the film, and spent hours on his computer generating many of the overlays that depict important radiographic features. Suzy Aller photographed most of the new images, meticulously taking and retaking each radiograph until she accomplished the desired result. In addition to being an author, Suzy is editor of this book and in that capacity, she addressed the task of reorganizing, rewriting, and finalizing each chapter. Our vision that the quality of the publication be paramount was due, in large part, to her tenacious dedication. Finally, we extend much appreciation to Jean Hawkins, DVM, MS, who served as reviewer and contributed her vast knowledge of and proficiency in the field of dental radiography.

Our own interests drove us, but the completion of the *Atlas of Canine and Feline Dental Radiography* would not have been possible without the support of our employees, friends, and family. We would first like to recognize those people who initially piqued our interest in the use of intraoral radiographs, including Don Ross, Gary Beard, and Peter Emily.

We extend our sincere appreciation to Veterinary Learning Systems and its staff for helping us to realize our dream of creating this resource for the veterinary profession. We would also like to recognize the staffs at our hospitals for all their support, especially in the development of many of the intraoral radiographs. Finally, we thank our family pets—Annie, Smudge, Killian, Guin, Bandit, Joshua, Marshmallow, Anna, and many others—for providing some of the best examples for this book.

—*Charles A. Williams, DVM*
Diplomate, American Veterinary Dental College

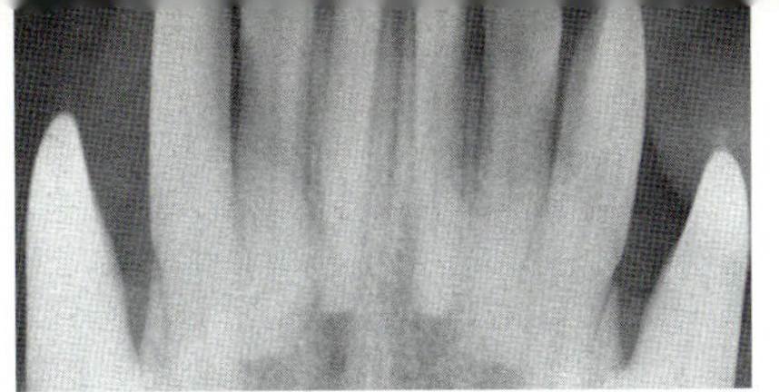

CONTENTS

Chapter One
BASIC EQUIPMENT NEEDS 7

Chapter Two
PROJECTION GEOMETRY 15

Chapter Three
EXTRAORAL IMAGING TECHNIQUES 23

Chapter Four
INTRAORAL IMAGING TECHNIQUES 27

Chapter Five
TECHNICAL ERRORS AND TROUBLESHOOTING 45

Chapter Six
PRINCIPLES OF IMAGE INTERPRETATION 65

Chapter Seven
NORMAL RADIOGRAPHIC ANATOMY 68

Chapter Eight
DEVELOPMENTAL PROBLEMS AND DENTAL ANOMALIES 91

Chapter Nine
INTERPRETATION OF PERIODONTAL DISEASE 104

Chapter Ten
INTERPRETATION OF ENDODONTIC DISEASE 124

Chapter Eleven
ACQUIRED DEFECTS: CARIES AND REGRESSIVE CHANGES 153

Chapter Twelve
TRAUMA 170

Chapter Thirteen
OSSEOUS LESIONS 184

Chapter Fourteen
DENTAL RADIOGRAPHY: A TEST OF YOUR KNOWLEDGE 204

GLOSSARY 231

INDEX 239

Chapter One

BASIC EQUIPMENT NEEDS

Most small animal clinics and hospitals are equipped with a stationary x-ray machine that operates on 300 milliamperes (mA) with a 150-kilovolt peak (kVp). Although this machine is practical and reliable for general-practice radiography, it is not recommended for dental radiography for several reasons. First, fine adjustments in the focal film distance, angulation, and collimation are required for dental radiography. These adjustments are difficult to make because of the manner in which the tubehead is mounted on standard equipment. Second, dental surgery is rarely performed in the same room that contains standard x-ray equipment. If radiography is required during dental surgery, patients must be moved from the treatment room—a process that is cumbersome and time-consuming and may require disconnecting anesthetic equipment.

Both problems can be resolved by installing a dental x-ray machine on the floor or wall of the dental treatment room. This machine is compact and features an adjustable arm that allows the user to maneuver the tubehead easily (Figure 1-1). In addition, the angle of the horizontal and vertical x-ray beams can be quickly modified (Figure 1-2A), and proper focal film distance and collimation of the beam are possible using the cylindrical device that fits over the x-ray port (Figure 1-2B). Equipping your practice with a dental x-ray machine not only improves the quality of dental services but can facilitate taking radiographs during dental surgery.

DIFFERENCES BETWEEN STANDARD AND DENTAL EQUIPMENT

Although dental x-ray machines essentially feature the same circuitry design as standard machines, some differences need to be noted. On standard machines, the operator can set the kilovolts peak, milliamperes, and exposure time in almost any combination. In comparison, the settings for most dental x-ray machines are very limited, with the milliamperes and kilovolts peak often being preset. On dental x-ray machines, the timer is usually the only control that is adjustable (Figure 1-3). The time is displayed in impulses or seconds; on some models, it may be preprogrammed according to the type of tooth being radiographed.

Many dental x-ray machines measure time as impulses because of the limitations of using alternating current (AC). The voltage of alternating current undergoes a cyclical change from a positive phase to zero and then to a negative phase 60 times per second. The polarity of the x-ray tube alternates with the same frequency and produces x-rays only during the positive half of the cycle. Thus, when an x-ray tube is powered with 60-cycle alternating current, 60 pulses of x-rays are generated each second, with each pulse having a duration of 1/120 of a second. If a timer device is calibrated in impulses, one impulse is defined as 1/60 of a second. For example:

$$30 \text{ impulses} = (30)(1/60) = 0.5 \text{ seconds}$$

When the alternating high voltage is applied directly across the x-ray tube, the x-ray tube itself functions as a rectifier of the current and eliminates the negative phase of the current. Therefore, most dental x-ray units function as self-rectified or half-wave rectified tubes by using only the positive half of the AC cycle to generate x-rays.[1,2]

FILM

The film used for dental radiography is a type of photographic film composed of an emulsion on a flexible polyester base material that is semitransparent. The emulsion contains silver-halide crystals in a gelatin matrix that is coated with a protective layer to withstand handling during processing. The film speed is an inherent property related to the size of the silver-halide grain in the emulsion and determines the amount of exposure required to produce an image. The larger the grain size, the faster the film speed. Although faster film requires less exposure to produce a useful radiographic image, the image may appear grainy and lack sharpness.[1,2]

Standard-size film, which is available for use in cassettes

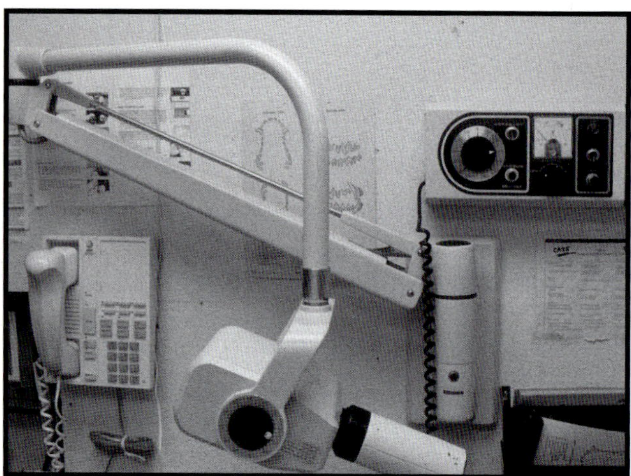

FIGURE 1-1

A wall-mounted dental x-ray machine.

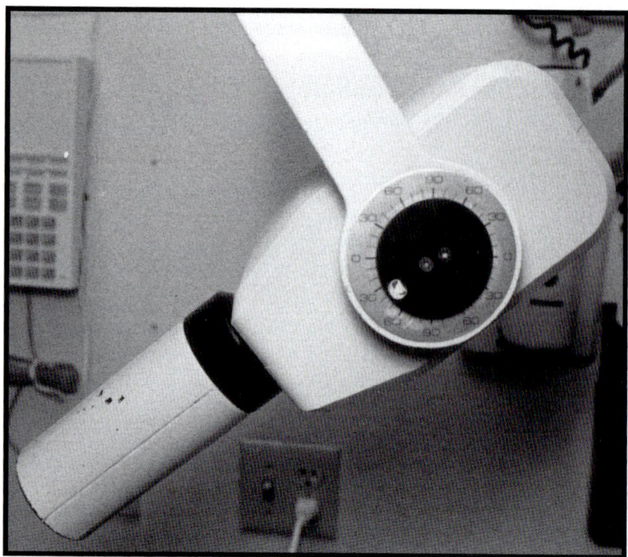

FIGURE 1-2A

FIGURE 1-2B

Figure 1-2A *The tubehead of a dental x-ray machine can be easily adjusted to any position. The dial on the side of the tubehead displays measurements for vertical angulation.* **Figure 1-2B** *In anesthetized animals, any required beam angulation is possible by adjusting the tubehead on the dental x-ray machine.*

used with or without intensifying screens, can be used for taking extraoral radiographs. The faster films and those that have intensifying screens decrease the radiation dose required to expose the film properly; however, radiographic details are usually sacrificed as a result.[3] Many screen-film systems do provide fairly good detail, especially those with slower-speed films designated for use in fine screen-film systems that radiograph the extremities, other orthopedic structures, or mammography systems.[2,4]

Intraoral film that is manufactured for human dentistry does not use intensifying screens. Various small sizes can be easily adapted to small animal patients (Figure 1-4), and the flexibility of the film packet allows intraoral positioning in most small animal patients. In small animal dentistry, sizes 2 and 4 are the most useful. Size 2 film works well for taking radiographs of specific teeth in all breeds of dogs and fits nicely in the restricted oral space of toy canine breeds and all domestic cats. Size 4 film is used for taking oral radiographs in large-breed dogs. Although these films can be exposed using standard equipment, we recommend a dental x-ray machine.

Each film packet has a single or double sheet of film and is wrapped to protect the packet from light and moisture. The film inside the packet is wrapped in a black paper envelope. A lead foil that protects the film from secondary radiation covers the side of the film to be positioned away from the x-ray tube (Figure 1-5). A dimple can be found on the corner of the film itself. The convex surface of the dimple always faces toward the x-ray tube during exposure and serves as an orientation marker on the radiograph for determining the left or right side of a patient (Figure 1-6).

The speed of dental x-ray film is designated by a letter. The fastest dental film currently available has a rating of E. The

most commonly used intraoral films are Kodak Ultra-speed film group D (Figure 1-7) and Kodak Ektaspeed film group E (Eastman Kodak—Rochester, New York). Although Ektaspeed film is faster and reduces patient exposure to radiation by about one half the time required for Ultra-speed film, some practitioners believe that the images produced by Ultra-speed film may be easier to read. Many studies, however, have not found a substantial difference in the quality of images produced by D- and E-speed films.[5-7] Ultra-speed film does produce useful images from a wider range of exposures and there-

FIGURE 1-3

Typical control panel on a dental x-ray machine. Note that the timer is the only adjustable control.

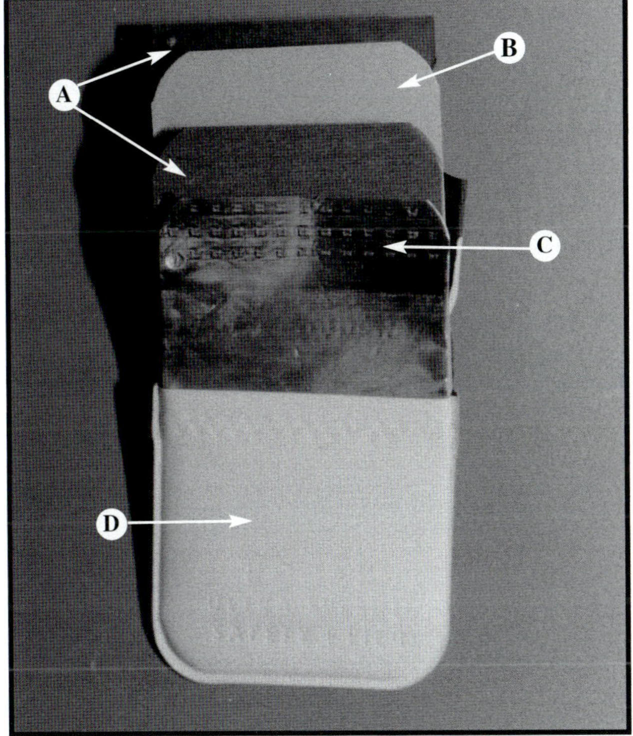

FIGURE 1-5

The inside of an intraoral film packet: (A) paper folder, (B) film, (C) lead foil, and (D) outside cover.

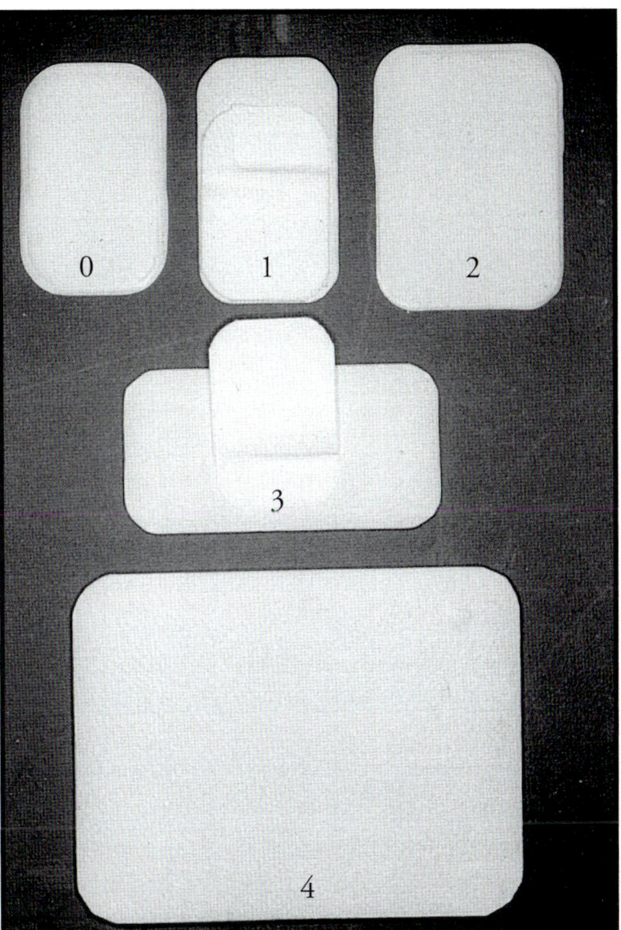

FIGURE 1-4

Intraoral film packets are available in sizes 0, 1, 2, 3, and 4.

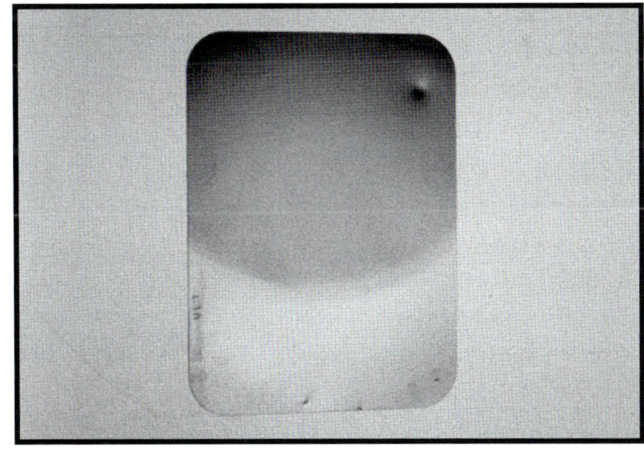

FIGURE 1-6

A dimple on the corner of each film is convex on one side and concave on the other side. The dimple provides spatial orientation for reading the radiograph.

fore is more forgiving if the exposure technique is inconsistent.[1,4,8]

PROCESSING EQUIPMENT

Standard-size x-ray film can be developed either manually or using large automatic processors in a darkroom. Smaller intraoral film can be developed in standard tanks by using the appropriate clips to hang the film (Figures 1-8A and 8B). Dental film can be developed in large automatic processors if it is taped to a larger film that serves as a leader by dragging the dental film through the processor (Figure 1-9A). Care is necessary to avoid

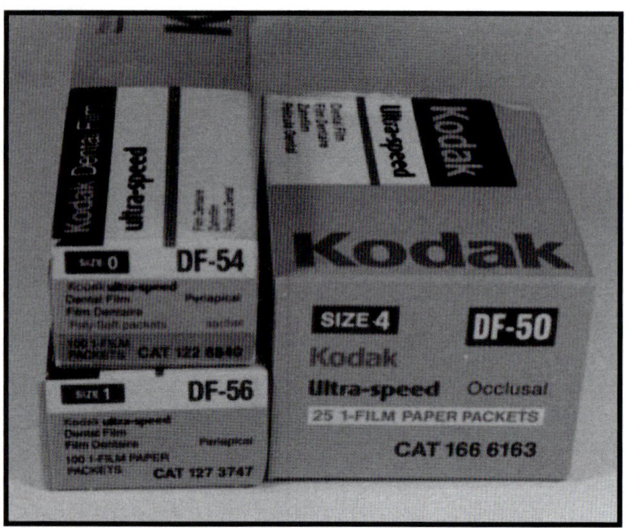

FIGURE 1-7

Kodak Ultra-speed film group D (Eastman Kodak) is one of the more commonly used films for dental radiography.

taping any area of the dental film that contains the radiographic image. We recommend using a leader tape that is designed for automatic processors (e.g., Pakor Tape®—Pakor Inc., Maple Grove, Minnesota). An alternative to the taping method is to use a carrier transport system (Figure 1-9B).

Automatic processors that are designed to handle dental films can deliver processed film in a fully fixed, dried condition within five minutes without requiring the use of a leader film (Figure 1-10). Ease of operation and the convenience of producing finished films with minimal effort are advantageous. These processors, however, can be cost prohibitive for the average veterinary practice, especially if their limitations are considered. Most models cannot process film larger than 8 inches by 10 inches. A plumbing connection may also be necessary. In addition, these processors require either a darkroom or a glovebox attachment that is used for chairside development.

Efficient chairside processing is possible in a well-lighted room by using rapid-processing chemicals in a small glovebox device called a chairside darkroom or chairside developer. D-speed film can be processed in one minute without staff leaving the surgical area. Film is placed in small containers of processing chemicals that can be visualized by the staff member responsible for

developing the film (Figure 1-11). The special transparent top also prevents environmental light from fogging the film.

Another convenient chairside-developing device (Figure 1-12) is the Veterinary Dental Film System (Hanskin Technical Laboratory, Ltd.—Osaka, Japan; distributed by Hawaii Mega-Cor, Inc., Aiea, Hawaii). This system has its own line of solutions and film. The film packet is specially constructed with a channel that introduces a combination developer–fixer solution into the packet without opening it. The monobath solution is injected in calibrated doses using a volumetric injector. After the filled packet has been agitated for 30 seconds, the film is removed from the packet, rinsed, and ready for viewing. A special chemical that cures the radiograph before final storage is provided.

The simplest, least-expensive, and most popular way to develop film is by hand. Hand development of film is traditionally done in four steps: developing, rinsing, fixing, and washing. Five small containers should be placed on a convenient shelf in the darkroom. These containers must have a smooth surface to minimize scratching the film as it is being swirled in chemical solutions. Containers with airtight lids, such as those designed for food storage, can protect the solutions (Figure 1-13).

Several manufacturers market rapid-processing chemicals that produce diagnostic film in about 90 seconds. These chemicals, however, are relatively expensive and degenerate quickly. Replenishment of rapid-processing chemicals is determined by the caseload.

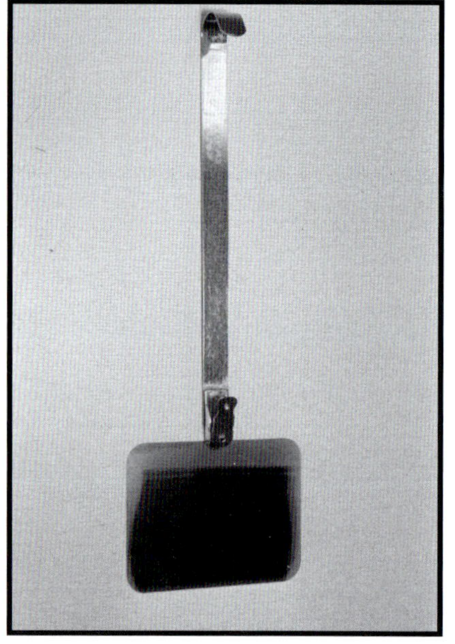

FIGURE 1-8A

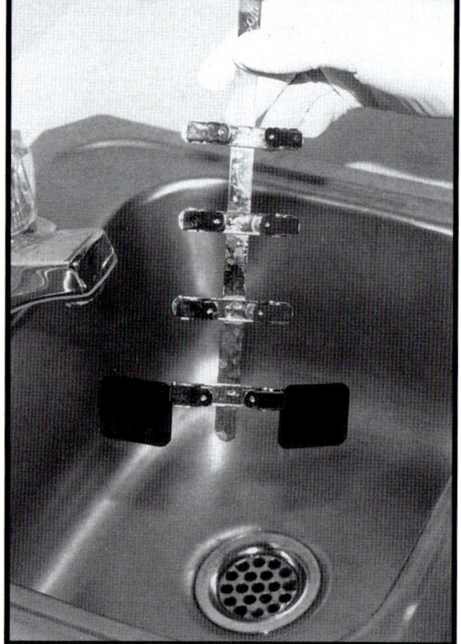

FIGURE 1-8B

Hanging clips can be used to dry (**Figure 1-8A**) *one film or* (**Figure 1-8B**) *multiple films during hand development.*

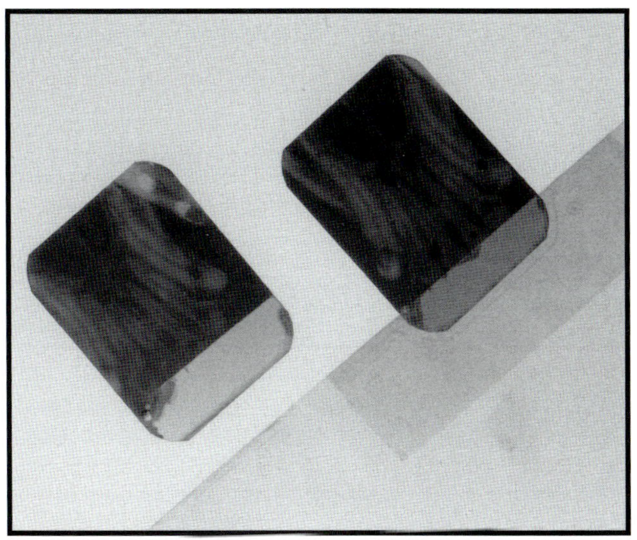

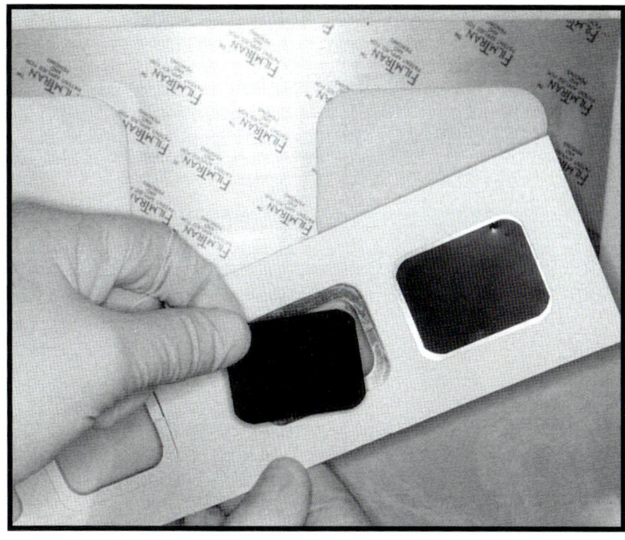

Figure 1-9A *Dental film can be taped onto a leader film for processing in a standard-size automatic processor. The taped area does not develop.* **Figure 1-9B** *A carrier transport system also can be used to process dental film in a standard automatic processor (FilmTran™—Bisco International Inc., Hillside, Illinois).*

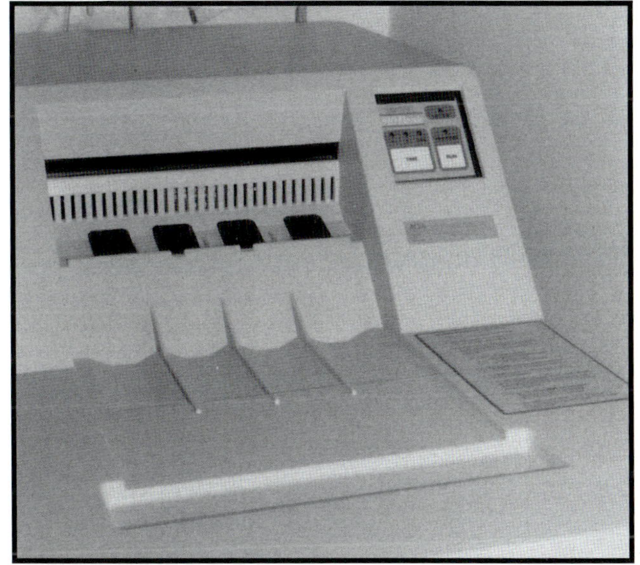

This automatic processor is designed specifically for intraoral film.

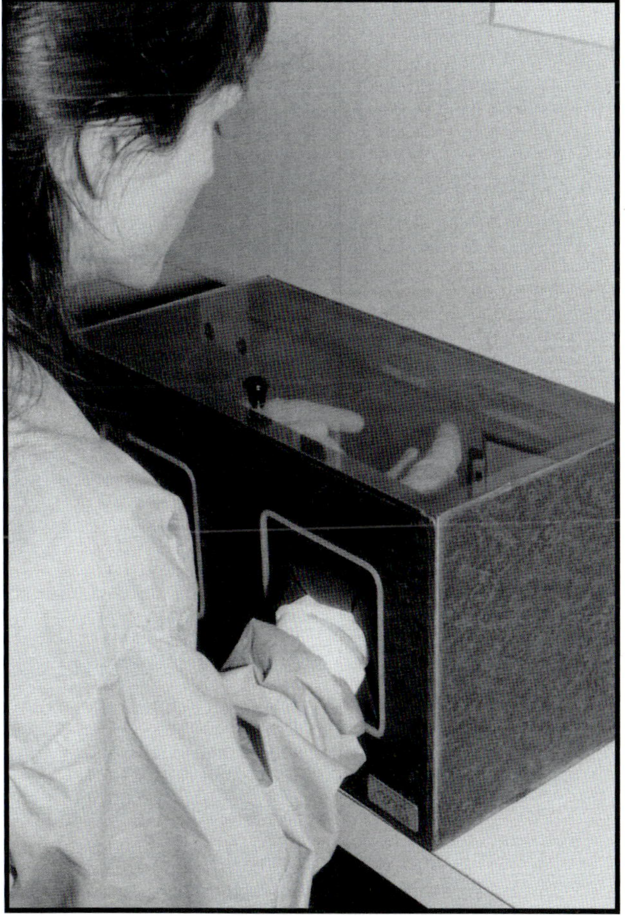

Chairside developers are small glovebox devices that allow the user to visualize film through a transparent top (Chairside Developer—Rinn Corp., Elgin, Illinois).

Although hand-developing techniques can be quick and efficient, the processed film may discolor with time if certain precautions are not taken. Dental film requires thorough rinsing after it is immersed in each chemical solution. Otherwise, the surface of the film retains a heavy scum or residue that can stain the radiograph. After developing but before fixing the film, it should be thoroughly swirled in two separate water rinses. The radiograph can be viewed immediately after processing but should then be placed in a conventional fixing

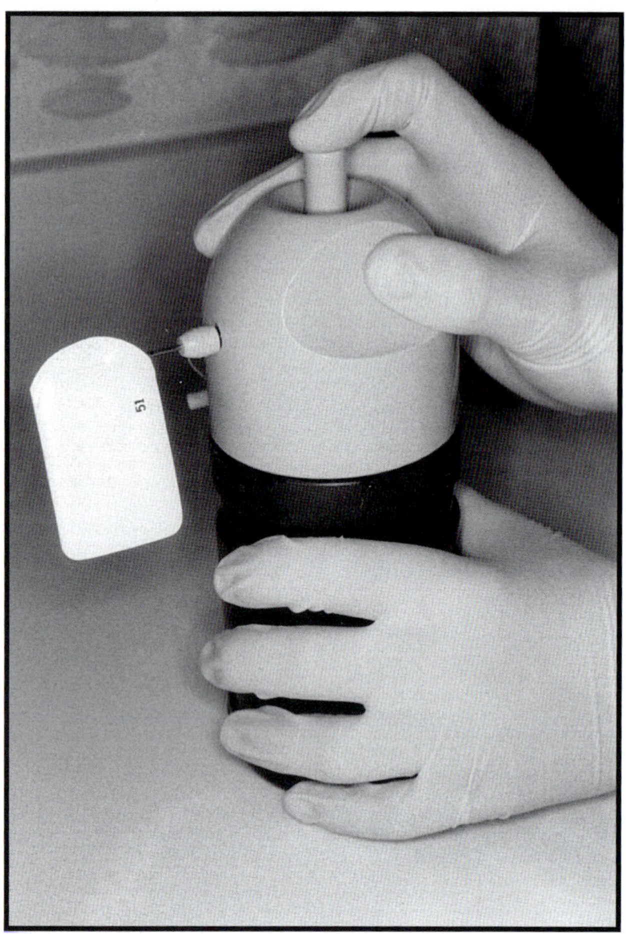

FIGURE 1-12
The Veterinary Dental Film System (Hanskin Technical Laboratory, Ltd.) is designed for chairside processing. Chemicals are injected directly into the film packet.

FIGURE 1-13
Containers with airtight lids should be used for hand-developing film.

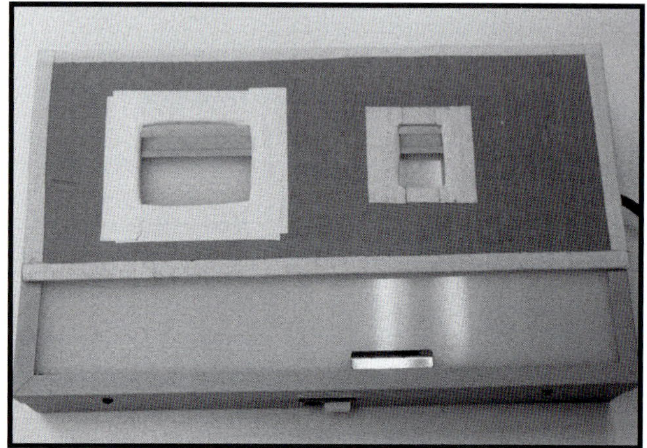

FIGURE 1-14
Example of a dental radiographic viewer with a viewing frame.

solution for 10 minutes and washed for 20 minutes before being dried. If this technique is observed, the quality of archived radiographs can be preserved.[2,8]

RADIOGRAPHIC VIEWERS

The traditional viewers used for dental radiographs are smaller than the standard viewers found in most veterinary practices. Although viewers designed for dental radiographs are convenient, they are not a necessity. Any radiographic viewer can be used.

Because intraoral film is small, considerable glare can be emitted from the edges of the radiograph being viewed—a tendency that can make it difficult to read. Glare may be most apparent when working with size 1 or 2 film. A simple solution is to make a cover from cardboard or another thin, light-blocking material. Holes that correspond to the size of the radiograph being viewed should be cut. When the cover is used as a viewing frame, all unwanted peripheral light is eliminated (Figure 1-14).

LABELING AND STORAGE

Cassette films are labeled using conventional marking systems. Small intraoral radiographs can be identified most easily by inserting the labeling information on plastic or cardboard mounts, plastic sleeves, or a carrier transport system (Figures 1-9B and 1-15). Placing radiopaque adhesive letters and numbers or using a marking pen to write directly on a radiograph can mar it or inadvertently obscure the image to be viewed.

To maintain the quality of archived radiographs, they should be placed into protective envelopes and stored in either a special storage cabinet for dental radiographs or into the patient's file folder. Of greater importance, however, is that the radiographs be thoroughly fixed and completely dried before they are stored; otherwise, they can deteriorate rapidly.

POSITIONING DEVICES

When taking dental radiographs (especially extraoral radi-

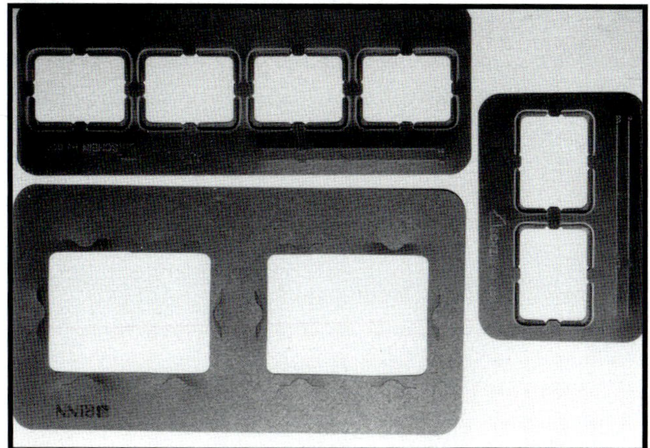

Plastic film mounts are available in various sizes and combinations.

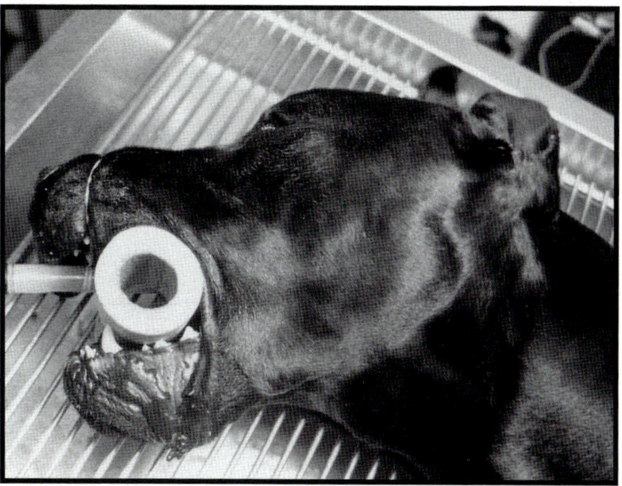

ographs), most practitioners use a device that keeps the patient's mouth open so that the upper and lower arcades are not superimposed on each other. The most commonly used device is a bite block. Practitioners can be very creative when selecting material to perform the function of a bite block. The only requirement is that the material be radiolucent. Foam rolls (which are sold as pipe insulators in building supply stores) work well in large-breed dogs (Figure 1-16A), whereas foam rubber hair rollers or cosmetic foam sponges are more appropriate for small-breed dogs and all domestic cats. Recycled syringe cases also are ideal because they can be cut to any length to accommodate even a very small mouth, are readily available, and are radiolucent (Figure 1-16B). In some instances, film-holding devices can serve as a bite block (Figure 1-16C). In addition to these makeshift devices, radiolucent veterinary bite blocks (Figure 1-17) are available on the market and traditional metal mouth gags (Figure 1-18) are still commonly used, although the radiopaque metal interferes with the image. Regardless of which device is chosen, it should be used carefully to avoid trauma to oral tissue, the temporomandibular joint, and the muscles of mastication.

RADIATION SAFETY EQUIPMENT

Regardless of the type of x-ray machine used, every attempt should be made to limit the amount of radiation exposure. Radiation safety begins with a good-quality x-ray machine that has been accurately collimated and is equipped with an accurate exposure timer. Most U.S. states now require inspection and certification of radiographic equipment at regular intervals to ensure that it is functioning properly.

Veterinary staff need to be protected against three sources of radiation: the primary beam, secondary radiation emitted

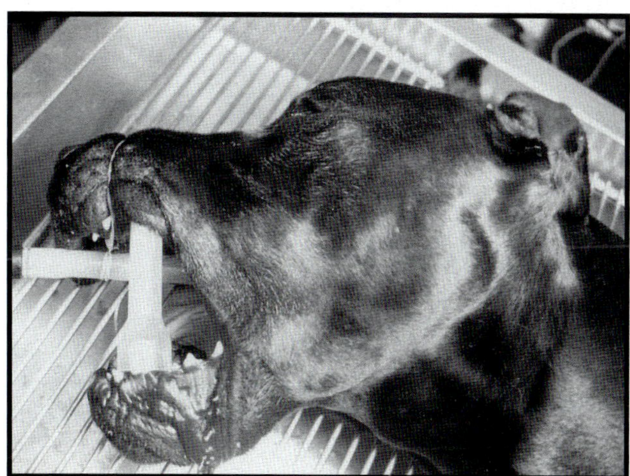

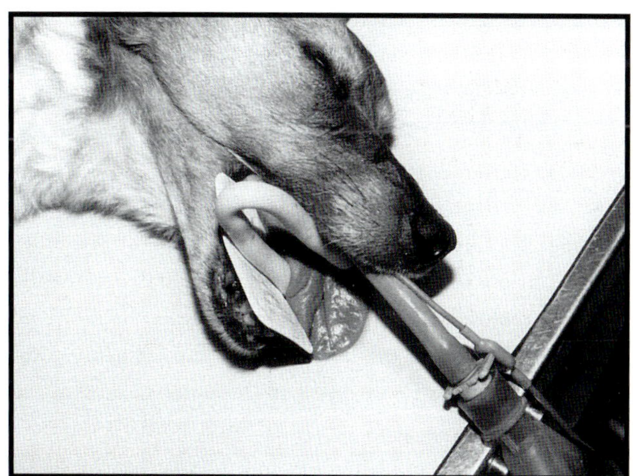

Various makeshift devices can be used as a bite block. **Figure 1-16A** *A foam roll works well in large-breed dogs.* **Figure 1-16B** *A syringe case serves as a bite block while deflecting the endotracheal tube.* **Figure 1-16C** *Film-holding devices (Flexi-film Holders™—Dr. Shipp's Laboratories, Beverly Hills, California) also can be used as a mouth prop.*

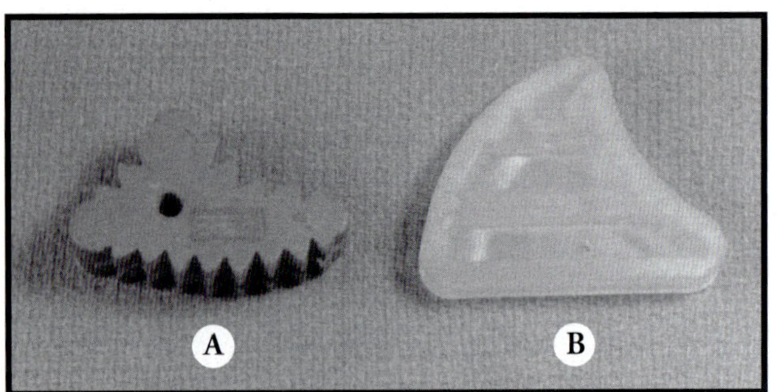

FIGURE 1-17

Commercial bite blocks include (A) *Mouth prop (Universal bite block—Henry Schein Inc., Port Washington, New York) and* (B) *the Wedge™ Veterinary Mouth Prop (Veterinary Devices, Inc., Wauwatosa, Wisconsin).*

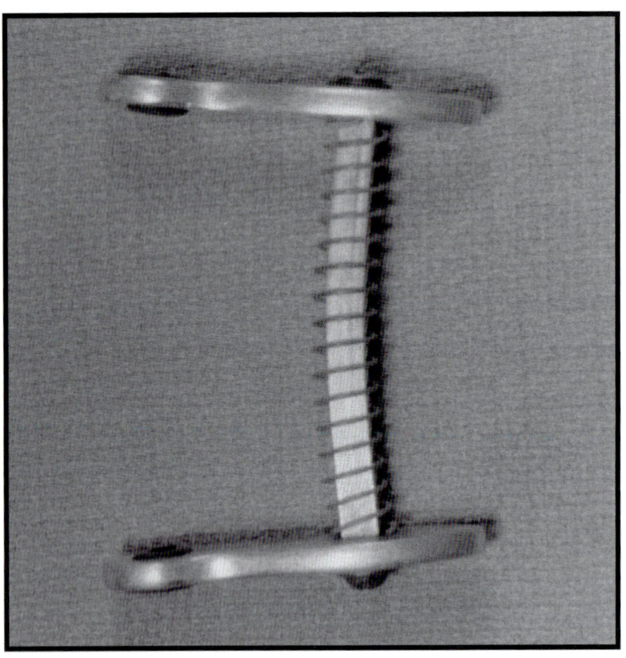

FIGURE 1-18

Although metal mouth gags continue to be used by some practitioners, we do not recommend them for oral radiography.

from the patient, and radiation leakage from the machine housing. Because virtually all veterinary patients are anesthetized before dental radiographs are taken, staff can usually leave the room during the procedure. Such film-holding and positioning devices as rubber bands, bite blocks, gauze pads, or paper towels can be used to stabilize intraoral film. If the room cannot be vacated, personnel should wear protective lead-lined gear or stand behind a protective barrier. If a barrier is not available, all personnel should stand at least six feet away from the patient and x-ray equipment. The operator must ensure that no one is standing in the path of the beam, even if it is at a point beyond the patient, but rather to the side of the patient and x-ray equipment, at an angle of 90° to 135°.

When the patient is on a stainless steel table, a lead sheet or mat should be placed under the animal to reduce scatter radiation. In addition, all staff should wear film badges that monitor occupational exposure to radiation.[1,8,9]

Patient exposure to radiation can be controlled by using fast film, limiting exposure to the least amount of time necessary to produce an acceptable radiograph, and adhering to proper techniques for positioning patients. Correct positioning minimizes the necessity to repeat the procedure.[1,2]

HAZARDOUS MATERIALS

The primary hazardous materials associated with dental radiography are the film developer and fixer solutions. Veterinary staff who handle these materials should be familiar with the information on the material safety data sheets provided by manufacturers. All processing solutions should be disposed of properly in accordance with each state's law.

In addition to following disposal instruction for film-processing solutions, veterinary staff should use appropriate caution when discarding the lead foil from film packets. The foil should be separated from each packet and stored for recycling.[1,2]

REFERENCES

1. Miles DA, Van Dis ML, Razmus TF: *Basic Principles of Oral and Maxillofacial Radiology.* Philadelphia, WB Saunders Co, 1992, pp 4–13, 45–47, 49–66.

2. Goaz PW, White SC: *Oral Radiology: Principles and Interpretation.* Philadelphia, CV Mosby Co, 1994, pp 6–12, 79–96, 119–121.

3. Miyabayashi T, Biller DS, Haider PR, Takiguchi M: Radiographic appearances of the nasal conchae in dogs using different screen-film systems: A postmortem study. *JAAHA* 30:382–388, 1994.

4. Zontine WJ: Canine dental radiology: Radiographic technic, development, and anatomy of the teeth. *J Am Vet Rad Soc* 16:75–82, 1975

5. Abramovitch K, Thomas LP: X-radiation: Potential risks and dose-reduction mechanisms. *Compend Contin Educ Dent* XIV(5): 642–648, 1993.

6. Gratt BM, White SC, Halse A: Clinical recommendations for the use of D-speed film, E-speed film, and xeroradiography. *J Am Dental Assoc* 117:609–614, 1988.

7. Hintze H, Christoffersen L, Wenzel A: In vitro comparison of Kodak Ultra-speed, Ekta-speed, and Ektaspeed Plus, and Agfa M2 Comfort dental x-ray films for the detection of caries. *Oral Surg Oral Med Oral Pathol Oral Radiol Endod* 81:240–244, 1996.

8. Eisner ER: Problems associated with veterinary dental radiography, in Manfra Marretta S (ed): *Problems in Veterinary Medicine—Dentistry,* vol 2, no. 1. Philadelphia, JB Lippincott, 1990, pp 46–84.

9. Zontine WJ: Dental radiographic technique and interpretation, in *The Veterinary Clinics of North America,* vol 4, no. 4. Philadelphia, WB Saunders Co, 1974, pp 741–762.

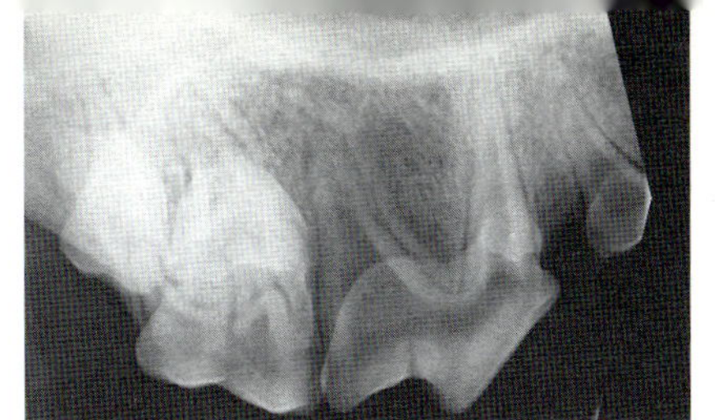

Chapter Two

PROJECTION GEOMETRY

The diagnostic value of a radiograph depends on its quality, and the degree of quality is determined by technique. Patient positioning and film exposure and processing collectively affect the value of a radiograph. Although basic radiographic principles apply to dental radiography, the equipment used in veterinary practices requires some modifications. The advantages and disadvantages associated with the different types of radiographic equipment are discussed in Chapter 1. Most of the radiographs that appear in this book were generated by using a dental x-ray machine and intraoral film.

In addition to equipment modifications, basic techniques of radiography often need to be adapted to dental radiology. For example, although parallel positioning offers the most accurate radiographic representation, the bisecting angle technique is used for most intraoral imaging. The following information covers specific techniques of projection geometry for intraoral imaging and represents the broadest application of dental radiology in veterinary practice. Regardless of the equipment or technique selected, however, the goal is a readable film that clearly shows the subject.

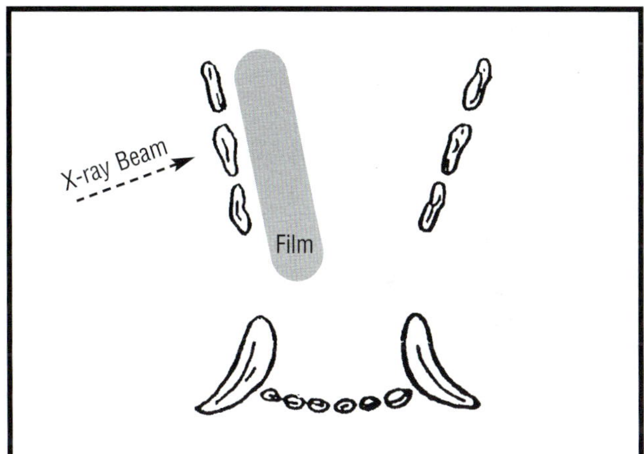

FIGURE 2-1B

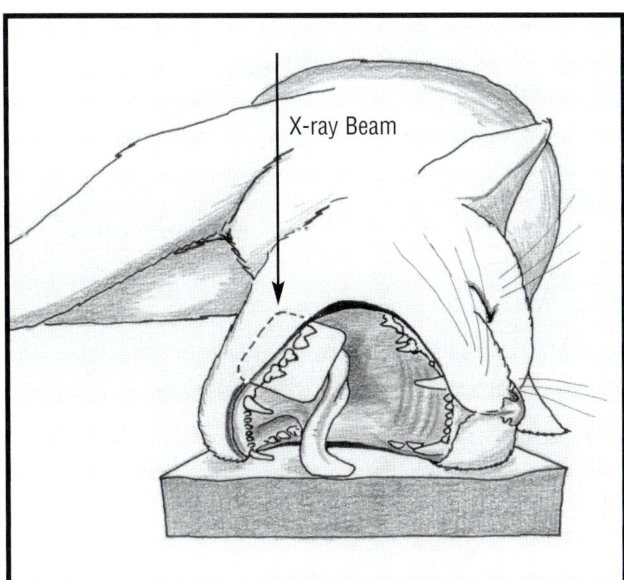

FIGURE 2-1A

Figures 2-1A and 2-1B *These two illustrations depict parallel positioning. The film and teeth are positioned parallel to each other, and the x-ray beam is directed perpendicular to both.* **Figure 2-1C** *This radiograph exemplifies the view that is captured when using parallel positioning.*

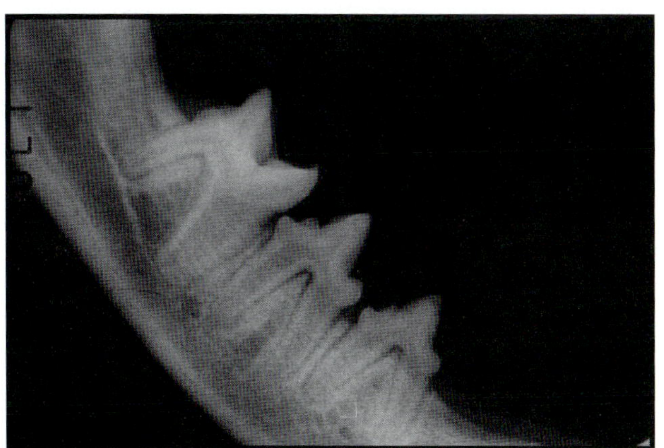

FIGURE 2-1C

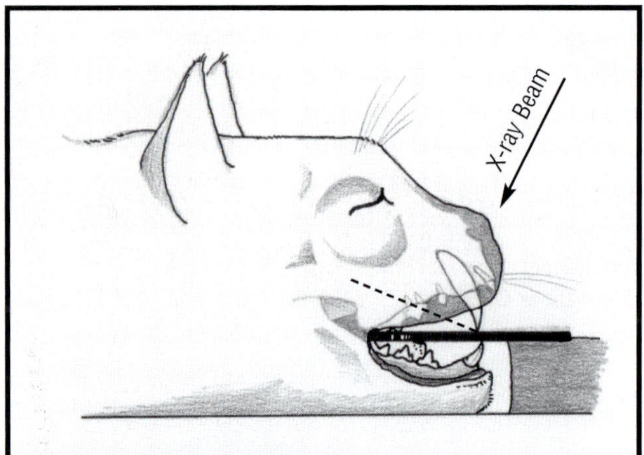

FIGURE 2-2A

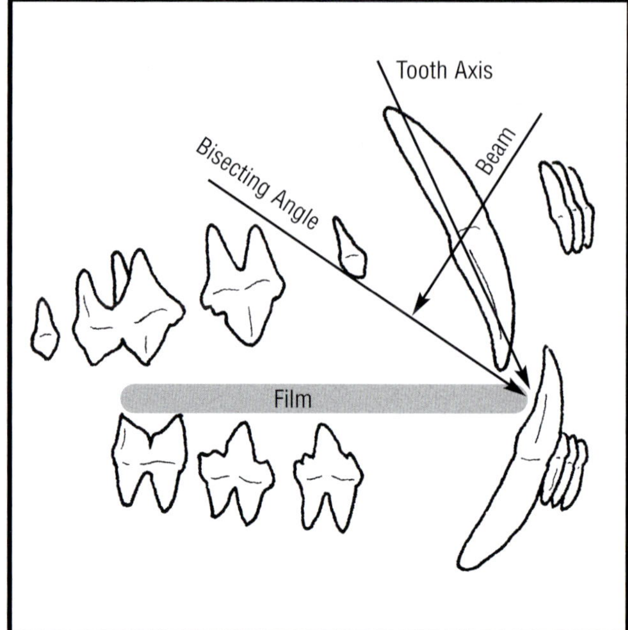

FIGURE 2-2B

Figures 2-2A and 2-2B *These two illustrations depict the bisecting angle technique. The subject and film are not positioned parallel to each other, and the x-ray beam is directed perpendicular to a plane between the subject and the film.* **Figure 2-2C** *This radiograph exemplifies the view that is captured when using the bisecting angle technique.*

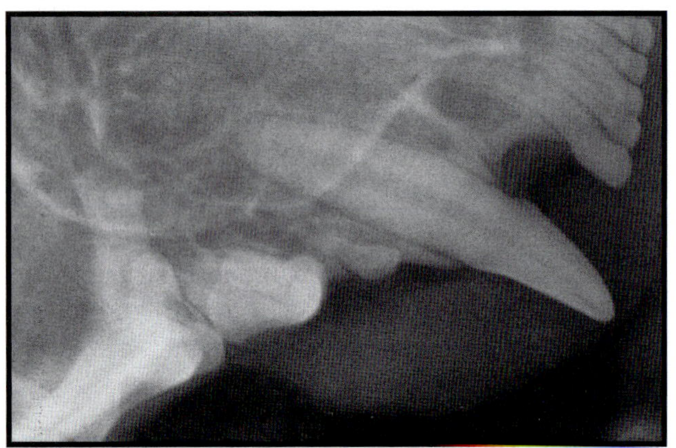

FIGURE 2-2C

FIGURE 2-3A

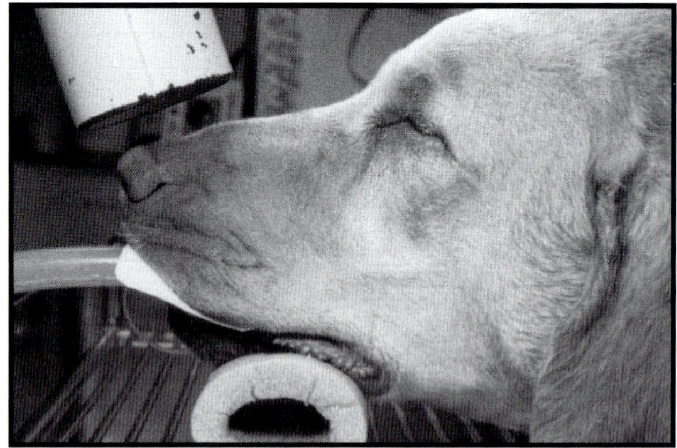

FIGURE 2-3B

Figures 2-3A and 2-3B *These two figures demonstrate how the vertical angle of the x-ray beam can be adjusted in relation to the position of the patient's head.*

PARALLEL TECHNIQUE

The technique known as parallel positioning produces the most consistently accurate image of an object. The film is positioned directly behind and parallel to the long axis of the area being radiographed, and the x-ray beam is directed per-pendicular to that area and the film (Figure 2-1). The film is placed as close as possible to the subject area to avoid magni-fication, and the beam is oriented perpendicular to the film and the subject area to avoid elongation or foreshortening of the image. To minimize blurriness of the image, the longest

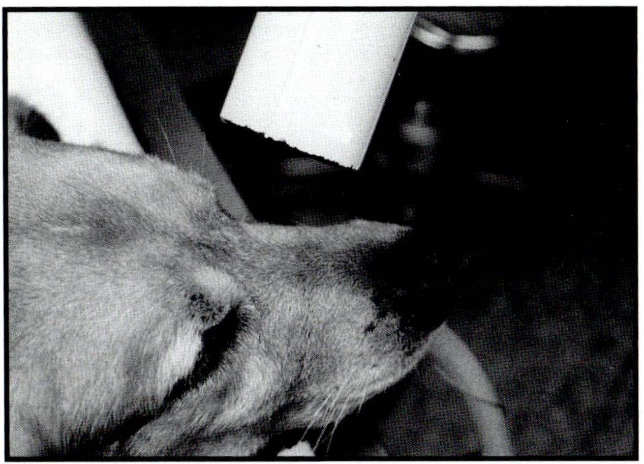

FIGURE 2-4A

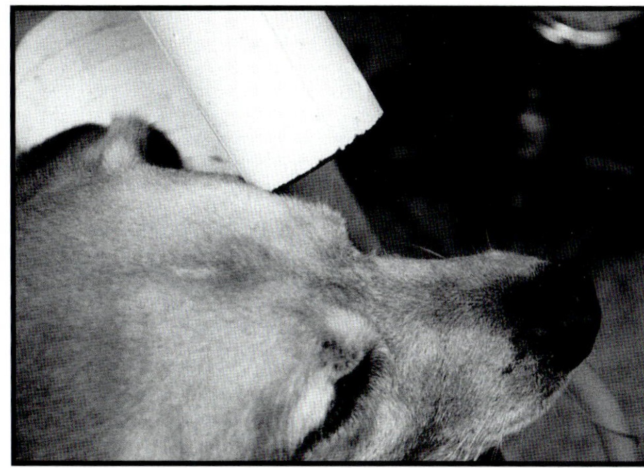

FIGURE 2-4B

Figures 2-4A and 2-4B *These two figures demonstrate how the horizontal angle of the x-ray beam can be adjusted in relation to the position of the patient's head.*

FIGURE 2-5A

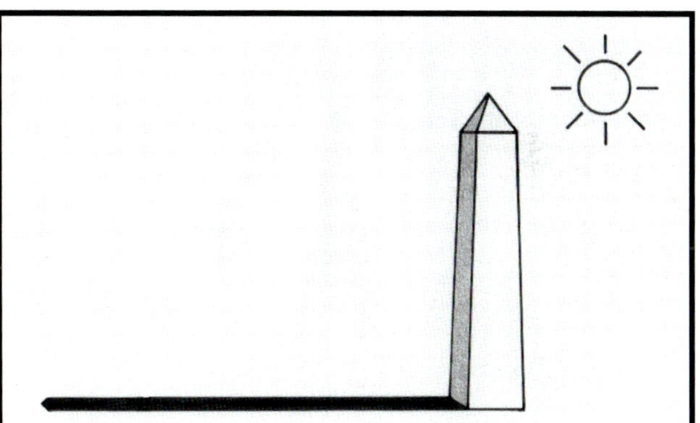

FIGURE 2-5B

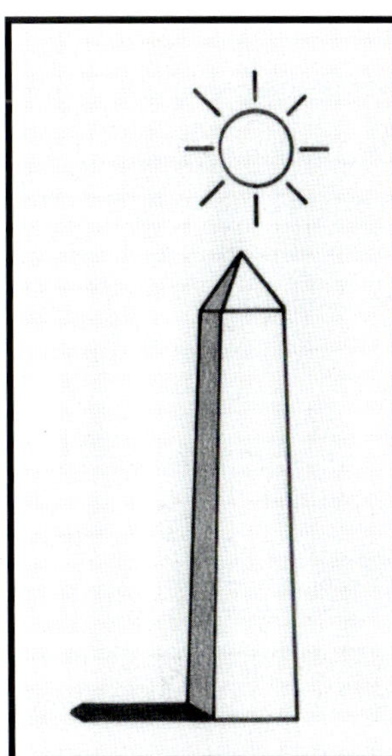

FIGURE 2-5C

Figure 2-5A *In this example of projection geometry, the sun's rays are striking the tower at an angle of 45°, that is, perpendicular to the bisecting plane between the ground and the tower. The resultant shadow (or image) is an accurate representation of the object.* **Figure 2-5B** *In this example of projection geometry, the sun's rays are striking the tower too close to the horizontal plane. The result is elongation of the shadow.* **Figure 2-5C** *In this example of projection geometry, the sun's rays are striking the tower too close to the vertical plane. The result is foreshortening of the shadow.*

focal film distance that is practical should be used. Following this technique minimizes image distortion and allows the radiographic image to reflect the area being radiographed as accurately as possible (that is, close to a 1:1 reproduction ratio of subject area to image).[1] Because of the oral anatomy of dogs and cats, the parallel technique in dental radiography is limited to imaging the mandibular premolars and molars.

BISECTING ANGLE TECHNIQUE

When the film and area being radiographed are not positioned parallel to each other, the resultant image will probably be elongated, foreshortened, unclear, or magnified. Adjusting the angle of the x-ray beam can correct such image distortion. A bisecting angle is an imaginary plane that equally divides the distance between the planes of the long axis of the area being radiographed and the film. The tubehead on the x-ray machine is positioned to allow the primary beam to be perpendicular to this imaginary bisecting plane (Figure 2-2). The resultant image is a fairly accurate depiction of the subject.[1]

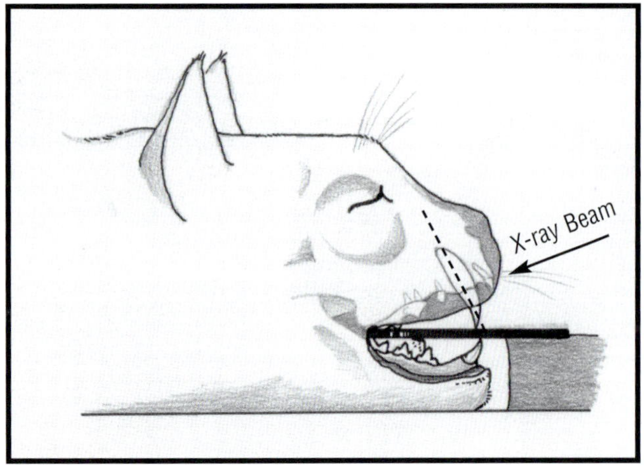

FIGURE 2-6A

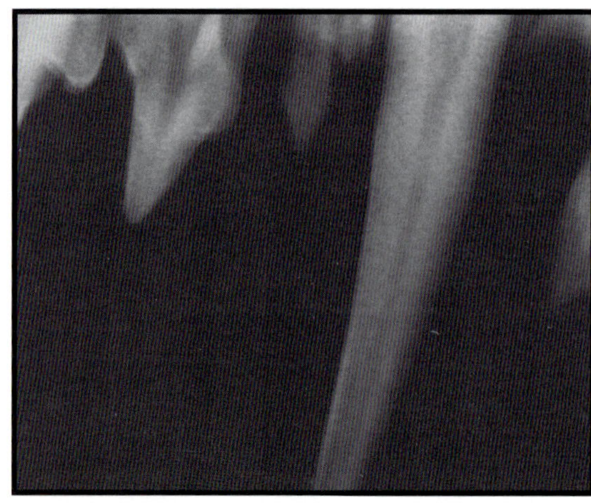

FIGURE 2-6B

Figure 2-6A *This illustration depicts the concept of elongation when using the bisecting angle technique. If the x-ray beam is directed perpendicular to the subject instead of to the bisector, the angle of the beam will be too close to the horizontal plane. The result would be elongation of the image.* **Figure 2-6B** *This radiograph exemplifies the image that is captured when the angle of the x-ray beam is positioned too close to the horizontal plane.*

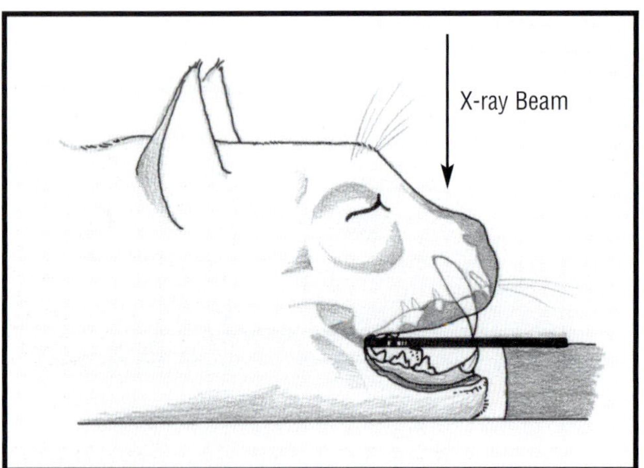

FIGURE 2-7A

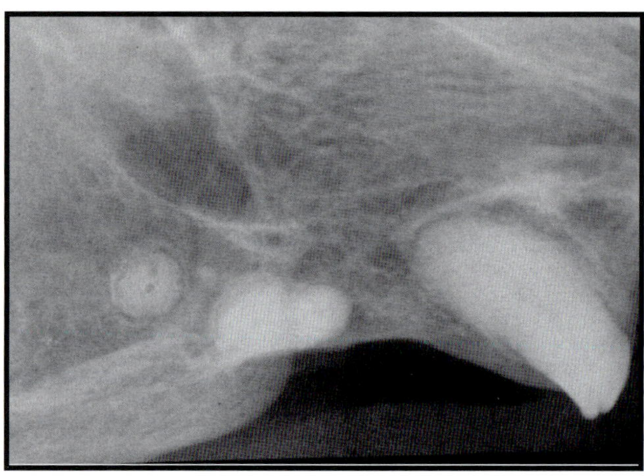

FIGURE 2-7B

Figure 2-7A *This illustration depicts the concept of foreshortening with the bisecting angle technique. If the x-ray beam is directed perpendicular to the film instead of to the bisector, the angle of the beam will be too close to the vertical plane. The result would be foreshortening of the image.* **Figure 2-7B** *This radiograph exemplifies the image that is captured when the x-ray beam is positioned too close to the vertical plane.*

CONE POSITIONING

Many errors in technique can cause image distortion when using dental x-ray machines. For example, the angle of the beam can be vertically adjusted along the long axis of the tooth being radiographed or horizontally adjusted along the dental arch (Figures 2-3 and 2-4). Proper placement of the x-ray cone and accurate beam angulation are essential to prevent elongation, foreshortening, or superimposition.

How the rules of projection geometry affect formation of an image can be demonstrated by visualizing a shadow that is cast by an object in a field. The sun's rays correspond to the x-ray beam, the object corresponds to the tooth being radiographed, the ground corresponds to the film surface, and the shadow that is cast corresponds to the radiographic image. The top of the object corresponds to the apex of the root, and that portion of the object on the ground corresponds to the crown of the tooth. The shape of the shadow depends on the angle of the sun's rays.[1,2] Figure 2-5 (which we refer to as the tower analogy) demonstrates this concept.

ERRORS IN VERTICAL ANGULATION
Elongation

If the x-ray beam is projected perpendicular to the long axis of the tooth rather than to the bisecting line, the angle of the beam will be incorrect. Likewise, if the tubehead is adjusted too far in the coronal direction and is excessively tilted toward

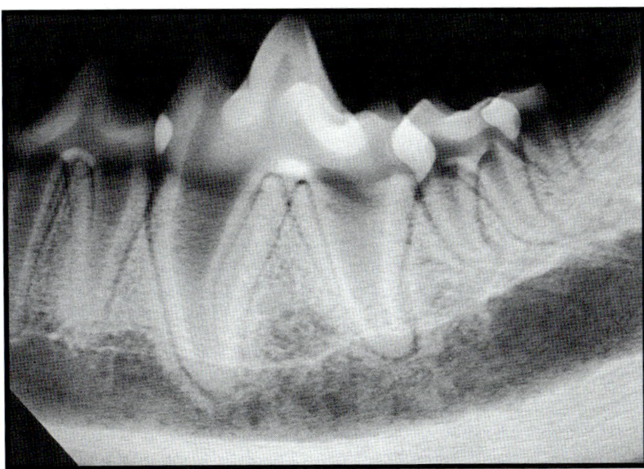

FIGURE 2-8A

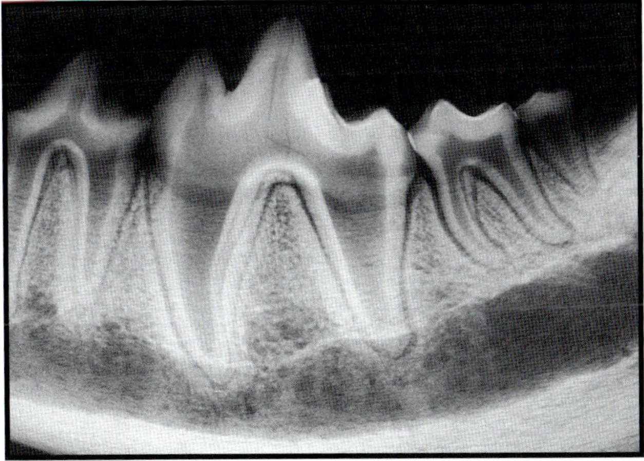

FIGURE 2-8B

Figure 2-8A *This radiograph demonstrates superimposition caused by an oblique x-ray beam; that is, the images of the teeth overlap when the horizontal angle of the x-ray beam is obliquely aligned.* **Figure 2-8B** *Superimposition is corrected when the x-ray beam is directed straight through the interproximal spaces.*

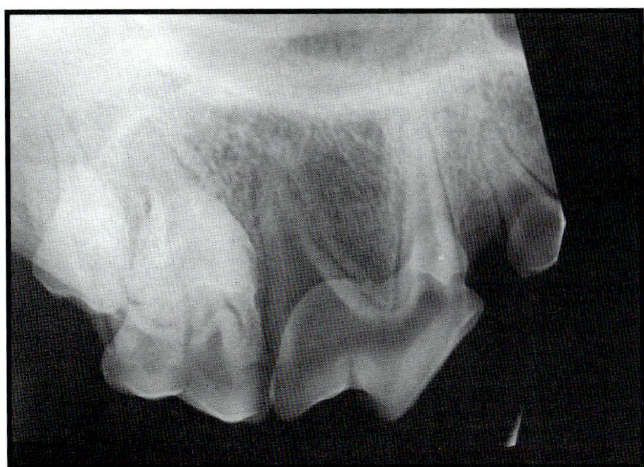

FIGURE 2-9A

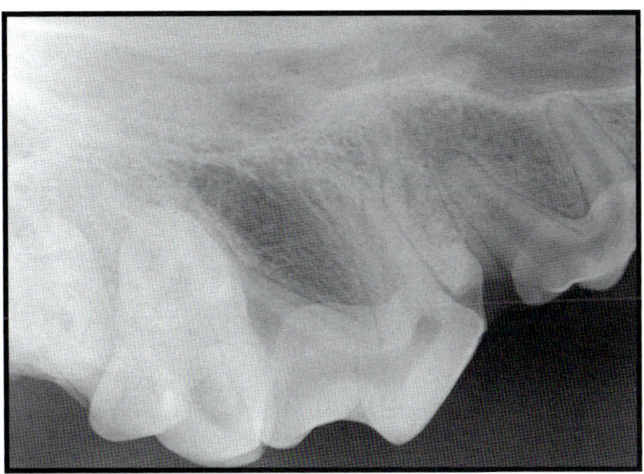

FIGURE 2-9B

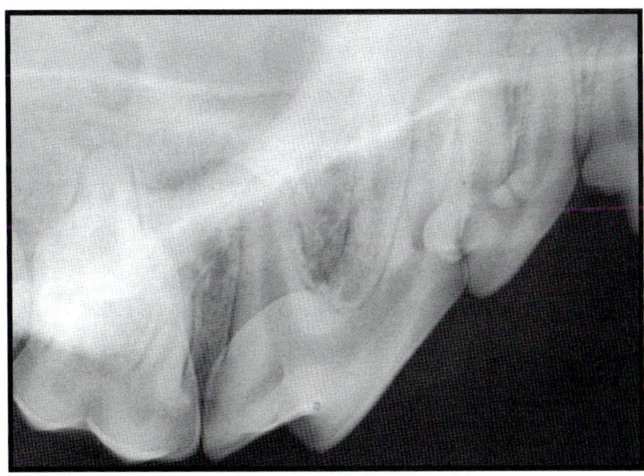

FIGURE 2-9C

Figure 2-9A *In this radiograph, the images of the mesial roots of the fourth premolar are superimposed in the lateral projection.* **Figure 2-9B** *The images of the mesial roots are separated in the mesiolateral oblique projection. Note that the mesiobuccal root is centered between the palatal and distal roots.* **Figure 2-9C** *This radiograph demonstrates distolateral oblique projection. The images of the mesial roots of the fourth premolar are separated, and the image of the distal root shifts away from the molar. Note that the palatal root is centered between the other two roots.*

the apex of the root, the angle of the x-ray beam will be too close to the horizontal plane, thereby making the angle incorrect. This error results in an image that is too long, which limits the diagnostic value of the film. On occasion, the image may be projected off the edge of the film (Figure 2-6).

Elongation is easy to visualize if the tooth is likened to the object in a field and a sun's ray is likened to the x-ray beam. The portion of the object closest to the ground corresponds to the crown of the tooth. If the sun is on the horizon, the shadow that is cast will be elongated (Figure 2-5).

Foreshortening

If the x-ray beam is projected perpendicular to the film rather than to the bisecting line, the angle of the beam will be incorrect. In addition, if the x-ray tube is adjusted too far in an

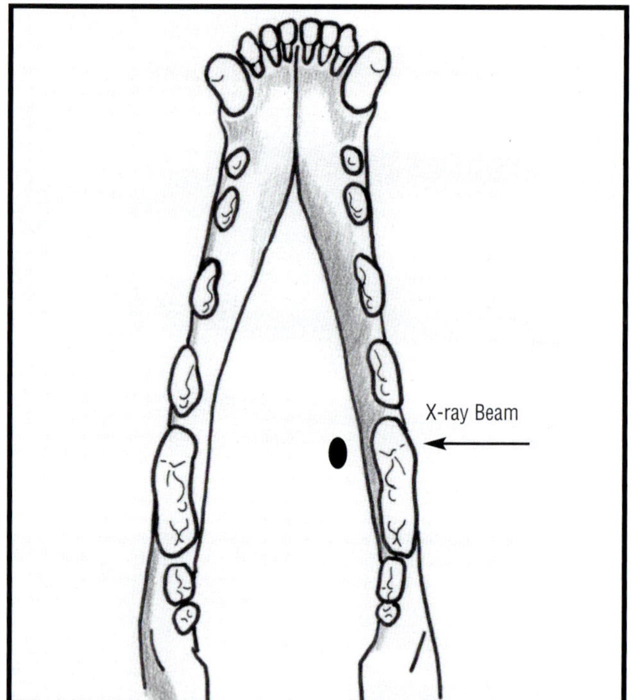

FIGURE 2-10A

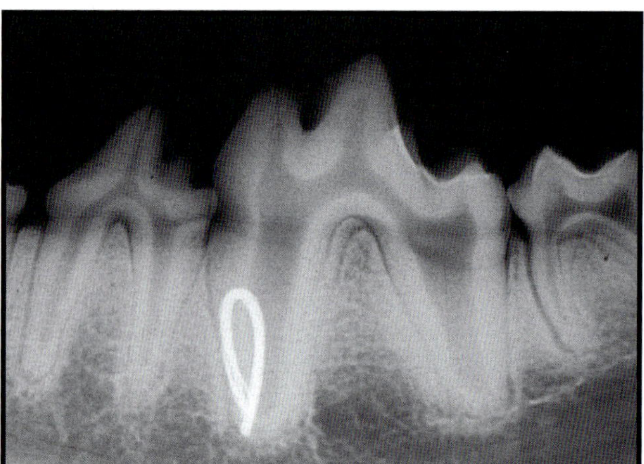

FIGURE 2-10B

Figure 2-10A *In this illustration, the mesial root of the mandibular molar represents the anatomic landmark selected as a reference point.*
Figure 2-10B *This radiograph reflects the resultant image.*

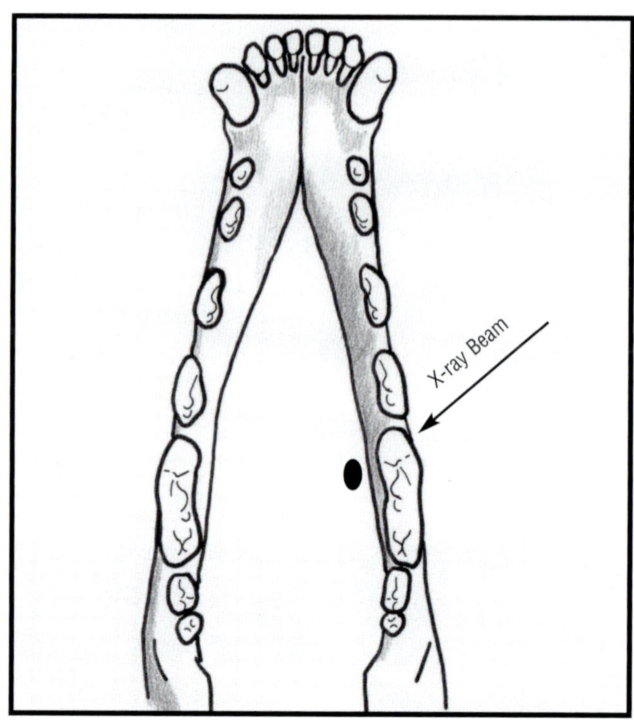

FIGURE 2-11A

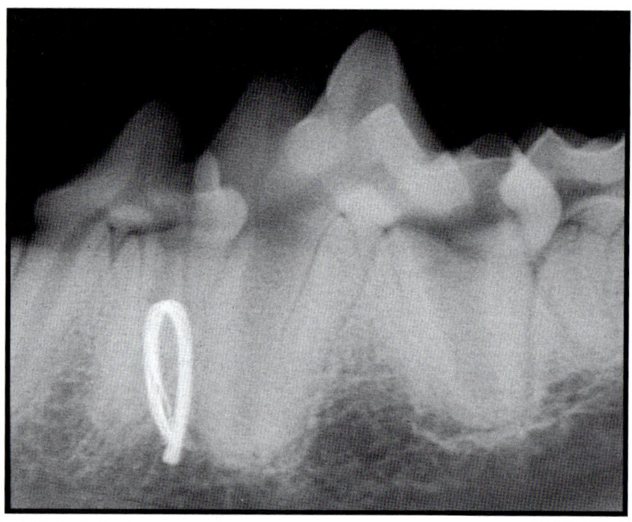

FIGURE 2-11B

Figure 2-11A *This illustration demonstrates the concept of* same-lingual, *and the radiograph in* **Figure 2-11B** *shows the resultant image. Note that the object shifted mesially to the root of the molar as the tubehead is moved in the same direction.*

apical direction and is tilted excessively toward the crown of the tooth, the angle of the x-ray beam will be too close to the vertical plane, thereby making the angle to the bisecting plane incorrect. This error results in an image that is too short and usually obliterates a diagnostic view of the apex of the root (Figure 2-7).

Using the tower analogy, if the sun is directly overhead (that is, 12:00 noon), the shadow will be foreshortened. The top of the tower corresponds to the apex of the tooth (Figure 2-5).

ERRORS IN HORIZONTAL ANGULATION
Superimposition

Because the tubehead on the x-ray machine moves in three dimensions, errors can also occur in the horizontal plane. If the x-ray beam is obliquely aligned in a mesial (nearer the center point of the dental arch) or distal (farther from the central point of the dental arch) direction, the interdental spaces appear narrow or superimposition occurs. When superimposi-

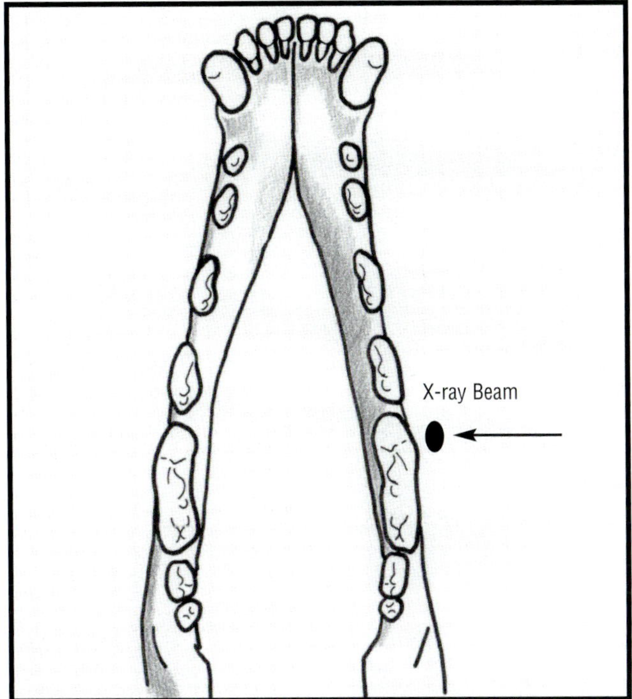

FIGURE 2-12A

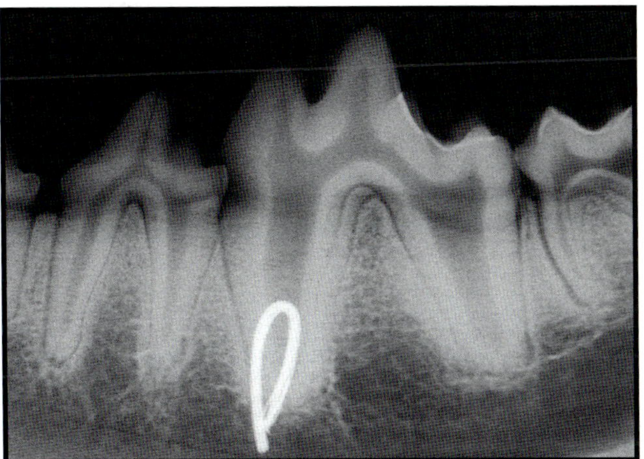

FIGURE 2-12B

Figure 2-12A *In this illustration, the mesial root of the mandibular molar represents the anatomic landmark selected as a reference point.* **Figure 2-12B** *This radiograph reflects the resultant image.*

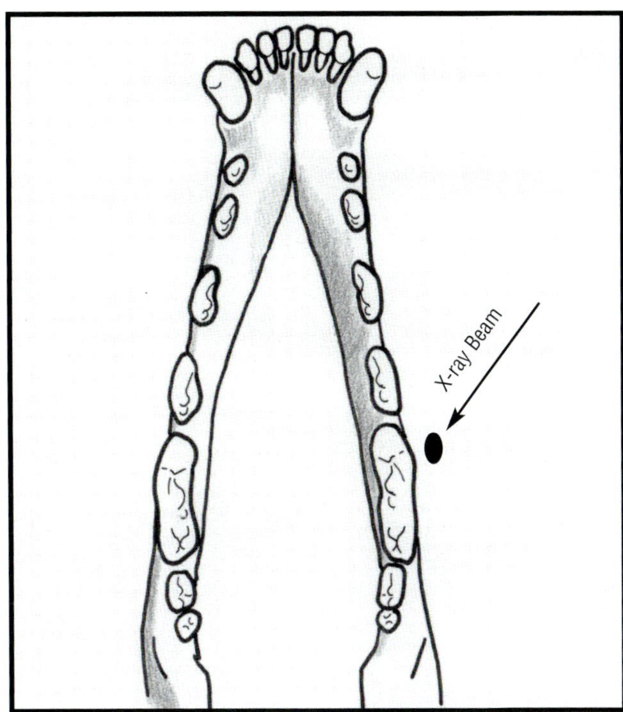

FIGURE 2-13A

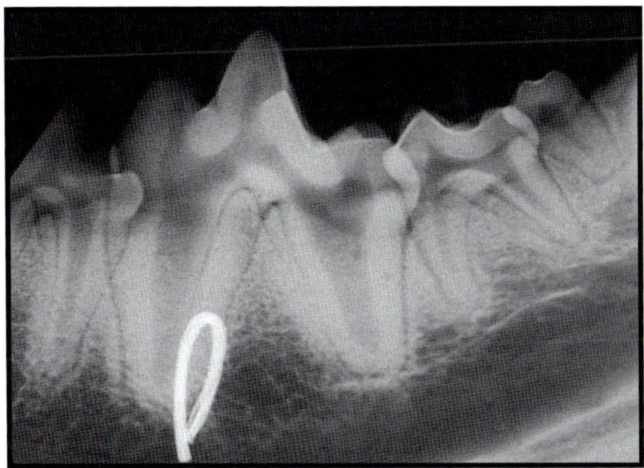

FIGURE 2-13B

Figure 2-13A *This illustration demonstrates the concept of opposite-buccal.* **Figure 2-13B** *This radiograph captures the resultant image of an object that has shifted in a distal direction as the tubehead is shifted mesially to the root of the molar; therefore, the object is buccal to the molar.*

tion occurs, the images are overlapped and the interproximal areas become obscured. To prevent superimposition, the angle of the x-ray beam should be aligned directly through the interproximal spaces[1,2] (Figure 2-8).

Three-Rooted Teeth

When the horizontal angle is correct, the two mesial roots of the fourth maxillary premolar are naturally superimposed (Figure 2-9). To separate the images of the mesial roots, the horizontal angle of the x-ray beam must be purposely adjust-ed obliquely toward the mesial or distal aspect.

When the tubehead is obliquely positioned at the mesial aspect, the images of the roots of the fourth maxillary premolar are separated; however, the distal root may be superimposed over the first molar. Of the three roots, the mesiobuccal root is in the center position. When the tubehead is obliquely positioned at the distal aspect, the images of the roots of the fourth maxillary premolar are again separated; however, the mesiobuccal root may sometimes be superimposed over the third premolar. Of the three roots, the palatal root is now the

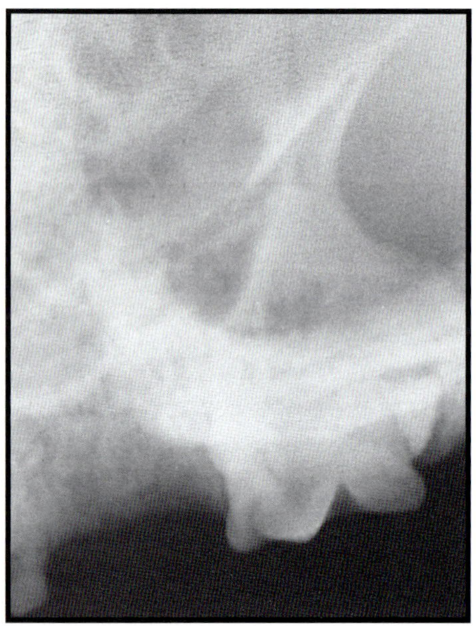

FIGURE 2-14A

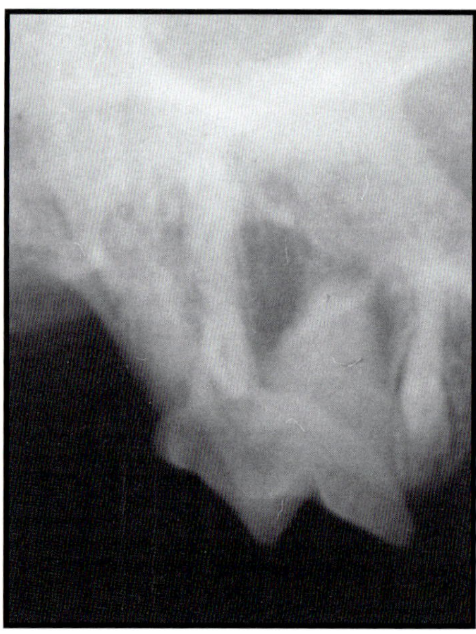

FIGURE 2-14B

Figure 2-14A *When the tubehead is shifted in a dorsal direction, the image of the zygomatic arch shifts toward the ventral aspect and obscures the image of the teeth.* **Figure 2-14B** *When the tubehead is shifted in a ventral direction, the image of the zygomatic arch shifts toward the dorsal aspect and the teeth can be better visualized.*

root in the center position. This radiographic view of the fourth premolar is usually preferred because the distal root is clearly visible for evaluation.

THE SLOB RULE AND OBJECT LOCALIZATION

The shifting of images in relation to the angle or direction of the x-ray is often referred to as the *tube-shift technique*. The phenomenon is also called the *buccal object rule* or *Clark's Rule*.[2]

Two radiographic views (preferably at right angles to each other) are necessary to obtain three-dimensional information from a radiograph. The oral anatomy of dogs and cats, however, usually limits right-angle projections. An alternative is to follow the tube-shift technique. Adjustments to beam angulation shift the projected images to provide a slightly different view of the subject.

The tube-shift technique can resolve problems with superimposition and can help to locate foreign bodies and impacted teeth. The concept of images shifting in relation to the angle of the x-ray beam can be remembered as SLOB: same-lingual, opposite-buccal. An anatomic landmark is selected as a reference point on the radiograph. How the position of the subject to be radiographed shifts in relation to that of the anatomic landmark is then observed according to the angle of the x-ray beam (Figures 2-10 through 2-13).

Same-Lingual. If the image of the subject to be radiographed moves in a mesial direction from the anatomic landmark when the tubehead is shifted toward the mesial aspect or moves in a distal direction when the tubehead is shifted toward the distal aspect, then the image is moving in the *same* direc-tion as the tubehead. An image that moves from the landmark reference in the *same* direction as the tubehead is located *lingual* to the known reference point[2] (Figures 2-10 and 2-11).

Opposite-Buccal. If the image of the subject to be radiographed moves in a mesial direction from the anatomic landmark when the tubehead is shifted toward the distal aspect or moves in a distal direction when the tubehead is shifted toward the mesial aspect, then the image is moving in the *opposite* direction as the tubehead. An image that moves from the anatomic landmark in the opposite direction as the tubehead is located *buccal* to the known reference point[2] (Figures 2-12 and 2-13).

If the image does not move with respect to the anatomic landmark, then the image is at the same depth as the reference point.[2] The SLOB rule also applies to changes in the vertical angle of the x-ray beam. For example, radiographic images of the zygomatic arch shift their position over the teeth according to the vertical angle of the beam. Because the zygomatic arch is buccal to the teeth, the resultant images move in a ventral direction over the crowns of the teeth as the tubehead is shifted toward the dorsal aspect (that is, *opposite-buccal*)[2,3] (Figure 2-14).

REFERENCES

1. Razmus TF, Williamson GF: *Current Oral and Maxillofacial Imaging.* Philadelphia, WB Saunders Co, 1996, pp 117–179.
2. Goaz PW, White SC: *Oral Radiology: Principles and Interpretation.* Philadelphia, CV Mosby Co, 1994, pp 97–105.
3. Pasler FA: Localization using various methods, in Rateitschak KH, Wolf HF (eds): *Color Atlas of Dental Medicine and Radiology.* New York, Thieme Medical Publishers, 1993, pp 83–87.

Chapter Three

EXTRAORAL IMAGING TECHNIQUES

Dental radiography involves two kinds of imaging: extraoral and intraoral. Intraoral imaging is presented in Chapter 4. Although most dental radiography relies on intraoral techniques, extraoral imaging can be advantageous when very large lesions need to be radiographed or when intraoral imaging is not feasible.

As discussed in Chapter 1, standard-size cassettes with or without intensifying screens are normally used for extraoral radiographs, although dental film can also be used. We prefer size 4 intraoral film for extraoral radiographs of cats.

The film cassette is placed in an extraoral position to the teeth being radiographed. The x-ray beam is generally angled perpendicular to the film, and the animal's head is rotated to prevent superimposition of other structures onto the teeth being radiographed.

Extraoral radiographs can appear to be magnified, to have a blurry image, or both because of the distance between the film cassette and the teeth being radiographed. In addition, some portions of extraoral radiographs may be over- or underexposed because skeletal structures surrounding the teeth vary in thickness and mass from animal to animal. Regardless of these limitations, extraoral imaging is often more practical. As mentioned, such imaging can be useful if a large lesion (such as a neoplasm or a lesion associated with metabolic bone disease) needs to be studied.

Quality extraoral radiographs of the teeth can be achieved if certain guidelines

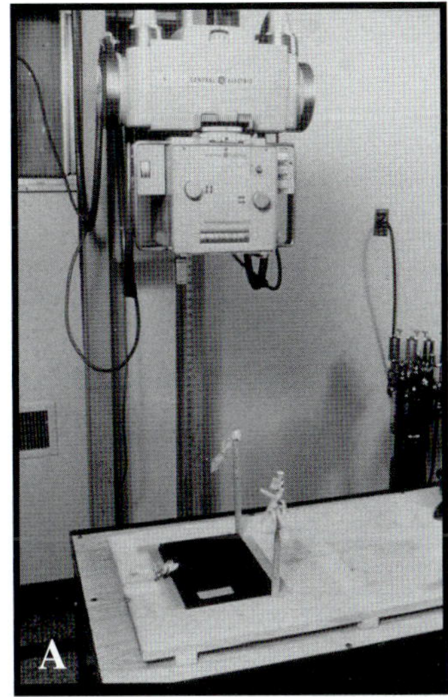
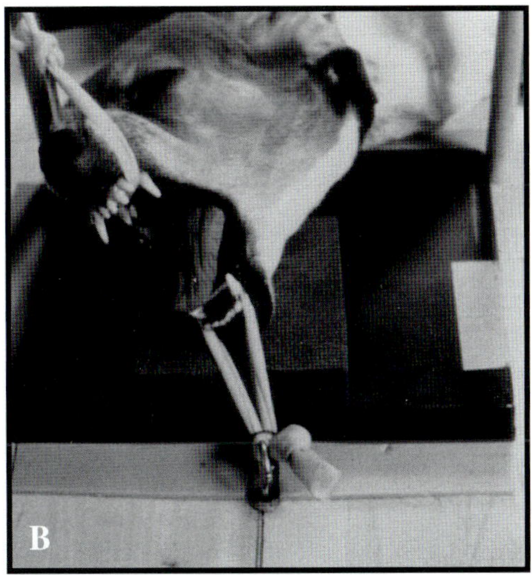
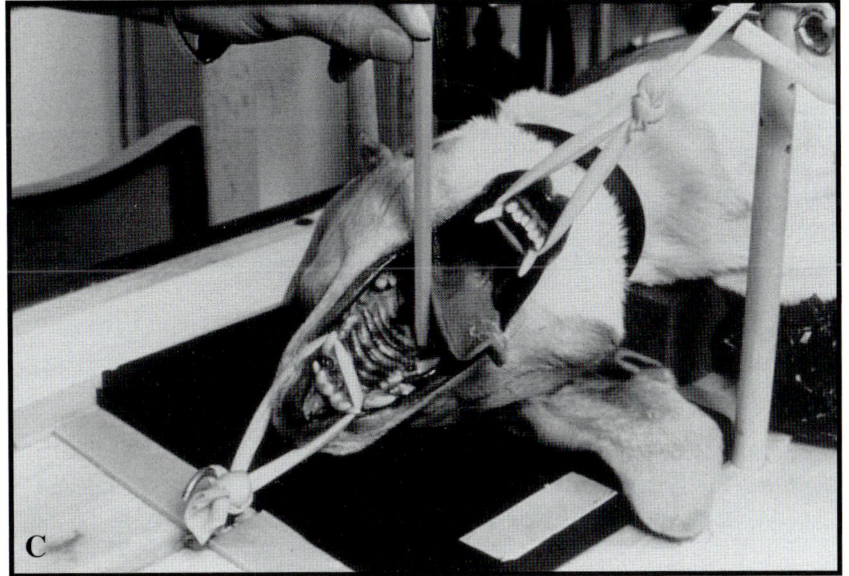

FIGURE 3-1

The device shown in this figure stabilizes the patient's head when extraoral radiographs of the mandible or maxilla are needed. (A) First, the device should be aligned under the tubehead as shown. (B) Elastic tubing can be used to hold the patient's head firmly in position. In this case, an extraoral radiograph from a lateral oblique view is being taken of the mandible. (C) The angle of the pencil reflects the angle of the primary x-ray beam when the patient is being positioned for a lateral oblique radiograph of the maxilla. (Photographs courtesy of Gregg Boring, DVM).

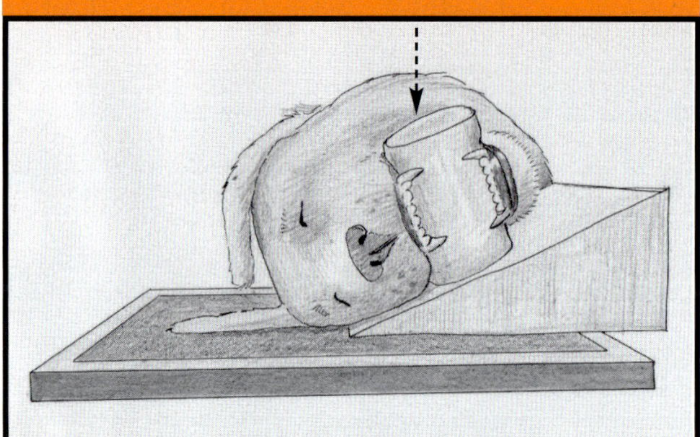

FIGURE 3-2A Patient Positioning and Film Placement

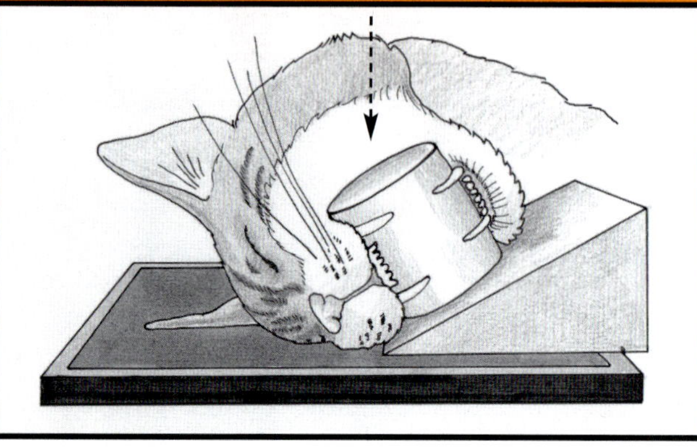

FIGURE 3-3A Patient Positioning and Film Placement

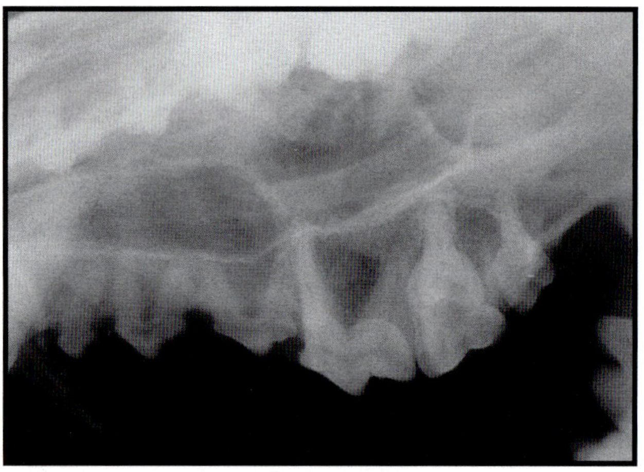

FIGURE 3-2B Radiographic Image

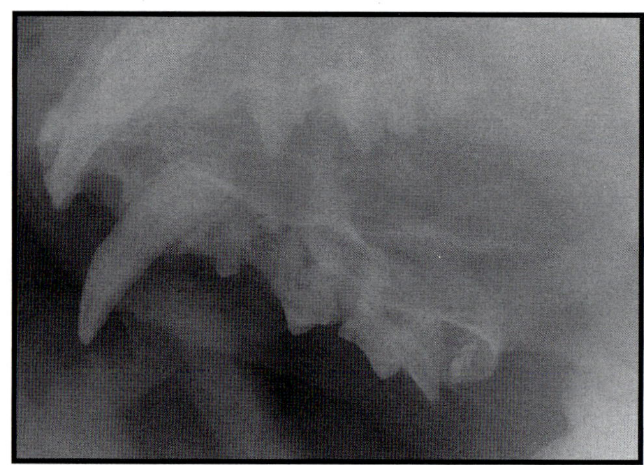

FIGURE 3-3B Radiographic Image

Figures 3-2A and 3-3A *The animal's head should be positioned in lateral recumbency and a foam pad inserted in an inclined position under the mandible to keep the maxilla within an oblique plane.[5,6] The head must be positioned at about 45° to ensure that the teeth to be radiographed are not superimposed over contralateral teeth. A foam spacer or syringe case can be used to prevent superimposition. The extraoral film plate should then be placed under the animal's head. The x-ray beam should be centered over the fourth premolar.* **Figures 3-2B and 3-3B** *The film should be positioned 90° to the primary x-ray beam without moving the animal from its oblique position. The entire upper lateral quadrant can be captured in one view. The image will reflect the maxillary dentition positioned closest to the film (referred to as the down side). The resultant image will not, however, be an exact reproduction (that is, will not be a 1:1 ratio); some magnification, blurriness, or both usually occur.*

are followed. As mentioned in Chapter 1, extraoral radiographs can be generated by using either a standard or dental x-ray machine. The x-ray beam, however, should be collimated so that only the specific structure being radiographed is captured on film (that is, the entire film area does not have to be exposed; rather, exposure is limited to the specific area of interest). Standard-size, slow-speed film and cassettes with a fine screen should be used for extraoral radiographs of large-breed dogs. In contrast, the ideal film to use for extraoral radiographs of very small breeds of dogs and all cats is size 4 intraoral film, which is placed in an extraoral position.

EXPOSURE TECHNIQUES

Exposure techniques differ for standard and dental x-ray equipment and according to the film being used on that equipment. In all instances, tabletop radiography is recommended because it limits the distance between the film and the teeth being radiographed.[1-3] When film cassettes with an intensifying screen are being used with a standard x-ray machine, tabletop radiography that adheres to the technique typically followed when using the equipment is recommended.[2,4] If nonscreen intraoral film (D speed) is being used with a standard x-ray machine, the focal film distance should be about 16 inches.

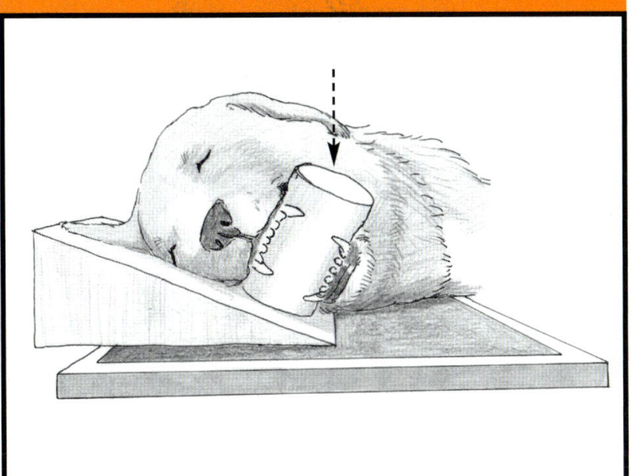

FIGURE 3-4A Patient Positioning and Film Placement

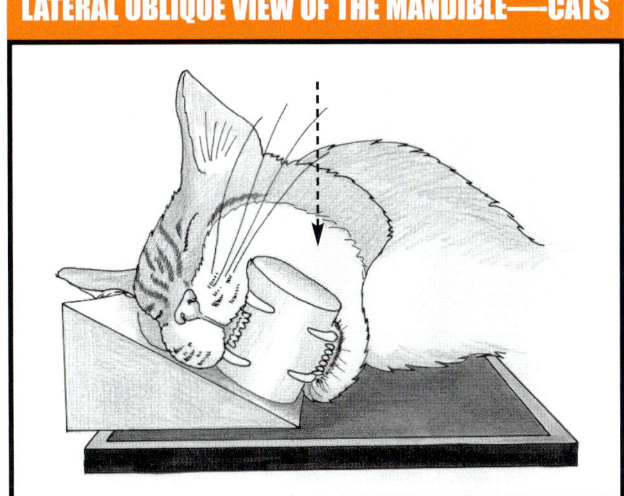

FIGURE 3-5A Patient Positioning and Film Placement

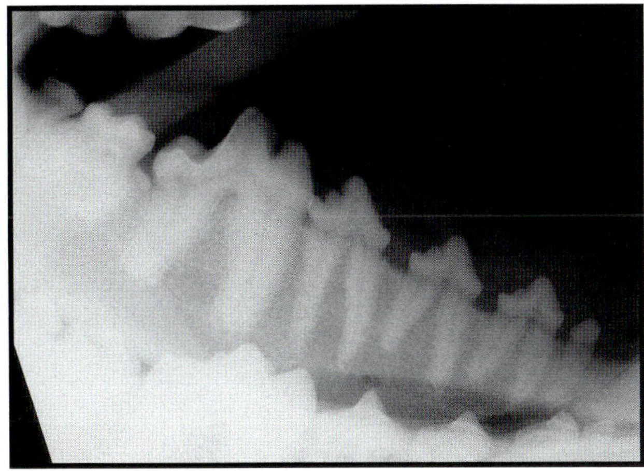

FIGURE 3-4B Radiographic View

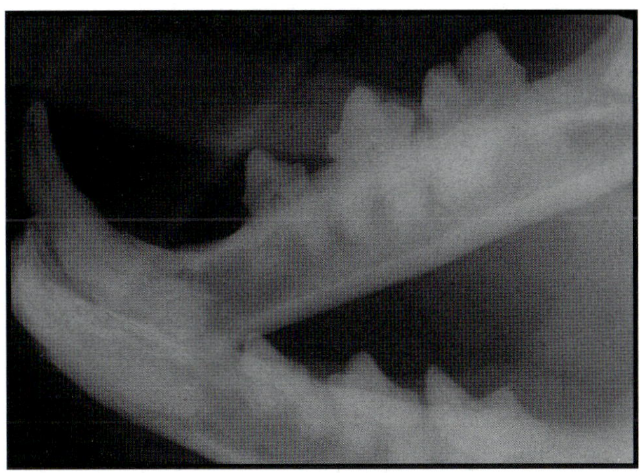

FIGURE 3-5B Radiographic View

Figures 3-4A and 3-5A The animal's head should be positioned in lateral recumbency and a foam pad inserted in an inclined position under the head to keep the mandible within an oblique plane.[5,6] The mandible must be positioned at about 45° to ensure that contralateral teeth are not superimposed over the teeth being radiographed. A foam spacer or syringe case can be used to prevent superimposition. The extraoral film plate should then be positioned under the animal's head. The x-ray beam should be centered over the first molar in dogs and over the fourth premolar and molar in cats. Figures 3-4B and 3-5B The film should be positioned 90° to the primary x-ray beam without moving the animal from its oblique position. The entire lower lateral quadrant can be captured in one view. In dogs, however, the caudal dentition may be obscured; whereas in cats, this view may be very difficult to capture if an endotracheal tube is present. The resultant view will not be an exact reproduction; some magnification, blurriness, or both usually occur.

The settings on the x-ray machine should be 100 milliamperes (mA) and 65 to 85 kilovolts peak (kVp), depending on the size of the patient. The exposure time should be set for ½ of a second for cats and ⅒ of a second for dogs.[3,4]

If cassette film with an intensifying screen is being used with a dental x-ray machine, the exposure will depend on the specific screen-film system and x-ray machine being used. Therefore, users should refer to the manufacturer's recommendations for film exposure and extrapolate information on exposure accordingly. If dental film is being used with a dental x-ray machine, the same procedures normally practiced for film exposure should be followed. Compared with the exposure time required for intraoral radiographs, however, users may need to increase the exposure time slightly for extraoral radiographs.

PATIENT POSITIONING

Most extraoral radiographs are generated using the open-mouth lateral oblique position in which the patient's mouth is opened wide and the head is rotated.[4,5] The animal must be anesthetized, and any device(s) used as a mouth prop should be inserted. The positioning device shown in Figure 3-1 keeps the patient's head stationary, is ideal for taking lateral oblique radiographs of the maxilla and mandible, and is inexpensive to construct. Materials required are two wooden dowels, a platform that can accommodate the film plate, a retaining ring, and elastic tubing. Other common positioning devices are discussed in Chapter 1.

Examples of patient positioning for taking extraoral radiographs of the maxilla and mandible of dogs and cats are shown in Figures 3-2 through 3-5, along with corresponding radiographic images.

ACKNOWLEDGMENT

The positioning device shown in Figure 3-1 was originally developed by Gregg Boring, DVM, MS, Professor, Biomedical Research Facility, College of Veterinary Medicine, Mississippi State University, Mississippi State, Mississippi.

REFERENCES

1. McNeel SV: Radiology of the skull and cervical spine, in Kealy JK (ed): *The Veterinary Clinics of North America. Small Animal Practice.* Philadelphia, WB Saunders Co, 1982, pp 259–279.
2. Burk RL: Oral radiology, in Bojrab MJ, Tholen M (eds): *Small Animal Oral Medicine and Surgery.* Philadelphia, Lea & Febiger, 1990, pp 56–74.
3. Eisner E: Problems associated with veterinary dental radiography, in Manfra Marretta S (ed): *Problems in Veterinary Medicine: Dentistry,* ed 2, no. 1. Philadelphia, JB Lippincott Co, 1990, pp 46–84.
4. Zontine WJ: Dental radiographic technique and interpretation, in Suter PF (ed): *The Veterinary Clinics of North America. Radiology,* ed 4, no. 4. Philadelphia, WB Saunders Co, 1974, pp 741–762.
5. Ticer JW: *Radiographic Technique in Veterinary Practice,* ed 2. Philadelphia, WB Saunders Co, 1984, pp 231–253.
6. Douglas SW, Herritage ME, Williamson HD: *Principles of Veterinary Radiography,* ed 4. Philadelphia, Balliere Tindall, 1987, pp 193–198.

Chapter Four

INTRAORAL IMAGING TECHNIQUES

The bisecting angle technique described in Chapter 2 is required to obtain quality intraoral radiographs because the shape and small size of the oral cavity in dogs and cats limit parallel positioning of the film. Film designed for intraoral radiography is very flexible (Figure 4-1), thereby accommodating the tight confinement of the oral cavity. Additional features of intraoral film are discussed in Chapter 1. Basic principles for positioning, exposure and processing, and projection geometry are addressed in Chapters 1 and 2.

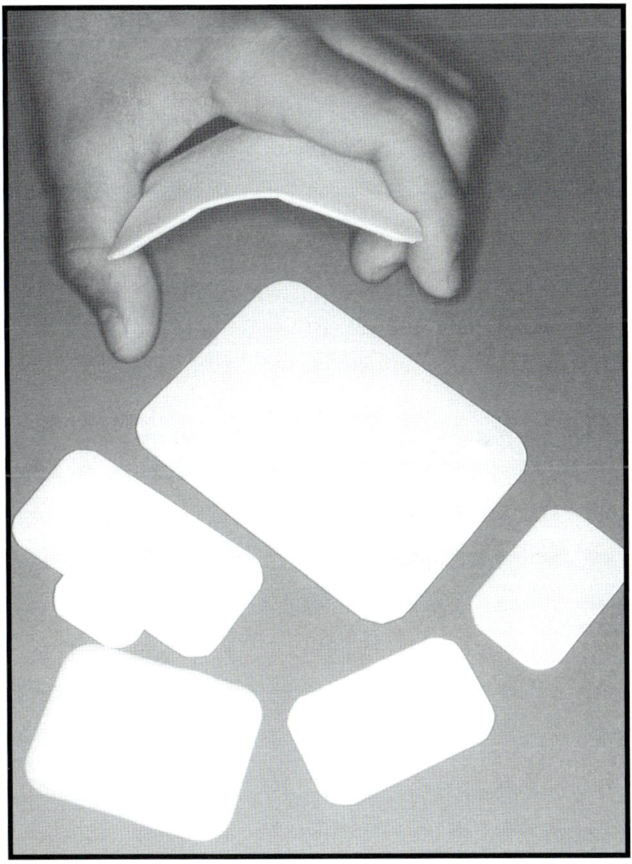

Figure 4-1 *This photograph demonstrates the flexibility of intraoral film. Flexibility is advantageous because of the small size of the oral cavity in dogs and cats.*

EXPOSURE TECHNIQUES

As mentioned in Chapter 1, when a dental x-ray unit and intraoral film are being used, the kilovolt peak (kVp) and milliamperes (mA) usually remain constant and the only variable would be the exposure time. Tables that indicate exposure specifications for intraoral radiographs of humans are provided by equipment manufacturers.[1] Pediatric values approximate the values for small-breed dogs and all cats, whereas adult human values approximate the values for large-breed dogs (Tables 4-1 and 4-2) and can serve as a basis for obtaining quality intraoral images when using D-speed film. E-speed film requires about one half the exposure time as that required for D-speed film.[2] Exposure guidelines vary between standard x-ray and dental x-ray units. On standard x-ray units, we recommend 7.5 to 10 mA-seconds (mA-sec) for all cats and small-breed dogs and 10 to 12.5 mA-sec for large-breed dogs.[3,4] If a standard x-ray machine and cassette film are being used, the standard technique chart for the equipment should be followed along with the concepts for applying bisecting angle and extraoral oblique techniques (see Chapter 2 for detailed discussions of these techniques). Additional informa-

TABLE 4 – 1
EXPOSURE GUIDELINES FOR USING D-SPEED FILM ON A DENTAL X-RAY MACHINE

Region	*Impulses*[a] *(70 kVp, 7 mA)*	*Seconds*[a] *(70 kVp, 7 mA)*
Mandibular Teeth		
Cats	9–11	0.15
Small-breed dogs	12	0.20
Large-breed dogs	14–16	0.25
Maxillary Teeth		
Cats	12–14	0.20
Small-breed dogs	14	0.25
Large-breed dogs	14–18	0.30

[a]Distance of 12 inches between source and film.

	TABLE 4 – 2			
EXPOSURE GUIDELINES FOR D-SPEED FILM ON A STANDARD X-RAY MACHINE[3,4]				
Patient Size	*Milliamperes*	*Kilovolt Peak*	*Seconds[a]*	*Seconds[b]*
Cats, Puppies, and Toy Breeds	100	50–60	0.1	0.4
Small-Breed Dogs (less than 30 pounds)	100	60	0.1	0.4
Medium-Sized dogs (30 to 60 pounds)	100	60–70	0.1	0.4
Large-Breed Dogs (more than 60 pounds)	100	70–85	0.1	0.4–0.6

[a]Focal film distance of 16 inches.
[b]Focal film distance of 36 inches. Note that accurately aligning the beam with the film is more difficult at this distance.

CONVERSIONS, ADAPTATIONS, AND TRICKS OF THE TRADE[2]

■ Time

To convert a technique chart for D-speed film to a chart for E-speed film, divide the exposure time by one half.

To convert seconds to impulses, multiply the number of seconds by 60:

$$\text{Exposure time in seconds} \times 60 = \text{exposure time in impulses}$$

To convert impulses to seconds, divide the number of impulses by 60:

$$\text{Exposure time in impulses} \div 60 = \text{exposure time in seconds}$$

■ Distance

If the distance between the source and film needs to be adjusted, the exposure technique needs to change accordingly because the Inverse Square Law states that beam intensity diminishes with distance in a predictable fashion.[2] The following rules apply when adjusting distances:

When the source-to-film distance is doubled, the original exposure time should be multiplied by four.

When the source-to-film distance is halved, the original exposure time should be one fourth of the original exposure time.

The mathematical formula is as follows:

$$\frac{\text{Old Exposure Time}}{\text{New Exposure Time}} = \frac{(\text{Old Distance})^2}{(\text{New Distance})^2}$$

■ Kilovolt Peak (kVp) Settings

A higher beam of 75 to 90 kVp produces an image that contains more shades of gray between the black and white range than a lower kVp beam. More shades of gray give a more gradual transition from black to white and is called *long-scale contrast*. *Short-scale contrast* (also referred to as high contrast) occurs when only a few shades of gray occur in the transition from black to white. If the contrast is changed without changing the overall film density, the following guidelines apply:

To increase contrast, decrease the original kilovolt peak setting by 15 kVp and double the exposure time.

To decrease contrast (increase the number of shades of gray), increase the original kilovolt peak setting by 15 kVp and use one half the original exposure time.

tion that is useful for determining exposure time and requirements is provided in the box entitled Conversions, Adaptations, and Tricks of the Trade.

FILM ORIENTATION

Because the characteristics of intraoral film are discussed in Chapter 1, only the major points are presented in this chapter. For film orientation, a dimple that is convex (raised area) on one side and concave (depressed area) on the other side is located on one corner of the film. The film is packaged so that the convex side always faces the x-ray tube (Figure 4-2).

When positioning intraoral film, the dimple should be facing the coronal aspect of a tooth rather than placed deep into the oral cavity. The dimple leaves a permanent mark on the

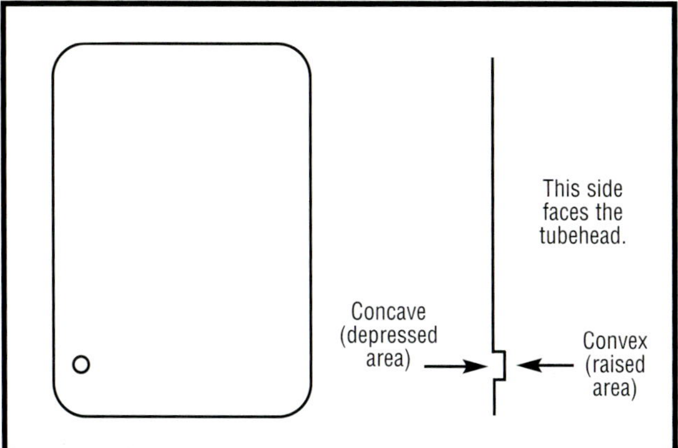

FIGURE 4-2A

FIGURE 4-2B

Figure 4-2A *illustrates the convex and concave aspects of the dimple located on intraoral film. In this illustration, the dimple appears in the lower left-hand corner.* **Figure 4-2B** *is a photograph of actual intraoral film. Note the positions of the dimple* (A and B arrows) *and its concave and convex aspects.*

radiograph, which could interfere with the image. When several intraoral radiographs are being viewed for one patient, determining which radiograph corresponds to the animal's left or right side may be difficult without knowing on which side of the film the dimple is located. Because the convex side always faces the tubehead, the right and left side of the radiograph can be identified by the permanent mark on the radiograph. When radiographs of the cheek teeth are being viewed, the left and right sides can be easily discerned by facing the concave side of the dimple. If the distal teeth (the molars) are on the left, then the radiograph is showing the left side of the arcade; whereas if the distal teeth are on the right, the radiograph is showing the right side (Figure 4-3).

PATIENT POSITIONING AND FILM PLACEMENT

Although the concept of the bisecting angle technique is

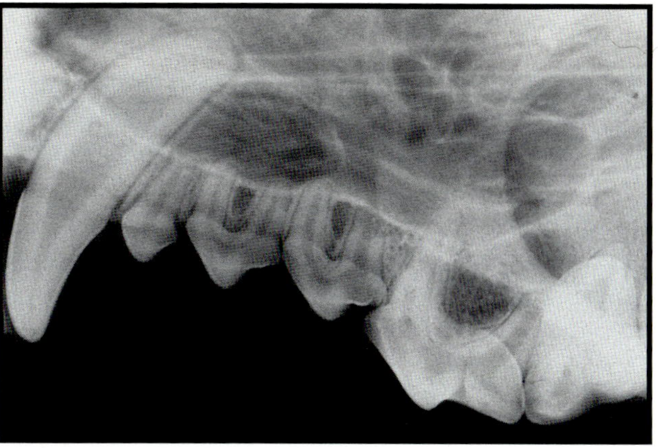

FIGURE 4-3

Figure 4-3 *This radiograph shows the upper right dentition when the image is viewed from the concave surface of the dimple.*

not difficult, its use does require some practice to produce consistently high-quality images of diagnostic value. Vertical angulation of the x-ray beam is indicated by the angle guide on the side of the tubehead. Most errors initially encountered include patient positioning, film placement, or angle of the x-ray cone. Errors that involve patient positioning and film placement can be eliminated by positioning the animal's head and body in the same manner and placing the film in the same location. Practitioners who have considerable experience taking radiographs use the palate instead of the tabletop as a reference for determining the bisecting angle. Figures 4-4 through 4-27 offer guidelines for patient positioning and film placement and show the resultant radiographic images. For oral radiographs of most breeds of dogs, size 4 intraoral film is used. Size 2 intraoral film may be more appropriate for all cats, pediatric patients, and toy and brachycephalic breeds of dogs.

Before studying Figures 4-4 through 4-27, we recommend that readers carefully review the intraoral bisecting angle technique outlined in Chapter 2 on Projection Geometry. Most of the figures depict this radiographic technique, whereas radiographs of the mandibular premolars and molars require parallel positioning.

REFERENCES

1. Exposure Guidelines for Kodak Ultra-speed and Ektaspeed Dental Films and Conversion Charts. Rochester, NY, Eastman Kodak Co, 1990.
2. Razmus TF, Williamson GF: *Current Oral and Maxillofacial Imaging.* Philadelphia, WB Saunders Co, 1996, pp 28-36, 103-104.
3. Zontine WJ: Dental radiographic technique and interpretation, in Suter PF (ed): *The Veterinary Clinics of North America*, ed 4. Philadelphia, WB Saunders Co, pp 741-762.
4. Eisner ER: Problems associated with veterinary dental radiography, in Manfra Marretta SM (ed): *Problems in Veterinary Medicine*, ed 2. Philadelphia, JB Lippincott Co, pp 46-84.

POSITIONING FOR INCISORS AND CANINE TEETH

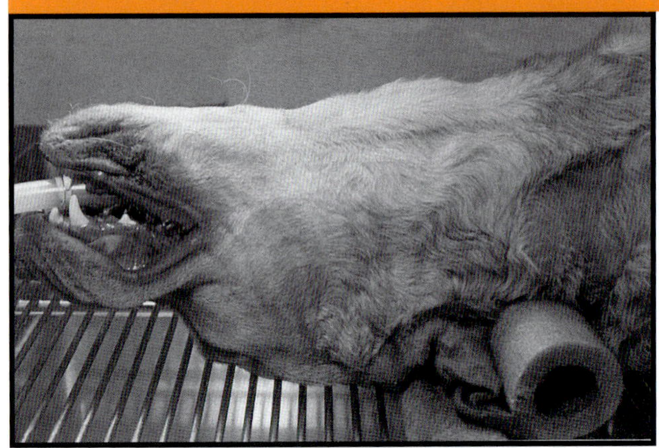

FIGURE 4-4A Patient Positioning—Mandibular Incisors and Canine Teeth

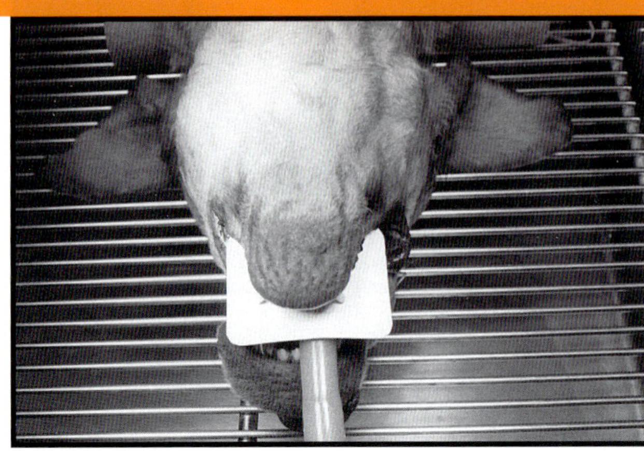

FIGURE 4-4B Film Placement—Mandibular Incisors and Canine Teeth

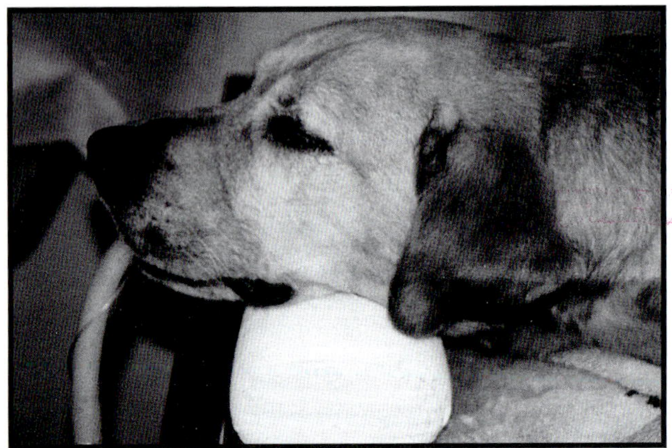

FIGURE 4-5A Patient Positioning—Maxillary Incisors and Canine Teeth

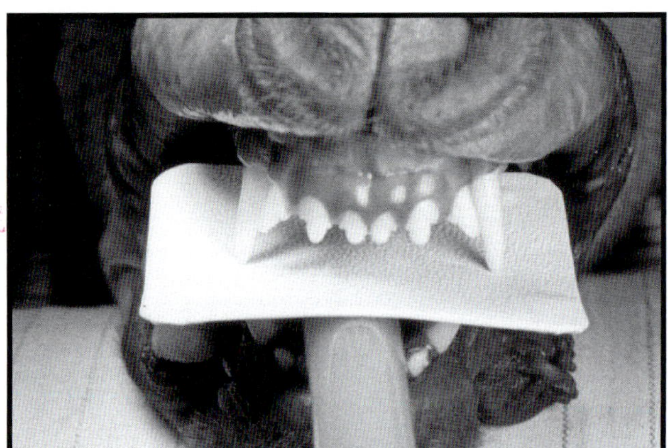

FIGURE 4-5B Film Placement—Maxillary Incisors and Canine Teeth

Figures 4-4A and 5A *The animal's head should be placed in dorsal recumbency for radiographs of the mandibular incisors and canine teeth and in ventral recumbency for radiographs of the maxillary incisors and canine teeth. A spacer can be placed under the neck to keep the hard palate equidistant to the same horizontal plane as the table top.* **Figures 4-4B and 5B** *Intraoral film is positioned under the incisors and canine teeth as shown.*

POSITIONING FOR PREMOLARS AND MOLARS

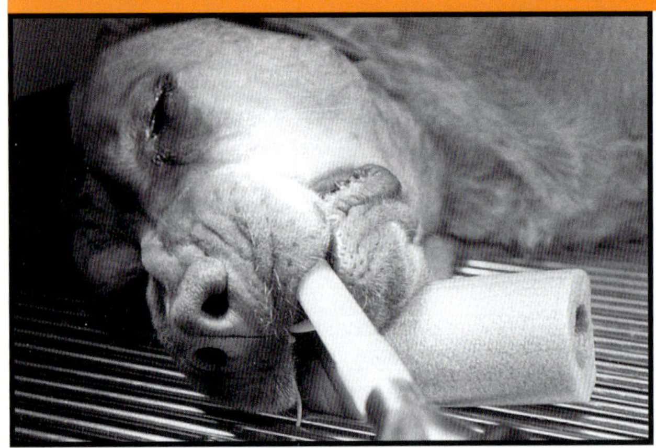

FIGURE 4-6A Patient Positioning—Mandibular Premolars and Molars

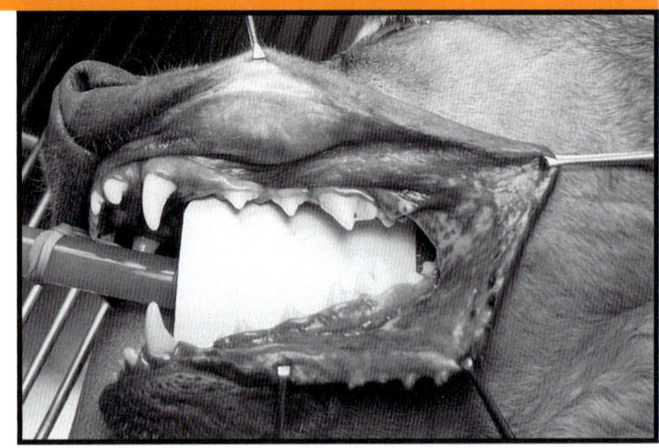

FIGURE 4-6B Film Placement—Mandibular Premolars and Molars

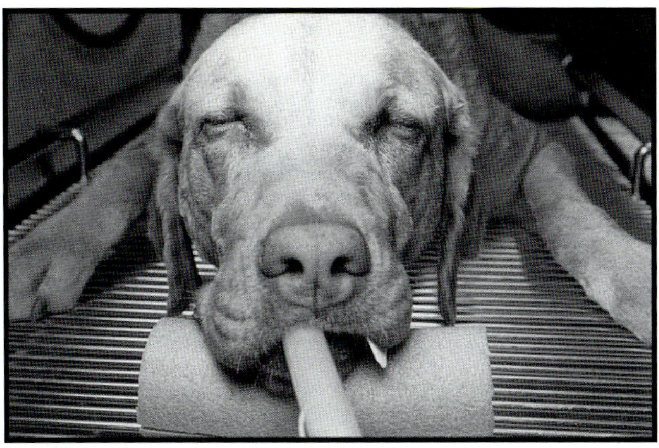

FIGURE 4-7A Patient Positioning—Maxillary Premolars and Molars

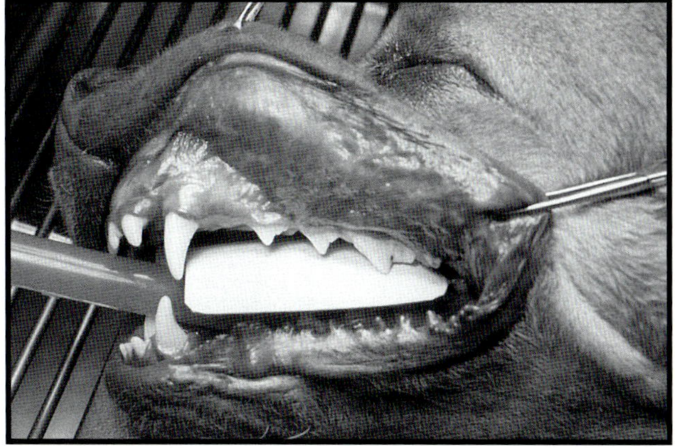

FIGURE 4-7B Film Placement—Maxillary Premolars and Molars

Figures 4-6A and 7A *The animal's head should be placed in lateral recumbency for radiographs of the mandibular premolars and molars and in ventral recumbency for radiographs of the maxillary premolars and molars. A spacer can be positioned under the jaw or chin to keep the hard palate perpendicular or equidistant to the same horizontal plane as the table top.* **Figures 4-6B and 7B** *When placing the film for radiographs of the mandibular premolars and molars, the film may have to be gently bent during insertion. For radiographs of the maxillary premolars and molars, the intraoral film should be placed under the upper dentition and adjacent to the hard palate. The film should be inserted until it is near the lingual aspect of the contralateral teeth.*

PROJECTION FOR MAXILLARY INCISORS—DOGS

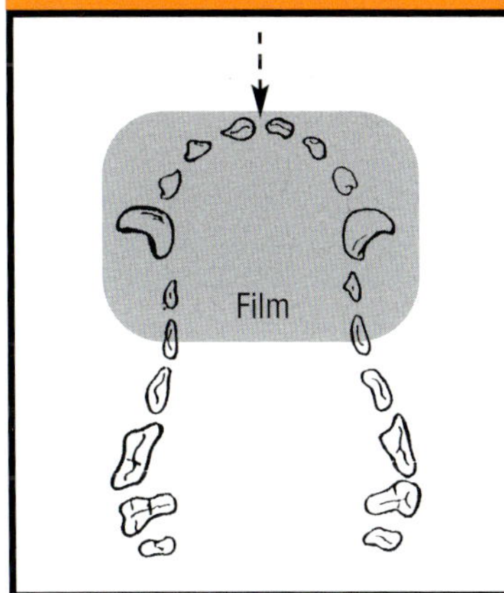

FIGURE 4-8A Image Field and Film Placement

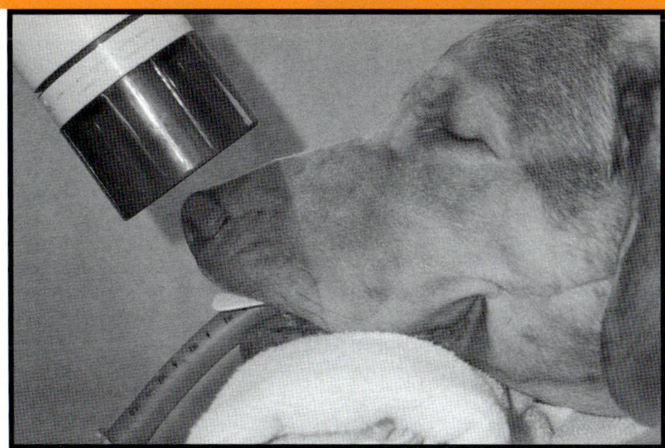

FIGURE 4-8B Patient Positioning

Figure 4-8A *The tubehead should be centered over the dog's nose. The arrow depicts the direction of the primary x-ray beam. Size 2 or 4 intraoral film can be used.* **Figure 4-8B** *The dog's head should be placed in ventral recumbency and a roll positioned under the jaw to keep the hard palate parallel to the table top. Size 2 or 4 intraoral film can then be placed below the incisors and canine teeth. The x-ray cone should be oriented downward at an approximate angle of 50° to 60° to the horizontal plane (or the hard palate). The primary x-ray beam will be almost perpendicular to the bisecting plane.* **Figure 4-8C** *The x-ray cone should be positioned perpendicular to the angle of the wing of the nares.* **Figure 4-8D** *The resultant image is shown in this radiograph.*

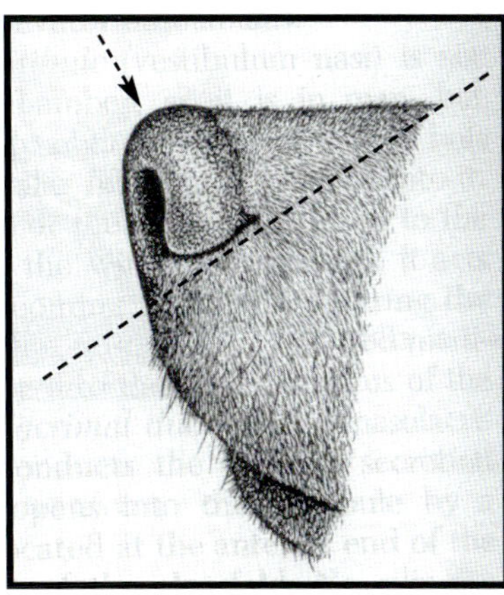

FIGURE 4-8C Positioning Hint

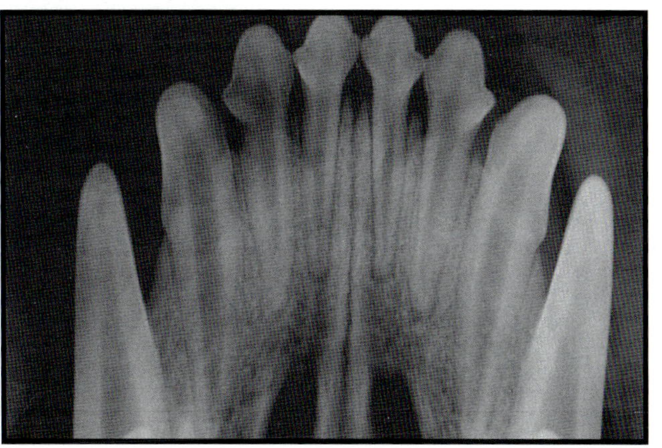

FIGURE 4-8D Radiographic Image

LATERAL OBLIQUE PROJECTION FOR MAXILLARY CANINE TEETH—DOGS

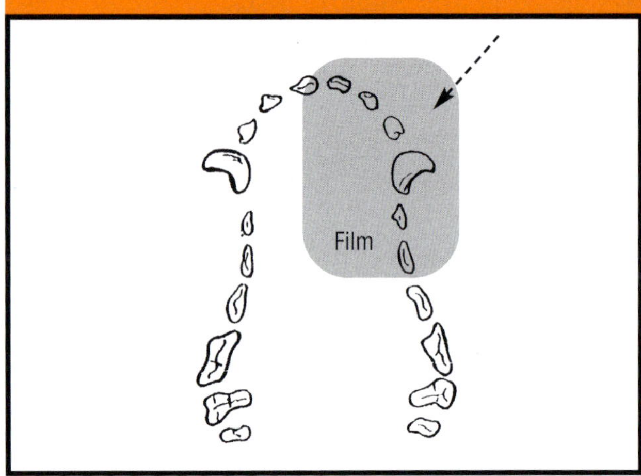

FIGURE 4-9A Image Field and Film Placement

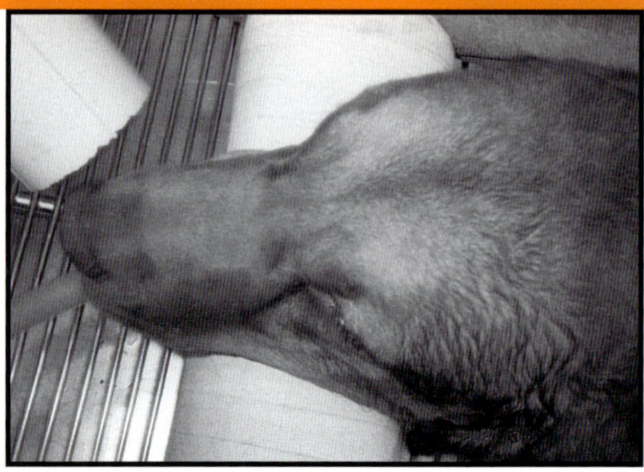

FIGURE 4-9B Patient Positioning

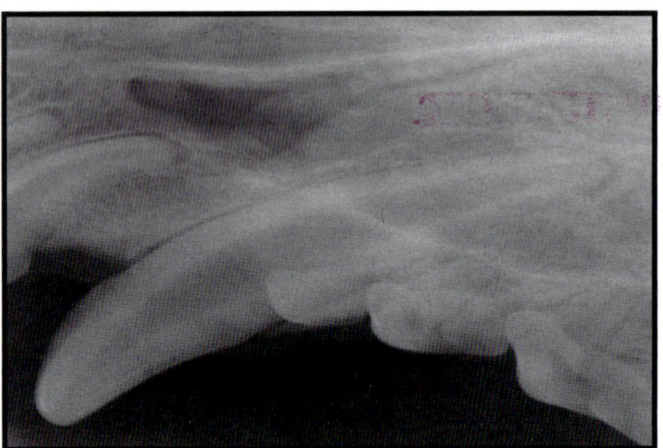

FIGURE 4-9C Radiographic Image

Each of the four canine teeth can be radiographed from three differing perspectives. The rostrocaudal projection produces an excellent image of the coronal aspect of the canine teeth, but the apex of the canine teeth will be superimposed over the premolars. The lateral projection offers a different view of the canine teeth than that provided by the lateral oblique projection. Therefore, both views are recommended. **Figure 4-9A** *The tubehead should be centered over the upper canine tooth. The* arrow *depicts the direction of the primary x-ray beam. Size 4 intraoral film is recommended for most breeds of dogs, although size 2 film may be more appropriate for toy and some brachycephalic breeds. The film can be gently bowed to accommodate placement.* **Figure 4-9B** *The dog's head should be placed in ventral recumbency and a roll placed under the jaw to keep the hard palate parallel to the table top. Size 2 or 4 intraoral film can then be placed under the canine tooth, extending under the incisors and premolars. The x-ray cone should be directed toward the midline at an angle of 45° as shown and tipped downward at an angle of 45° to 60° to the hard palate. The primary x-ray beam will be almost perpendicular to the bisecting plane.* **Figure 4-9C** *This radiograph shows the resultant image.*

LATERAL PROJECTION FOR MAXILLARY CANINE TEETH—DOGS

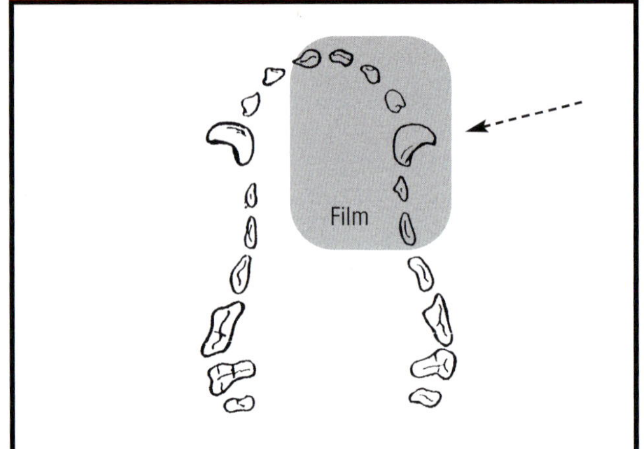

FIGURE 4-10A Image Field and Film Placement

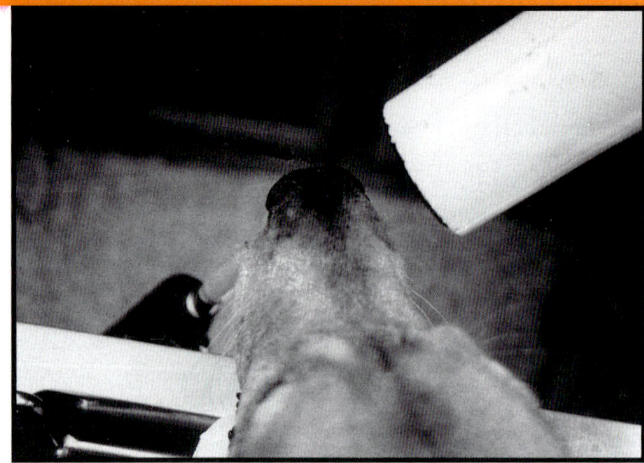

FIGURE 4-10B Patient Positioning

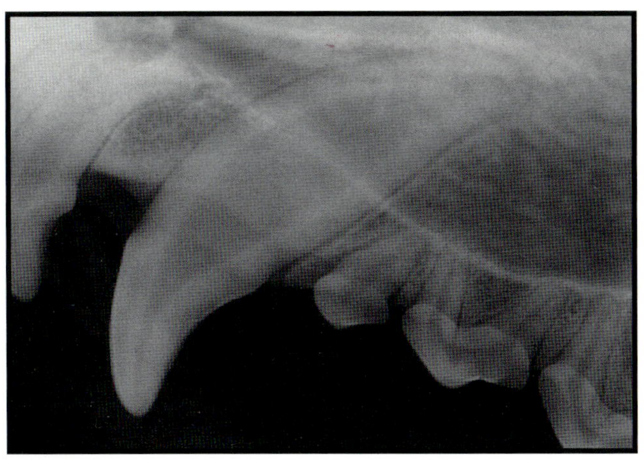

FIGURE 4-10C Radiographic Image

Both lateral and lateral oblique projections are recommended when radiographing the canine teeth. **Figure 4-10A** *The tubehead should be centered over the upper canine tooth and adjacent teeth at least as far as the second premolar. The arrow depicts the direction of the primary x-ray beam. Although size 4 intraoral film is recommended for most breeds of dogs, size 2 film may be more appropriate for radiographs of toy breeds and some brachycephalic breeds. The film can be gently bowed to accommodate placement.* **Figure 4-10B** *The dog's head should be placed in ventral recumbency and a roll positioned under the jaw to keep the hard palate parallel to the table top. Size 2 or 4 intraoral film can then be placed adjacent to the canine tooth toward the lingual aspect, extending under the incisors and second premolar. The x-ray cone should be directed at an approximate angle of 80° to 90° to the median plane or midline and tipped downward at an angle of 45° to the hard palate. The primary beam will be almost perpendicular to the bisecting plane.* **Figure 4-10C** *This radiograph shows the resultant image.*

LATERAL PROJECTION FOR MAXILLARY PREMOLARS—DOGS

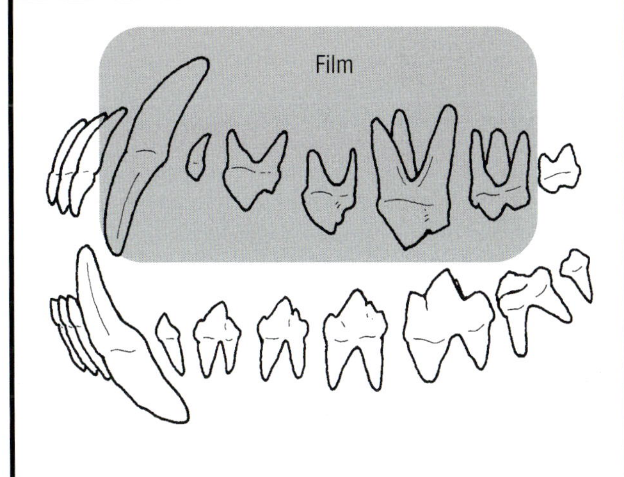

FIGURE 4-11A Image Field and Film Placement

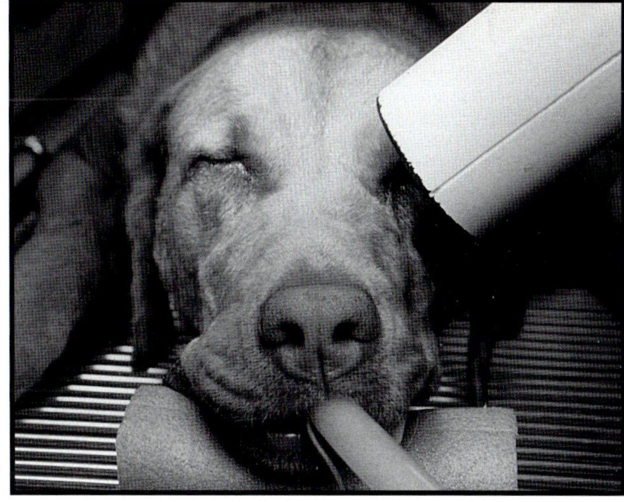

FIGURE 4-11B Patient Positioning

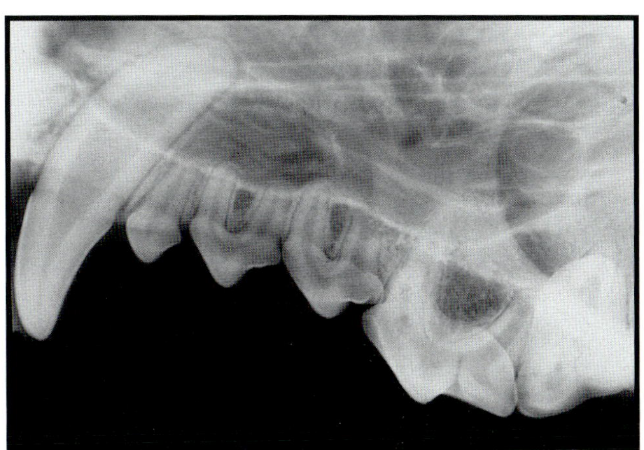

FIGURE 4-11C Radiographic Image

Figure 4-11A *The tubehead should be positioned over the premolars to be radiographed. Size 4 intraoral film is recommended for most breeds of dogs; however, size 2 film may be more appropriate for toy breeds and some brachycephalic breeds.* **Figure 4-11B** *The dog's head should be placed in ventral recumbency and a roll positioned under the jaw to keep the hard palate parallel to the table top. Intraoral film can then be placed adjacent to the hard palate and inserted until it is near the lingual aspect of the contralateral teeth. The x-ray cone should be positioned in a lateral direction at an angle of 45° to the hard palate. The primary x-ray beam will be almost perpendicular to the bisecting plane.* **Figure 4-11C** *This radiograph shows the resultant image.*

LATERAL PROJECTION FOR MAXILLARY PREMOLARS AND MOLARS—DOGS

FIGURE 4-12A Image Field and Film Placement

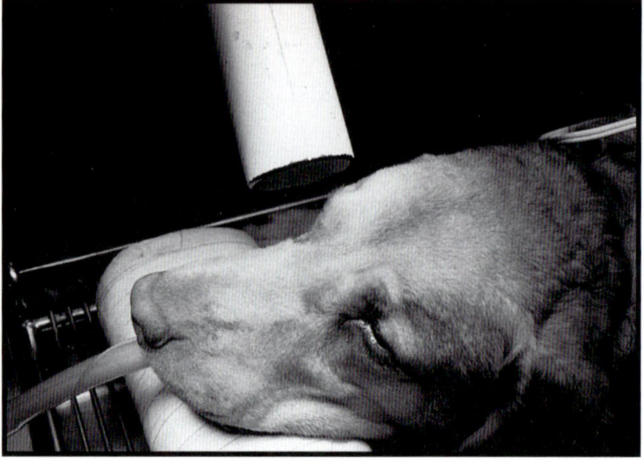

FIGURE 4-12B Patient Positioning

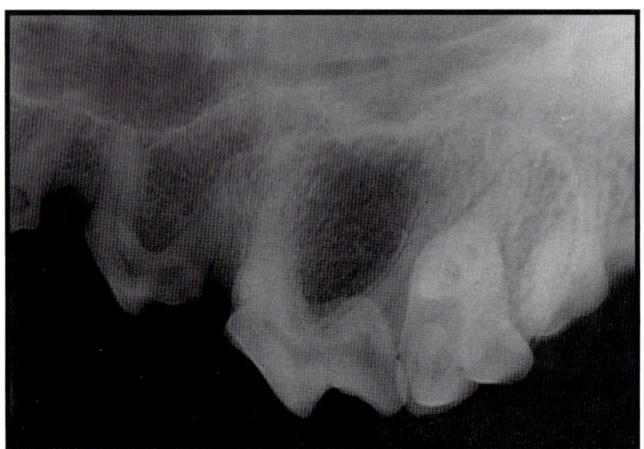

FIGURE 4-12C Radiographic Image

Figure 4-12A *The tubehead should be positioned over the premolars and molars to be radiographed. Although size 4 intraoral film is recommended for most breeds of dogs, size 2 film may be more appropriate if the space within the oral cavity is restricted. The film can be gently bowed to accommodate placement.* **Figure 4-12B** *The dog's head should be placed in ventral recumbency and a roll positioned under the jaw to keep the hard palate parallel to the table top. Intraoral film can then be placed adjacent to the hard palate and inserted until it is near the lingual aspect of the contralateral teeth. The x-ray cone should be positioned lateral to the dentition at the lateral canthus and directed at an angle of 45° to the hard palate. The primary x-ray beam will be almost perpendicular to the bisecting plane.* **Figure 4-12C** *This radiograph shows the resultant image.*

MESIOLATERAL OBLIQUE PROJECTION FOR FOURTH MAXILLARY PREMOLAR—DOGS

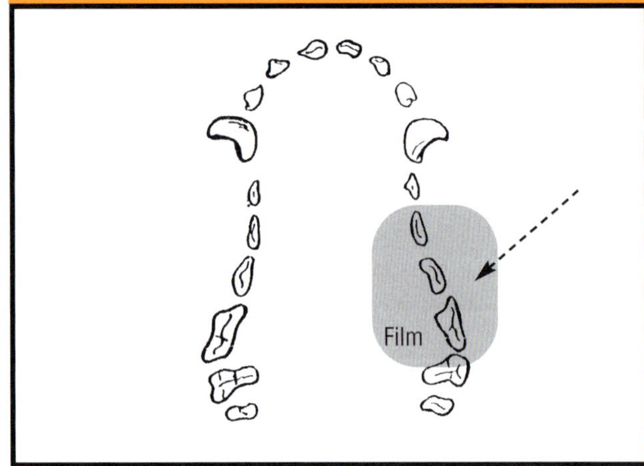

FIGURE 4-13A Image Field and Film Placement

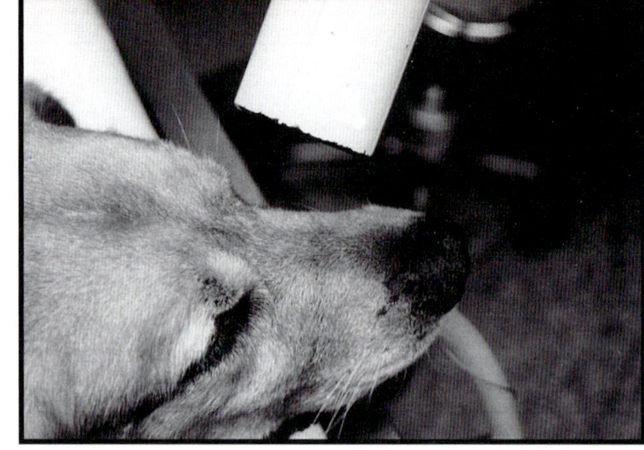

FIGURE 4-13B Patient Positioning

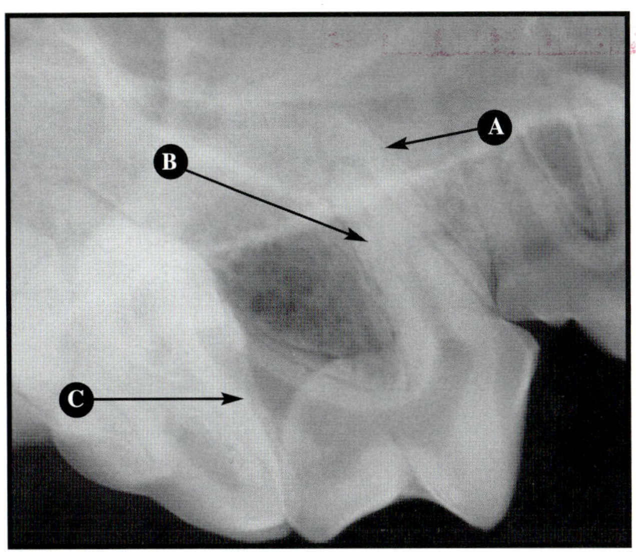

FIGURE 4-13C Radiographic Image

The mesiolateral oblique projection is primarily used when the upper fourth premolar needs to be evaluated for signs of endodontic involvement. **Figure 4-13A** *The tubehead should be positioned over the upper third and fourth premolars. The* arrow *depicts the rostrocaudal angle of the primary x-ray beam. Size 4 intraoral film is recommended for most breeds of dogs; however, size 2 film may be more appropriate for toy and some brachycephalic breeds. The film can be gently bowed to accommodate placement.* **Figure 4-13B** *The dog's head should be placed in ventral recumbency and a roll positioned under the jaw to keep the hard palate parallel to the table top. Intraoral film can then be placed adjacent to the hard palate and inserted until it is near the lingual aspect of the contralateral teeth. The x-ray cone should be vertically positioned at an angle of 45° to the palate and then shifted toward the mesial aspect as shown. The primary x-ray beam will be almost perpendicular to the bisecting plane. If the primary x-ray beam intersects the tooth at a lateral angle, the image of the mesiobuccal and palatal roots will be superimposed. To avoid such superimposition, the tubehead can be shifted mesially so that the beam obliquely strikes the fourth premolar from the mesial to distal aspects. The images of the two mesial roots will in turn shift to prevent superimposition.* **Figure 4-13C** *This radiograph shows the resultant image. (A) The root in the mesial aspect is the palatal root. (B) The image of the mesiobuccal root is centered between that of the other two roots. (C) The first molar is often superimposed over the distal root of the premolar.*

DISTOLATERAL OBLIQUE PROJECTION FOR MAXILLARY FOURTH PREMOLAR—DOGS

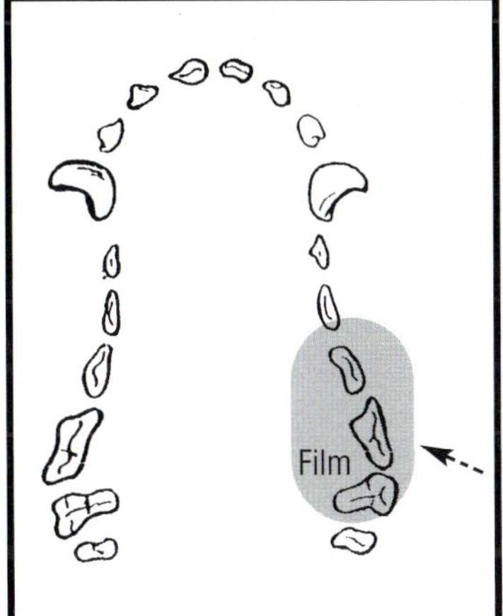

FIGURE 4-14A Image Field and Film Placement

FIGURE 4-14B Patient Positioning

The distolateral oblique projection is primarily used when the upper fourth premolar needs to be evaluated for signs of endodontic involvement. The view shown in this figure is more advantageous than the mesiolateral oblique view because a portion of the distal root of the premolar typically is not superimposed over that of the first molar. **Figure 4-14A** *The tubehead should be positioned over the upper fourth premolar and first molar. The* arrow *depicts the distolateral oblique angle of the primary x-ray beam. Size 4 intraoral film is recommended for most breeds of dogs; however, size 2 film may be more appropriate for toy and some brachycephalic breeds.*

Figure 4-14B The dog's head should be placed in ventral recumbency with a roll positioned under the jaw to keep the hard palate parallel to the table top. Intraoral film can then be placed against the hard palate and inserted until it is near the lingual aspect of the contralateral teeth. The x-ray cone should be positioned at a standard angle of 45° and then obliquely directed from the distal to mesial direction as shown. The primary x-ray beam will be almost perpendicular to the bisecting plane. If the beam intersects the tooth at a lateral angle, the image of the mesiobuccal and palatine roots will be superimposed. To prevent such superimposition, the tubehead should be shifted distally so that the beam obliquely strikes the fourth premolar from a distal to mesial direction. The image of the two mesial roots will shift to prevent superimposition. **Figure 4-14C** *This radiograph shows the resultant image. (A) The image of the palatal root is centered between that of the other two roots.*

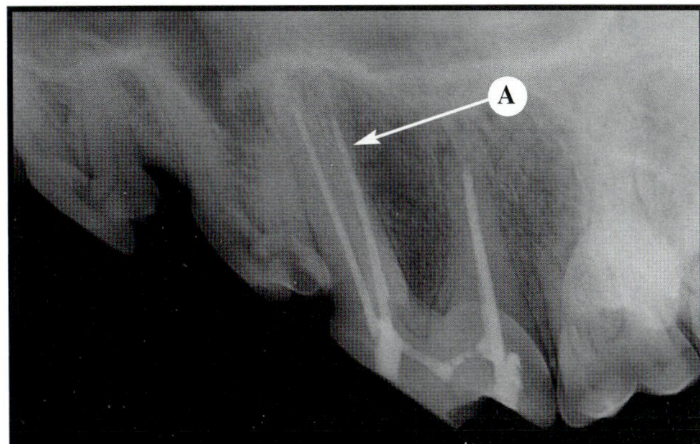

FIGURE 4-14C Radiographic Image

PROJECTION FOR MANDIBULAR INCISORS AND CANINE TEETH—DOGS

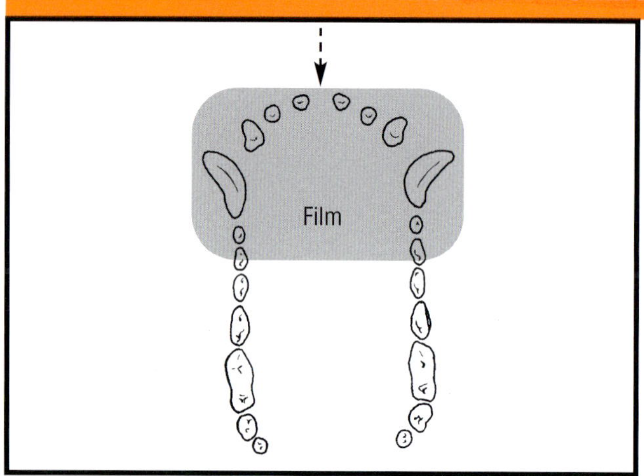

FIGURE 4-15A Image Field and Film Placement

FIGURE 4-15B Patient Positioning

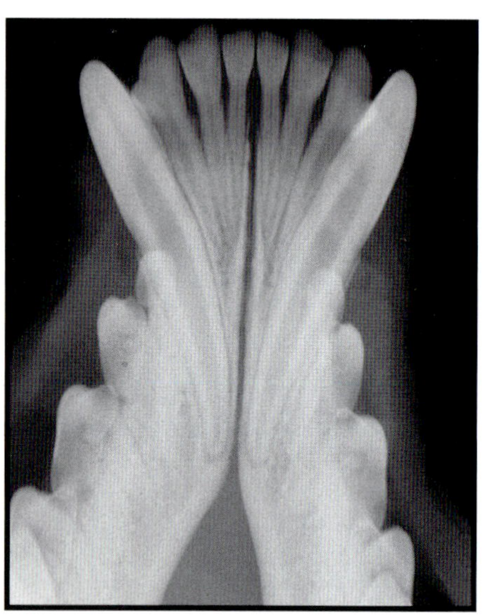

FIGURE 4-15C Radiographic Image

Figure 4-15A *The tubehead should be centered over the mandibular incisors. The arrow depicts the direction of the primary x-ray beam. Although size 4 intraoral film is recommended to obtain an image from the lower dentition to the level of the first or second premolar, size 2 film may be appropriate if the image concentrates on specific teeth.* **Figure 4-15B** *The dog's head should be placed in dorsal recumbency and a small roll inserted under the neck to keep the hard palate parallel to the table top. Size 4 or 2 intraoral film can then be placed below the incisors and canine teeth. The x-ray cone should be directed downward at an approximate angle of 60° to the horizontal plane of the hard palate. The primary x-ray beam will be almost perpendicular to the bisecting plane.* **Figure 4-15C** *This radiograph shows the resultant image. An image of greater diagnostic value for evaluating specific incisors can usually be captured by using the following oblique projection.*

LATERAL OBLIQUE PROJECTION FOR MANDIBULAR CANINE TEETH—DOGS

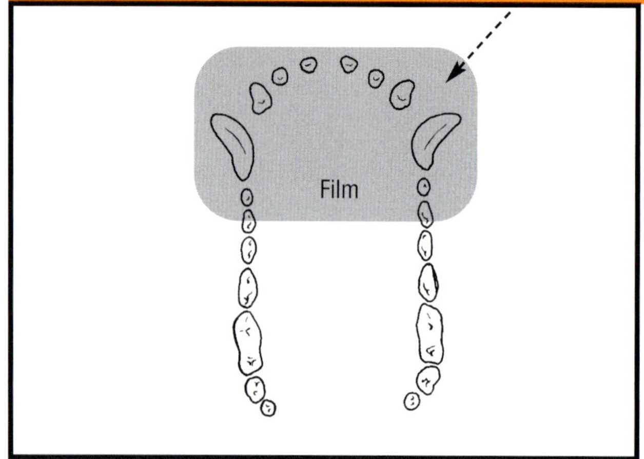

FIGURE 4-16A Image Field and Film Placement

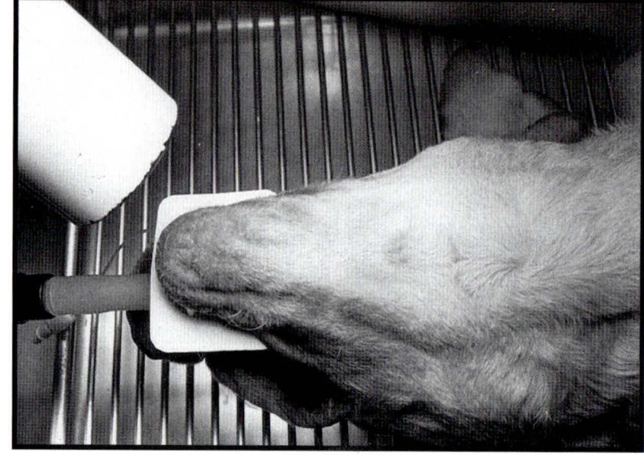

FIGURE 4-16B Patient Positioning

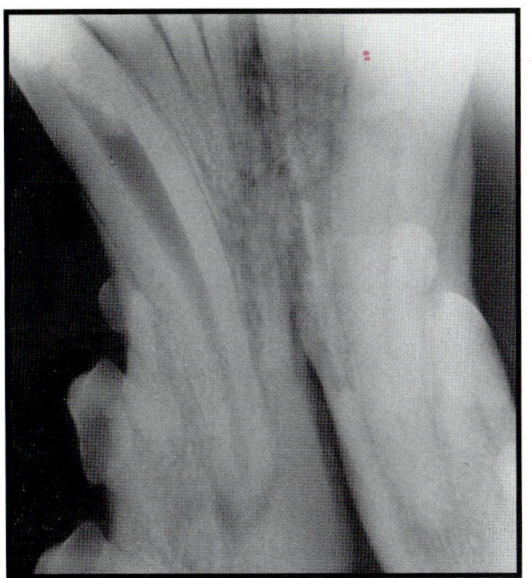

FIGURE 4-16C Radiographic Image

Figure 4-16A *The tubehead should be centered over the lower canine tooth and adjacent incisors, covering to at least the third premolar. The* arrow *depicts the direction of the primary x-ray beam. Size 4 intraoral film is recommended for most breeds of dogs; however, size 2 film may be more appropriate for toy and some brachycephalic breeds. The film can be gently bowed to accommodate placement.* **Figure 4-16B** *The dog's head should be placed in dorsal recumbency and a roll positioned under the neck to keep the hard palate parallel to the table top. Size 4 or 2 intraoral film can then be placed under the canine tooth and distal second premolar. The x-ray cone should be directed toward the interproximal space of the lateral incisor and canine tooth and tipped downward at an angle of 45° to 60° to the hard palate. The apices of the lower canine teeth are usually located at the mesial root of the second premolar. The primary x-ray beam will be almost perpendicular to the bisecting plane.* **Figure 4-16C** *This radiograph shows the resultant image, which is often preferred to viewing the incisors in the same quadrant as the canine tooth.*

LATERAL PROJECTION FOR MANDIBULAR CANINE TEETH—DOGS

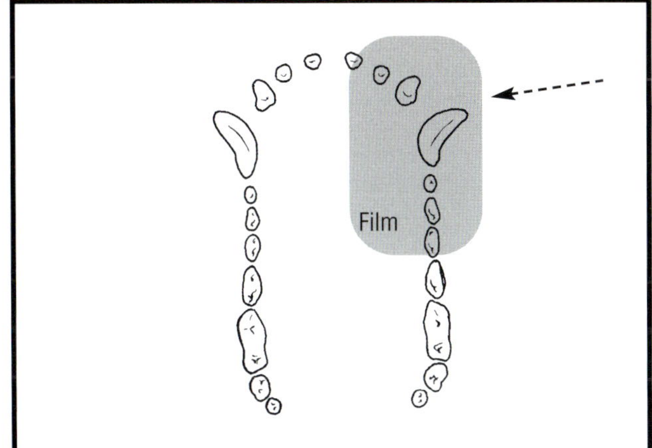

FIGURE 4-17A Image Field and Film Placement

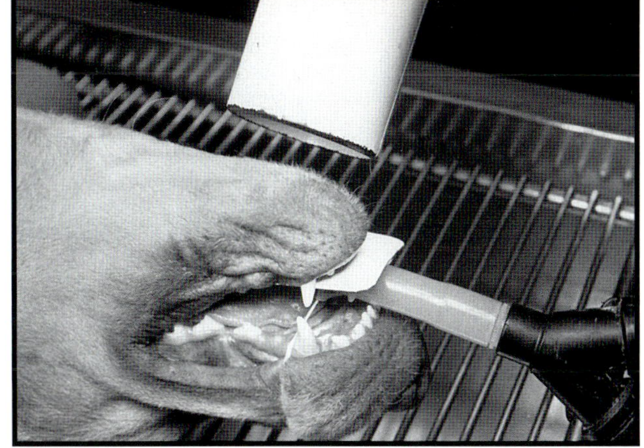

FIGURE 4-17B Patient Positioning

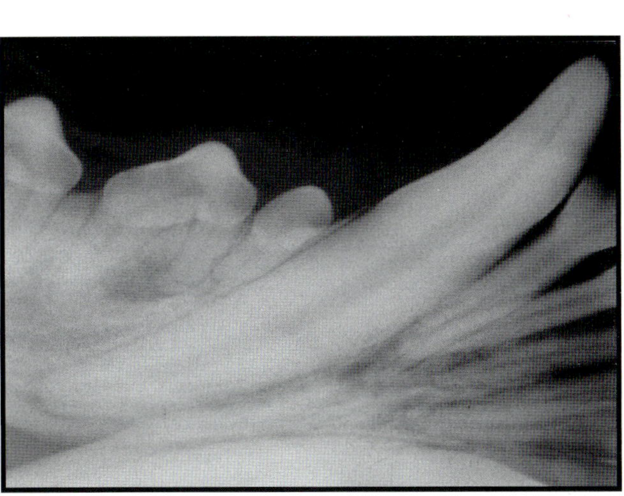

FIGURE 4-17C Radiographic Image

Figure 4-17A *The tubehead should be centered over the lower canine tooth and adjacent premolars. The* arrow *depicts the direction of the primary x-ray beam. Size 4 intraoral film is recommended for most breeds of dogs; however, size 2 film may be more appropriate if the canine tooth only is to be viewed in toy and some brachycephalic breeds. The film can be gently bowed to accommodate placement.* **Figure 4-17B** *The dog's head should be placed in dorsal recumbency and a roll positioned to keep the hard palate parallel to the table top. Size 4 or 2 intraoral film can then be placed under the canine tooth extending toward the third premolar. The x-ray cone should be laterally directed at the ventral border of the mandible as shown and tipped downward at an angle of 45° to 60° to the hard palate. The primary x-ray beam will be almost perpendicular to the bisecting plane.* **Figure 4-17C** *This radiograph shows the resultant image.*

PROJECTION FOR MANDIBULAR PREMOLARS AND MOLARS (PARALLEL TECHNIQUE)—DOGS

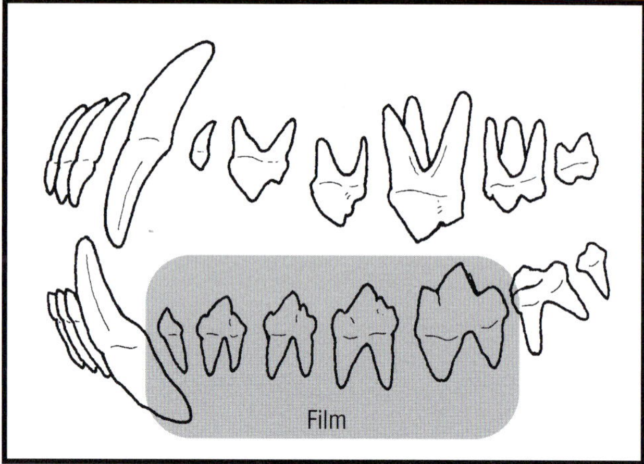

FIGURE 4-18A Image Field and Film Placement

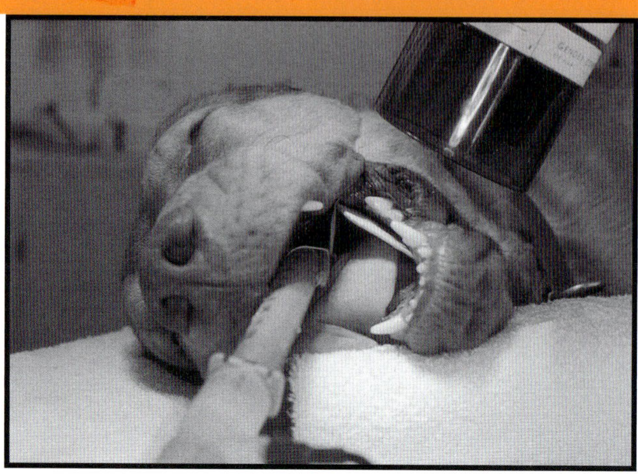

FIGURE 4-18B Patient Positioning

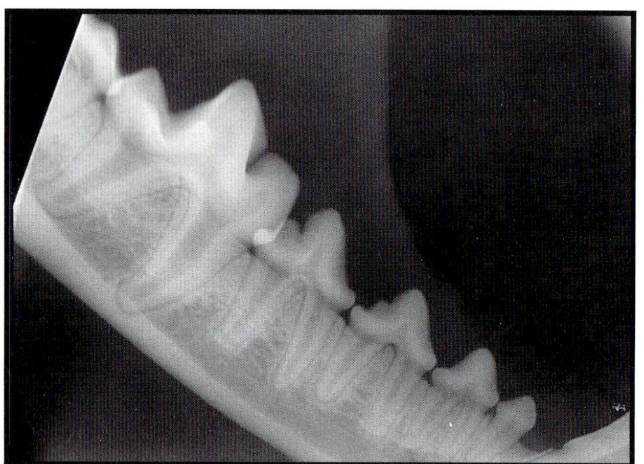

FIGURE 4-18C Radiographic Image

Figure 4-18A *The tubehead should be centered over the mandibular premolars. Size 4 intraoral film is recommended for most breeds of dogs; however, size 2 film may be more appropriate for toy and some brachycephalic breeds. The film can be gently folded to accommodate placement.*
Figure 4-18B *The dog's head should be placed in lateral recumbency and a spacer inserted under the jaw to keep the hard palate perpendicular to the plane of the table top. Size 4 or 2 intraoral film can then be placed between the tongue and the mandible parallel to the long axis of the teeth. The x-ray cone should be perpendicular to the teeth and film at an approximate angle of 60° to the horizontal plane of the table top. The primary x-ray beam will be almost perpendicular to the film and teeth.*
Figure 4-18C *This radiograph shows the resultant image.*

PROJECTION FOR MANDIBULAR MOLARS (PARALLEL TECHNIQUE)—DOGS

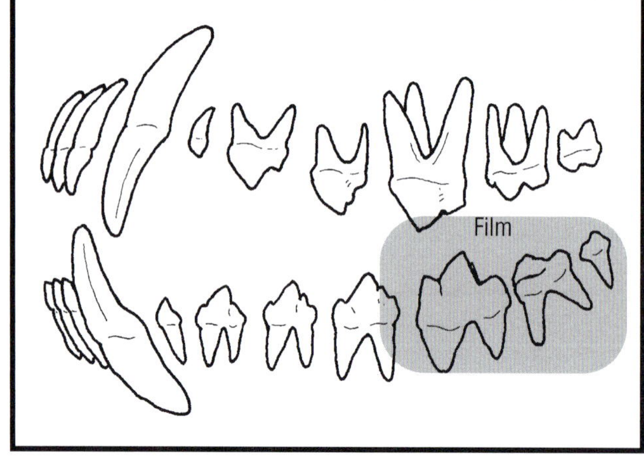

FIGURE 4-19A Image Field and Film Placement

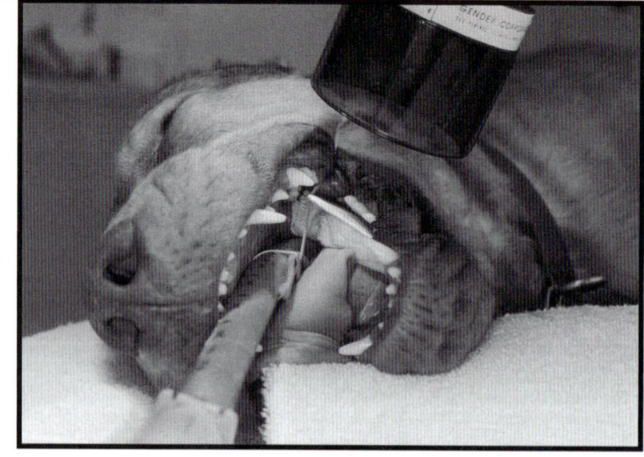

FIGURE 4-19B Patient Positioning

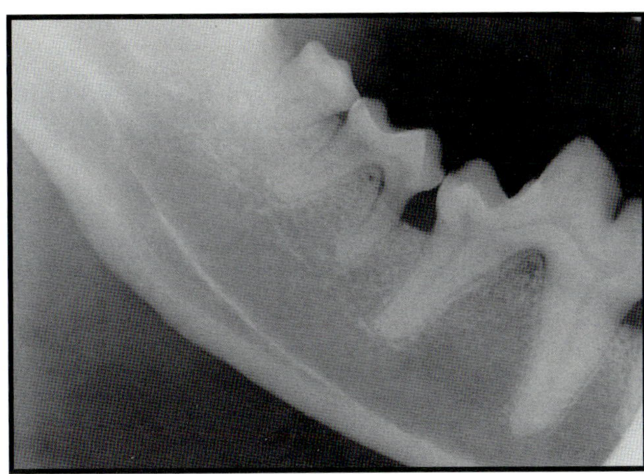

FIGURE 4-19C Radiographic Image

Figure 4-19A *The tubehead should be centered over the mandibular molars, making sure that the third molar will be captured on film. Because of the restricted space, size 2 intraoral film may be more appropriate than size 4 film. The film can be gently folded to accommodate placement.* **Figure 4-19B** *The dog's head should be placed in lateral recumbency and a spacer inserted under the jaw to keep the hard palate perpendicular to the plane of the table top. Size 2 or 4 intraoral film can then be placed between the tongue and mandible parallel to the long axis of the teeth. The x-ray cone should be positioned perpendicular to the film and teeth at an approximate angle of 60° to the horizontal plane of the table top. Gauze sponges can be used to stabilize the film in the back of the mouth. All sponges and the film should be removed at the same time.* **Figure 4-19C** *This radiograph shows the resultant image.*

PROJECTION FOR MAXILLARY INCISORS—CATS

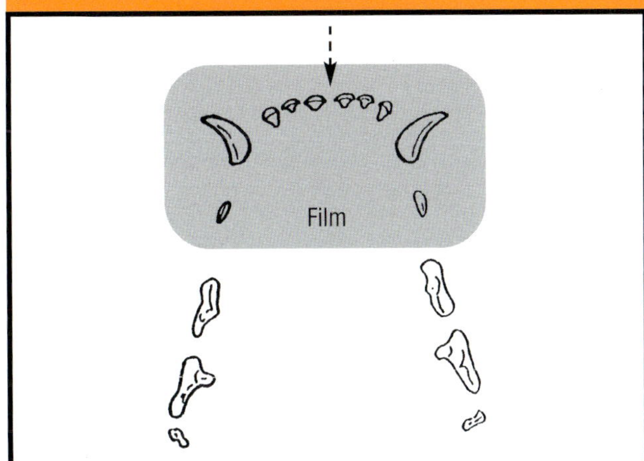

FIGURE 4-20A Image Field and Film Placement

FIGURE 4-20B Patient Positioning

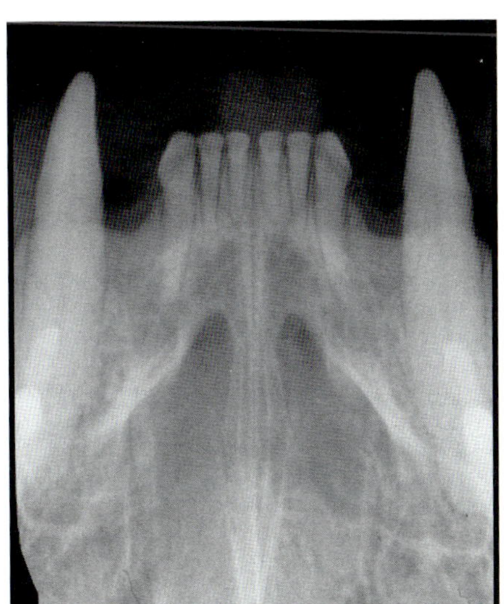

FIGURE 4-20C Radiographic Image

Figure 4-20A *The tubehead should be centered over the upper incisors. The arrow depicts the angle of the primary x-ray beam. Size 2 film is recommended for viewing all of the upper dentition to the level of the first premolar. Size 4 film can also fit in a cat's mouth if preferred.* **Figure 4-20B** *The cat's head should be placed in ventral recumbency with a small roll inserted under the chin to keep the hard palate parallel to the table top. Size 2 or 4 intraoral film can then be placed below the incisors and canine teeth. The x-ray cone should be directed downward at an angle of 60° to the horizontal plane or the hard palate. The primary x-ray beam will be almost perpendicular to the bisecting plane.* **Figure 4-20C** *This radiograph shows the resultant image.*

LATERAL OBLIQUE PROJECTION FOR MAXILLARY CANINE TEETH—CATS

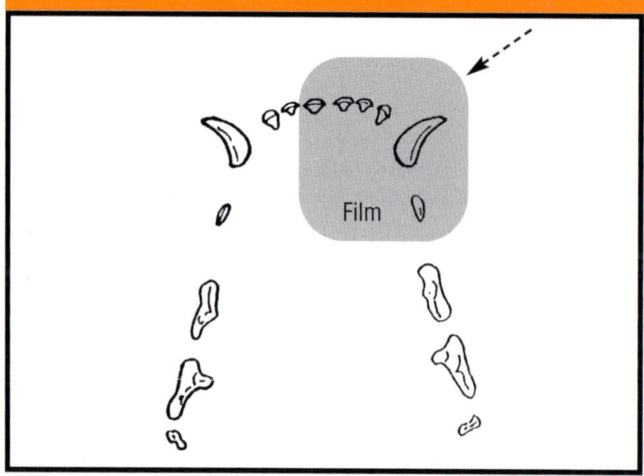

FIGURE 4-21A Image Field and Film Placement

FIGURE 4-21B Patient Positioning

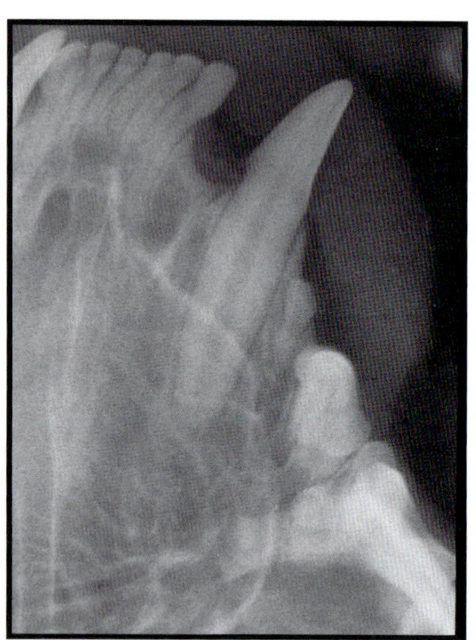

FIGURE 4-21C Radiographic Image

Figure 4-21A *The tubehead should be centered over the upper canine tooth and adjacent premolars. The* arrow *depicts the direction of the primary x-ray beam. Size 2 intraoral film is recommended for cats.* **Figure 4-21B** *The cat's head should be placed in ventral recumbency and a roll positioned under the jaw to keep the hard palate parallel to the table top. Size 2 intraoral film can then be placed under the canine tooth and premolars. The x-ray cone should be horizontally directed at an angle of 45° as shown and tipped downward at an angle of 45° to 60° to the horizontal plane of the hard palate. The primary x-ray beam will be almost perpendicular to the bisecting plane.* **Figure 4-21C** *This radiograph shows the resultant image.*

LATERAL PROJECTION FOR MAXILLARY CANINE TEETH AND PREMOLARS—CATS

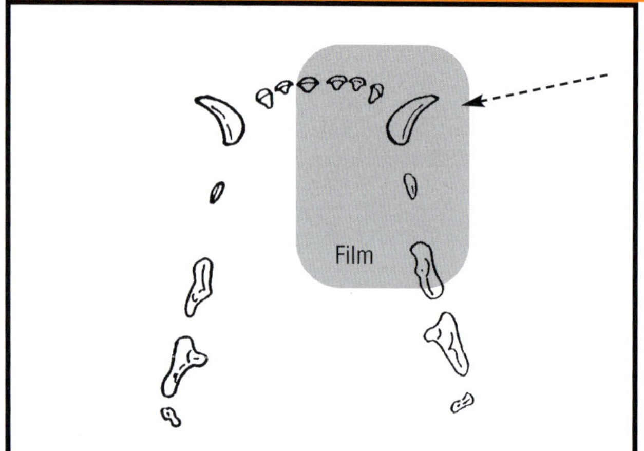

FIGURE 4-22A Image Field and Film Placement

FIGURE 4-22B Patient Positioning

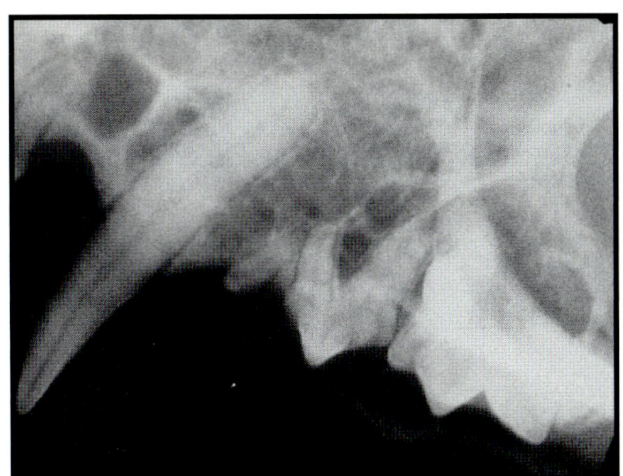

FIGURE 4-22C Radiographic Image

Figure 4-22A *The tubehead should be centered over the upper canine tooth and second premolar. The* arrow *depicts the direction of the primary x-ray beam. Size 2 intraoral film is recommended for viewing all upper dentition to the third premolar. Size 4 intraoral film can also be used if preferred.* **Figure 4-22B** *The cat's head should be placed in ventral recumbency and a roll positioned under the jaw to keep the hard palate parallel to the table top. Intraoral film can then be placed under the canine tooth, hard palate, and premolars. The x-ray cone should be directed toward the lateral aspect of the cat's muzzle as shown and tipped downward at an angle of 45° to the horizontal plane of the hard palate. The primary x-ray beam will be almost perpendicular to the bisecting plane.* **Figure 4-22C** *This radiograph shows the resultant image.*

PROJECTION FOR MAXILLARY PREMOLARS AND MOLAR—CATS

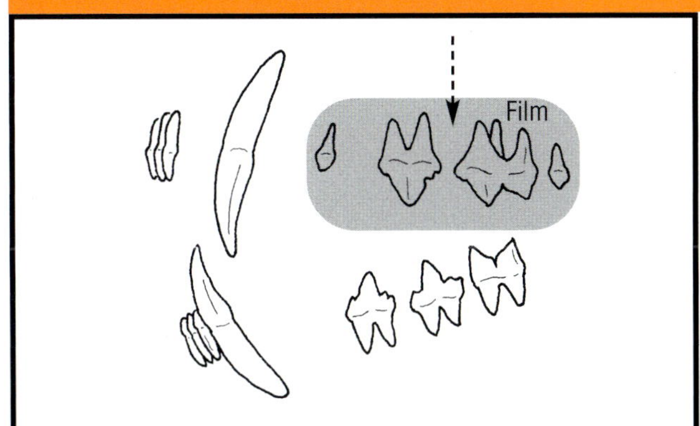

FIGURE 4-23A Image Field and Film Placement

FIGURE 4-23B Patient Positioning

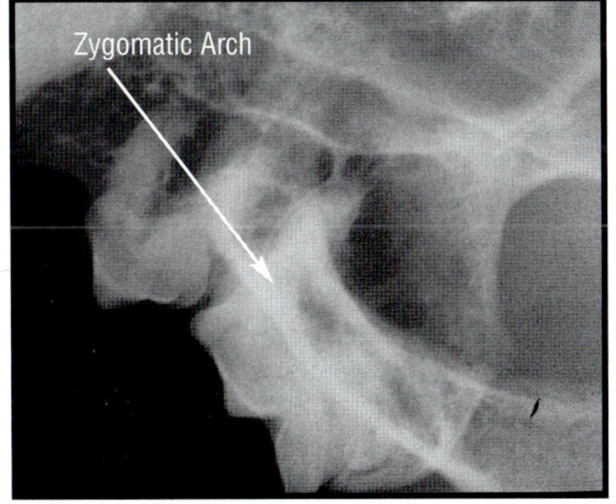

FIGURE 4-23C Radiographic Image at 45°

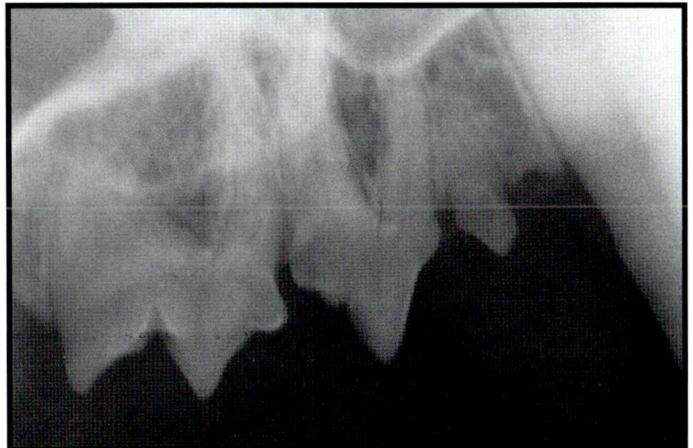

FIGURE 4-23D Radiographic Image at 30°

Figure 4-23A *The tubehead should be centered over the side of the cat's face just below the lateral canthus. The mesial edge of the cone is near the medial canthus, and the distal edge of the cone is at the base of the ear. The* arrow *depicts the direction of the primary x-ray beam. Size 2 or 4 intraoral film can be used to view the entire upper lateral dentition.* **Figure 4-23B** *The cat's head should be placed in ventral recumbency and a small roll inserted under the neck to keep the hard palate parallel to the table top. Size 2 or 4 intraoral film can then be placed under the hard palate. The x-ray cone should be directed downward at an approximate angle of 30° to the horizontal plane of the table top or hard palate to minimize superimposition of the zygomatic arch. The primary x-ray beam should not be perpendicular to the bisecting plane.* **Figure 4-23C** *The central x-ray beam must be directed at an angle of 30° to the cat's head and not at an angle of 45°. Such positioning is necessary because the zygomatic arch in cats superimposes the upper premolars and molars (see the description of the tube shift technique in Chapter 2).* **Figure 4-23D** *This radiograph shows the resultant image when the x-ray beam is positioned at an angle of 30° because the patient is a cat. The image is slightly elongated because of the unique positioning required.*

PROJECTION FOR MANDIBULAR INCISORS AND CANINE TEETH—CATS

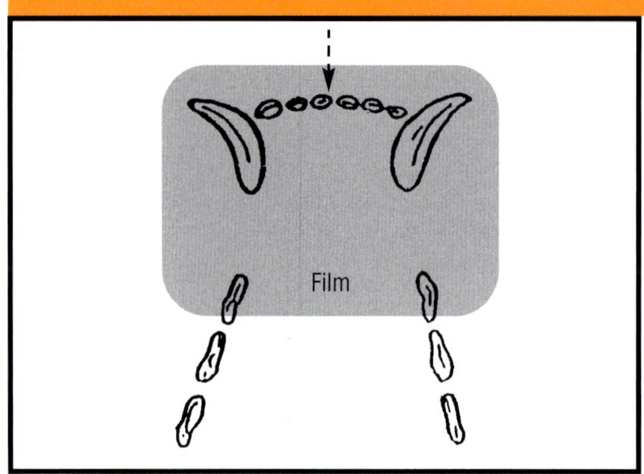

FIGURE 4-24A Image Field and Film Placement

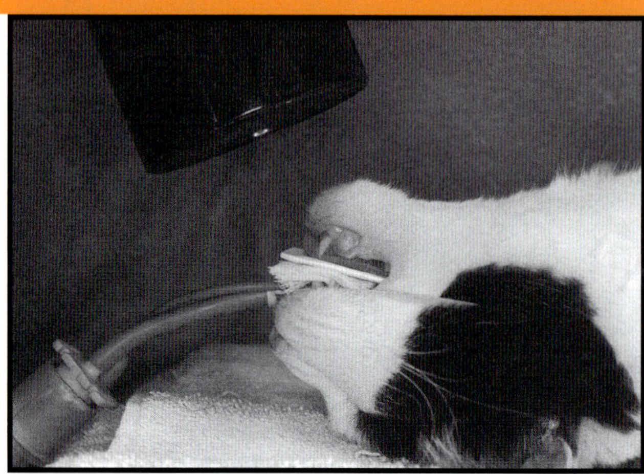

FIGURE 4-24B Patient Positioning

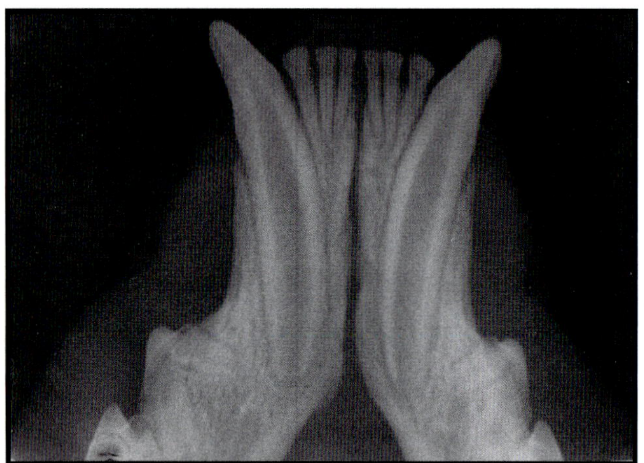

FIGURE 4-24C Radiographic Image

Figure 4-24A *The tubehead should be centered over the lower incisors. The arrow depicts the direction of the primary x-ray beam. Size 2 intraoral film is recommended for radiographs of the entire lower dentition to the canine teeth; however, size 4 film can be used if preferred.* **Figure 4-24B** *The cat's head should be positioned in dorsal recumbency and a small roll inserted under the neck to keep the hard palate parallel to the table top. Intraoral film can then be placed below the incisors and canine teeth. The x-ray cone should be directed at an approximate angle of 60° to the horizontal plane of the hard palate or table top. The primary x-ray beam will be almost perpendicular to the bisecting angle.* **Figure 4-24C** *This radiograph shows the resultant image.*

LATERAL OBLIQUE PROJECTION FOR MANDIBULAR CANINE TEETH—CATS

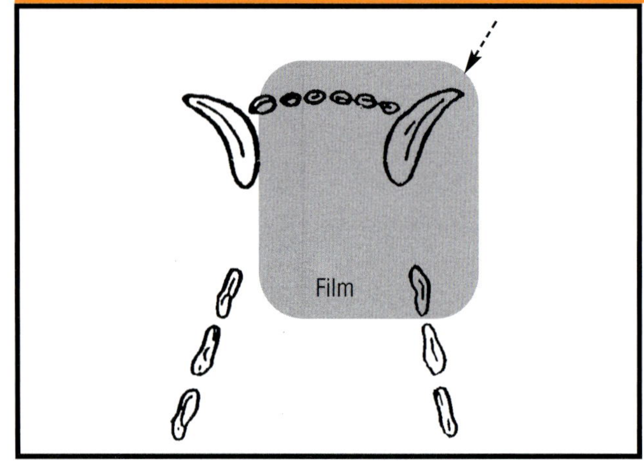

FIGURE 4-25A Image Field and Film Placement

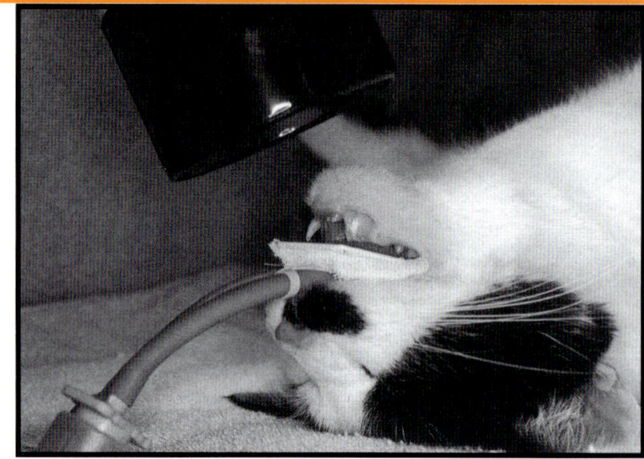

FIGURE 4-25B Patient Positioning

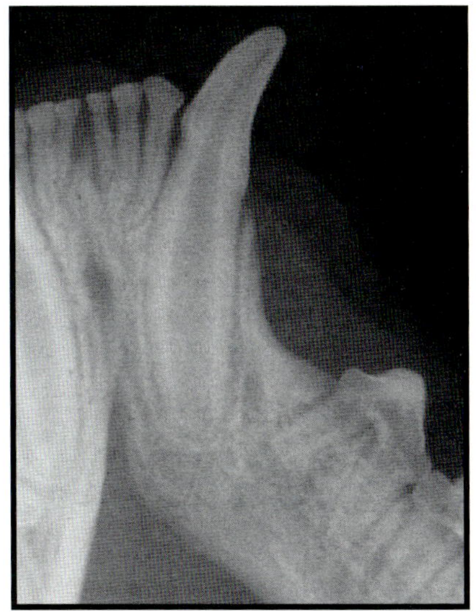

FIGURE 4-25C Radiographic Image

Figure 4-25A *The tubehead should be centered over the mandibular canine tooth. The* arrow *depicts the direction of the x-ray beam. Unlike oblique projection in dogs, the premolars and canine teeth do not usually become superimposed for radiographs of cats. The view shows the interdental (interproximal) bone of the canine tooth and third premolar. Size 2 intraoral film is recommended.* **Figure 4-25B** *The cat's head should be placed in dorsal recumbency and a roll positioned under the neck to keep the hard palate parallel to the table top. Size 2 intraoral film can then be placed under the canine tooth and third premolar. The x-ray cone should be directed toward the interproximal space of the lateral incisor and canine tooth as shown and tipped downward at an angle of 45° to 60° to the hard palate. The primary x-ray beam will be almost perpendicular to the bisecting plane. Gauze sponges can be used to stabilize the film in the back of the mouth. All sponges and the film should be removed at the same time.* **Figure 4-25C** *This radiograph shows the resultant image.*

LATERAL PROJECTION FOR MANDIBULAR PREMOLARS AND MOLAR (PARALLEL TECHNIQUE)—CATS

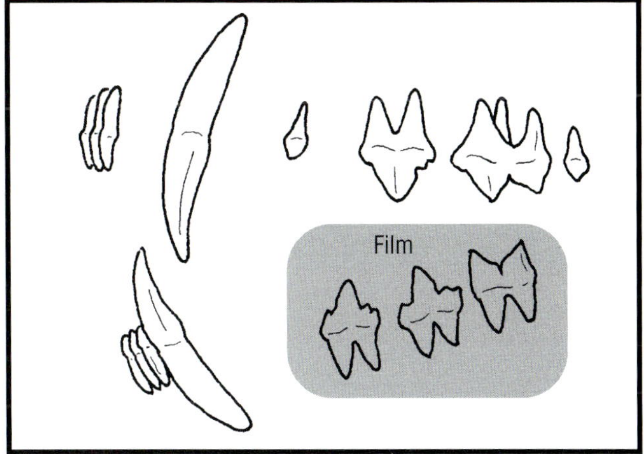

FIGURE 4-26A Image Field and Film Placement

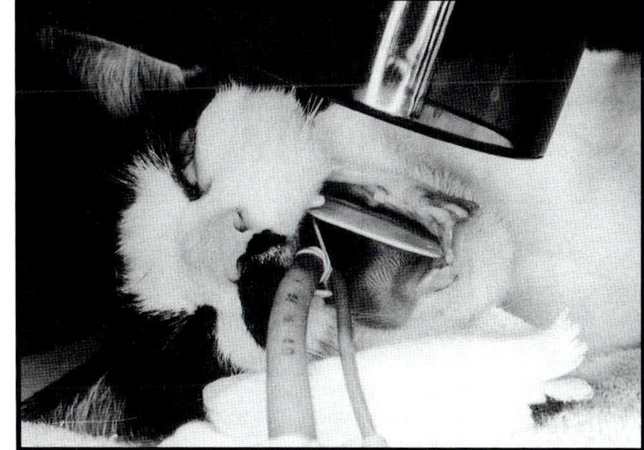

FIGURE 4-26B Patient Positioning

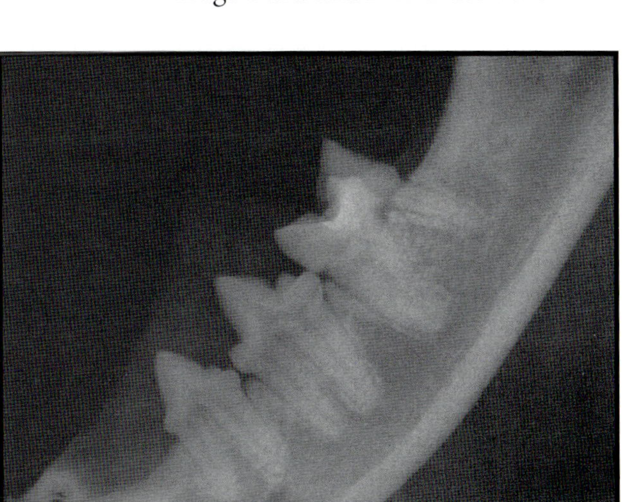

FIGURE 4-26C Radiographic Image

Figure 4-26A *The tubehead should be centered over the mandibular premolars and molar. Size 2 intraoral film is recommended.* **Figure 4-26B** *The cat's head should be placed in lateral recumbency and a spacer inserted under the jaw to keep the hard palate perpendicular to the plane of the table top. Size 2 intraoral film can then be placed in the vestibule between the tongue and mandible, making sure that the film does not traumatize the symphyseal area during insertion and removal. The x-ray cone should be positioned perpendicular to the film and teeth at an angle of about 60° to the horizontal plane or table top. The primary x-ray beam will be almost perpendicular to the film and teeth.* **Figure 4-26C** *This radiograph shows the resultant image.*

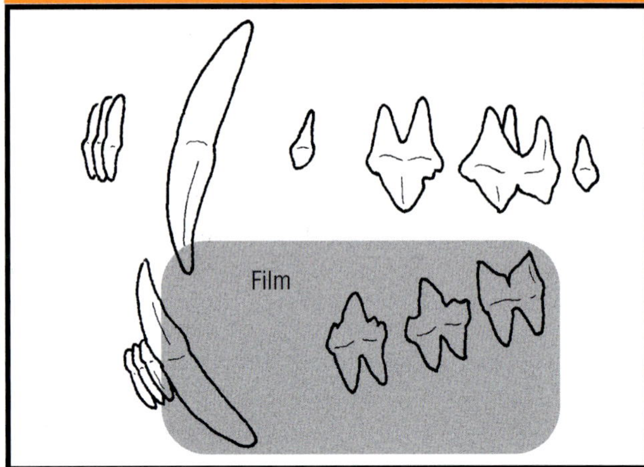

FIGURE 4-27A Image Field and Film Placement

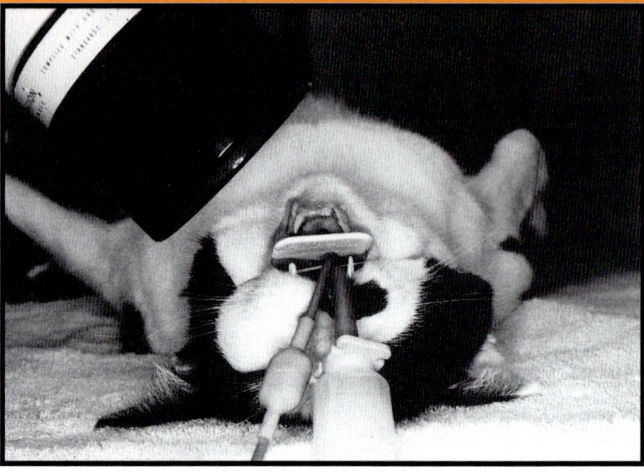

FIGURE 4-27B Patient Positioning

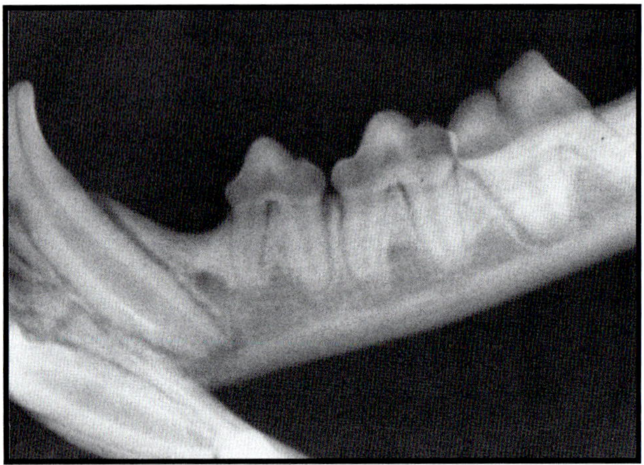

FIGURE 4-27C Radiographic Image

Figure 4-27A *The tubehead should be laterally centered over the hemimandible. This view can capture the area between the mandibular canine tooth and third premolar. Lateral views often miss this area because the symphysis limits film placement. Size 2 or 3 intraoral film is recommended.* **Figure 4-27B** *The cat's head should be placed in dorsal recumbency and a small roll inserted under the neck to keep the hard palate parallel to the table top. Size 2 or 3 intraoral film can then be placed below the tongue and hemimandible. The x-ray cone should be directed toward the border of the mandible at an angle of 45° to the horizontal plane (table top). The primary x-ray beam will be almost perpendicular to the bisecting angle.* **Figure 4-27C** *This radiograph shows the resultant image.*

TECHNICAL ERRORS AND TROUBLESHOOTING

Technical errors can occur at any stage: film placement, patient positioning, angle of the x-ray beam, exposure, processing, storage, or any combination of these stages. The result will be a radiograph of substandard quality. By being able to recognize in which stage the error most likely occurred, the problem can be quickly corrected.

PROBLEMS WITH THE RADIOGRAPHIC IMAGE

Substandard radiographic images result from lack of clarity, magnification, and distortion.[1-3] Lack of clarify (i.e., a blurry outline) and magnification can be prevented by placing the film close to the subject, using the furthest focal film distance that is practical, and limiting movement while the radiograph is being taken.[1-3]

When using the bisecting angle technique, the most common errors occur when the angle of the x-ray beam is incorrect. Incorrect angle can cause elongation, foreshortening, and superimposition (see Chapter 2). Elongation and foreshortening occur when the angle is vertically incorrect, whereas superimposition occurs when the angle is horizontally incorrect.[1] The bisecting angle technique can be challenging because of the diverse anatomy of dogs and cats, numerous film positions, and varied orientations when setting the x-ray beam. By following the techniques described in Chapter 4, technical problems with the radiographic image can be minimized. If patient positioning and film placement are correct but the radiographic image is substandard, adjusting the angle of the x-ray beam should resolve the problem (Table 5-1).

TABLE 5 – 1[1-4]
PROBLEMS WITH INTRAORAL IMAGES

Problem	Cause	Correction
Blurred Image	Patient, machine, or film moved during exposure	Retake radiograph
Double Image	Double exposure	Retake radiograph using fresh film and depress hand switch only once
Partial Image	Head of x-ray cone misaligned or film not completely immersed in developer	Check alignment of x-ray cone and replenish fluid levels
Image Distortion	Excessive bending of film during positioning	Keep film as flat as possible
Light or Blank Film	Hand switch depressed incorrectly or x-ray unit malfunctioning	Depress the switch completely for full exposure time
Inconsistent Film Density	Hand switch depressed incorrectly	Depress the switch completely for full exposure time
Short, Stubby, or Dense Teeth	X-ray beam aligned too close to the vertical aspect (foreshortening)	Realign x-ray beam toward the horizontal plane
Abnormally Long Teeth	X-ray beam aligned too close to the horizontal aspect (elongation)	Realign x-ray beam toward the vertical plane
Cervical Burnout	Too much radiation on the tooth neck	Reduce exposure settings (altering the horizontal angle of the x-ray beam may help)
Fuzzy, Indistinct Roots	Excessive horizontal angle of x-ray beam	Adjust horizontal angle
Blurry, Magnified Image	Film too far from subject (short x-ray cone)	Place film as close to subject as possible and use longer x-ray cone
Grainy Appearance	Low-intensity x-ray beam or elevated developer temperature	Alter exposure settings and check chemical temperature
Superimposition	X-ray beam aligned too far in mesial or distal direction	Realign x-ray beam

TABLE 5 – 2[1-4]
LIGHT RADIOGRAPHS WITH POOR CONTRAST

Problem	Cause	Correction
Underdevelopment	Low chemical temperature	Adjust temperature and verify accuracy of thermometer
	Developing time too short	Lengthen developing time
	Depleted or contaminated developer	Change developer
	Diluted developer	Change developer
	Overfixation	Shorten fixation time
Underexposure	Film positioned backward	Correct film positioning
	Insufficient exposure time	Increase exposure time
	Low milliamperes or kilovolt peak	Adjust exposure settings
	Excessive distance from film to tube	Adjust technique settings or correct distance
	Hand switch not depressed properly	Engage switch for the entire exposure time

TABLE 5 – 3[1-4]
DARK RADIOGRAPHS WITH POOR CONTRAST

Problem	Cause	Correction
Overdevelopment	High temperature	Adjust temperature
	Developing time too long	Adjust time
	Concentrated developer	Change developer
Light Exposure	Darkroom leaks or improper safelight	Evaluate integrity of darkroom and evaluate safelight
Radiation Exposure	Unexposed film stored too close to x-ray beam	Relocate unexposed film to radiation-safe area or container
	Excessive milliamperes, kilovolt peak, or time	Adjust settings
Overexposure	X-ray tube too close to subject	Increase distance from subject to tube

FILM DENSITY

When a radiograph is too dark or light or the contrast is poor, film exposure is typically the culprit. A dark radiograph generally indicates overexposure, whereas a light radiograph indicates underexposure. The parameters that govern exposure are kilovolt peak (kVp), milliamperes (mA), time, and distance.[1]

Three of the parameters (kilovolt peak, milliamperes, and time) are machine related. If a dental x-ray machine with pre-set kilovolt peak and milliampers is being used, only the time can be adjusted. Decreasing the exposure time corrects overexposure and increasing the time corrects underexposure (Tables 5-2 and 5-3).

If all three parameters can be adjusted on the x-ray machine, adjustments can be made to correct underexposed or overexposed film or change the contrast. If the contrast in the radiograph is high and more shades of gray are needed to correct the problem, the kVp setting should be higher and the exposure time decreased. If the contrast is low, the kVp setting should be lower and the exposure time increased.[2]

The distance between the x-ray source and the area being radiographed can also affect the resultant image because the intensity of the x-ray beam decreases with distance (Inverse Square Law). Magnification and blurry images can be eliminated by having the x-ray source as far as possible from the area being radiographed. A substantial change in focal film distance, however, requires an appropriate change in exposure technique.[3]

FILM PROCESSING AND STORAGE

If the radiograph is too dark or light and the exposure technique is correct, the cause is most likely an error in processing the film. The steps in processing are wetting, developing, rinsing, fixing, washing, and drying. The condition and temperature of the chemicals are very important. The performance of processing chemicals is markedly affected by temperature and use. Because the chemicals degrade quickly, they should be restored daily with replenishing solutions provided by the manufacturer and periodically replaced completely. If the chemicals become contaminated, they must be immediately replaced. Likewise, the condition of the water used for washing and rinsing is important. Fresh, distilled water should always be used for hand processing.

Additional factors that can affect the resultant radiograph-

TABLE 5 – 4[1-4]
GRAY RADIOGRAPHS WITH POOR DETAIL AND POOR CONTRAST

Problem	Cause	Correction
Safelight Problems	Damaged or improper filter	Check filter Film may require a red filter instead of orange filter
	High bulb wattage	Bulb should be no more than 15 watts
	Safelight too close to work area	Safelight should be farther than 4 feet from work area
	Excessive exposure to the safelight	Work more efficiently
Light Leaks	Defective safelight filter	Check filter
	Cracks around door or glovebox in chairside developer	Check darkroom
Radiation Exposure	Film stored near radiation area	Store unexposed film in radiation-safe area or containers
Overdevelopment	Temperature too high	Adjust chemical temperature
Chemical Problem	Developer contaminated with fixer	Clean tank thoroughly Replace developer
	Depleted solutions	Replace developer and fixer
Bad Film	Film stored at high temperature or humidity	Change storage environment
	Old film	Check expiration date

TABLE 5 – 5[1-4]
RADIOGRAPHS WITH A GLAZED OR FROSTED APPEARANCE

Problem	Cause	Correction
Inadequate Fixation	Insufficient time	Adjust fixation time
	Incorrect temperature	Check temperature
	Exhausted solution	Replace solution
Inadequate washing	Insufficient time	Adjust wash time
	Insufficient water level	Check water level and water lines and check the drain plugs
Automatic Processor Problem	Contaminated rollers	Clean rollers
	Malfunctioning rollers	Repair rollers

ic image include film handling, darkroom conditions, and condition of the tanks used for manual processing or the rollers on an automatic processor[1-3] (Tables 5-2 to 5-7).

Common problems that can arise when film is hand processed include the presence of spots, stains, scratches, or mars. In addition, environmental deterioration can occur from improper storage.

Spotted and stained films are usually the result of incomplete rinsing or failing to isolate the film in a clean, dry environment between processing and storing. The most common reason for the presence of spots, however, is keeping processed film in the proximity of film that is being developed in the darkroom; splashed chemicals can contaminate the work space or film that has already been processed.

The amount of scratches and mars on film depends on how carefully it is handled during processing and how well it is protected during the drying period. Because film emulsion is soft, the film can be easily damaged until it has dried for at least 12 hours. We recommend drying film for 12 to 24 hours before storage to prevent it from sticking to other film or materials during storage.

Film deterioration is usually attributed to inadequate fixation, inadequate rinsing or washing, or a poorly controlled storage environment. The radiographic image can deterioriate if the film has not been properly fixed. In addition, the film can gradually turn yellow or brown if chemical residues have not been adequately rinsed from the film. Film that is rapidly processed should be placed in conventional fixative for 10 minutes and washed for 20 minutes before it is allowed to dry. The storage area should be free of sunlight, humidity, and extreme ambient temperature[1-4] (Table 5-6).

If intraoral film is used, it should be stored differently than standard-size film. Mounts for labeling and small envelopes can protect archived intraoral film. In addition, a separate filing system should be used.

QUALITY CONTROL

Quality control can be effective if general guidelines are

TABLE 5 – 6[1-4]
RADIOGRAPHS WITH SPOTS, STAINS, AND MARKS

Problem	Cause	Correction
Dark Spots, Lines, or Smudges	*Predeveloping contamination of film with developer solution*	*Careful technique required*
	Paper film wrapper stuck to film during processing	*Be sure all remnants of wrapper are removed before processing film*
	Film in contact with tank or other film during fixation	*Process films separately*
Dark Lines or Crescent Marks	*Film bent or creased*	*Handle film carefully and keep it flat*
White Spots	*Film contamination with fixer before processing*	*Careful technique required*
	Film in contact with tank or other film during development	*Process film separately*
Fingerprints	*Touching film surface with fingers*	*Process films with clean, dry hands and touch on edges only*
Nonuniform Development	*Trapped air bubbles prevented contact with developer*	*Agitate film while processing*
	Depleted developer	*Replace developer*
	Depleted fixer	*Replace fixer*
Yellow or Brown Stains	*Inadequate fixation*	*Adjust fixation time*
	Inadequate washing (incomplete removal of fixer)	*Use fresh water* *Increase washing time*
	Contaminated solutions or rollers	*Change chemicals* *Clean rollers*
Streaks	*Inadequate washing*	*Increase washing time*
	Contaminated wash water	*Change wash water*
	Contaminated hanger clips	*Use clean hanger clips only*
	Uneven concentration of developer	*Agitate film immediately on entry into developer*
Black Spiderweb-like Lines, Lightning Streaks, or Wooly Worms	*Static electricity generated by opening packet too rapidly*	*Open packet more slowly*

first film, the coin should be removed from the second film and the film processed. If an image of the coin appears on the second film, light leaks are present in the darkroom. Such leaks can affect the quality of radiographs.[2,3]

Common errors that result from improper patient positioning and film placement, inaccurate angle of the x-ray beam, and poor processing and storage techniques are presented in Figures 5-3 through 5-53. Figures 5-3 through 5-8 apply to the bisecting angle technique, whereas the errors shown in Figures 9 and 10 can occur with improper parallel positioning. Readers should note that the bisecting angle technique may require some practice to produce quality film of diagnostic value (see Chapter 4). Although errors rarely occur when using parallel positioning, they are possible. Most errors occur in patient positioning and film placement. Consistently following the techniques demonstrated in Chapter 4 can minimize errors associated with positioning. Figures 5-11 through 5-27 show the most common errors. Incorrect exposure, errors made in the darkroom, and improper handling and storage are shown in Figures 5-28 through 5-53.

followed. To ensure that proper techniques are being used, good-quality film should be used as a control for comparison. The condition of processing chemicals can be monitored by periodically completing a step-wedge test[2,3] as shown in Figure 5-1. In addition, an easy coin test can be done to check for light leaks in the darkroom (Figure 5-2). First, a radiograph should be taken with a double packet of film. While one film is being processed, the other film should be set aside and a coin (usually a penny) placed on top of it. After processing the

REFERENCES

1. Goaz PW, White SW: *Oral Radiology: Principles and Interpretation*, ed 3. Philadelphia, CV Mosby Co, 1994, pp 97–125.
2. Miles DA, Van Dis ML, Raxmus TF: *Basic Principles of Oral and Maxillofacial Radiology*. Philadelphia, WB Saunders Co, 1992, pp 49–71, 155–184.
3. Razmus TF, Williamson GF: *Current Oral and Maxillofacial Imaging*. Philadelphia, WB Saunders Co, 1996, pp 50–78, 197–232.
4. Langlais RP, Kasle MJ: *Exercises in Oral Radiographic Interpretation*, ed 3. Philadelphia, WB Saunders Co, pp 19–37, 208–211.

TABLE 5 – 7[1-4]
AUTOMATIC PROCESSOR PROBLEMS

Problem	Cause	Correction
Light Radiograph	Developer problems: low temperature, depleted developer, developer contaminated with fixer, no agitation	Check temperature, clean tank, replace developer, check agitator paddles
Dark Radiograph	High temperature or water supply off	Check water supply, check temperature of developer and water supply
Gray Radiograph	High temperature, developer contaminated with fixer, bad film, light leaks, safelight problems	Check temperature, clean tank, replace developer, check for light leaks, check safelight, check film storage
Cloudy or Frosty Radiograph	Depleted fixer, insufficient washing, roller problems	Replace fixer, check water supply, check the plug in the wash tank, clean rollers
Damp Films	Depleted fixer, insufficient washing, low dryer temperature	Replace fixer, check water supply, adjust dryer temperature
Marks on Film	Roller problems, depleted fixer, high temperature	Check temperature, replace fixer, clean and repair rollers, add film hardeners
Lost Film	Roller problems, leader tape problem, bent film	Check rollers and assembly, securely affix leader tape to film
Scratched Film	Rough handling, roller problems	Handle film gently and by the edges, check rollers
Films Stuck Together	Rapid insertion	Insert films separately and more slowly

FIGURE 5-1A

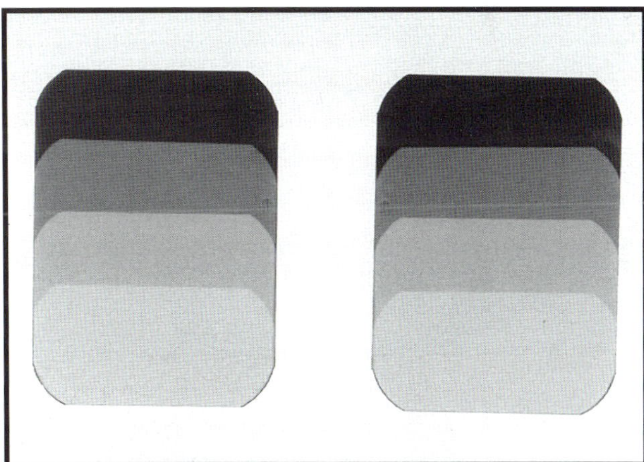

FIGURE 5-1B

Problems with technique can be identified by completing a step-wedge test. **Figure 5-1A** *A custom step wedge can be made by layering discarded lead foils over a piece of film.* **Figure 5-1B** *Exposed step-wedge radiographs. One radiograph serves as a control and the other as test film for comparison.*

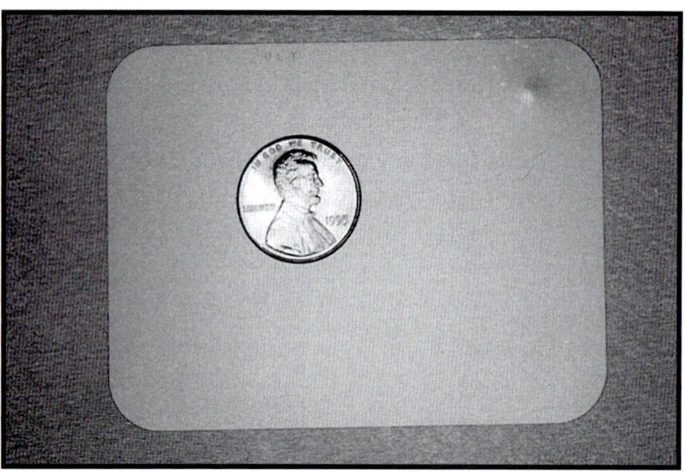

FIGURE 5-2A

A simple coin test can reveal whether light is leaking into the darkroom. **Figure 5-2A** *First, a radiograph is taken using a double-film packet. While processing one film, a coin is placed on the second film and the film is set aside. After removing the coin, the second film should be processed.* **Figure 5-2B** *If an image of the coin appears on the second film, light is leaking into the darkroom.*

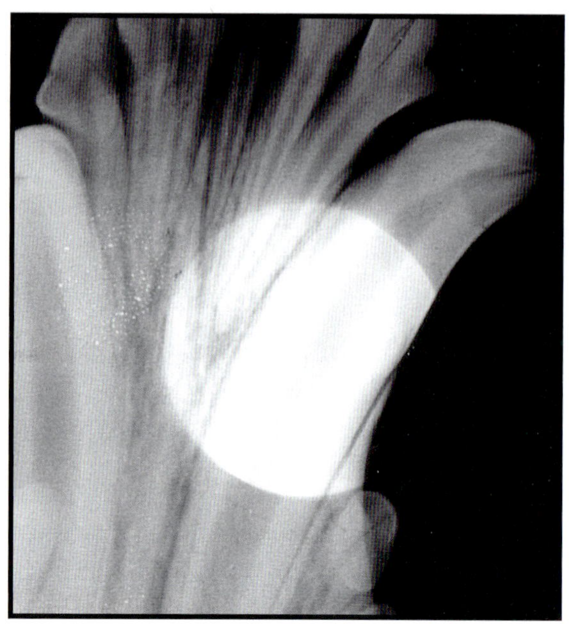

FIGURE 5-2B

ERRORS IN BISECTING ANGLE TECHNIQUE—ELONGATION

FIGURE 5-3A Decreased Vertical Angulation— Upper Rostral Dentition

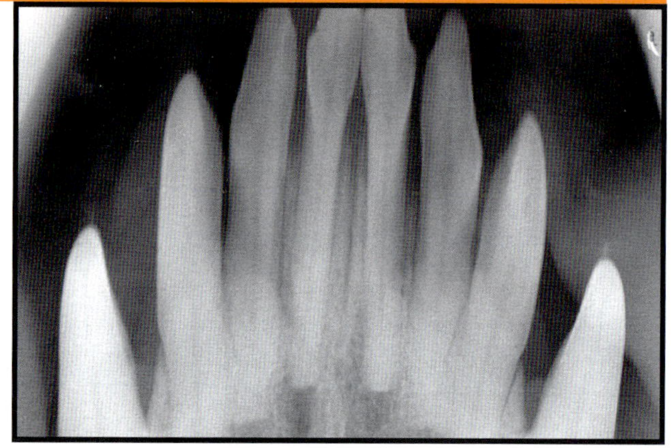

FIGURE 5-3B Radiographic Image

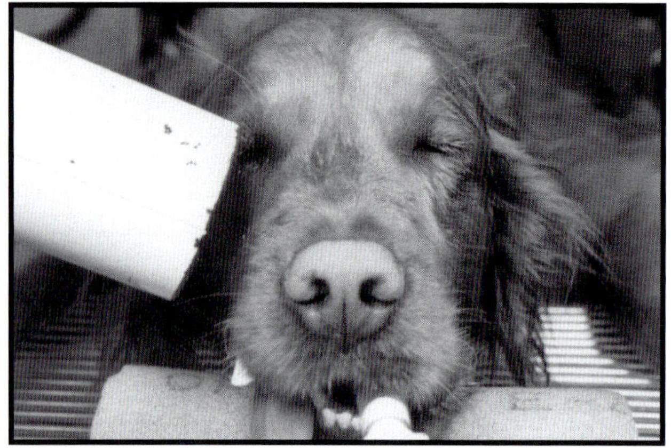

FIGURE 5-4A Decreased Vertical Angulation— Lateral View of Upper Dentition

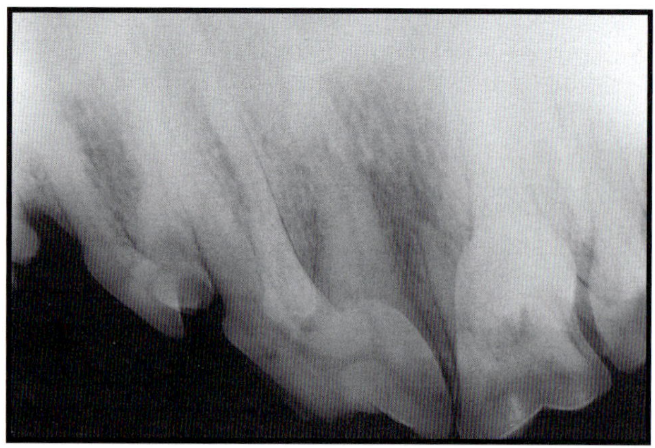

FIGURE 5-4B Radiographic Image

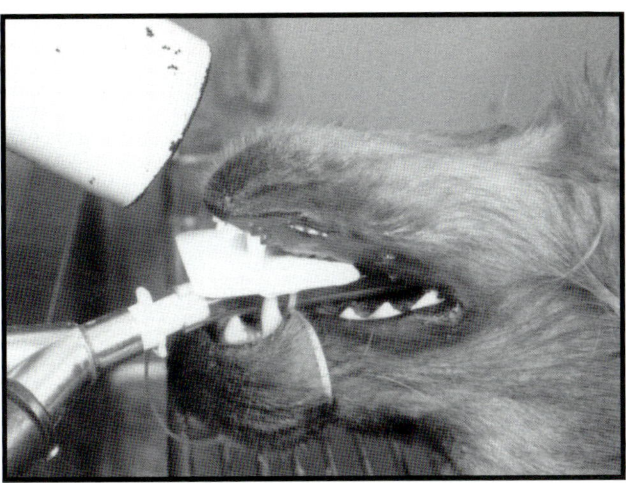

FIGURE 5-5A Decreased Vertical Angulation—
Lower Rostral Dentition

FIGURE 5-5B Radiographic Image

Figures 5-3A, 4A, and 5A *If the x-ray cone is positioned perpendicular to the long axis of the teeth, the teeth will appear elongated as shown in* **Figures 5-3B, 4B, and 5B.** *If elongation of the upper dentition (Figure 5-4B) is not severe, the film can be used; however, it will not provide a true representation. To capture a true representation, the x-ray beam must be positioned closer to an angle of 45° for radiographing premolars and 60° for radiographing incisors. (Note: Elongation of the lower dentition is shown in Figure 5-9 on parallel positioning.)*

ERRORS IN BISECTING ANGLE TECHNIQUE—FORESHORTENING

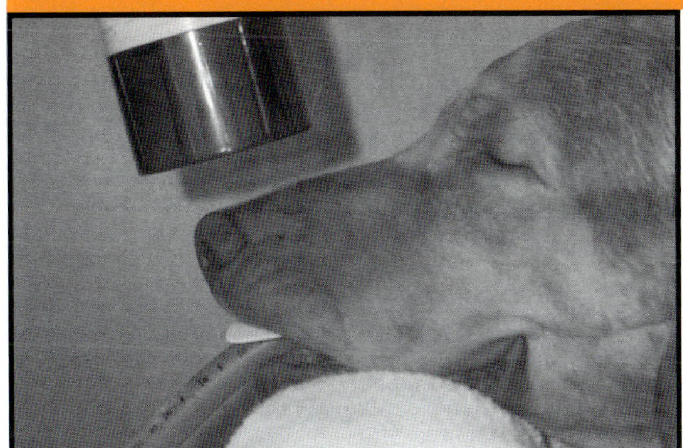

FIGURE 5-6A Increased Vertical Angulation—
Upper Rostral Dentition

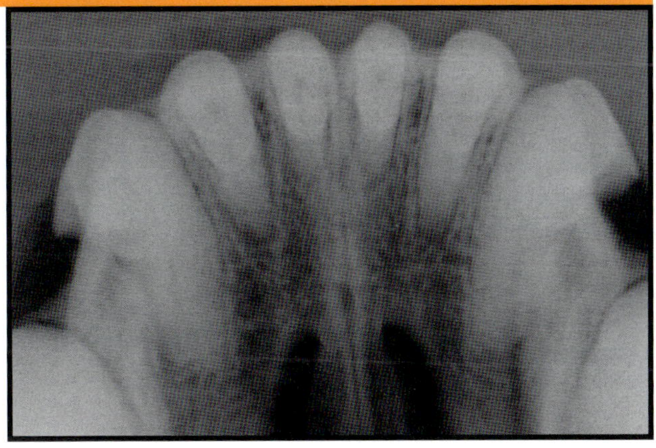

FIGURE 5-6B Radiographic Image

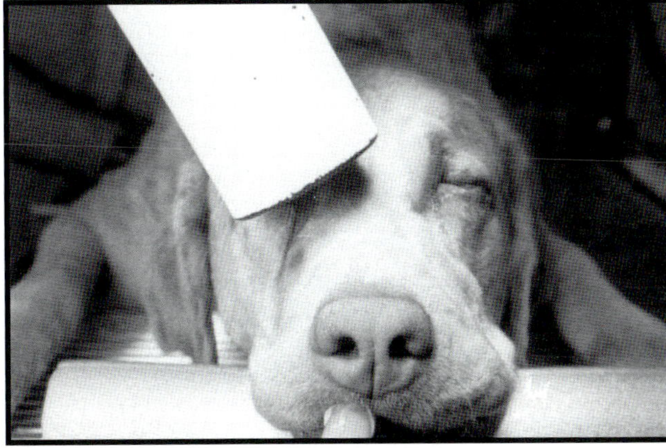

FIGURE 5-7A Increased Vertical Angulation—
Lateral View of Upper Dentition

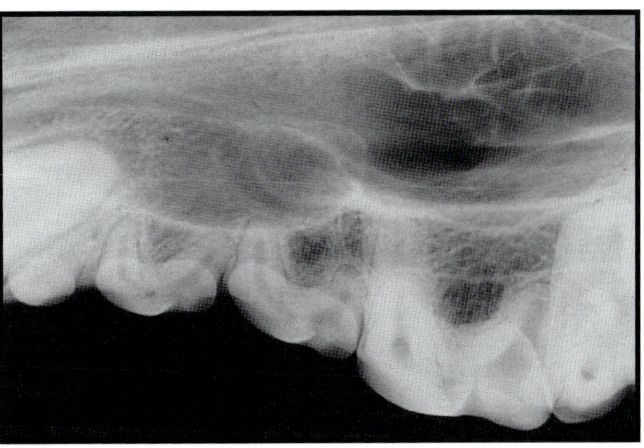

FIGURE 5-7B Radiographic Image

(Foreshortening figures and legend continue on next page)

ERRORS IN BISECTING ANGLE TECHNIQUE—FORESHORTENING *(continued)*

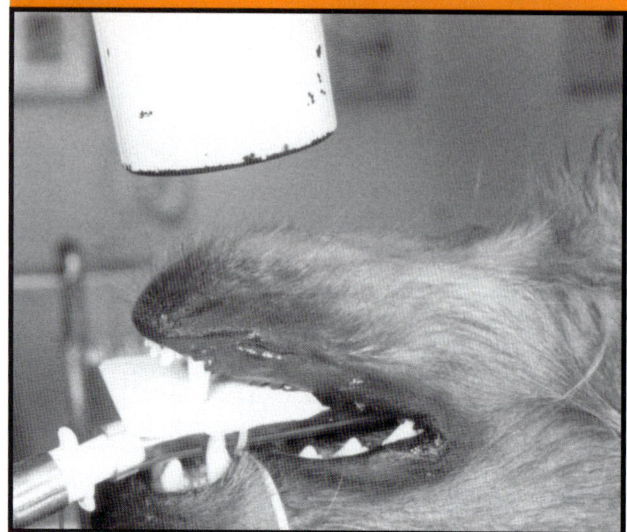

FIGURE 5-8A Increased Vertical Angulation—Lower Rostral Dentition

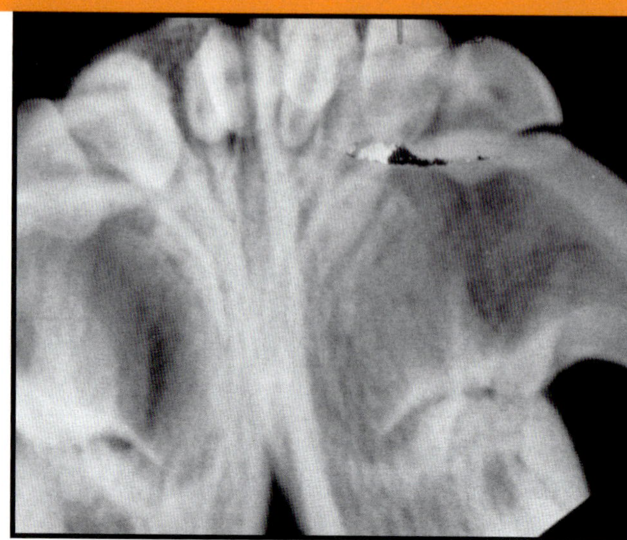

FIGURE 5-8B Radiographic Image

(continued from page 51)

Figure 5-6A *If the x-ray cone is positioned to the film as shown, the resultant image in* **Figure 5-6B** *will make the teeth appear short and stubby (foreshortening).* **Figure 5-7A** *If the x-ray cone is positioned to the film as shown over the dorsal aspect of the muzzle, the x-ray beam will travel down the long axis of the tooth being radiographed and the resultant image in* **Figure 5-7B** *will make the tooth appear shorter than it is.* **Figure 5-8A** *If the x-ray cone is positioned to the film as shown, the lower rostral dentition will appear short and stubby, as occurs in* **Figure 5-8B**. *To correct the error, the x-ray cone should be positioned at an angle of 45° to 60° to the horizontal plane. (Note: Foreshortening of the lower dentition is shown in Figure 5-10 on parallel positioning.)*

ERRORS IN PARALLEL POSITIONING—ELONGATION

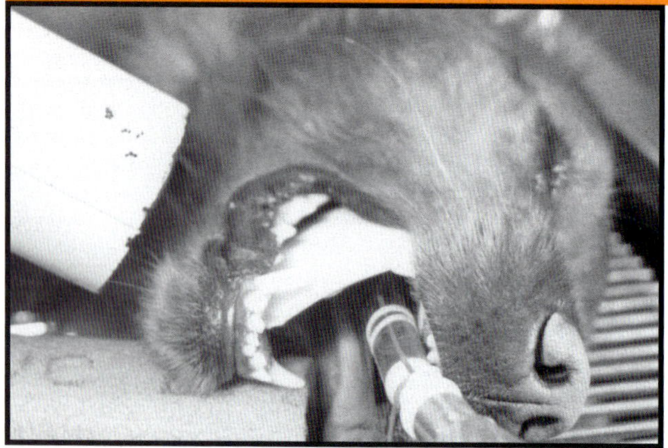

FIGURE 5-9A Decreased Vertical Angulation—Lateral View of Lower Dentition

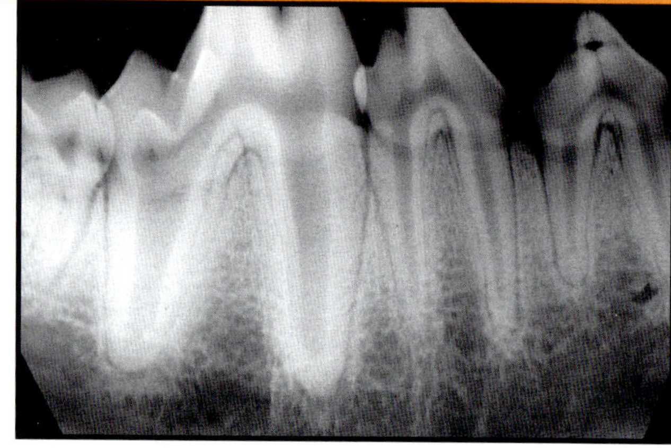

FIGURE 5-9B Radiographic Image

Figure 5-9A *If the x-ray cone is positioned incorrectly at the ventral border of the mandible, the radiographic image shown in* **Figure 5-9B** *will make the teeth appear longer than they are.*

ERRORS IN PARALLEL POSITIONING—FORESHORTENING

FIGURE 5-10A Increased Vertical Angulation—Lateral View of Lower Dentition

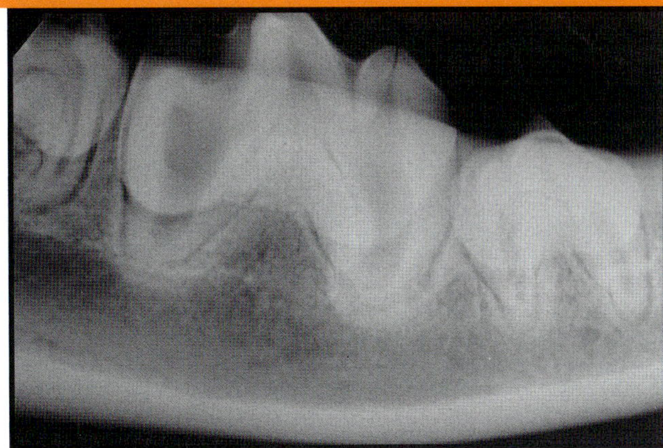

FIGURE 5-10B Radiographic Image

Figure 5-10A *If the x-ray cone is directed toward the occlusal surfaces of the teeth, the resultant image shown in* **Figure 5-10B** *will make the teeth appear short and stubby.*

COMMON POSITIONING ERRORS—SUPERIMPOSITION USING INTRAORAL FILM

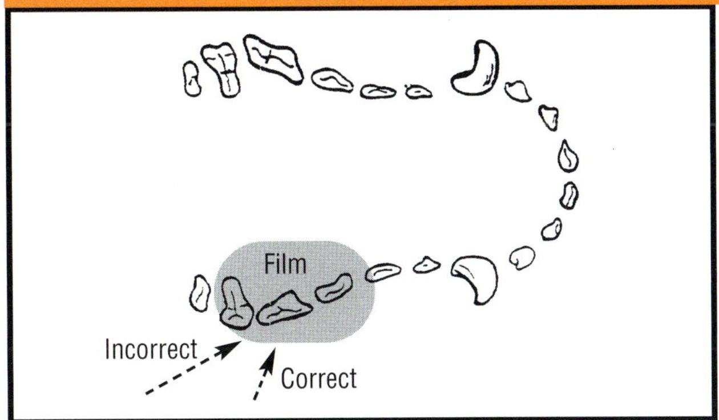

FIGURE 5-11A Excessive Horizontal Angulation

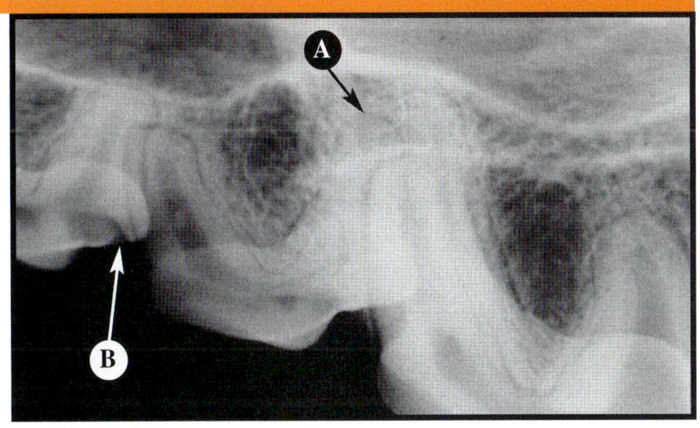

FIGURE 5-11B Radiographic Image

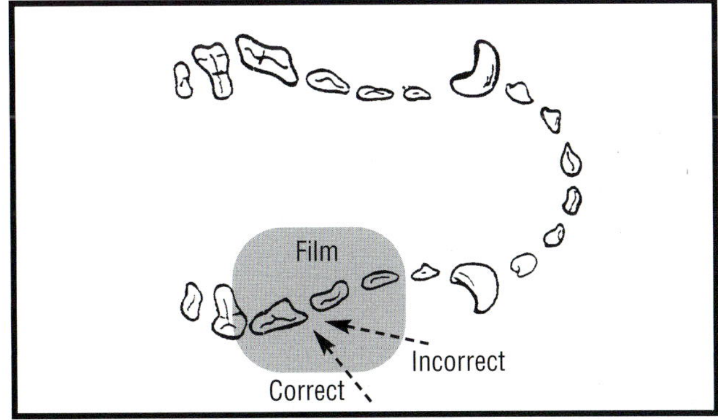

FIGURE 5-11C Excessive Horizontal Angulation

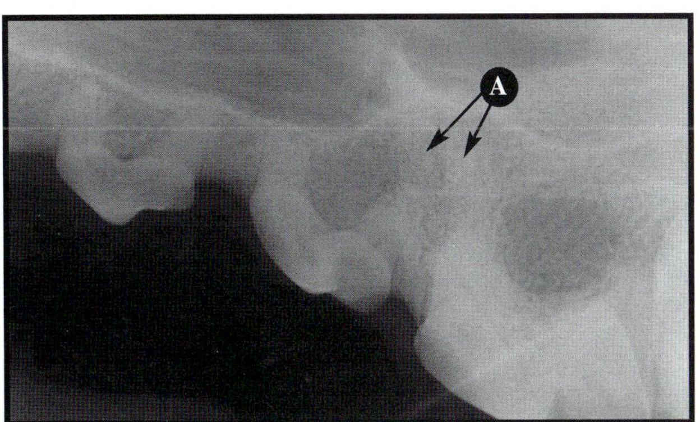

FIGURE 5-11D Radiographic Image

If the cone is horizontally positioned in an excessive oblique angle from the mesial or distal aspect, the angle of the x-ray beam will be incorrect and the interdental spaces will appear to be too narrow, the teeth may overlap, the image of the roots may be fuzzy or disappear, or the cervical area will appear to be burned out. **Figure 5-11A** *If the cone is horizontally directed from an excessive distal angle, the images of the roots of the third and fourth premolars are superimposed.* **Figure 5-11B** *This radiograph shows (A) how the images of the mesial roots of the upper fourth premolar and distal aspect of the upper third premolar are superimposed. The other interdental spaces (B) are also superimposed and appear more narrow than they are.* **Figure 5-11C** *If the cone is horizontally directed from an excessive mesial angle, the roots of the fourth premolar will be superimposed in a mesial direction on the third premolar and in a distal direction on the molar. The images of the mesial roots of the fourth premolar become less distinct.* **Figure 5-11D** *This radiograph shows (A) how changes in the horizontal angle of the x-ray beam can cause the image of the roots to appear fuzzy, especially in the fourth premolar, thereby mimicking resorption.*

COMMON POSITIONING ERRORS—EXTRAORAL PROJECTION

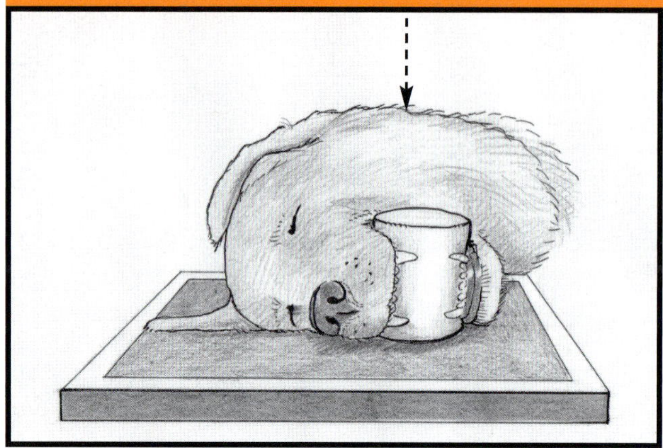

FIGURE 5-12A Lateral Projection

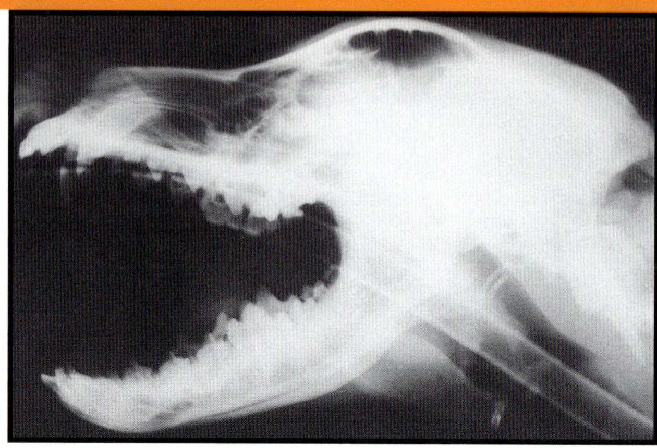

FIGURE 5-12B Radiographic Image

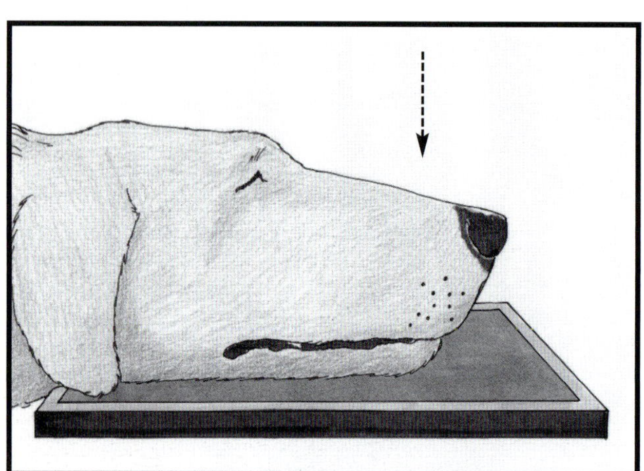

FIGURE 5-13A Dorsoventral Projection

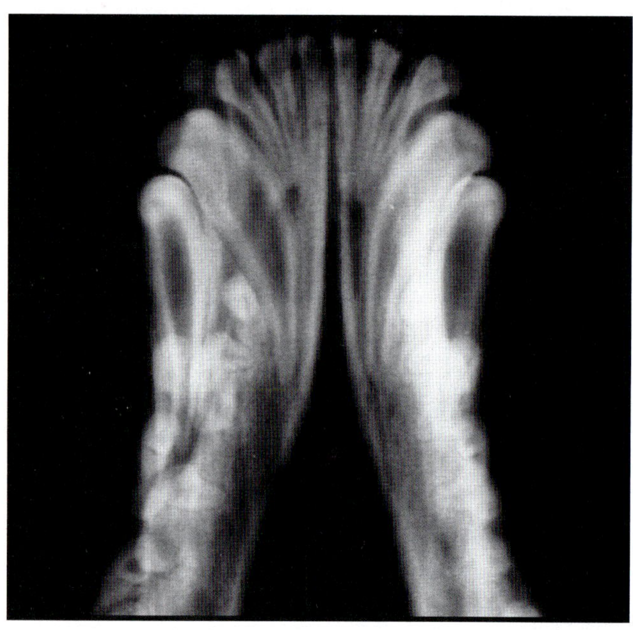

FIGURE 5-13B Radiographic Image

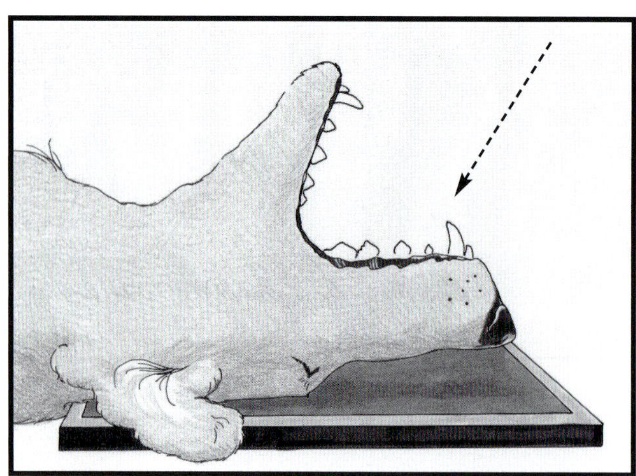

FIGURE 5-14A Open-Mouth Ventrodorsal Projection of Maxilla

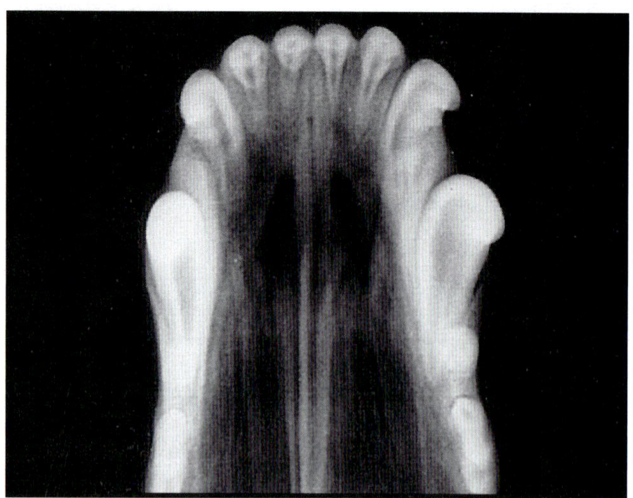

FIGURE 5-14B Radiographic Image

FIGURE 5-15A Open-Mouth Dorsoventral Projection of Mandible

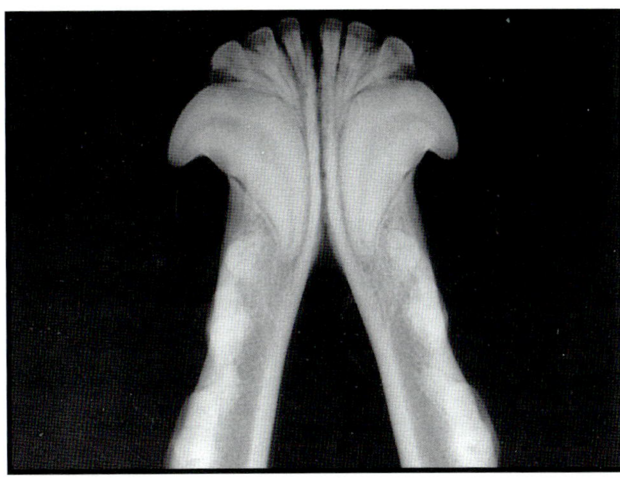

FIGURE 5-15B Radiographic Image

Figure 5-12A *A lateral view without adequate rotation of the patient's head can cause superimposition of the teeth.* **Figure 5-12B** *This radiograph shows the resultant image.* **Figure 5-13A** *If the x-ray beam is directed down the long axis of the teeth for a dorsoventral view, foreshortening and superimposition of the upper and lower arcades can occur. The error is often compounded if an endotracheal tube is not removed before exposure.* **Figure 5-13B** *This radiograph shows the resultant image.* **Figure 5-14A** *If the angle of the x-ray beam is too acute for the open-mouth ventrodorsal view, a foreshortened image will occur.* **Figure 5-14B** *This radiograph shows the resultant image.* **Figure 5-15A** *Although this dorsoventral view of the mandible is excellent for studying some lesions of the mandible, the view can cause foreshortening of the mandibular teeth.* **Figure 5-15B** *This radiograph shows the resultant image.*

COMMON POSITIONING ERRORS—PLACEMENT OF X-RAY CONE

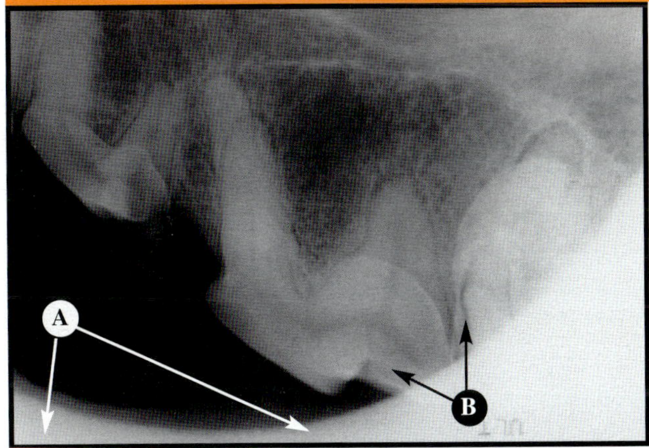

FIGURE 5-16

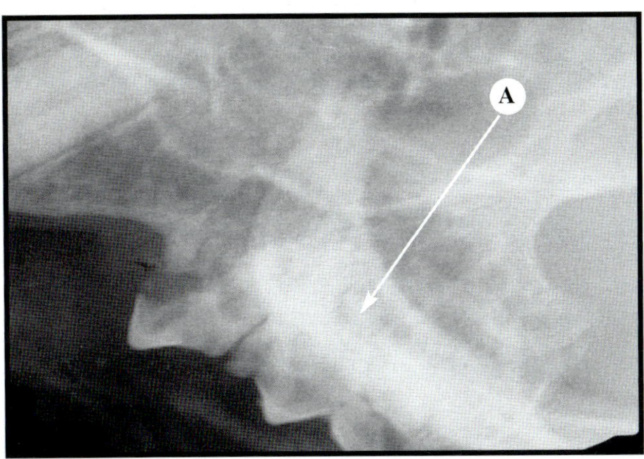

FIGURE 5-17

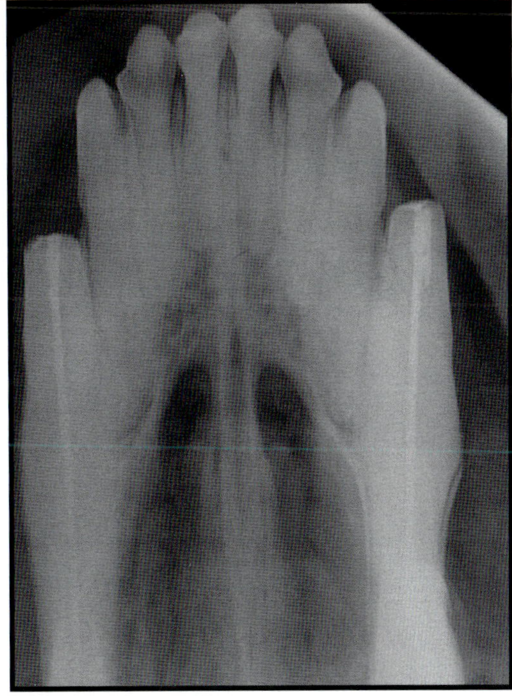

FIGURE 5-18

Figure 5-16 (A) *This area of the film is unexposed, whereas* (B) *a loss of image occurred because the x-ray cone (primary x-ray beam) was not centered over the film.* **Figure 5-17** *In this radiograph, the* (A) *zygomatic arch is superimposed over the premolars. The error can be corrected by following the tube-shift technique; that is by keeping the same angle, moving the cone down, or changing the angle of the cone.* **Figure 5-18** *In this radiograph, the apices of the canine teeth are superimposed over the premolars. The error can be corrected by changing the horizontal angle of the primary x-ray beam.*

ERRORS IN HANDLING THE FILM—BENDING AND CRIMPING

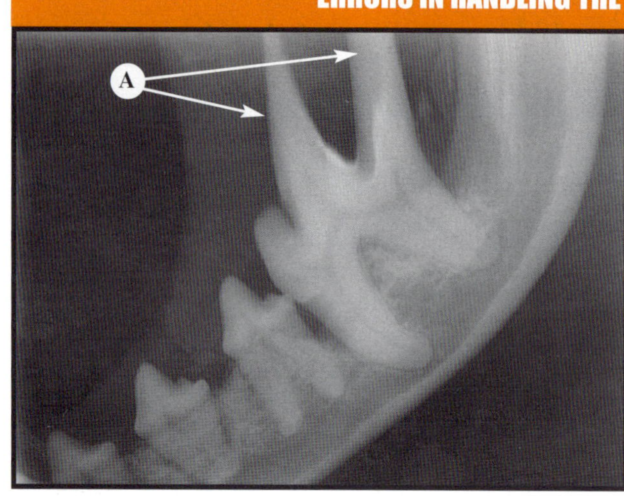

FIGURE 5-19 Bending the Film

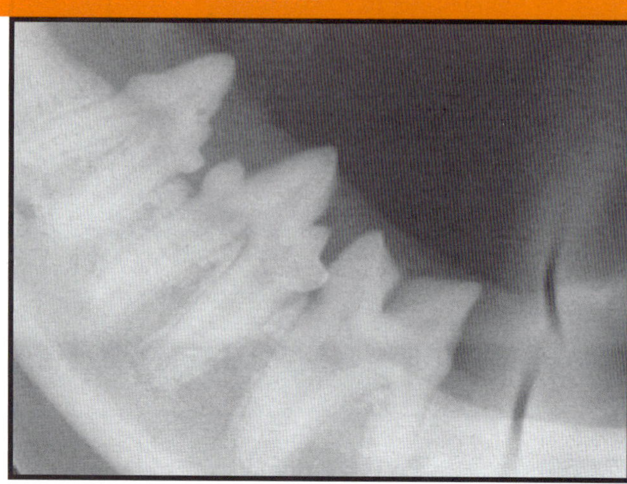

FIGURE 5-20 Crimping the Film

Figure 5-19 *This radiograph shows a distorted image of the crown because the film was bent. Bending or curving the film too much can cause such distortion.* **Figure 5-20** *The black crescent marks apparent on this radiograph were caused by crimping the film. Film must be handled carefully when it is being inserted into or removed from the patient's mouth and during processing.*

ERRORS IN HANDLING THE FILM—ARTIFACTS

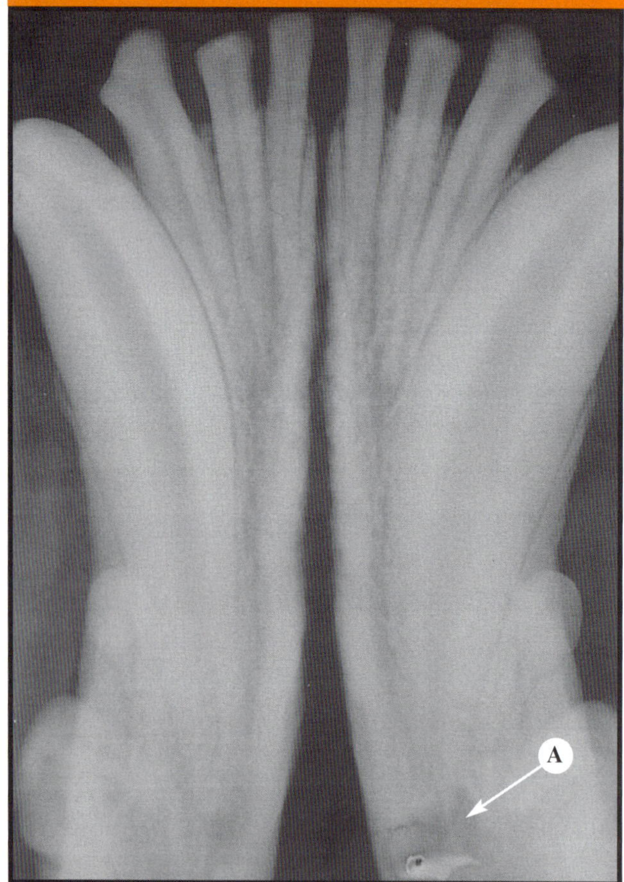

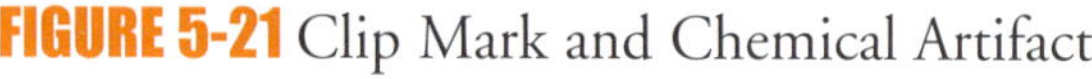

FIGURE 5-21 Clip Mark and Chemical Artifact

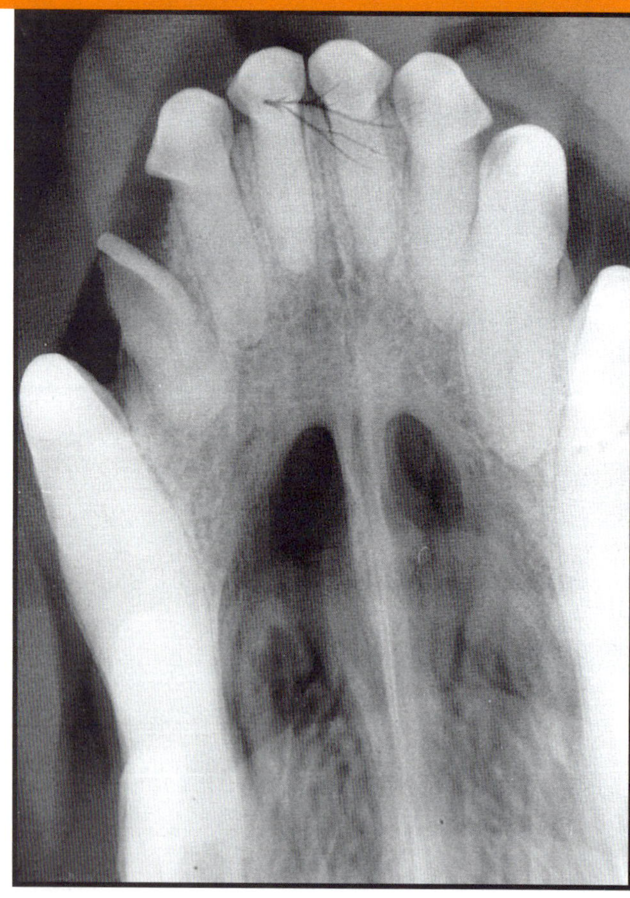

FIGURE 5-22 Static Electricity

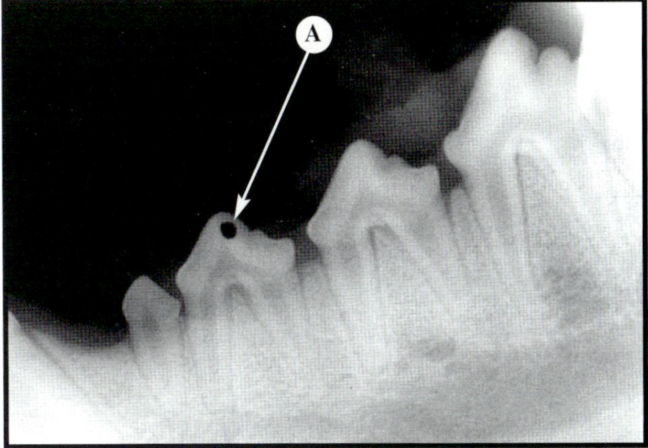

FIGURE 5-23 Bite Marks

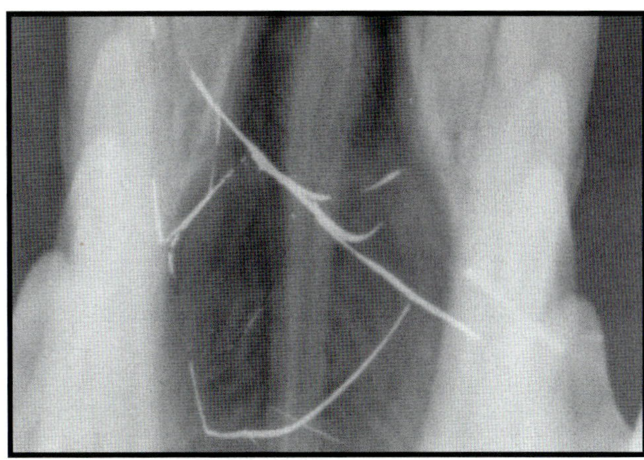

FIGURE 5-24 Scratched Film

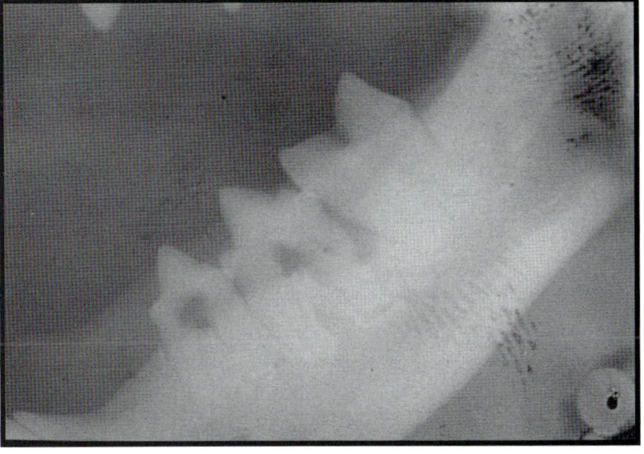

FIGURE 5-25A Black Fingerprints

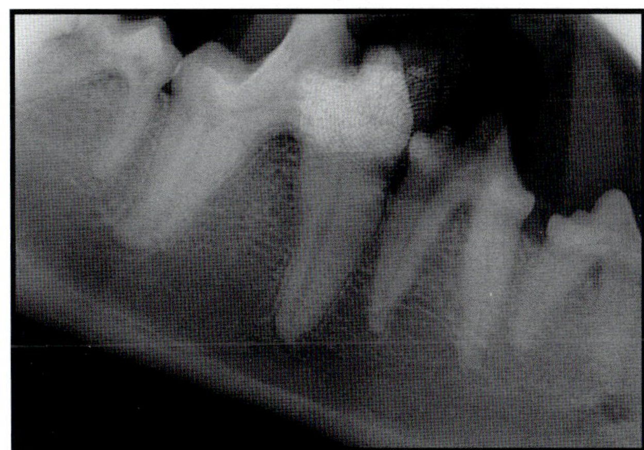

FIGURE 5-25B White Fingerprints

Figure 5-21 *A clip mark and chemical artifact* (A) *visible at the apex of the canine tooth in this radiograph mimic an apical lesion (or obscure it). This error can be prevented by using clean, dry clips and by clipping the film away from the subject area.* **Figure 5-22** *On this radiograph, static electricity is visible over the incisor. The error was caused by opening the film packet too rapidly.* **Figure 5-23** *A bite mark is visible* (A) *on this radiograph because the film was pressed against a tooth or other structure. The error can be corrected by handling the film gently, using bite blocks as needed, and keeping the animal anesthetized.* **Figure 5-24** *The scratches on this radiograph probably occurred from the film being dragged over a rough surface while the emulsion was still soft, a reminder that the film must be carefully handled throughout the procedure.* **Figures 5-25A and 25B** *Black fingerprints can occur if the film is touched with hands that have been soiled with fluoride or developer, whereas white fingerprints mean that the hands have been soiled with fixer.*

ERRORS IN FILM PLACEMENT

Figure 5-26 *A dimple* (A) *visible at the apex of the distal fourth premolar indicates that the film was not correctly oriented. The dimple on the film packet indicates that the film should be placed away from the area to be radiographed. Also note the clip mark* (B), *which occurred because the film was not clipped away from the subject area.*

FIGURE 5-26

ERRORS IN FILM PLACEMENT *(continued)*

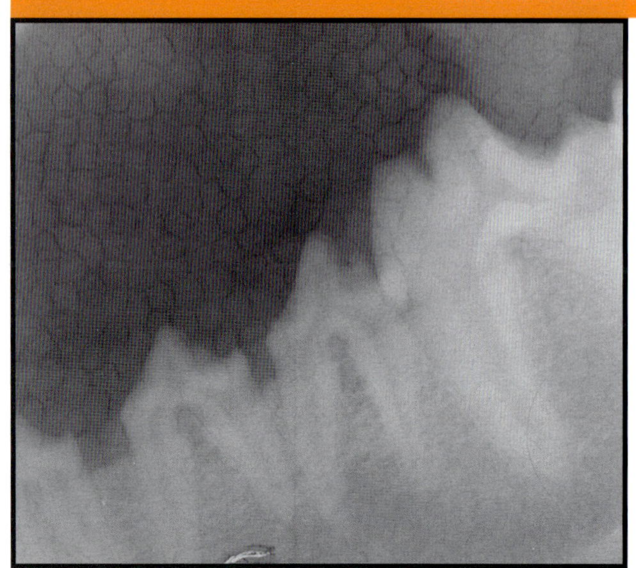

FIGURE 5-27A

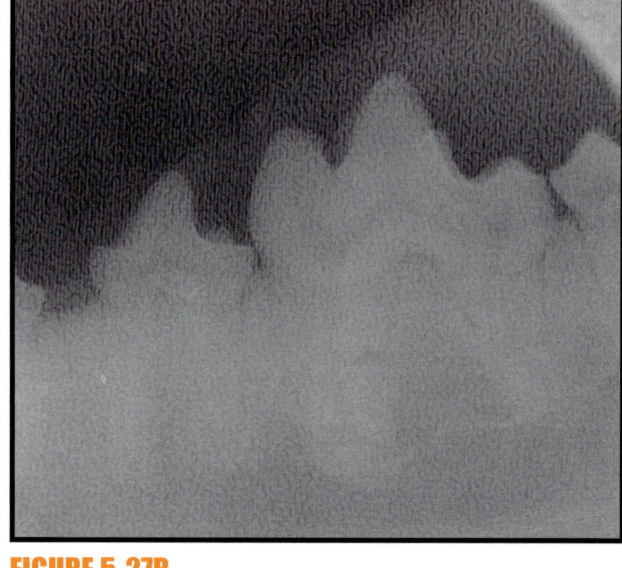

FIGURE 5-27B

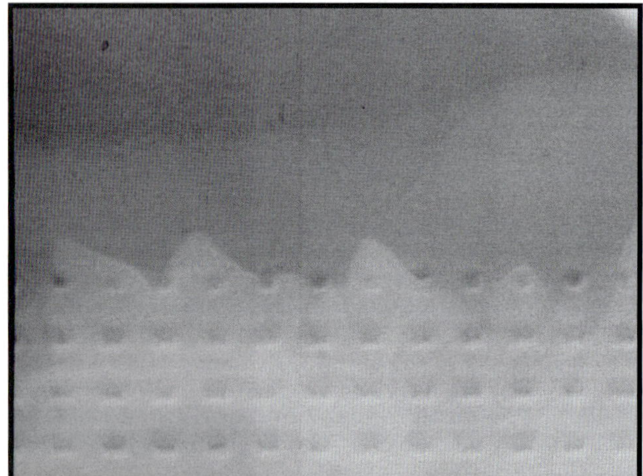

FIGURE 5-27C

Figures 5-27A, 27B, and 27C *The pebbled or stippled appearance of these radiographs occurred because the film was positioned backward. The x-ray beam is passing through the lead foil, thereby causing superimposition of the foil and the subject area.*

ERRORS IN EXPOSURE—DOUBLE EXPOSURE

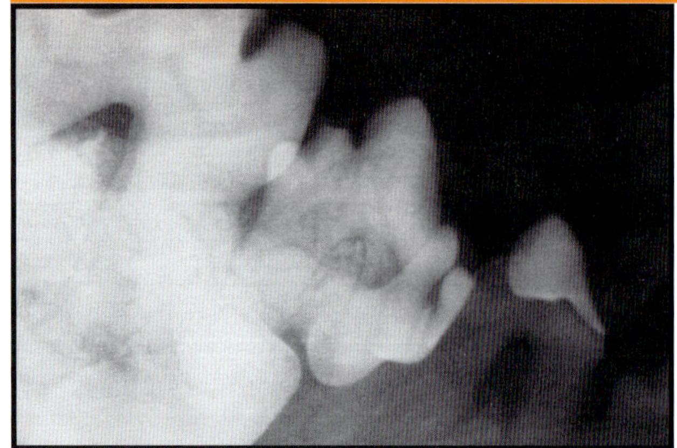

FIGURE 5-28A

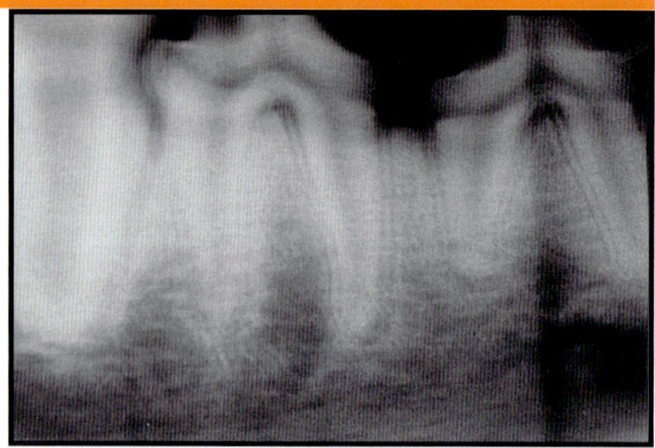

FIGURE 5-28B

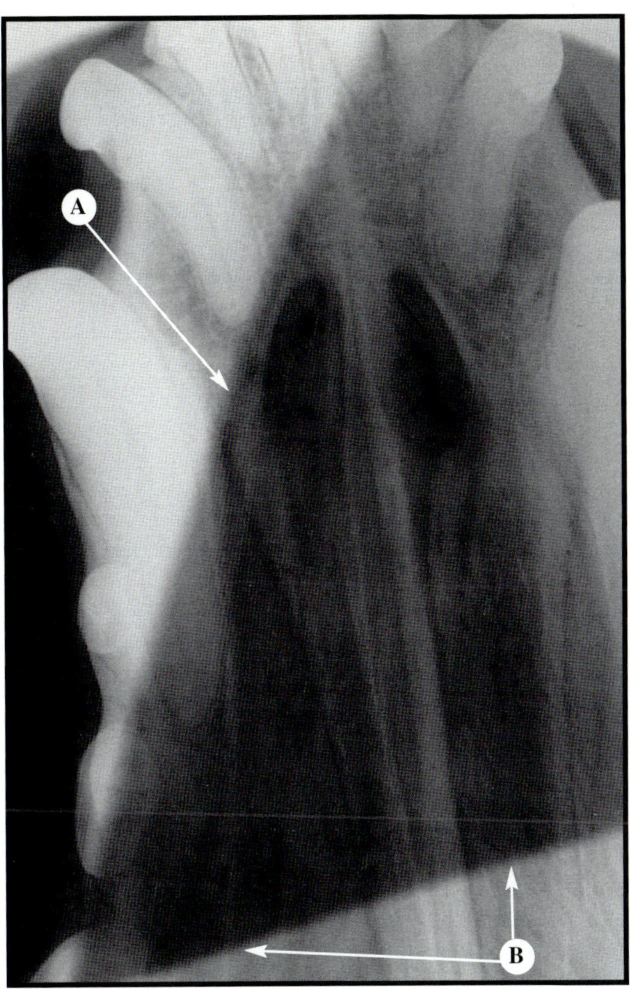

FIGURE 5-29

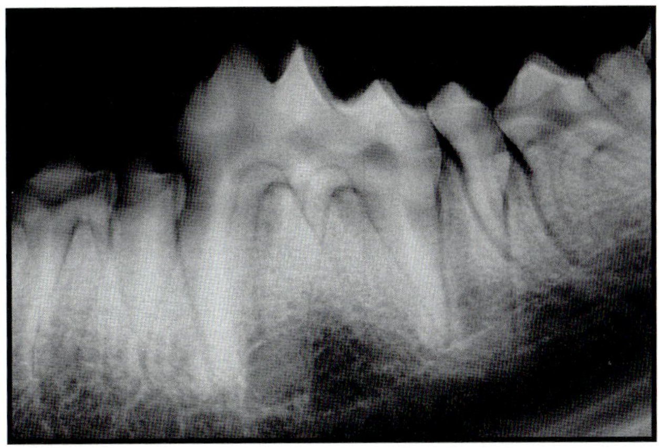

Figures 5-28A and 28B *The film in Figure 5-28A was used twice in two different regions of the mouth, whereas the film in Figure 5-28B was used twice in the same region of the mouth. Both radiographs are reminders that film must be removed immediately and taken to the area designated for processing.* **Figure 5-29** *A triangular-shaped dark shadow (A and B) was caused by inadvertent exposure to radiation. The film had been placed on a table too close to the primary x-ray beam while additional radiographs were being taken.* **Figure 5-30** *The double exposure visible in this radiograph occurred because the switch had been depressed twice and because of a slight shift in the position of the film. Caution should be taken to ensure that the switch is depressed only once and that the patient, film, and x-ray cone remain stationary throughout the procedure.*

ERRORS IN EXPOSURE—UNDEREXPOSURE AND OVEREXPOSURE

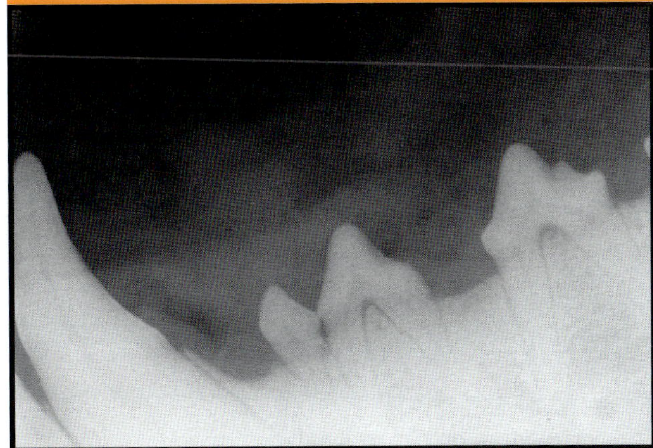

FIGURE 5-31 Underexposed Film

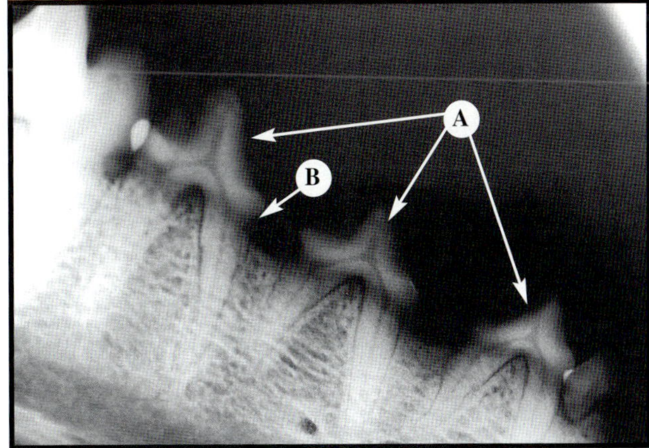

FIGURE 5-32 Overexposed Film

Figure 5-31 *This radiograph has a white appearance and the images of the teeth and bone seem to be coalescing, which occurred because the film was underexposed. The error can be corrected by increasing the exposure time and centering the area to be radiographed on the film.* **Figure 5-32** *The image in this radiograph is (A) too dark, lacks contrast, and has a ghost-like appearance. (B) Cervical burnout is also apparent. These errors in exposure can be corrected by decreasing the exposure time.*

ERRORS CAUSED BY PLACEMENT OF ENDOTRACHEAL TUBE

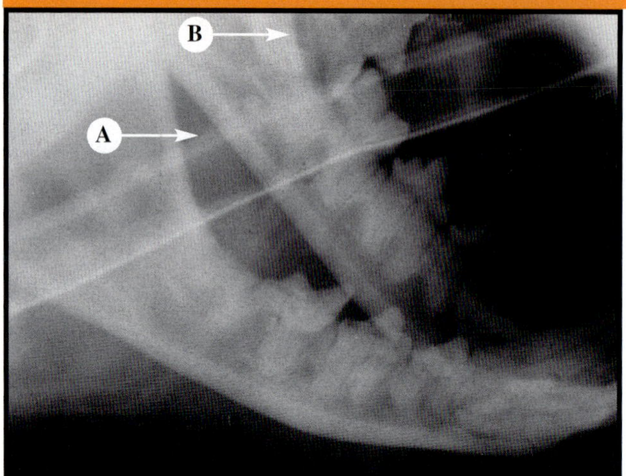

FIGURE 5-33

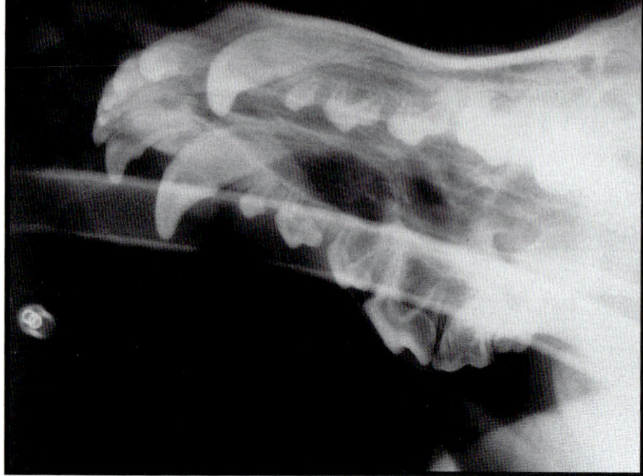

FIGURE 5-34

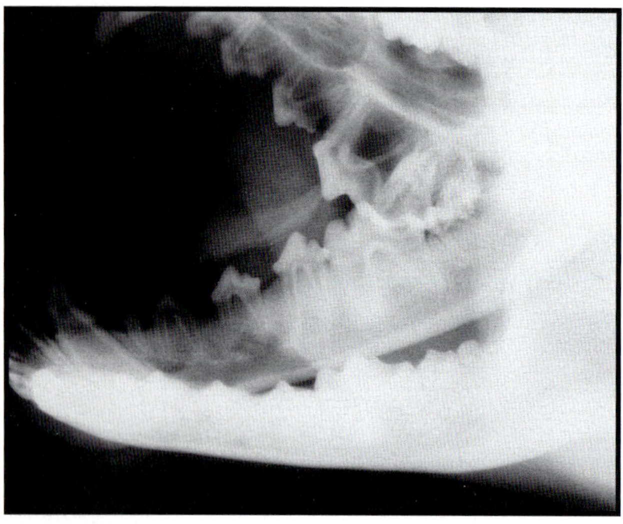

FIGURE 5-35

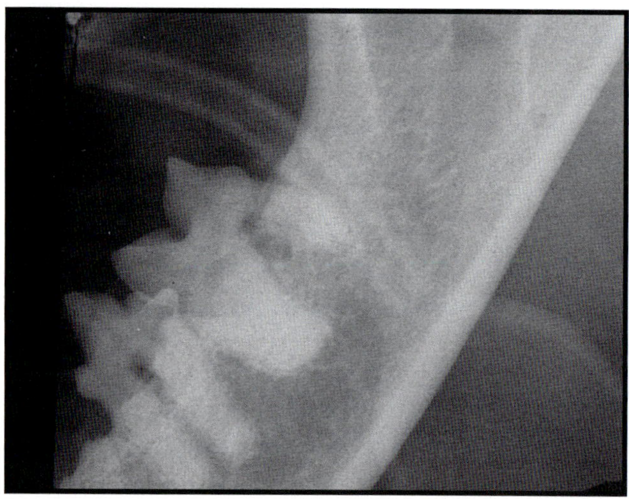

FIGURE 5-36

Figure 5-33 *An obscured image and superimposition are apparent in this radiograph.* (A) *The endotracheal tube is obscuring the image of the molars, and* (B) *the maxillary teeth are superimposed over the tube. The first error can be corrected by removing the endotracheal tube when extraoral radiographs are being taken in cats or small-breed dogs. The superimposition can be resolved by opening the animal's mouth wider or changing the degree of rotation of the animal's head.* **Figure 5-34** *In this radiograph, the endotracheal tube is overlapping the entire subject area. This error can be resolved by stabilizing the tube away from the subject area when taking extraoral radiographs.* **Figure 5-35** *Superimposition of the upper and lower molars is apparent and can be corrected by opening the animal's mouth wider or changing the degree of rotation of the animal's head.* **Figure 5-36** *The inflation bladder and tubing of the endotracheal tube are superimposed over the mandible, which can be corrected by moving the tube away from the subject area.*

PROCESSING ERRORS—OVERDEVELOPED OR UNDERDEVELOPED FILM

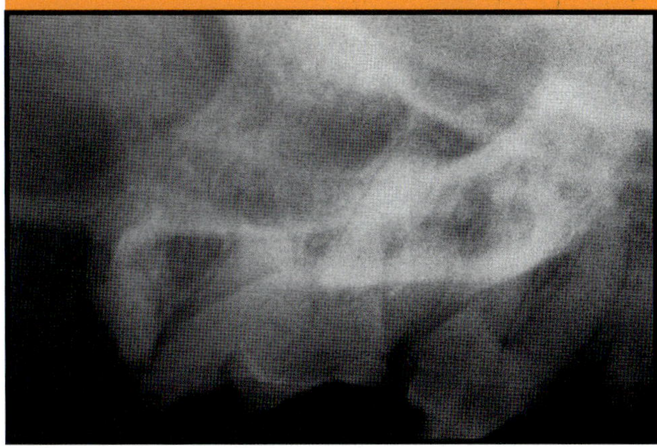

FIGURE 5-37 Overdeveloped Film

Figure 5-37 *This radiograph is the result of overdeveloped film, which remained in the developer too long. The result is overall darkening of the film. When hand-developing film, it should be removed from the developing solution and rinsed immediately after the film turns dark (see Table 5-3).* **Figure 5-38** *This radiograph is an example of underdeveloped film, which occurred because the developing solution was too old. The error can be corrected by replenishing chemicals as recommended (see Table 5-2).*

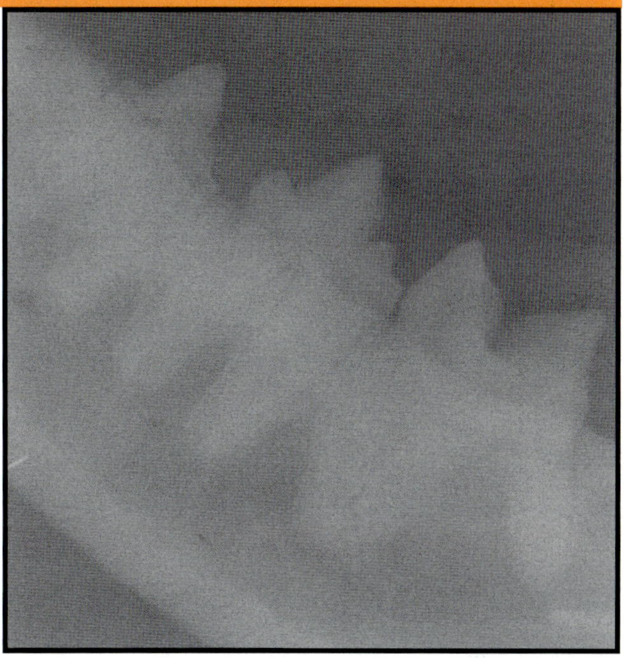

FIGURE 5-38 Underdeveloped Film

PROCESSING ERRORS—DARK ARTIFACTS

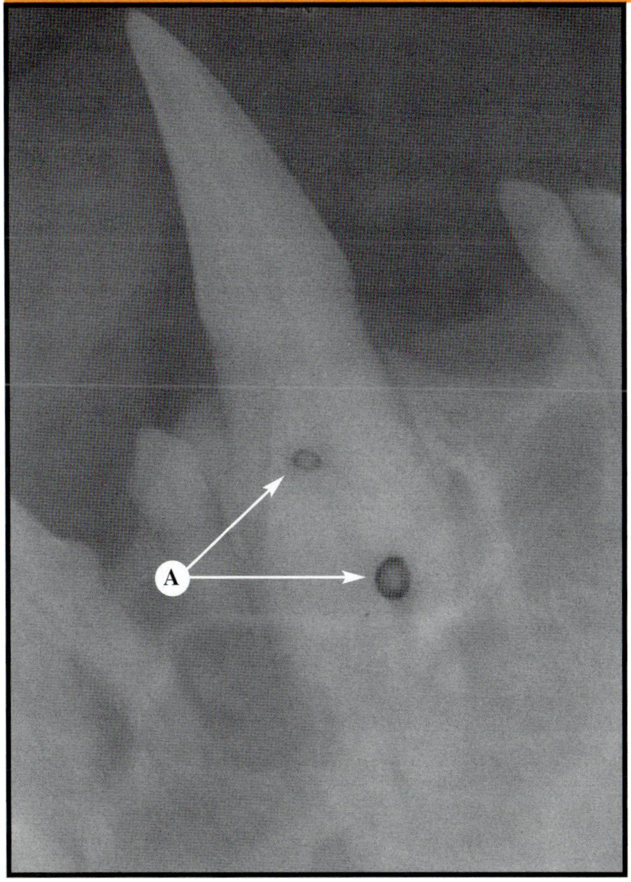

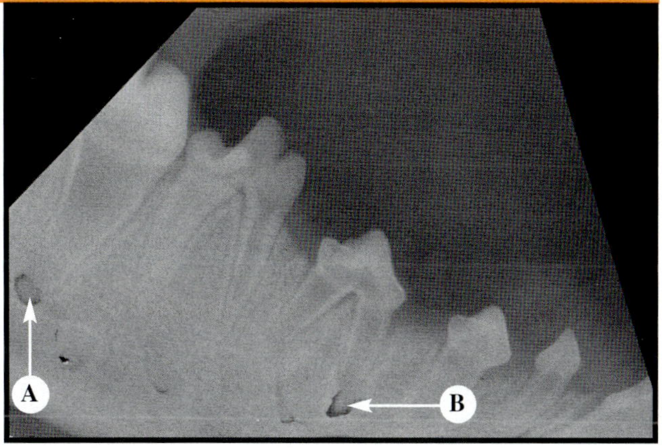

FIGURE 5-40 Chemical Artifacts and Gray Hue

Figure 5-39 *The dark spots (A) visible on this radiograph are the result of developing solution being splashed on the radiograph before processing. In addition, the radiograph is very gray. Contamination can be eliminated by working in a clean environment, and the gray appearance can be resolved by keeping the chemicals fresh.* **Figure 5-40** *The chemical artifacts (A and B) apparent on this radiograph also occurred when developer was splashed on the film before processing. Note how the spots mimic periapical lesions. In addition, the film is very gray.*

FIGURE 5-39 Dark Spots and Gray Hue

PROCESSING ERRORS—DARK ARTIFACTS *(continued)*

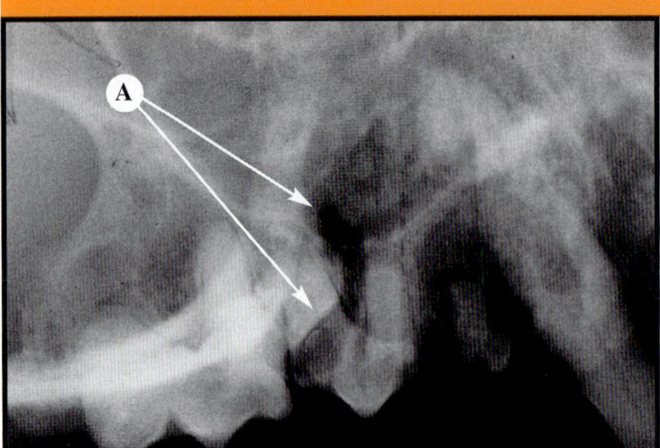

FIGURE 5-41 Developer Artifact

Figure 5-41 *The chemical artifacts (A) on this radiograph likewise were caused by developer being spilled on the film before processing. In this case, however, the spots are mimicking radiolucent, lytic, or resorptive lesions. The grainy appearance could be the result of temperature, especially if the temperature of the developer is elevated.*

PROCESSING ERRORS—WHITE ARTIFACTS

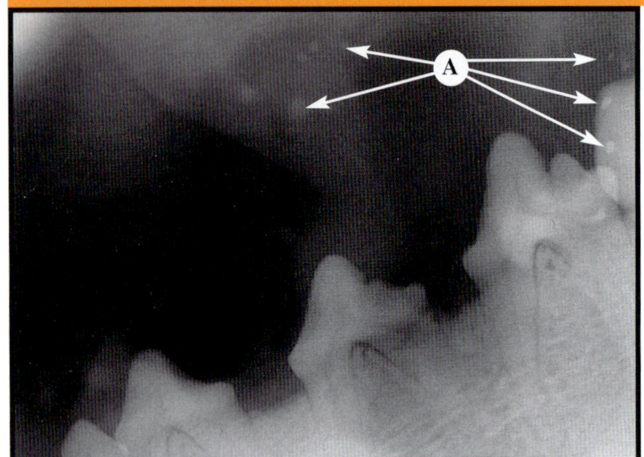

FIGURE 5-42 White Spots from Fixer

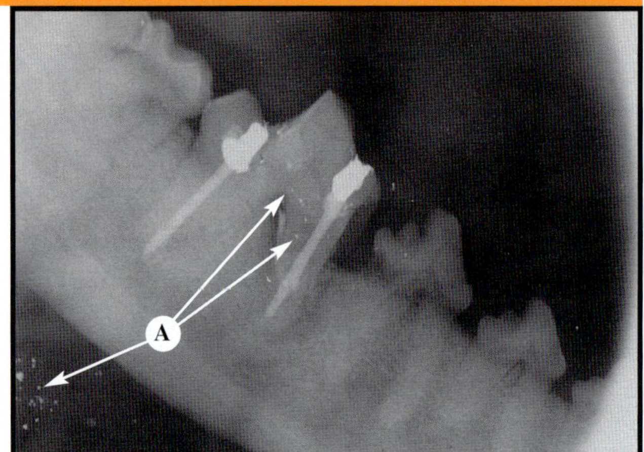

FIGURE 5-43 White Spots from Amalgam Debris

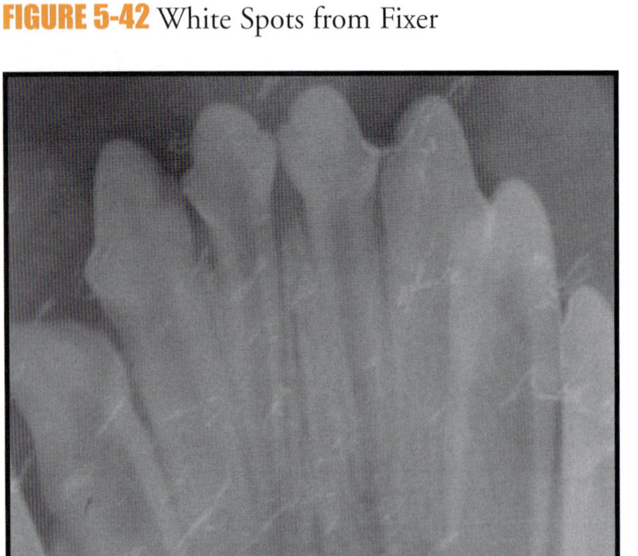

FIGURE 5-44 White Streaks

Figure 5-42 *The white spots (A) on this radiograph are caused by fixer being spilled on the film before it was processed. This error can be corrected by keeping the work area clean and dry.* **Figure 5-43** *The white spots (A) on this radiograph were caused by residual amalgam (which can mimic fixer contamination of undeveloped film) or from the fixer being spilled. In addition, the radiograph is mottled. Oftentimes, small pieces of amalgam not visible to the human eye can cause an image artifact. The mottled appearance is a processing artifact.* **Figure 5-44** *The white streaks apparent on this radiograph are caused by contaminated fixer solution or wash water or by inadequate washing. The film should be washed and rinsed thoroughly during processing; distilled water is recommended.*

PROCESSING ERRORS—MOTTLED FILM

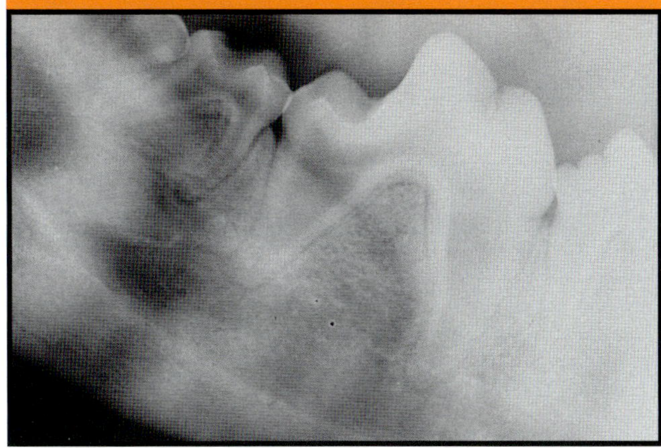

FIGURE 5-45

Figure 5-45 *The mottled appearance of this radiograph occurred because of uneven development, improper fixation, and contamination. Discoloration will progress with time. The rinse was inadequate, and the fixer must have been contaminated. Because the radiograph has no diagnostic value, new film should be used. This problem usually occurs during hand processing with rapid chemicals.* **Figure 5-46** *This radiograph exemplifies the frosty, glazed look that occurs when the film has not been properly fixed (see Table 5-5). The problem can be corrected by ensuring adequate fixation during automatic processing.*

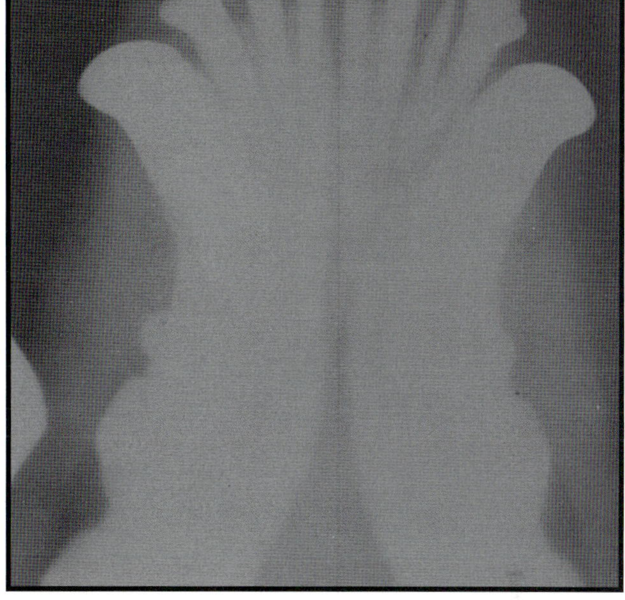

FIGURE 5-46

PROCESSING ERRORS—CONTAMINATED CLIPS

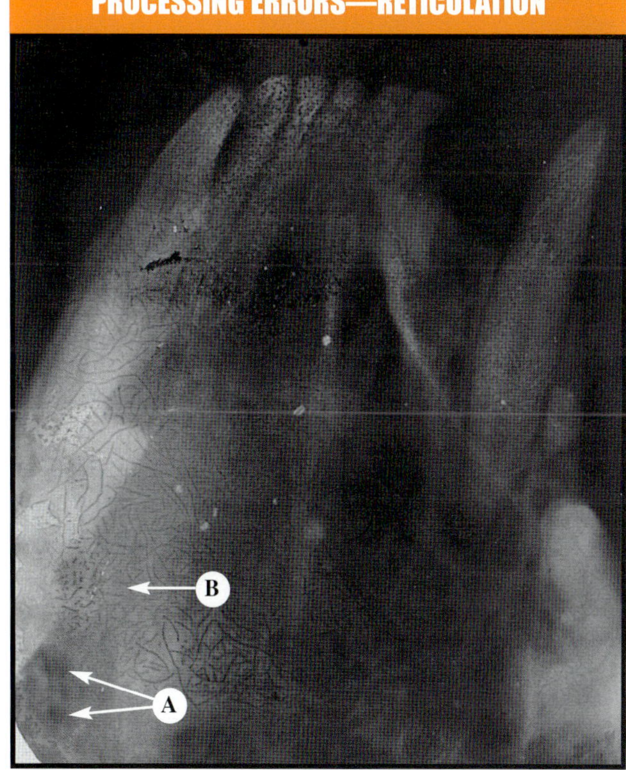

FIGURE 5-47

The contamination apparent on this radiograph occurred because a film clip had not been cleaned; chemicals on the clip dripped onto the film. Clean, dry clips should always be used for each film during processing. The clip should be immersed with the film during washing and rinsing to prevent buildup of chemicals.

PROCESSING ERRORS—RETICULATION

FIGURE 5-48

Cracked emulsion caused this reticulation-like pattern on the radiographic image. Cracked emulsions occur during processing when immersion into warm solutions is followed by washing in very cold water. If chemicals are left on the film, the residue will crack when drying. (A) Reticulated emulsion and (B) cracked residue.

AUTOMATIC PROCESSING ERRORS

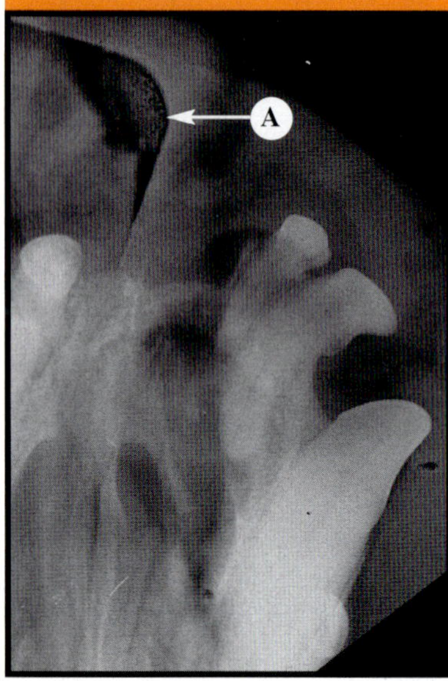

FIGURE 5-49 Overlapped Film

Figure 5-49 *If films are overlapped during processing, a mark will appear on the radiographic image (A). The films must have entered the automatic processor too quickly, or a double-film packet was not separated.* **Figure 5-50** *Uniform bands on processed film can be caused by contaminated rollers or by the rollers pressing onto or bending the film. The roller assembly should be cleaned and checked.* **Figure 5-51** *The scratches and torn emulsion on this radiograph occurred because the roller was dirty or became stalled during processing. Again, the entire roller assembly should be cleaned and checked.*

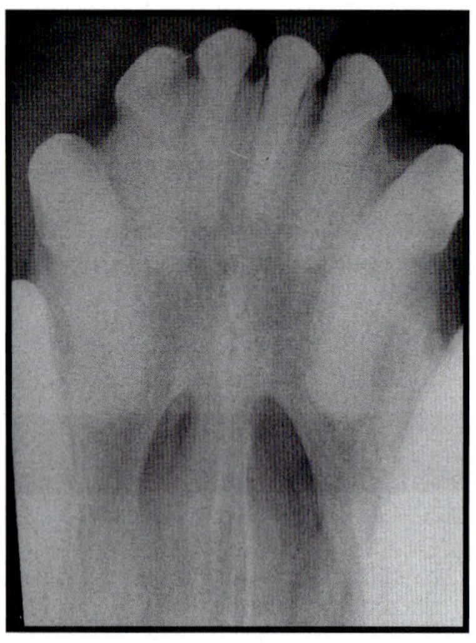

FIGURE 5-50 Roller Marks

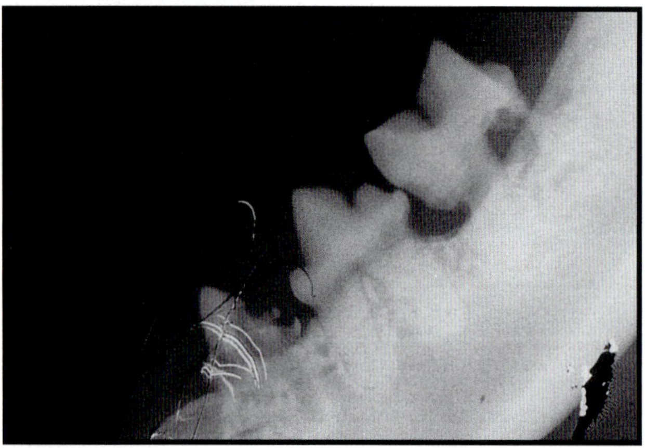

FIGURE 5-51 Scratches

STORAGE ERRORS

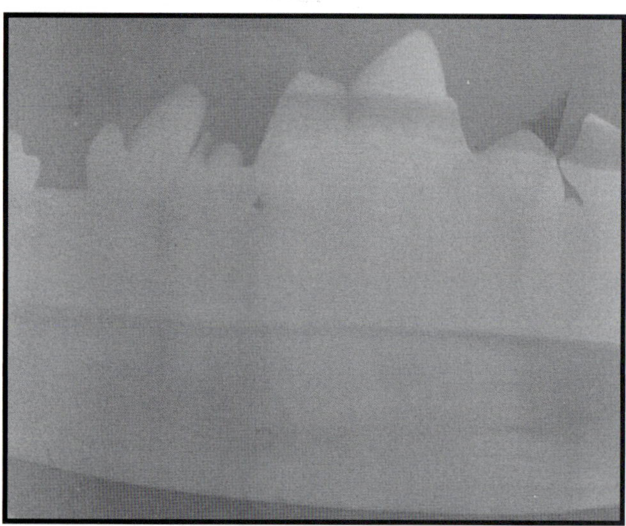

FIGURE 5-52 Premature Film Storage

FIGURE 5-53 Yellow or Brown Film

Figure 5-52 *One of these films was stored while still damp from processing. All the films stuck together and were ruined when subsequently separated. Film should be allowed to dry for 12 to 24 hours before being stored.* **Figure 5-53** *This film turned yellow and then eventually brown during storage. Such discoloration is usually caused by inadequate washing after fixation.*

Chapter Six

PRINCIPLES OF IMAGE INTERPRETATION

Because interpretation of the appearance of oral structures in radiographs is not always pathognomonic, a definitive diagnosis must also rely on a thorough case history, clinical examination, and the results of laboratory tests[1,2] (Table 6-1). When radiographs of lesions are required, the type of radiographs to be taken depends on the size and placement of the lesion. Intraoral film is usually adequate for studying most dental lesions found in dogs and cats. If large lesions or lesions with extensive mass are to be radiographed, an extraoral technique using large-size film is recommended to obtain the best evaluation (see Chapter 1). Although the ideal evaluation is made by obtaining two radiographic views taken at right angles to each other, the unique oral anatomy of dogs and cats often prevents capturing such views. Instead, small animal practitioners are often forced to be satisfied with one radiographic view or two oblique views.[1,2]

VIEWING

When using chairside viewing to evaluate radiographs, ambient light should be avoided. A bright viewing box with a background frame that eliminates extraneous light and a quality hand-magnifying glass are recommended (Figure 6-1).

As discussed in Chapter 1, all dental radiographs should be properly mounted and labeled. Each radiograph contains a dot that corresponds to a dimple on the film packet. This dot is useful in determining the left and right sides of the patient. Because the convex surface of the dimple on the film always faces toward the tubehead during exposure, the right or left side of the area being radiographed can be easily ascertained according to the area's relationship to the position of the dot (Figure 6-2).

When radiographs are being examined, they can be held or placed in the viewing box with the teeth on the patient's right side being viewed on the same side as that corresponding to the right side of the practitioner (i.e., as if the practitioner is sitting on the patient's tongue looking out the mouth). When the radiographs are being viewed, the concave surface of the dot

would be facing the practitioner. As an alternative, the radiographs can be displayed so that the patient's right side corresponds to the practitioner's left side (i.e., as if the practitioner is looking at the patient's face). The decision on how to use the orientation marker provided on the film is at the discretion of individual practitioners; however, to avoid confusion, all radiographs should be viewed following the same orientation (see Chapter 4).

NORMAL AND ABNORMAL CONDITIONS

The appearance of normal conditions varies considerably on radiographs of cats and dogs as well as among the many breeds. In general, bilateral symmetry suggests that the practitioner may be viewing a normal variant or a genetic or metabolic condition. A stable condition that does not change between radiographs taken at long intervals would support a conservative approach.[1]

IMAGE EVALUATION

When evaluating a radiograph, a specific protocol should be systematically followed. For example, despite discovery of a dramatic lesion, the practitioner should continue to examine the rest of the radiograph thoroughly. Likewise, if a lesion that satisfactorily answers the diagnostic queries surrounding a case

TABLE 6–1[1-3]
THE DIAGNOSTIC AND TREATMENT SEQUENCE

History and Clinical Examination
Radiographic Examination
Evaluation and Interpretation of Radiograph
Formulation of a Working Diagnosis
Determination of Treatment Protocol
Initiation of Treatment

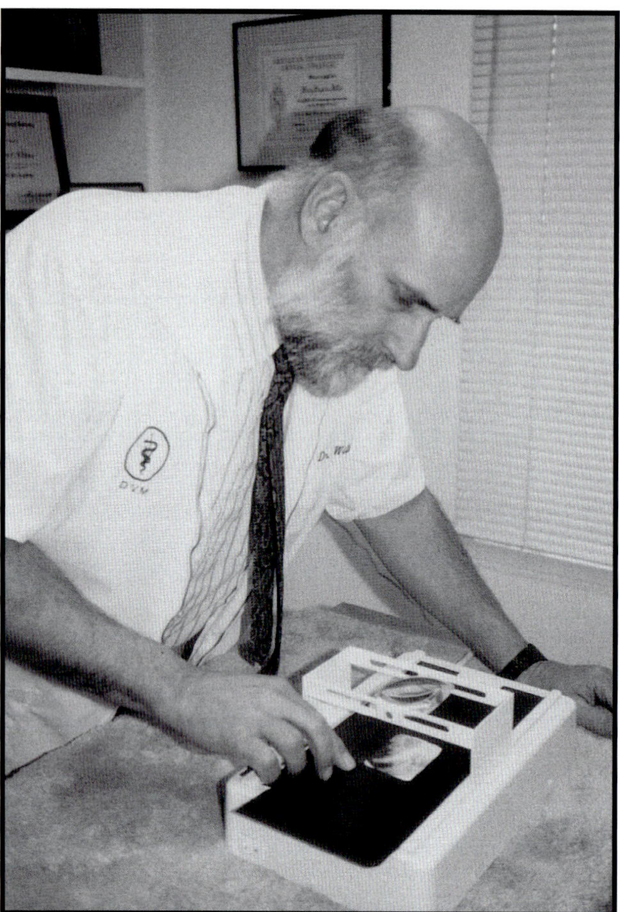

FIGURE 6-1 Film Evaluation

Figure 6-1 *Proper lighting and magnification are critical when interpreting radiographic images.*

is discovered, the practitioner should not prematurely stop viewing the rest of the radiograph.[3] The following list can serve as a guideline:

1. Practitioners should start at one end of the radiograph and view the teeth that were captured in the image from one side to the other. Each tooth structure should be evaluated for defects in contour or for changes in radiopacity to rule out caries or resorptive lesions.
2. The level of periodontal bone should be checked at all interproximal crests and furcations and the presence of any calculus noted.
3. The pulp chamber and periodontal space of each tooth should be carefully evaluated.
4. The bone surrounding the teeth should be checked. The overall density and trabeculation of the bone and the integrity of the lamina dura should be noted.
5. Normal anatomic landmarks should be identified.

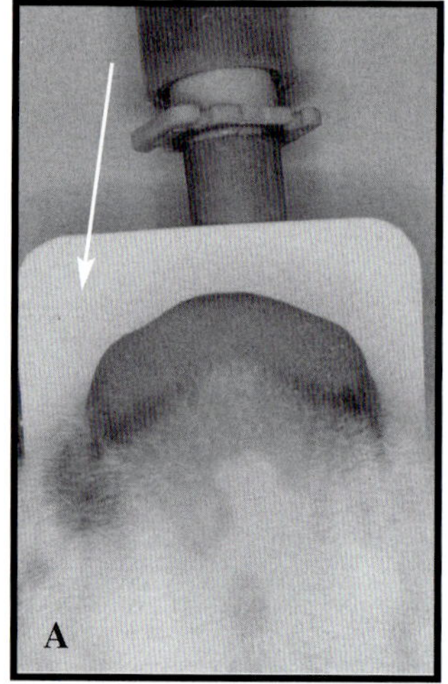
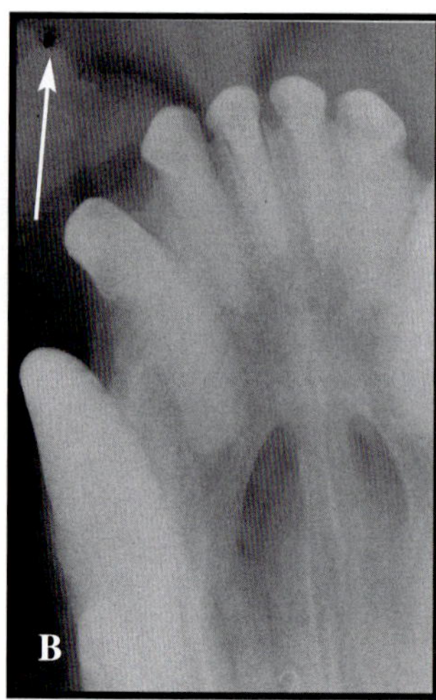

FIGURE 6-2 Orientation of Radiographs

Figure 6-2 *The dot (A and B) that appears on radiographs serves as an orientation marker. The convex surface of the dot corresponds to the convex side of the film packet and always faces the source of radiation (i.e., the tubehead). The side of the patient can easily be identified by the position of the dot and thereby used to recognize the teeth being captured in the radiographic image.*

CHARACTERISTICS OF JAW LESIONS

A lesion in the bone should be characterized according to its site, size, shape, symmetry, borders, contents, and association with and effects on other structures[1-5] (Table 6-2).

Site. The position and extent of the lesion and whether it is solitary, multiple, or generalized should be determined. Generalized bone involvement may indicate metabolic or nutritional disorders.[5]

Size. The size of the lesion should be measured in millimeters according to how it appears on the radiograph.

Shape. The shape of the lesion should be evaluated; to establish shape, radiographs taken from different views are recommended. Lesions tend to develop in the path of least resistance, and studying various views can establish overall shape.[1]

Symmetry. Asymmetry may suggest a medical problem. As already mentioned, bilateral symmetry usually suggests that a normal variant, metabolic condition, or inherited condition exists.[1]

Border. A substantial change in bone density, caused by either resorption or deposition, must occur for a lesion in the bone to be detected on radiographs. The borders of a lesion may have considerable diagnostic value because activity tends to occur at the growing edge. Such characteristics as demarcation, presence of sclerosis, or whether the border is continuous or interrupted should be identified. The silhouette principle is important to remember when evaluating margins. If two

TABLE 6 – 2[1,3–5]
COMMON DESCRIPTORS

Lesion Characteristics	*Descriptors*
Size	*Small, medium, or large*
Number	*Solitary or multiple*
Border	*Ill-defined, well-defined, or sclerotic; continuous or interrupted*
Shape	*Unilocular, multilocular, or nonlocular*
Density	*Radiolucent, radiopaque, or mixed; focal or diffuse*
Effect on Adjacent Structures	*Displacement, resorption, or unaffected*
Effect on Bone	*New bone formation; erosion, perforation, expansion, or lysis*
Association with Teeth	*Teeth associated or independent*

TABLE 6– 3[1,5]
BROAD CLASSIFICATION OF DISEASE

Developmental Anomaly or Congenital Condition
Degenerative Disease
Metabolic, Endocrinal, or Nutritional Disorder
Infectious or Inflammatory Disease
Cystic Disease
Neoplasia: Benign or Malignant Lesion
Trauma

FIGURE 6-3 Summation Effect

Figure 6-3 *This radiograph of a cluster of grapes demonstrates how the opacity of an object is altered by overlapping structures.*

of the bone, whereas a radiopaque lesion suggests mineralization. Patterns of lysis and mineralization, however, vary according to the type and stage of a lesion. Some lesions display a mixture of densities.[1]

Association with Other Structures. Oral lesions may or may not affect nearby teeth. A lesion can have two effects on the teeth: displacement or resorption. If neither is present, the lesion is likely to have no effect.[1,2,4]

CONCLUSION

By matching a radiographic pattern with a set of diagnostic differentials and establishing a correlation between that pattern and existing clinical signs, practitioners can classify the disease or condition and formulate a working diagnosis (Table 6-3). Careful interpretation of quality radiographs assists practitioners in determining whether additional tests need to be done, in selecting the correct treatment protocol, or in disseminating professional advice based on well-qualified knowledge.[1,2]

structures of the same radiopacity are in contact, their margins cannot be distinguished. On the other hand, if two structures of the same radiopacity are not in contact and are separated by a substance with different radiopacity, the borders can be ascertained on radiographs.[3] In addition, unless benign conditions are inflamed, they tend to be well defined and corticated.[1]

Contents. The density of the contents of a lesion should be judged, remembering that the radiolucent and radiopaque qualities are relative to the density and thickness of adjacent structures and the radiographic technique being used.[1,3] Summation effects from overlapping structures can determine the apparent radiopacity or radiolucency of a lesion[3] (Figure 6-3). The actual appearance of a lesion can also vary with the stage of pathogenesis. In general, a radiolucent lesion suggests lysis

REFERENCES

1. Goaz PW, White SC: *Oral Radiology: Principles and Interpretation.* Philadelphia, CV Mosby Co, 1994, pp 291–305.
2. Langlais RP, Rodriguez IE, Maselle I: Principles of radiographic selection and interpretation, in Miles DA, Van Dis ML (eds): *The Dental Clinics of North America,* vol 38, no. 1. Philadelphia, WB Saunders Co, 1994.
3. Thrall DE: Introduction to radiographic interpretation, in Thrall DE (ed): *Textbook of Veterinary Diagnostic Radiology,* ed 2. Philadelphia, WB Saunders Co, 1994, pp 1–14.
4. Miles DA, Van Dis ML, Kaugars GE, Lovas JGL: *Oral & Maxillofacial Radiology: Radiologic/Pathologic Correlations.* Philadelphia, WB Saunders Co, 1991, pp 1–5.
5. Konde JE: Aggressive versus nonaggressive bone lesions, in Thrall DE (ed): *Textbook of Veterinary Diagnostic Radiology,* ed 2. Philadelphia, WB Saunders Co, 1994, pp 15–22.

Chapter Seven

NORMAL RADIOGRAPHIC ANATOMY

Familiarity with the appearance of normal oral structures is necessary to diagnose abnormalities accurately. This chapter addresses the normal radiographic anatomy of the dentition of dogs and cats from preeruption development to maturation and aging. Because the eruption and maturation of teeth can vary among breeds, the age ranges cited are approximate.

BASIC DENTAL ANATOMY

Teeth are divided topographically into regions: the crown, neck or cervical line, and root. The radiopaque structures of the tooth are primarily dentin, with the crown being covered by enamel and the root by cementum. The neck region is the cementoenamel junction. The internal radiolucent structure of teeth is the pulp, which occupies the pulp chamber and root canal and often exhibits extensions (pulp horns) into the cusps or crests of the tubercles in the crown. In teeth with multiple roots, the area where the roots branch apart is referred to as the furcation. The tooth resides in a bony alveolar socket, which is suspended by the periodontal ligament that connects the root cementum to surrounding bone. The periodontal ligament space appears as a uniformly thin radiolucent space between the root and alveolar bone. The alveolus is lined with specialized cortical bone that appears on radiographs as a radiopaque line of fairly uniform thickness known as the lamina dura, which is contiguous with the crestal bone. Ideally, the alveolar bone should extend to the cementoenamel junction[1-3] (Figures 7-1 to 7-3).

ANATOMIC LANDMARKS

Several normal anatomic structures can serve as landmarks or produce artifacts on dental radiographs (Figures 7-4 to 7-11). Radiolucent structures of the mandible include the mandibular canal, mental foramina, and mandibular symphysis. The mandibular canal, which houses the vessels and nerves supplying the mandibular teeth, is a radiolucent tubular structure that parallels the ventral border of the mandible. The roots of the mandibular teeth sometimes seem to project into or across the canal. The mental foramina can be seen as radiolucent defects in the mandible that sometimes overlay the roots. The most cranial mental foramen is located in the incisor area, the middle mental foramen is at the level of the mandibular canine root, and the most caudal mental foramen is usually at the level of the third premolar. When viewing dental radiographs, a foramen could be mistaken for a pathologic periapical lesion if it is superimposed over a root. If a second radiograph is viewed from a different angle, the position of the foramen shifts in relation to the tooth. In contrast, the position of a true periapical lesion does not change in relation to the position of the root. The mandibular symphysis is a fibrous joint that joins the two halves of the mandible and presents as a thin, sometimes irregular, radiolucent line in animals.[1-7]

Radiolucent structures of the maxilla include the palatine fissures, palatine foramen, incisive canal, infraorbital foramen, and infraorbital canal. The palatine fissures are large paired openings caudal to the third incisors. The palatine foramina are near the fourth premolar. The incisive canal is a small opening located at the palatine suture line just caudal to the central incisors. The infraorbital foramen is located at the junction where the zygomatic arch joins the maxilla and can be seen as a radiolucent structure overlying the distal root of the third upper premolar or the mesial roots of the fourth premolar. The infraorbital canal is a tubular structure that extends caudally from the infraorbital foramen in dogs and is visible in some radiographs of the fourth premolar and molars.[5,6,8]

Radiopaque structures of the maxilla include the zygomatic arch, nasal septum, and floor of the nasal sinus. The zygomatic arch is often superimposed over the fourth premolar of cats and sometimes over the molars of dogs. Such superimposition can obscure anatomic details. To visualize the fourth upper premolar apices in cats, the radiographic image often needs to be elongated. In lateral radiographic views, the lateral aspect of the floor of the nasal cavity, where the bones of the palate meet the bones of the maxilla, is seen as a thin radiopaque undulating line superimposed over the roots of the maxillary teeth. The position of the line can change slightly according to the angle of the primary x-ray beam. The nasal

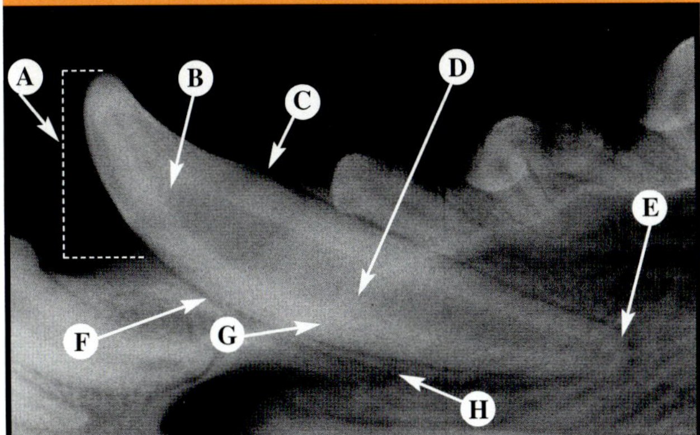

FIGURE 7-1 Single-Rooted Teeth

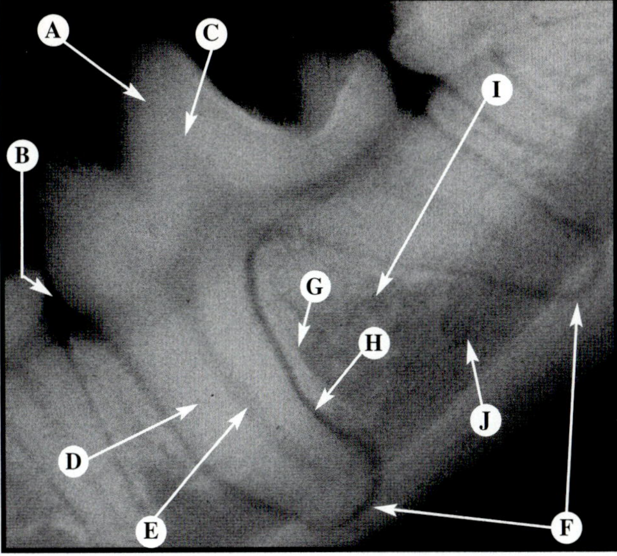

FIGURE 7-2A Teeth with Two Roots

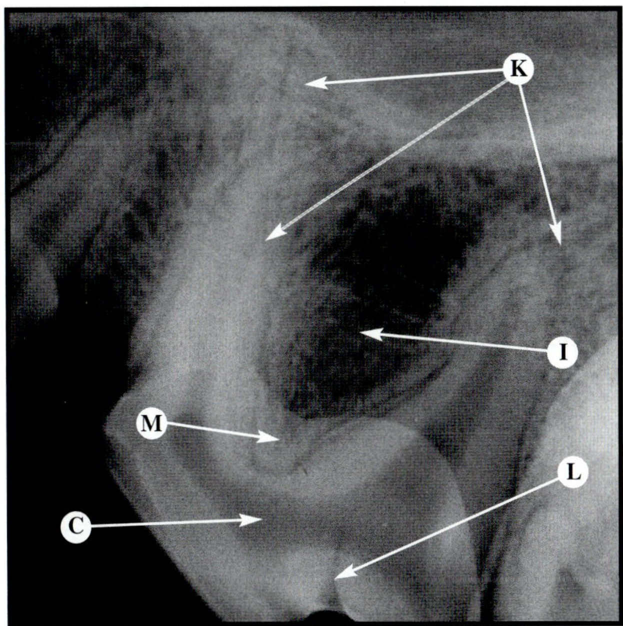

FIGURE 7-2B Teeth with Three Roots

Figure 7-1 *This radiograph shows the normal anatomy of a tooth with a single root. (A) Crown, (B) pulp chamber, (C) cervical line (cementoenamel junction, (D) root canal, (E) apex, (F) periodontal space, (G) dentinal wall, and (H) lamina dura.* **Figures 7-2A and 2B** *These radiographs show the normal anatomy of a tooth with two or three roots. (A) Crown, (B) cervical line (cementoenamel junction), (C) pulp chamber and pulp horn, (D) dentinal wall, (E) root canal, (F) apex, (G) lamina dura, (H) periodontal space, (I) interradicular bone, (J) mandibular canal, (K) three roots, (L) developmental groove, and (M) furcation.*

septum is seen as a radiopaque line that bisects the nasal cavity in dorsoventral views[4,7,8] (Figures 7-4 to 7-11).

THE DENTITION

Dogs and cats are diphyodonts; that is, they have two sets of teeth. The first set (the deciduous or baby teeth) serve the animal early in life. The second set (the permanent teeth) serve the animal in adulthood and erupt as the deciduous teeth undergo resorption and exfoliation. Deciduous teeth are smaller and slimmer than the permanent teeth and, when viewed on radiographs, have large pulp chambers and root canals with comparatively long, thin roots. The first premolar

of dogs and all the molars of dogs and cats have no deciduous precursors.[9]

The number and classification of teeth for dogs and cats are

DENTAL FORMULA FOR CATS

Deciduous Teeth

$$2 \times (I\tfrac{3}{3} \quad C\tfrac{1}{1} \quad P\tfrac{3}{2}) = 26$$

Permanent Teeth

$$2 \times (I\tfrac{3}{3} \quad C\tfrac{1}{1} \quad P\tfrac{3}{2} \quad M\tfrac{1}{1}) = 30$$

DENTAL FORMULA FOR DOGS

Deciduous Teeth

$$2 \times (I\tfrac{3}{3} \quad C\tfrac{1}{1} \quad P\tfrac{3}{3}) = 28$$

Permanent Teeth

$$2 \times (I\tfrac{3}{3} \quad C\tfrac{1}{1} \quad P\tfrac{4}{4} \quad M\tfrac{2}{2}) = 42$$

Legend: I = incisors, C = canine teeth, P = premolars, M = molars.

NORMAL ANATOMY OF MAXILLARY STRUCTURES—DOGS

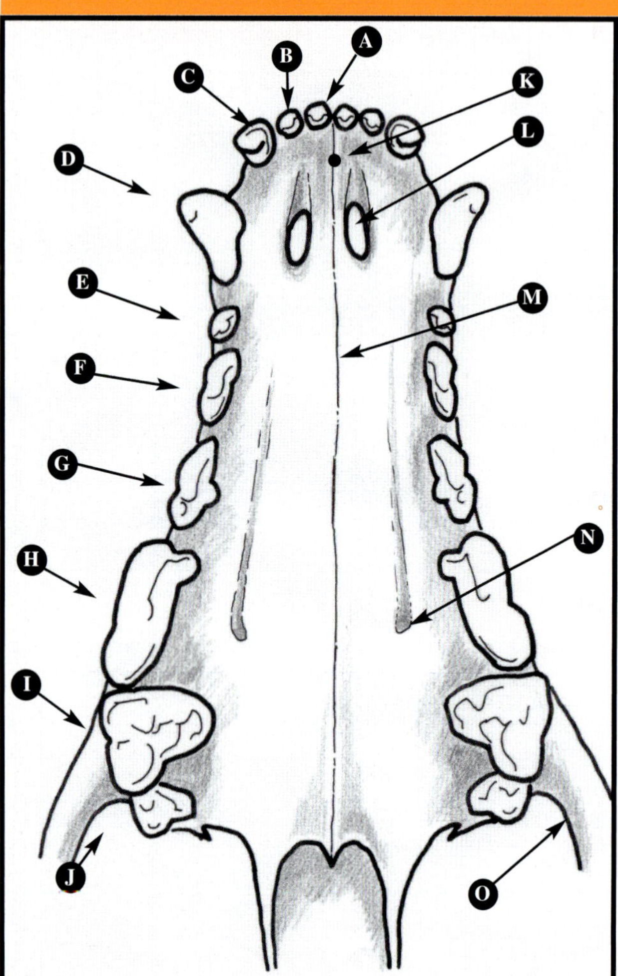

FIGURE 7-3A Structural Anatomy

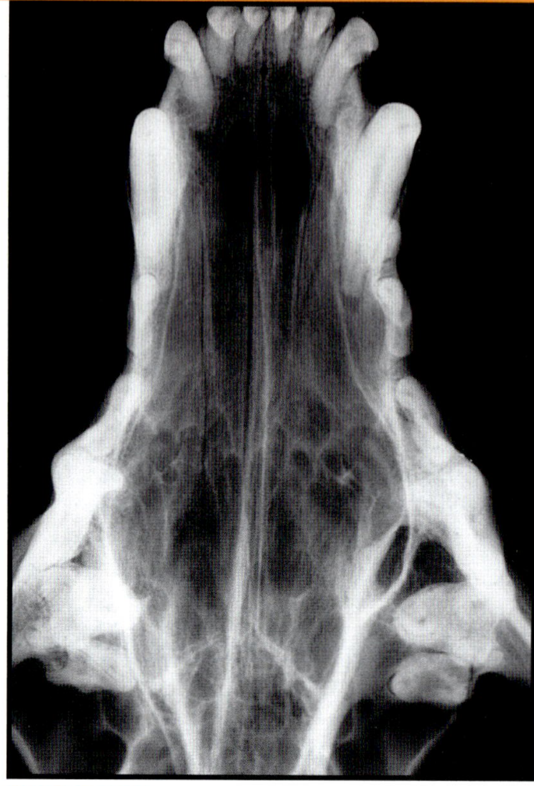

FIGURE 7-3B Radiographic Image

Figure 7-3A *This illustration identifies the major structures in the maxilla of dogs. (A) Central incisor 1, (B) intermediate incisor 2, (C) corner incisor 3, (D) canine tooth, (E) premolar 1, (F) premolar 2, (G) premolar 3, (H) premolar 4 (I) molar 1, (J) molar 2, (K) incisive canal, (L) palatine fissure, (M) palatine suture and nasal septum, (N) palatine foramen, and (O) zygomatic arch.* **Figure 7-3B** *This radiograph captures the images of the structures.*

NORMAL ANATOMY OF THE MAXILLA FROM EXTRAORAL OBLIQUE VIEW—DOGS

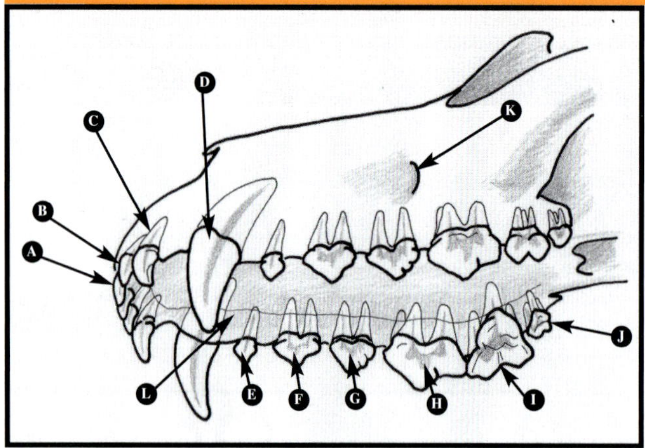

FIGURE 7-4A Structural Anatomy

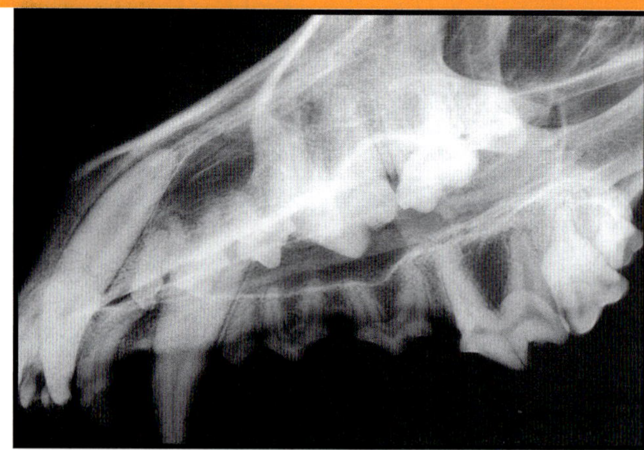

FIGURE 7-4B Radiographic Image

Figure 7-4A *This illustration identifies the major structures of the maxilla when radiographed from an oblique view. (A) Central incisor 1, (B) intermediate incisor 2, (C) corner incisor 3, (D) canine tooth, (E) premolar 1, (F) premolar 2, (G) premolar 3, (H) premolar 4, (I) molar 1, (J) molar 2, (K) infraorbital foramen, and (L) confluence of bones of the palate and maxilla.* **Figure 7-4B** *This radiograph captures the images of the structures.*

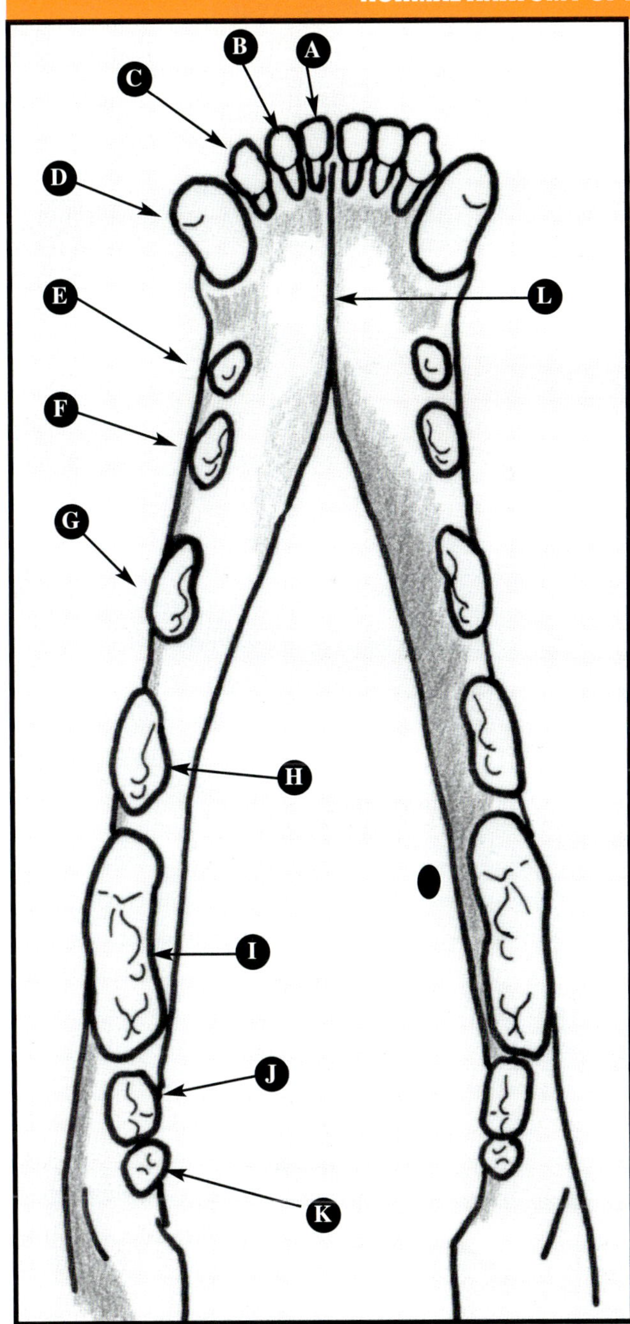

FIGURE 7-5A Structural Anatomy

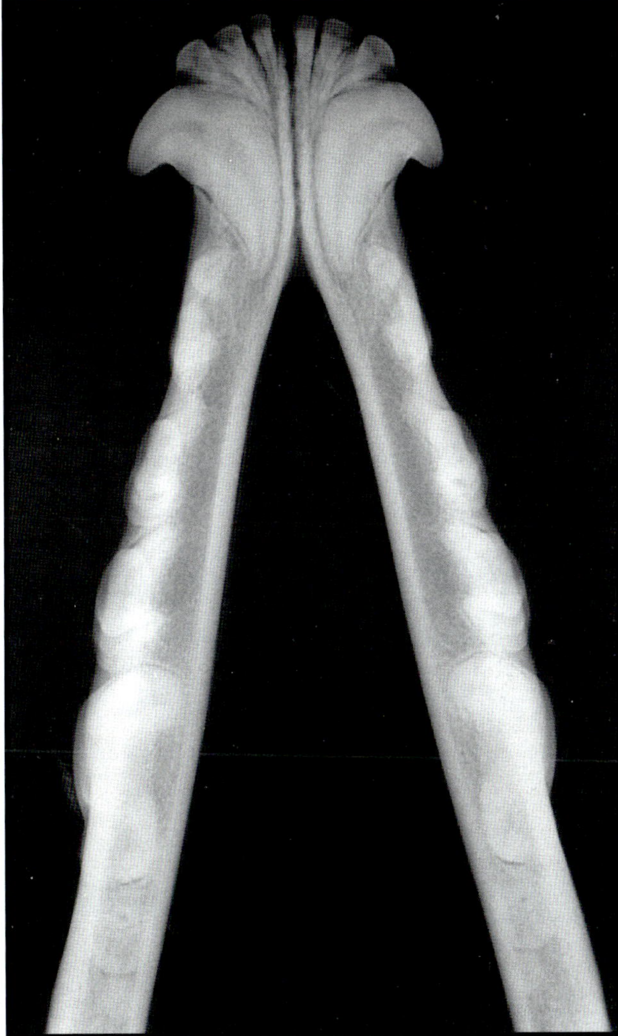

FIGURE 7-5B Radiographic Image

Figure 7-5A *This illustration identifies the major structures shown in a ventrodorsal view of the mandibular teeth.* (A) *Central incisor 1,* (B) *intermediate incisor 2,* (C) *corner incisor 3,* (D) *canine tooth,* (E) *premolar 1,* (F) *premolar 2,* (G) *premolar 3,* (H) *premolar 4,* (I) *molar 1,* (J) *molar 2,* (K) *molar 3, and* (L) *symphysis.* **Figure 7-5B** *This radiograph captures the image of the structures.*

conveniently represented by formulas (see Dental Formula for Dogs and Dental Formula for Cats). Dogs normally have 42 permanent teeth. Of these, 22 are single rooted: 12 incisors, 4 canine teeth, 4 first premolars, and 2 lower last molars. The upper fourth premolar and upper molars have three roots. The rest of the teeth have 2 roots.[9-11] Cats have 30 permanent teeth. Of these, 18 are single rooted: 12 incisors, 4 canine teeth, and 2 upper premolars. The upper fourth premolar has 3 roots, whereas all the remaining teeth have 2 roots (although the upper molar sometimes has a single fused root).[6,9,11]

The multiple roots of teeth are not necessarily symmetrical in size, shape, or length. For example, the distal root of the upper fourth premolar is often wider in dogs and cats than the tooth's two mesial roots. Likewise, the mesial root of the lower molar in cats is wider and often longer than the distal root.

ERUPTION AND DEVELOPMENT

In dogs, the crowns of deciduous teeth and first permanent molar have some calcification by day 55 of gestation. The calcification is radiographically evident at birth. The formation of

TABLE 7 – 1[9]
ERUPTION OF TEETH IN DOGS

Type of Tooth	Deciduous Teeth (age in weeks)	Permanent Teeth (age in months)
Incisors	3–4	3–5
Canine Teeth	3–5	4–6
Premolars	4–12	4–6
Molars	Not applicable	5–7

TABLE 7 – 2[9]
ERUPTION OF TEETH IN CATS

Type of Tooth	Deciduous Teeth (age in weeks)	Permanent Teeth (age in months)
Incisors	2–4	3–4
Canine Teeth	3–4	4–5
Premolars	3–6	4–6
Molars	Not applicable	4–5

TABLE 7 – 3
STAGES OF TOOTH ERUPTION VISIBLE ON RADIOGRAPHS[2,13,14]

Root Formation of Permanent Teeth
Formation of Gubernacular Canal
Resorption of Overlying Bone
Resorption of Deciduous Roots
Exfoliation of Deciduous Teeth
Formation of Alveolar Bone

TABLE 7 – 4
RADIOGRAPHIC FEATURES OF MATURATION AND AGING[1,2,16]

Changes to Tooth Structure
Root becomes elongated
Apical closure
Thickening of the dentinal wall
Constriction and shortening of the pulp and root canal
Changes to Bone Structure
Increasing coarseness and density of bone
Indistinct lamina dura
Slight regression of the alveolar crest

all deciduous crowns and the roots is almost complete by day 40 to 45 after birth, and formation of the deciduous canine tooth is complete by about day 55 after birth. Clinical emergence of the deciduous teeth usually occurs between 3 and 12 weeks of age in dogs and 2 to 6 weeks of age in cats. The permanent teeth develop lingual and palatal to deciduous teeth and are somewhat calcified by 3 months of age. For that reason, radiographic screening for missing permanent teeth should be done at 3 months of age if missing teeth are a concern.[2,3,9,12]

The deciduous teeth are usually exfoliated by 6 months of age and the permanent teeth clinically evident by 6 to 7 months of age in dogs and 6 months of age in cats. The initial clinical emergence into the oral cavity and total eruption time to occlusion vary[2,3,9] (Tables 7-1 and 7-2).

Tooth eruption is a complex process that can be captured on radiographs. Developing teeth reside within a dental follicle that plays an essential role in tooth formation and eruption. When root formation and alveolar bone deposition are initiated, tooth eruption is usually impending. The eruption rate can be initially slow but then accelerates as the roots become elongated. A narrow bony channel, known as the gubernacular canal, is located at the occlusal aspect of the follicle. Enlargement of the canal results in resorption of overlying bone and formation of an eruption pathway. Ideally, resorption of deciduous roots coincides with eruption of the succedent teeth. When the tooth emerges into the oral cavity, often about one half or more of the root length has been formed[2,13,15] (Table 7-3 and Figures 7-11 and 7-12).

MATURATION AND AGING

After the teeth have erupted, completion of root formation occurs. First the root becomes elongated, and then the apex closes to form an apical delta for egression and ingression of pulp vessels and nerves. The apical delta is not normally visible on radiographs unless it is delineated by endodontic filling material. Maturation and apical closure vary but usually occur from 8 to 18 months of age in dogs and 8 to 12 months in cats. Because the odontoblasts that line the pulp chamber and root canal continue to deposit dentin internally during the life of the animal,

NORMAL ANATOMY OF THE MANDIBLE FROM THE EXTRAORAL OBLIQUE VIEW—DOGS

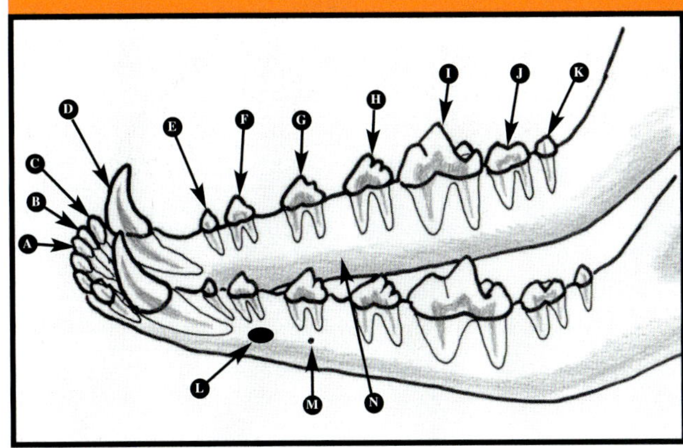

FIGURE 7-6A Structural Anatomy

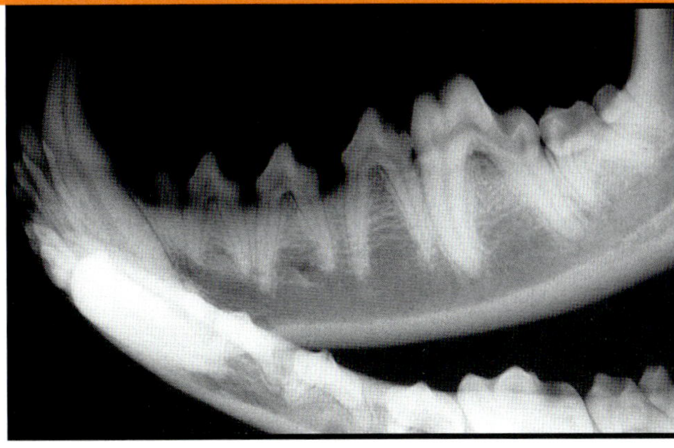

FIGURE 7-6B Radiographic Image

Figure 7-6A *This illustration identifies the major structures of the mandible when radiographed from an extraoral oblique view.* (A) *Central incisor 1,* (B) *intermediate incisor 2,* (C) *corner incisor 3,* (D) *canine tooth,* (E) *premolar 1,* (F) *premolar 2,* (G) *premolar 3,* (H) *premolar 4,* (I) *molar 1,* (J) *molar 2,* (K) *molar 3,* (L) *middle mental foramen,* (M) *caudal mental foramen, and* (N) *region of the mandibular canal.* **Figure 7-6B** *This radiograph captures the image of these structures.*

NORMAL ANATOMY OF THE MAXILLARY TEETH—CATS

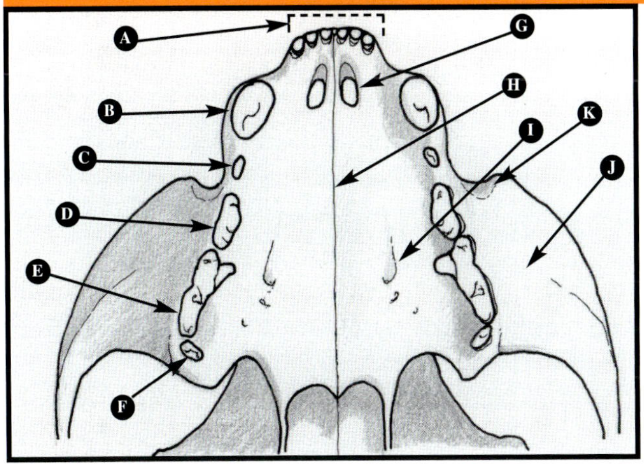

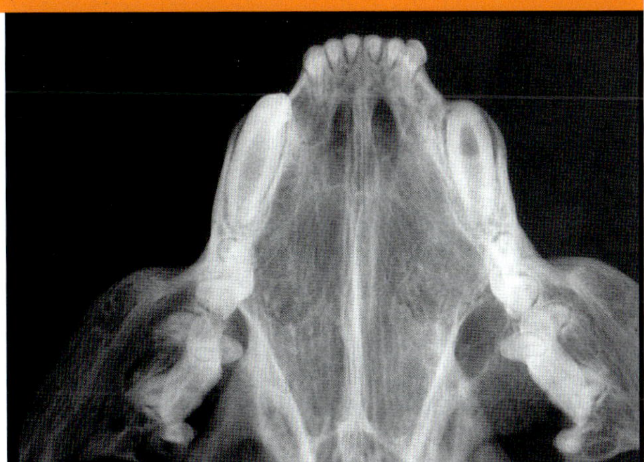

FIGURE 7-7A Structural Anatomy

FIGURE 7-7B Radiographic Image

Figure 7-7A *This illustration identifies the structures shown in a dorsoventral view of the maxillary teeth.* (A) *Incisors,* (B) *canine tooth,* (C) *premolar 2,* (D) *premolar 3,* (E) *premolar 4,* (F) *molar,* (G) *palatine fissure,* (H) *palatine suture and nasal septum,* (I) *palatine arterial fossa,* (J) *zygomatic arch, and* (K) *infraorbital foramen.* **Figure 7-7B** *This radiograph captures the image of these structures.*

the pulp chamber and root canal become progressively more narrow and shorter with time. As an animal ages, the trabecular bone that surrounds the teeth becomes more dense and on radiographs presents a more coarse and increasingly radiopaque pattern that can obscure the lamina dura. Some slight regression of the crestal bone is also considered normal with age and should not be confused with periodontal disease[1,2,16] (Table 7-4).

The age of an animal can often be estimated by observing the maturation of teeth. For example, examination of the canine tooth might be helpful in estimating the age of a cat, whereas the incisors in cats are too small to evaluate.

The radiographs in Figures 7-13 through 7-77 show the distinguishing characteristics and structural features of the normal dental anatomy of dogs and cats.

REFERENCES

1. Zontine WJ: Dental radiographic technique and interpretation, in Suter P (ed): *The Veterinary Clinics of North America.* Philadelphia, WB Saunders Co, 1974, pp 741-762.
2. Zontine WJ: Canine dental radiology: Radiographic technic, development, and anatomy of the teeth. *J Am Vet Radiol Soc* 16:75-82, 1975.
3. Orsini P, Hennet P: Anatomy of the mouth and teeth of the cat, in Harvey CE (ed): *The Veterinary Clinics of North America.*

(References continue on page 90)

NORMAL ANATOMY OF THE MAXILLA FROM THE EXTRAORAL OBLIQUE VIEW—CATS

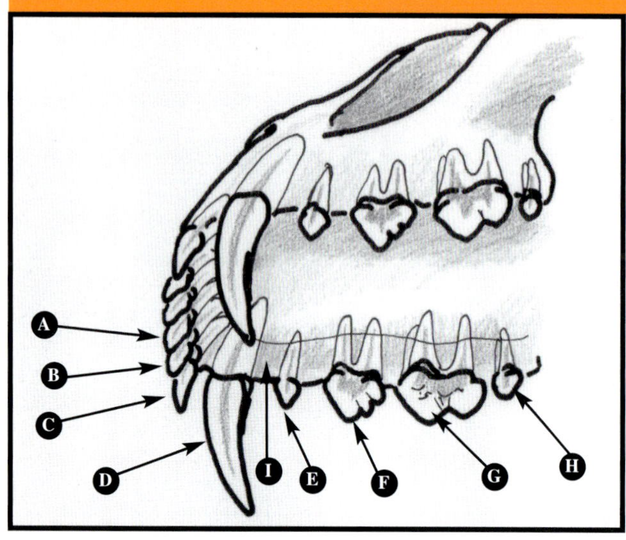

FIGURE 7-8A Structural Anatomy

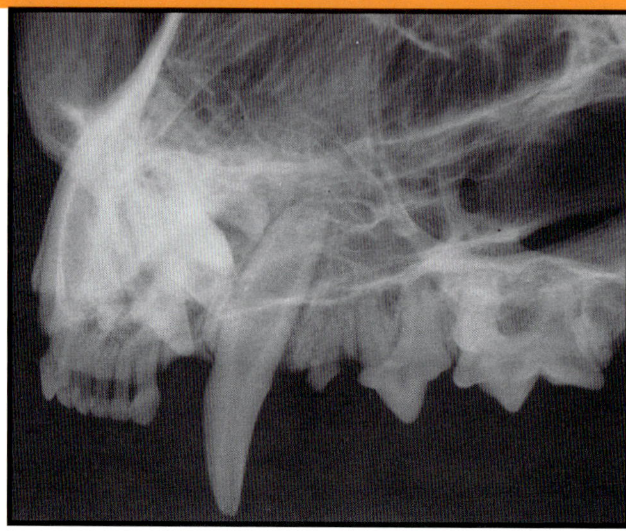

FIGURE 7-8B Radiographic Image

Figure 7-8A *This illustration identifies the structures of the maxilla when radiographed from an extraoral oblique view. (A) Central incisor 1, (B) intermediate incisor 2, (C) corner incisor 3, (D) canine tooth, (E) premolar 2, (F) premolar 3, (G) premolar 4, (H) molar, and (I) confluence of the bones of the palate and maxilla.* **Figure 7-8B** *This radiograph captures the image of these structures.*

NORMAL ANATOMY OF THE MANDIBULAR TEETH—CATS

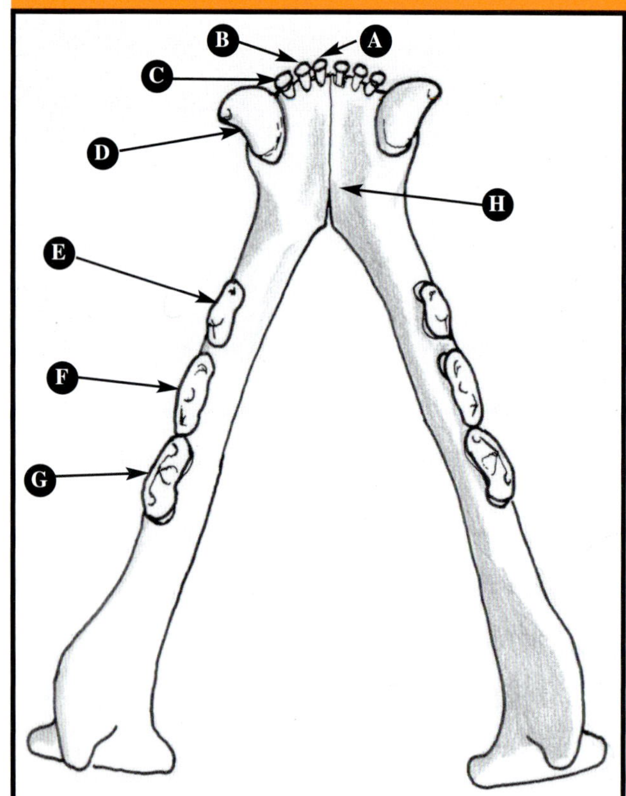

FIGURE 7-9A Structural Anatomy

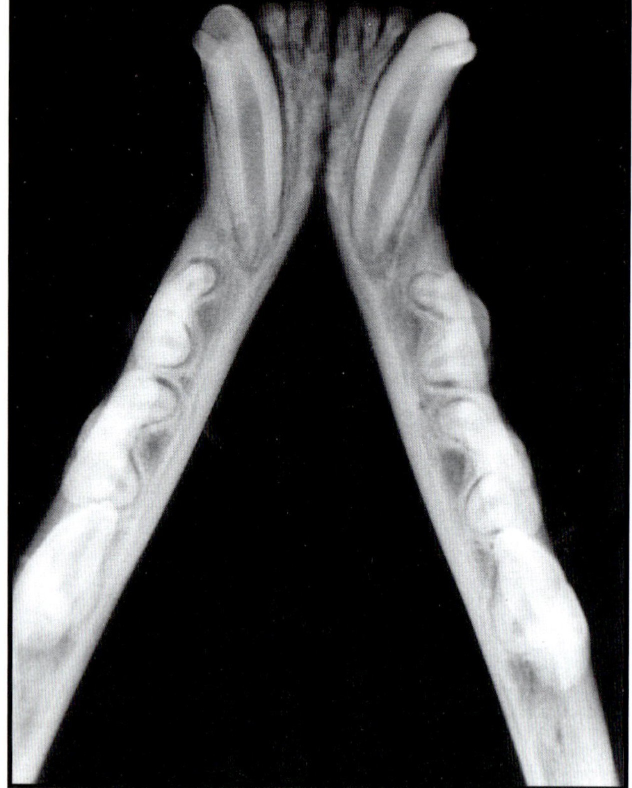

FIGURE 7-9B Radiographic Image

Figure 7-9A *This illustration identifies the structures of the mandibular teeth when radiographed from a ventrodorsal view. (A) Central incisor 1, (B) intermediate incisor 2, (C) corner incisor 3, (D) canine tooth, (E) premolar 3, (F) premolar 4, (G) molar, and (H) symphysis.* **Figure 7-9B** *This radiograph captures the image of these structures.*

NORMAL ANATOMY OF THE MANDIBLE FROM THE EXTRAORAL OBLIQUE VIEW—CATS

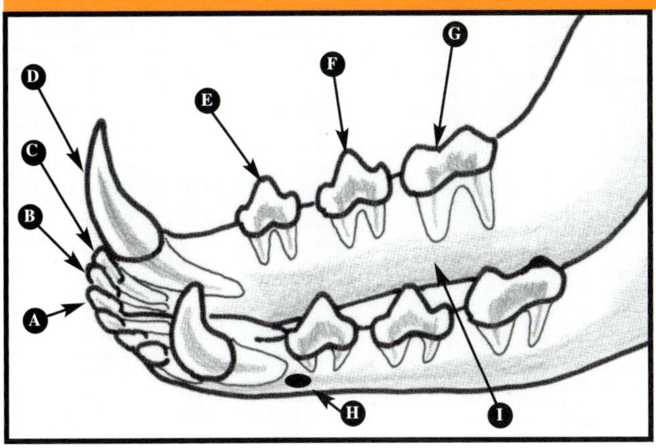

FIGURE 7-10A Structural Anatomy

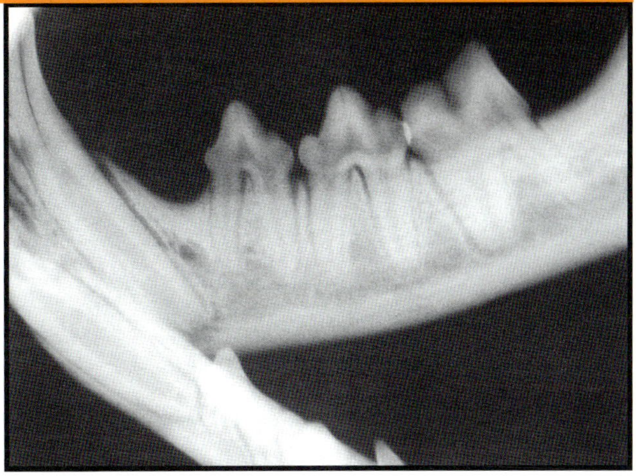

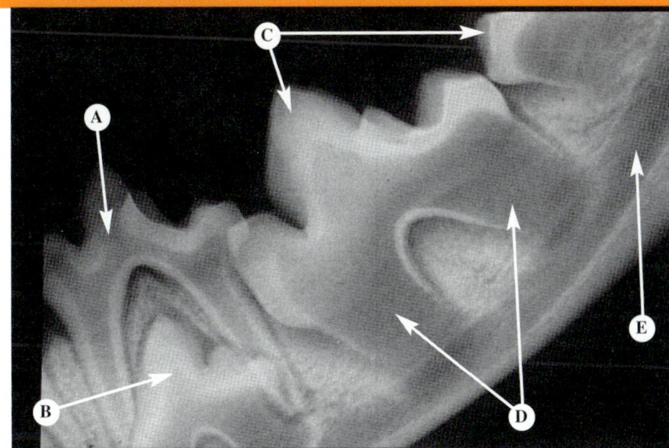

FIGURE 7-10B Radiographic Image

Figure 7-10A *This illustration identifies the structures of the mandible radiographed from an extraoral oblique view.* (A) *Central incisor 1,* (B) *intermediate incisor 2,* (C) *corner incisor 3,* (D) *canine tooth,* (E) *premolar 3,* (F) *premolar 4,* (G) *molar 1,* (H) *middle mental foramen, and* (I) *region of the mandibular canal.* **Figure 7-10B** *This radiograph captures the image of these structures.*

NORMAL ANATOMY OF ERUPTING TEETH—DOGS

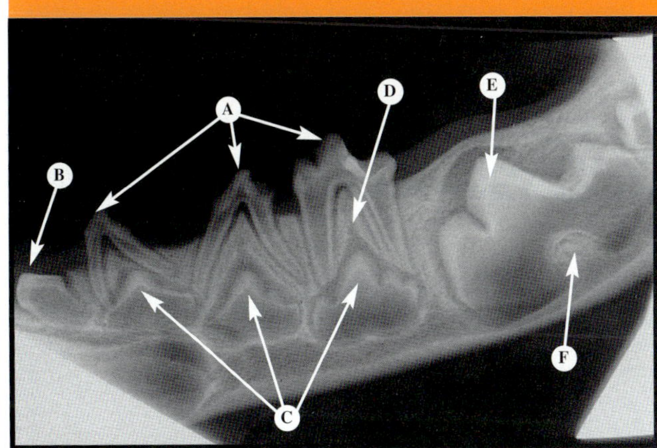

FIGURE 7-11 Premolars at 12 Weeks of Age

FIGURE 7-12 Molars at 5 Months of Age

Figure 7-11 *The anatomy of the premolars in dogs at 12 weeks of age. Note that the morphology of the third deciduous premolar is that of the molar even though it is a precursor of the fourth premolar.* (A) *Deciduous premolars,* (B) *permanent premolar 1,* (C) *permanent premolars 2, 3, and 4,* (D) *gubernacular canal,* (E) *first molar, and* (F) *furcation.* **Figure 7-12** *This radiograph shows how the molar emerges in a 5-month-old dog before root formation is complete.* (A) *Deciduous premolar,* (B) *permanent premolar 4,* (C) *molars,* (D) *formation of the root, and* (E) *mandibular canal.*

NORMAL ANATOMY OF MATURATION AND AGING—DOGS

FIGURE 7-13 Structural Anatomy at 18 Months of Age

FIGURE 7-14 Structural Anatomy at 8 Years of Age

Figure 7-13 *This radiograph shows the thickness of the dentinal wall, length of the root, and apical closure in an 18-month-old dog.* (A) *Thick dentinal wall and* (B) *root length complete with apical closure.* **Figure 7-14** *As a comparison, this radiograph shows how structural anatomy is affected by aging.* (A) *Thicker dentinal wall,* (B) *constricted root canal,* (C) *blunted alveolar crest, and* (D) *bone density obscuring the lamina dura.*

NORMAL ANATOMY OF THE MANDIBULAR INCISORS AND CANINE TEETH—DOGS

FIGURE 7-15 Structural Anatomy at 6 to 8 Weeks of Age

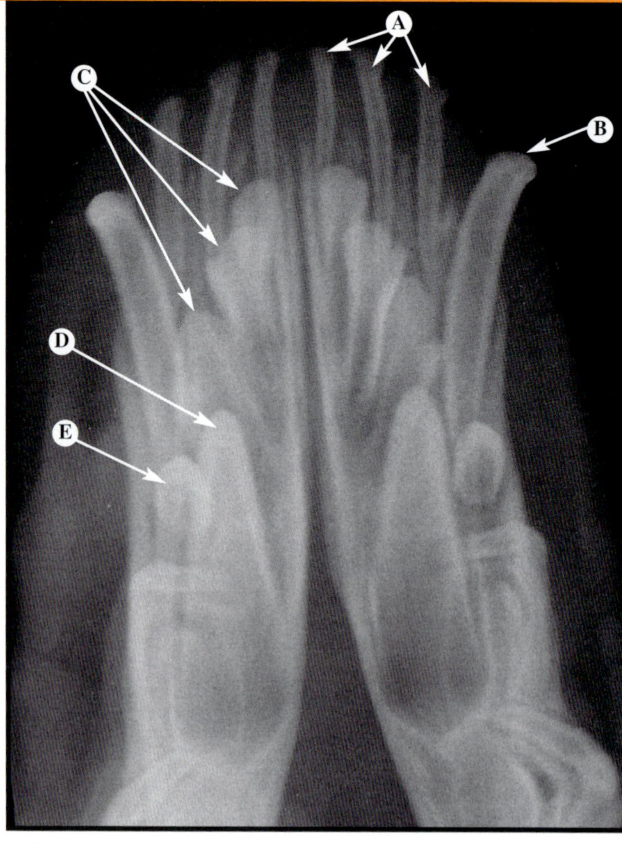

FIGURE 7-16 Structural Anatomy at 12 to 14 Weeks of Age

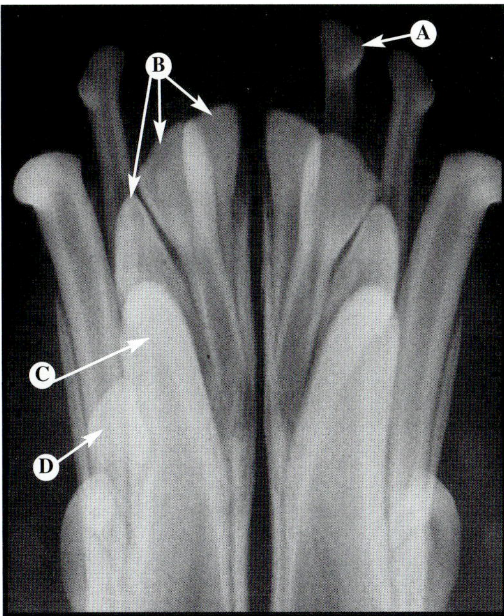

FIGURE 7-17 Structural Anatomy at 14 to 16 Weeks of Age

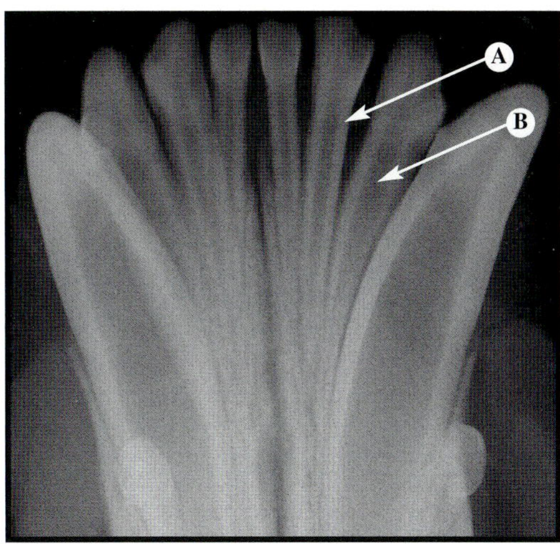

FIGURE 7-18 Structural Anatomy at 6 Months of Age

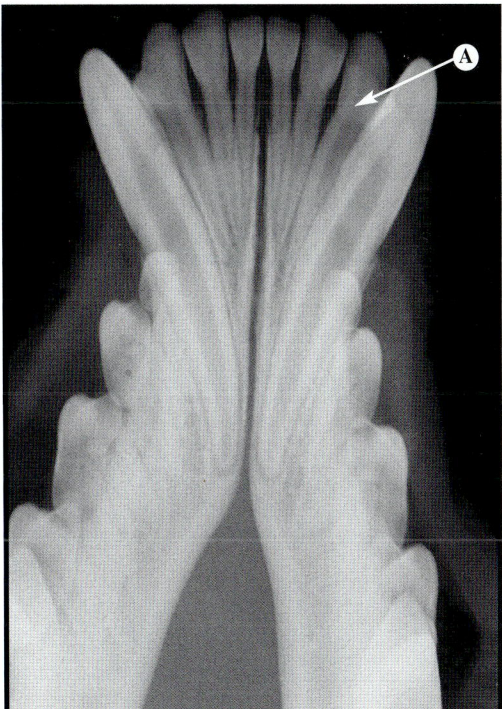

FIGURE 7-19 Structural Anatomy at 18 Months of Age

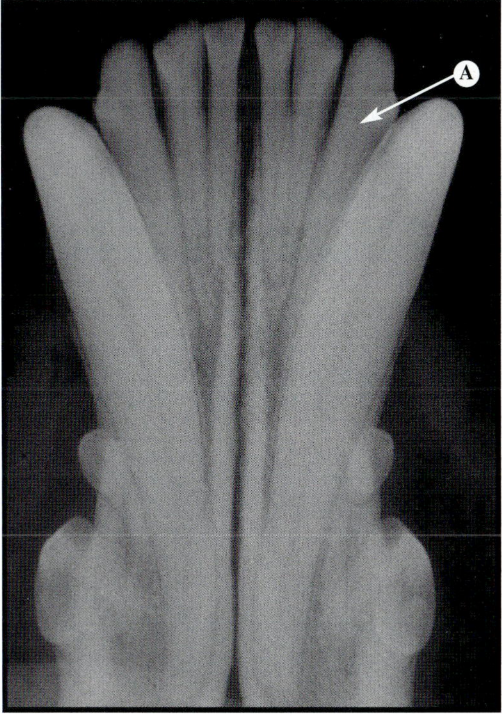

FIGURE 7-20 Structural Anatomy of a Geriatric Dog

Figure 7-15 *The mandibular incisors and canine teeth in a 6- to 8-week-old puppy. (A) Deciduous incisors, (B) deciduous canine tooth, (C) unerupted permanent incisors, and (D) unerupted permanent canine tooth.* **Figure 7-16** *This radiograph shows the mandibular incisors and canine teeth in a 12- to 14-week-old puppy. (A) Deciduous incisors, (B) canine tooth, (C) unerupted permanent incisors, (D) unerupted permanent canine tooth, and (E) first premolar.* **Figure 7-17** *This radiograph shows the mandibular incisors and canine teeth in a 14- to 16-week-old puppy. (A) Deciduous incisor undergoing resorption, (B) unerupted permanent incisors, (C) unerupted permanent canine tooth, and (D) first premolar.* **Figure 7-18** *This radiograph shows the mandibular incisors and canine teeth in a 6-month-old dog. (A) Thin dentinal wall and (B) wide lumen in the root canal.* **Figure 7-19** *This radiograph shows the mandibular incisors and canine teeth in an 18-month-old dog. Note the (A) thick dentinal wall and narrowing lumen in the root canal.* **Figure 7-20** *This radiograph shows the mandibular incisors and canine teeth in a geriatric dog. In comparison to Figure 7-19, note the (A) thick dentinal wall and constricted lumen in the root canal.*

NORMAL ANATOMY OF THE MAXILLARY INCISORS—DOGS

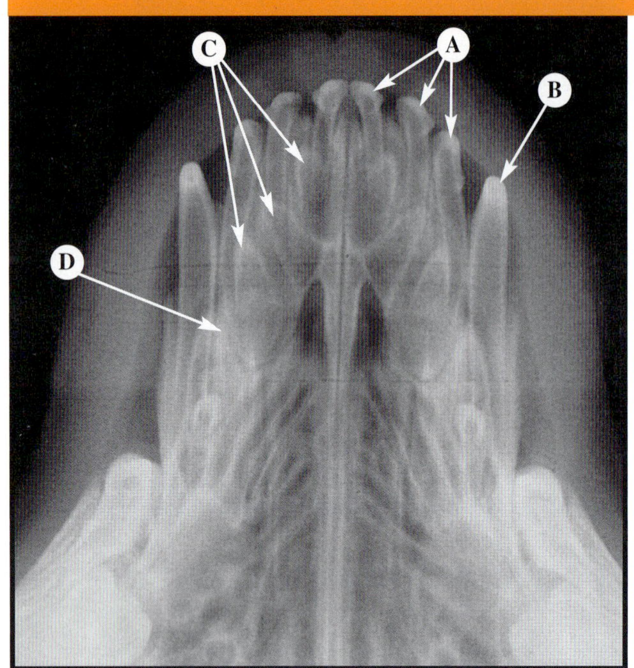

FIGURE 7-21 Structural Anatomy at 6 to 8 Weeks of Age

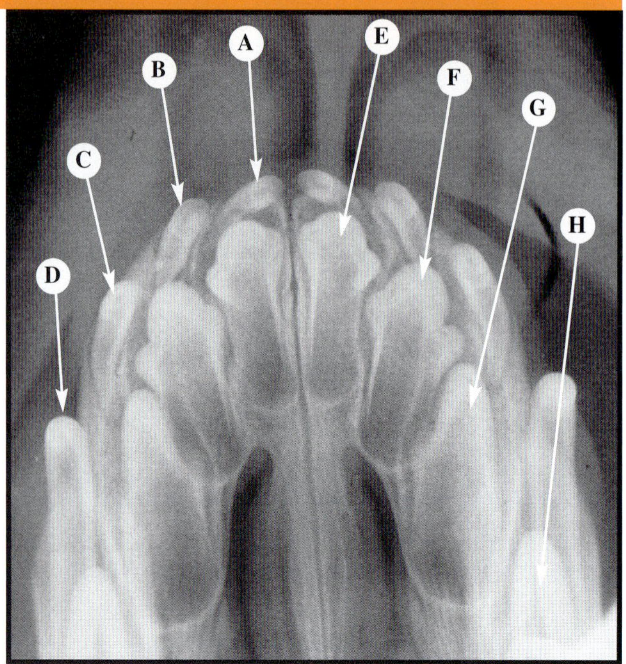

FIGURE 7-22 Structural Anatomy at 12 to 14 Weeks of Age

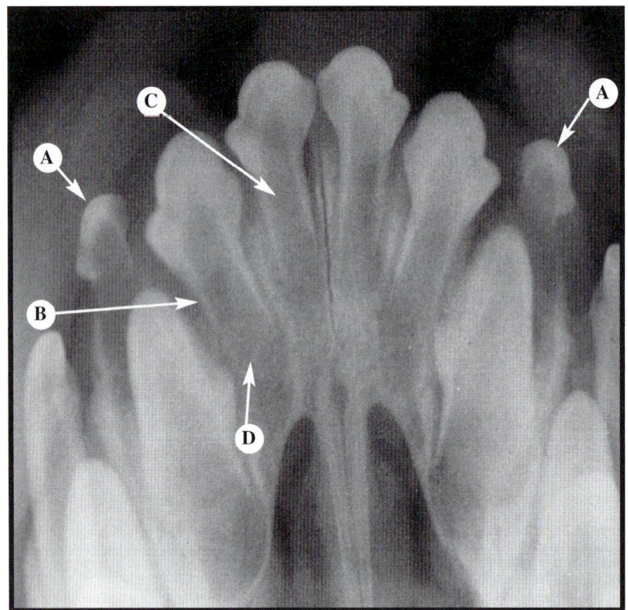

FIGURE 7-23 Structural Anatomy at 14 to 16 Weeks of Age

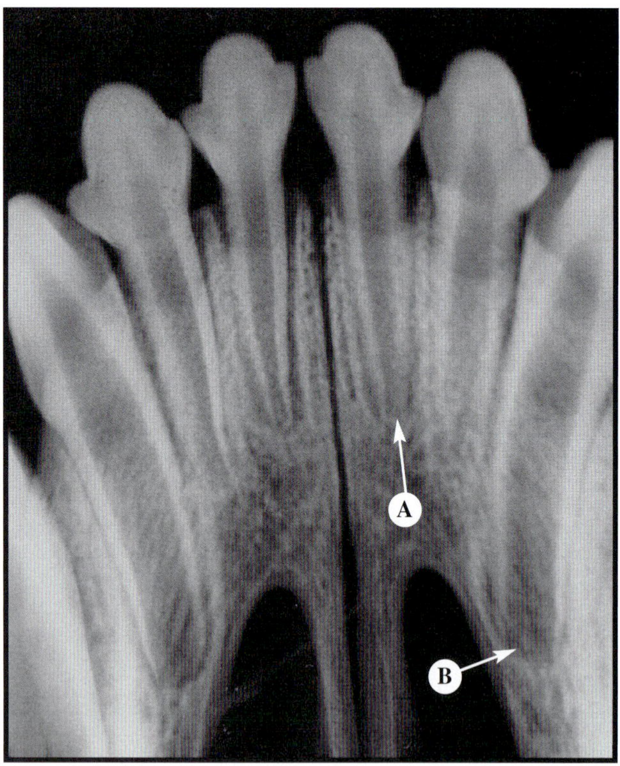

FIGURE 7-24 Structural Anatomy at 5 Months of Age

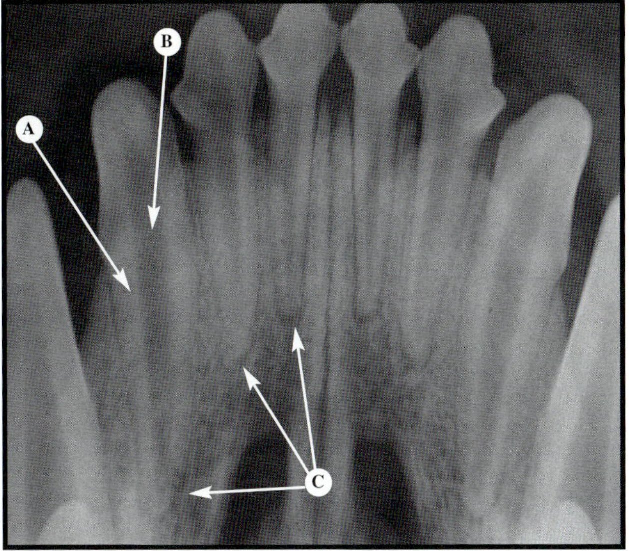

FIGURE 7-25 Structural Anatomy at 8 Months of Age

FIGURE 7-26 Structural Anatomy of a Geriatric Dog

Figure 7-21 *The maxillary incisors in a 6- to 8-week-old puppy. (A) Deciduous incisors, (B) deciduous canine tooth, (C) unerupted permanent incisors, and (D) unerupted permanent canine tooth.* **Figure 7-22** *This radiograph shows the maxillary incisors in a 12- to 14-week-old puppy. (A) Deciduous central incisor, (B) deciduous intermediate incisor, (C) deciduous corner incisor, (D) deciduous canine tooth, (E) unerupted permanent incisor 1, (F) unerupted permanent incisor 2, (G) unerupted permanent incisor 3, and (H) unerupted permanent canine tooth.* **Figure 7-23** *This radiograph shows the maxillary incisors and dental structures in a 14- to 16-week-old puppy. (A) Deciduous incisors, (B) thin dentinal walls, (C) wide lumen in the root canal, and (D) incomplete root length and open apex.* **Figure 7-24** *This radiograph shows the structures surrounding the incisors in a 5-month-old dog. Note the (A) apex closing and (B) open apex. Teeth that erupt first and those with shorter roots undergo apical closure first.* **Figure 7-25** *This radiograph shows the permanent incisors in an 8-month-old dog. (A) Narrow dentinal walls, (B) wide lumen, and (C) apical closure.* **Figure 7-26** *This radiograph shows the permanent incisors in a geriatric dog. Note the (A) constriction of the lumen in the root canal, (B) complete apical closure, and (C) wide dentinal walls.*

NORMAL ANATOMY OF CANINE TEETH—DOGS

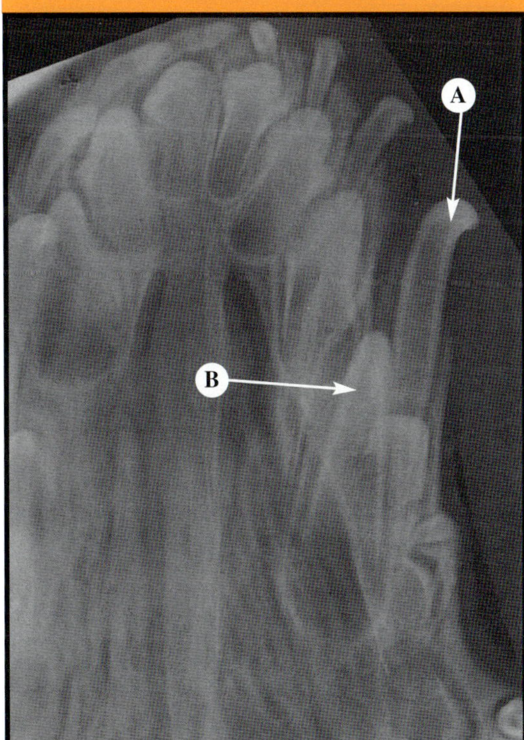

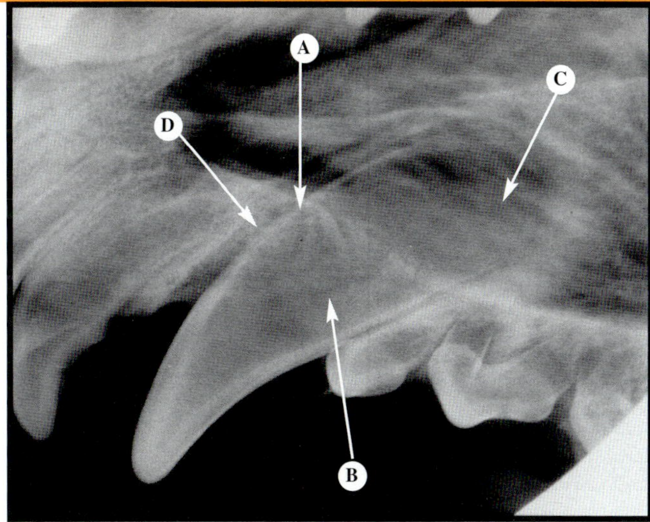

FIGURE 7-28 Structural Anatomy at 6 Months of Age

FIGURE 7-27 Structural Anatomy at 14 Weeks of Age

Figure 7-27 *The canine teeth in a 14-week-old dog. (A) Deciduous canine tooth and (B) unerupted permanent canine tooth.* **Figure 7-28** *This radiograph shows the permanent canine teeth and surrounding structures in a 6-month-old dog. Note the (A) narrow dentinal walls, (B) wide lumen, (C) incomplete lengthening of the root, and (D) wide periodontal space.*

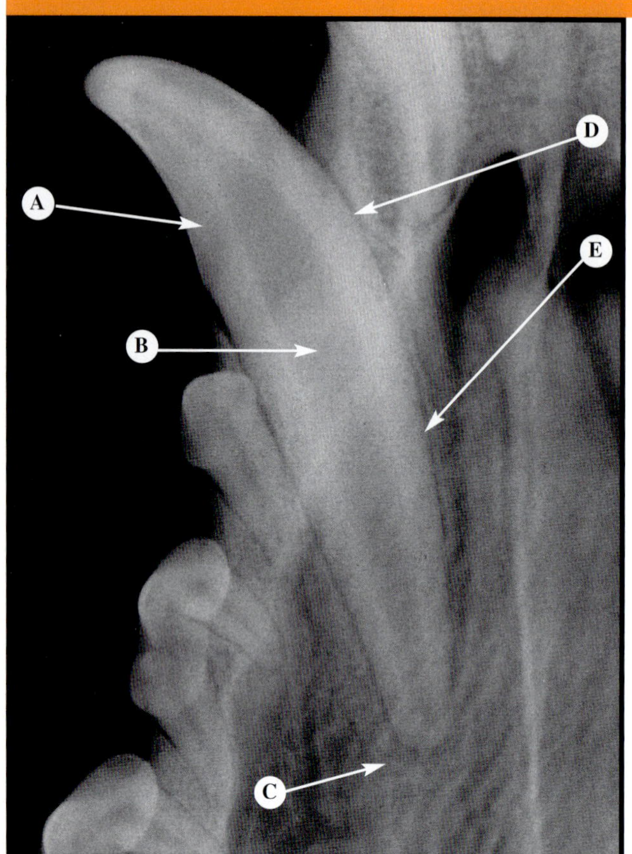

FIGURE 7-29 Structural Anatomy After 8 Months of Age

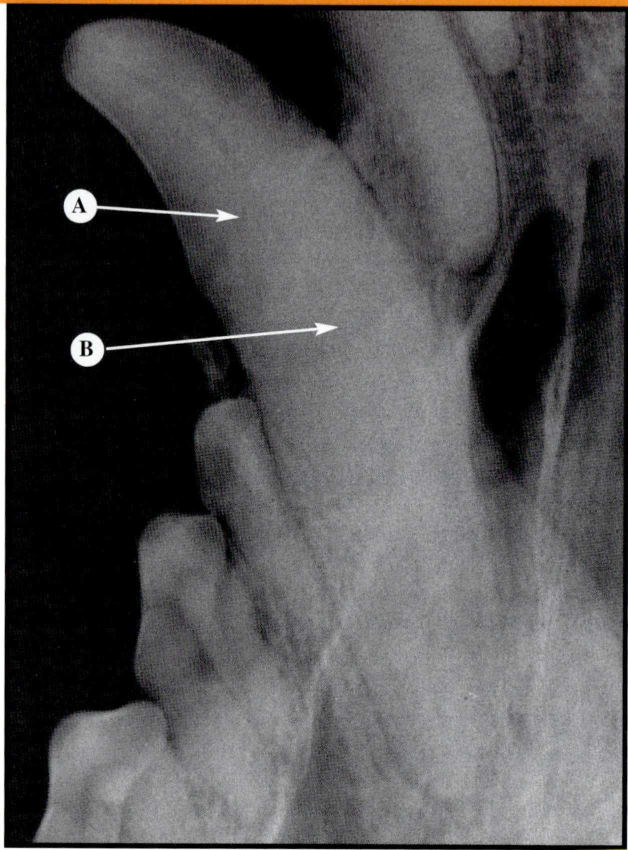

FIGURE 7-31 Structural Anatomy of a Geriatric Dog

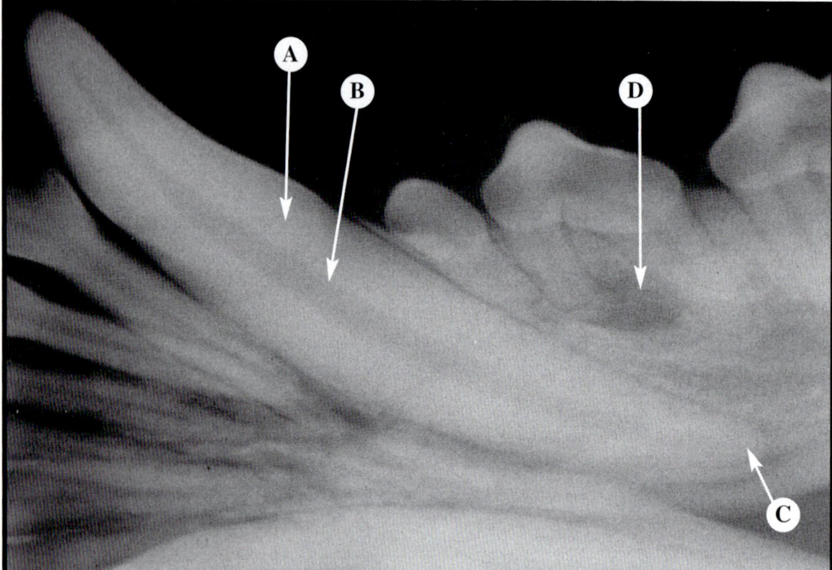

FIGURE 7-30 Structural Anatomy at 2 Years of Age

Figure 7-29 *This radiograph shows the permanent canine teeth in a dog older than 8 months of age. In comparison with Figure 7-28, note the (A) wider dentinal walls, (B) decreasing lumen, (C) completed length of the root and the apical closure, (D) size of the periodontal space, and (E) lamina dura.* **Figure 7-30** *This radiograph shows the permanent canine teeth and surrounding structures in a 2-year-old dog. Note the (A) thick dentinal walls, (B) constricting lumen, (C) closed apex, and (D) mental foramen.* **Figure 7-31** *This radiograph shows the canine tooth and surrounding structures in a geriatric dog. In comparison with Figures 7-28 and 7-29, note the (A) wide dentinal walls and (B) constricted pulp chamber and root canal.*

NORMAL ANATOMY OF THE MAXILLARY PREMOLARS—DOGS

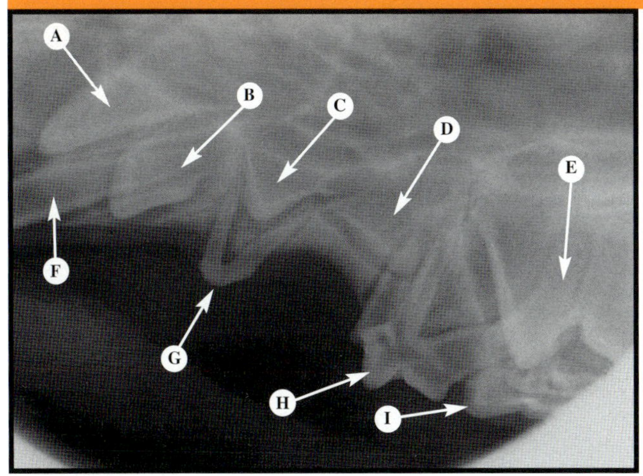

FIGURE 7-32 Structural Anatomy at 12 to 14 Weeks of Age

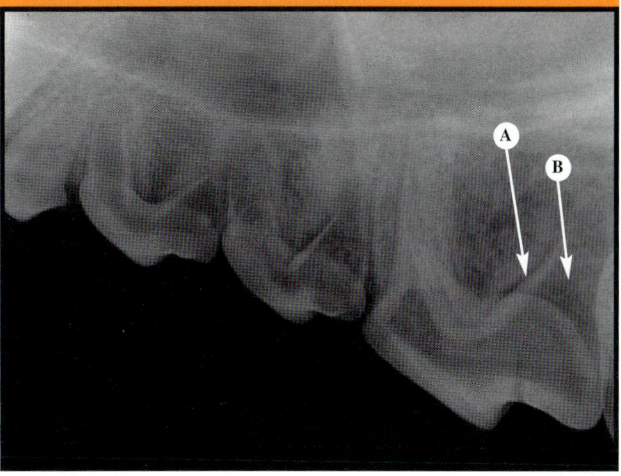

FIGURE 7-33 Structural Anatomy at 6 Months of Age

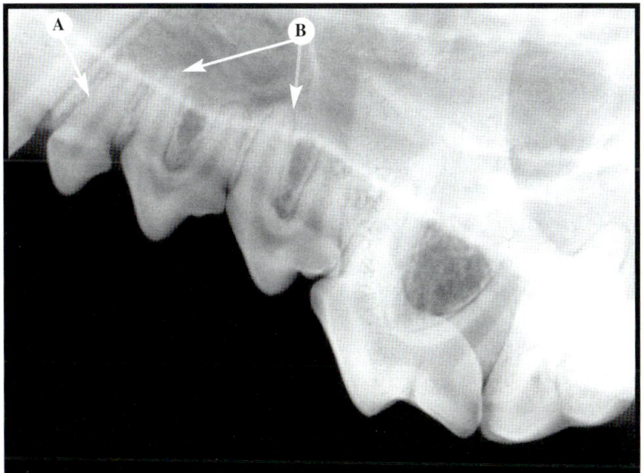

FIGURE 7-34 Structural Anatomy at 8 Months of Age

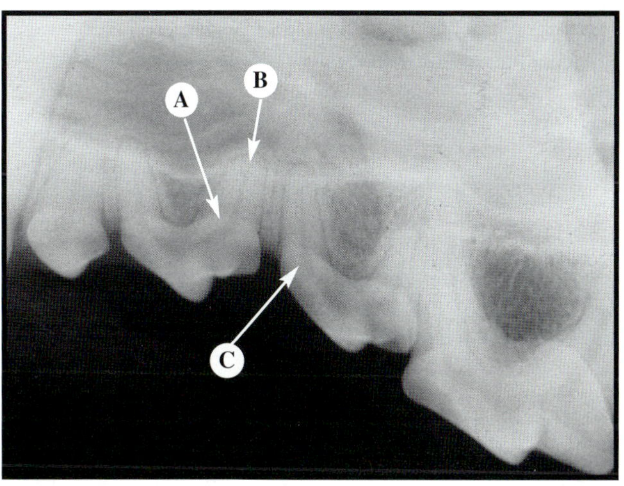

FIGURE 7-35 Structural Anatomy at 2 Years of Age

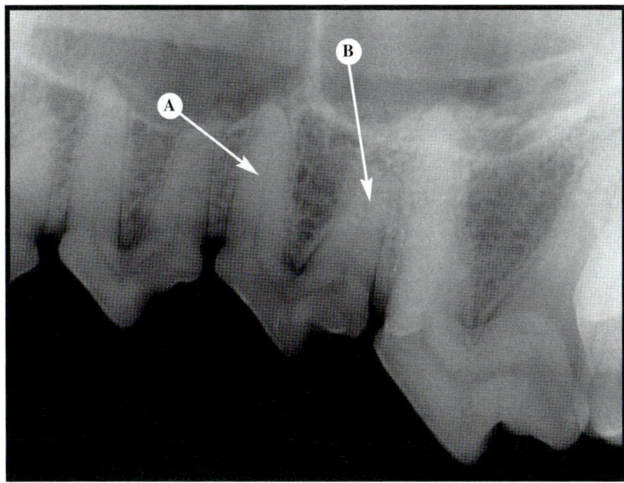

FIGURE 7-36 Structural Anatomy of a Geriatric Dog

Figure 7-32 *The maxillary premolars and surrounding structures in a 12- to 14-week-old puppy. Note that the deciduous precursor for the third permanent premolar has three roots, as does the permanent fourth premolar. (A) Unerupted permanent canine tooth, (B) unerupted premolar 1, (C) unerupted permanent premolar 2, (D) unerupted permanent premolar 3, (E) unerupted permanent premolar 4, (F) deciduous canine tooth, (G) deciduous precursor for premolar 2, (H) deciduous precursor for premolar 3, and (I) deciduous precursor for premolar 4.* **Figure 7-33** *This radiograph shows the permanent premolars and surrounding structures in a 6-month-old dog. Note the (A) narrow dentinal walls and (B) wide lumen and incomplete apical closure.* **Figure 7-34** *This radiograph shows the permanent premolars in an 8-month-old dog. In comparison with Figure 7-33, note the (A) thicker dentinal walls and (B) closing apices.* **Figure 7-35** *This radiograph shows the permanent premolars and surrounding structures in a 2-year-old dog. Note the (A) thick dentinal wall, (B) apical closure, and (C) constricting lumen.* **Figure 7-36** *This radiograph shows the maxillary premolars in a geriatric dog. In comparison with Figures 7-34 and 7-35, note the (A) wide dentinal walls and (B) constricted root canal.*

NORMAL ANATOMY OF THE MANDIBULAR PREMOLARS—DOGS

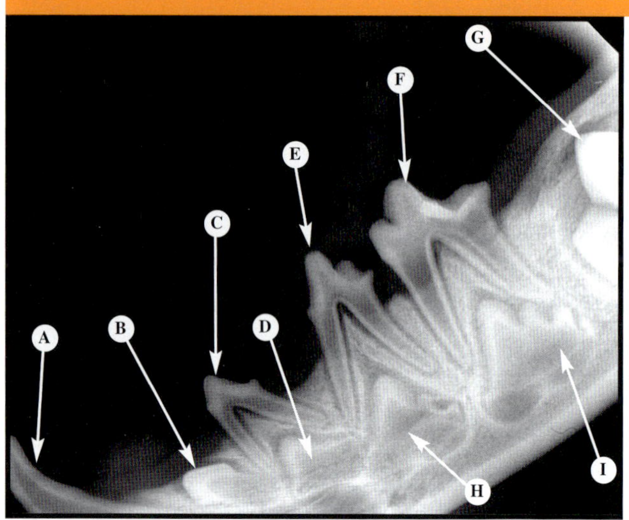

FIGURE 7-37 Structural Anatomy at 12 to 14 Weeks of Age

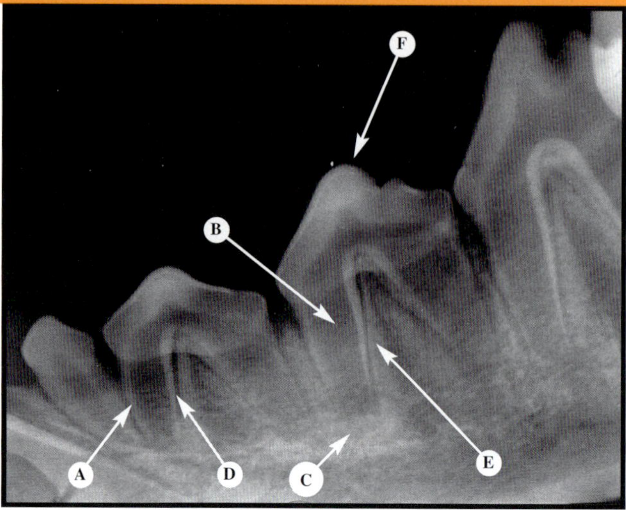

FIGURE 7-38 Structural Anatomy at 5 to 6 Months of Age

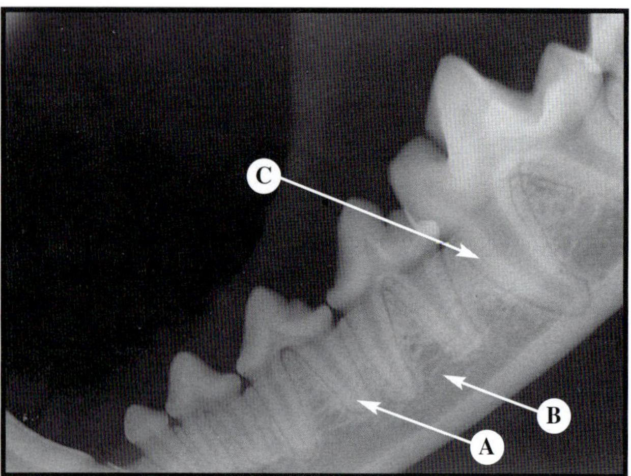

FIGURE 7-39 Structural Anatomy at 2 Years of Age

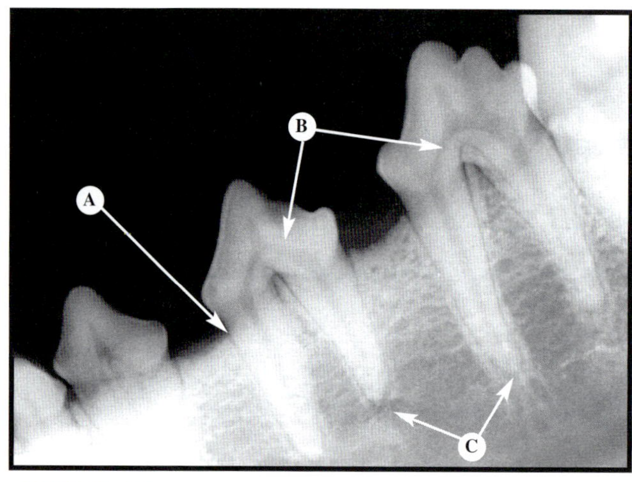

FIGURE 7-40 Structural Anatomy at 4 Years of Age

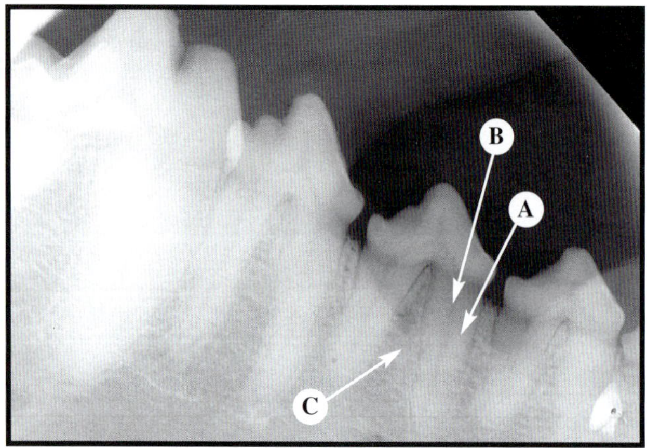

FIGURE 7-41 Structural Anatomy of a Geriatric Dog

Figure 7-37 *Mandibular premolars in a 12- to 14-week-old puppy. (A) Deciduous canine tooth, (B) first premolar, (C) deciduous premolar, (D) unerupted permanent premolar 2, (E) deciduous premolar, (F) deciduous premolar, (G) unerupted first permanent molar, (H) unerupted permanent premolar 3, and (I) unerupted permanent premolar 4.* **Figure 7-38** *Permanent premolars and surrounding structures in a 5- to 6-month-old dog. Note the (A) narrow dentinal walls, (B) wide lumen, (C) incomplete apex, (D) wide periodontal space, (E) prominent lamina dura, and (F) lack of coronal attrition.* **Figure 7-39** *Permanent premolars in a 2-year-old dog. Note the (A) complete root formation, (B) mandibular canal, and (C) thick dentinal walls.* **Figure 7-40** *Permanent premolars in a 4-year-old dog. Note the (A) thick dentinal walls, (B) narrow lumen in the pulp chamber, and (C) complete apical closure.* **Figure 7-41** *Premolars in a geriatric dog. In comparison with Figures 7-37 to 7-40, note the (A) wide dentinal walls, (B) constricted lumen, and (C) indistinct lamina dura.*

NORMAL ANATOMY OF THE MAXILLARY MOLARS—DOGS

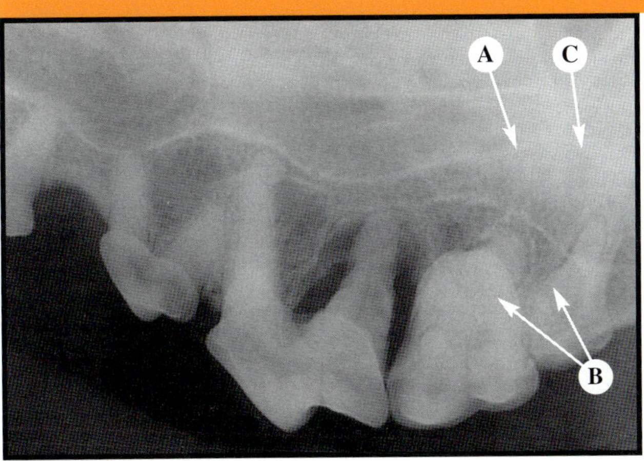

FIGURE 7-42 Mesial Projection in a Mature Adult

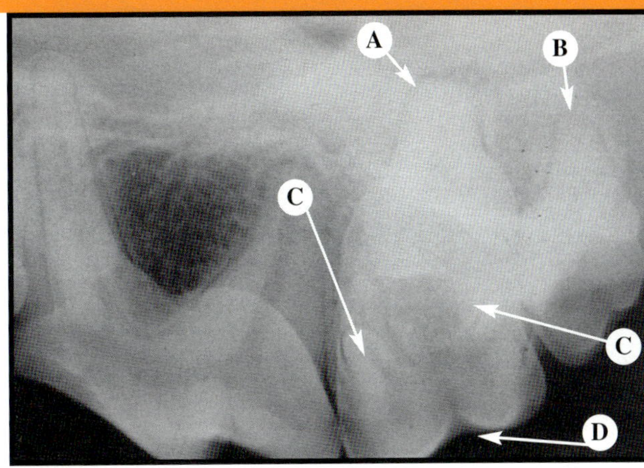

FIGURE 7-43 Distal Projection in a Mature Adult

Figure 7-42 *The mesial projection of the molars in a mature adult. (A) Palatal root of the first molar, (B) buccal roots of the first molar, and (C) palatal root of the second molar.* **Figure 7-43** *The distal projection of the molars in a mature adult. (A) Palatal root of the first molar, (B) palatal root of the second molar, (C) buccal roots, and (D) developmental groove.*

NORMAL ANATOMY OF THE MANDIBULAR INCISORS AND CANINE TEETH—CATS

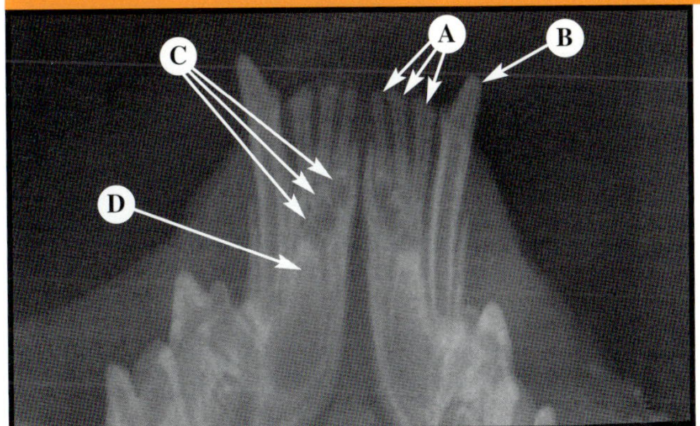

FIGURE 7-44 Structural Anatomy at 6 to 8 Weeks of Age

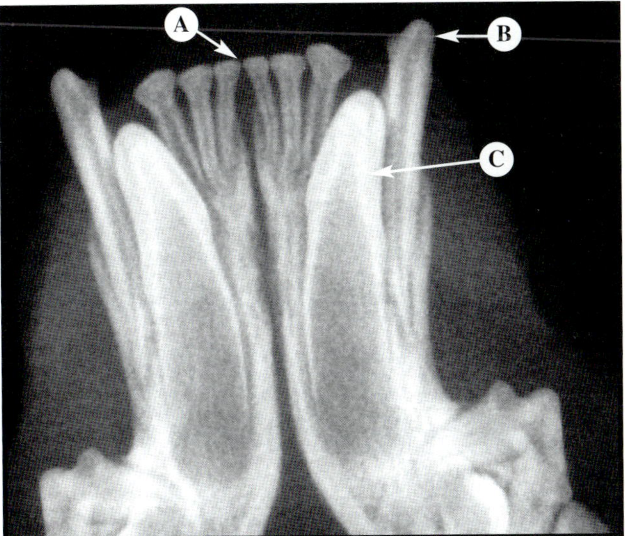

FIGURE 7-46 Structural Anatomy at 4 Months of Age

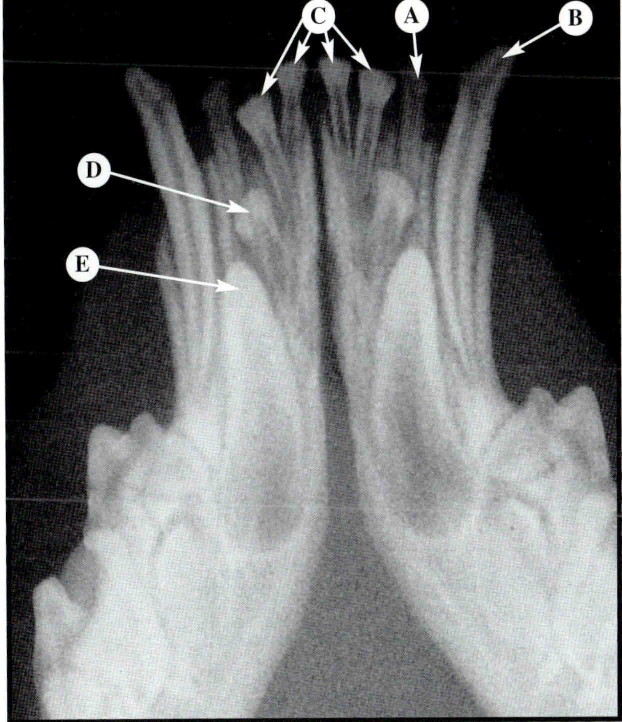

FIGURE 7-45 Structural Anatomy at 12 to 14 Weeks of Age

Figure 7-44 *Mandibular incisors and canine teeth in a 6- to 8-week-old kitten. (A) Deciduous incisors, (B) deciduous canine tooth, (C) unerupted permanent incisors, and (D) unerupted permanent canine tooth.* **Figure 7-45** *Mandibular incisors and canine teeth in a 12- to 14-week-old kitten. (A) Deciduous incisor, (B) deciduous canine tooth, (C) erupting permanent incisors, (D) unerupted permanent incisor, and (E) unerupted permanent canine tooth.* **Figure 7-46** *Mandibular incisors and canine teeth in a 4-month-old kitten. (A) permanent incisors, (B) deciduous canine tooth, and (C) permanent canine tooth.*

NORMAL ANATOMY OF THE MANDIBULAR INCISORS AND CANINE TEETH—CATS *(continued)*

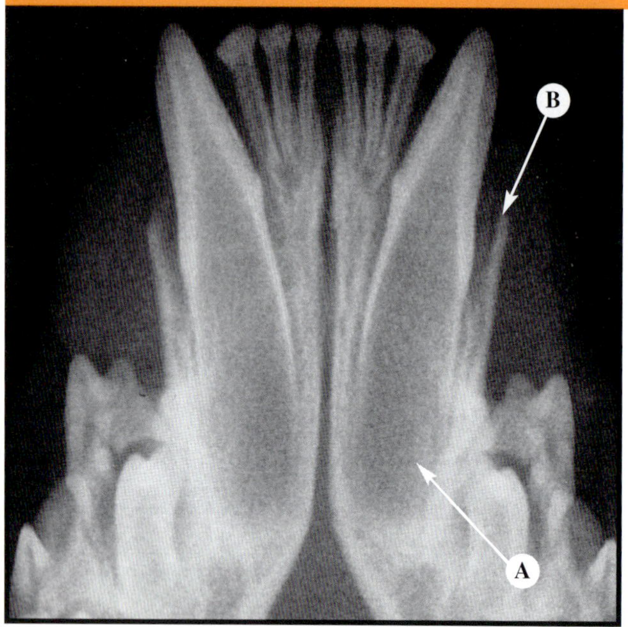

FIGURE 7-47 Structural Anatomy at 5 Months of Age

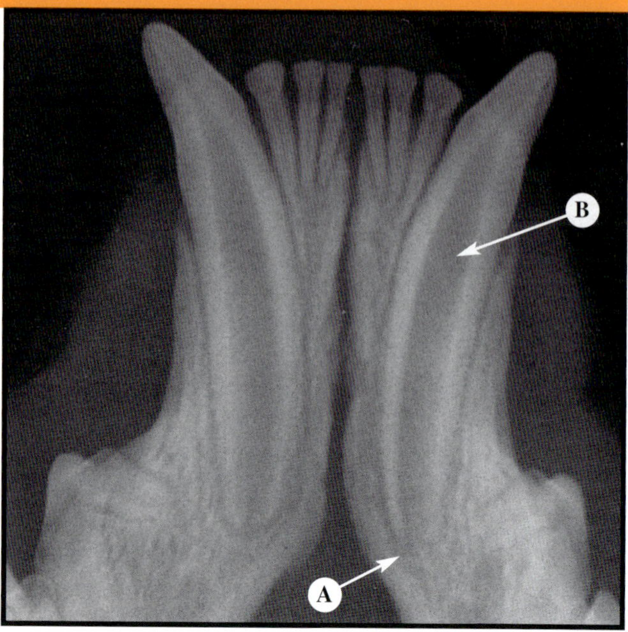

FIGURE 7-48 Structural Anatomy at 8 Months of Age

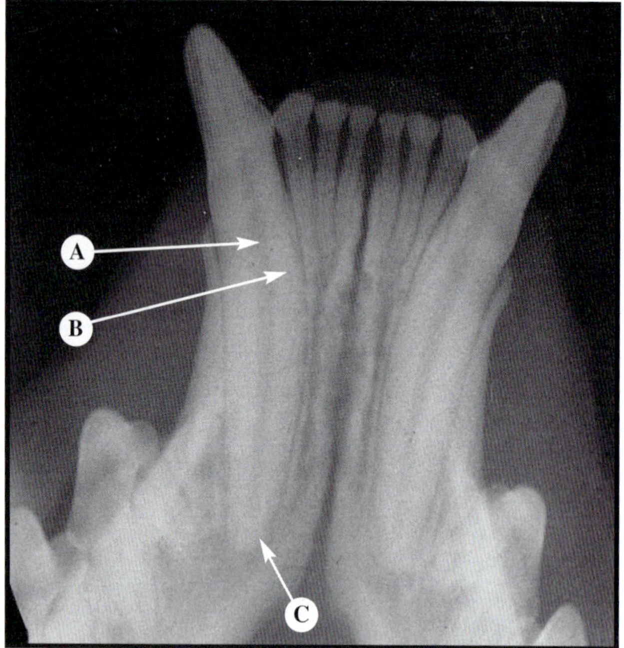

FIGURE 7-49 Structural Anatomy at 2 Years of Age

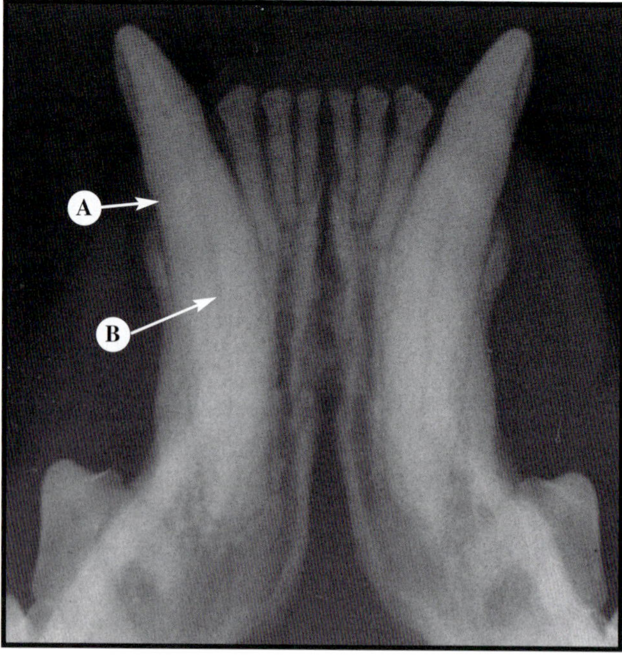

FIGURE 7-50 Structural Anatomy of a Geriatric Cat

Figure 7-47 *Mandibular incisors, canine teeth, and surrounding structures in a 5-month-old kitten. As in Figure 7-46, active eruption of the canine teeth continues to occur. Note the (A) open apex and incomplete length of the roots and (B) wide periodontal space.* **Figure 7-48** *Permanent incisors, canine teeth, and surrounding structures in an 8-month-old cat. Note that the (A) root has lengthened and the apex is open. In addition, the lumen (B) is wide.* **Figure 7-49** *Permanent incisors, canine teeth, and surrounding structures in a 2-year-old cat. Note the (A) constricting lumen, (B) thick dentinal walls, and (C) apical closure.* **Figure 7-50** *Mandibular incisors and canine teeth in a geriatric cat. Note the (A) thicker dentinal walls and (B) narrow lumen.*

NORMAL ANATOMY OF THE MAXILLARY INCISORS—CATS

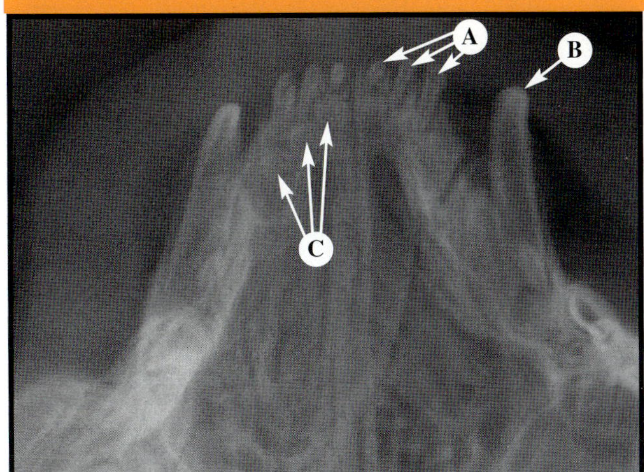

FIGURE 7-51 Structural Anatomy at 6 to 8 Weeks of Age

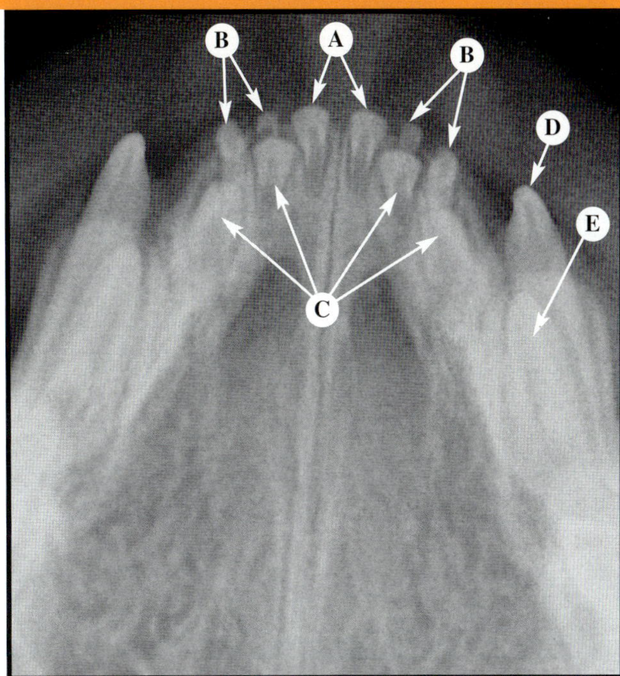

FIGURE 7-52 Structural Anatomy at 10 to 12 Weeks of Age

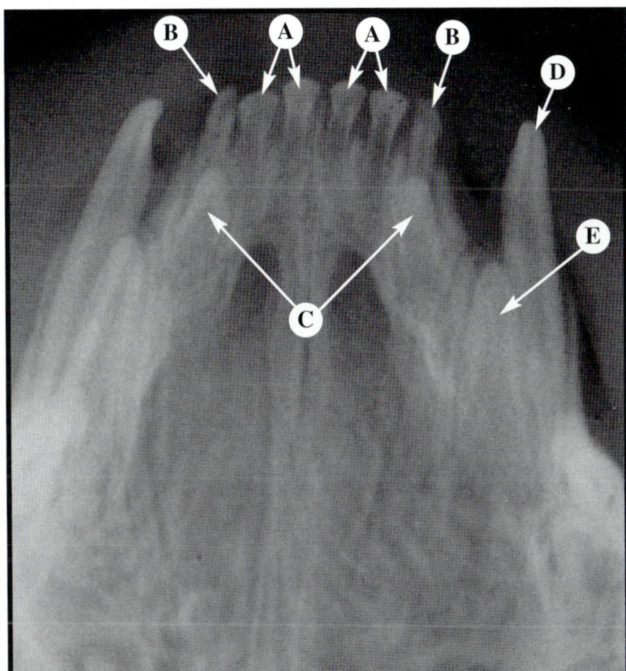

FIGURE 7-53 Structural Anatomy at 12 to 14 Weeks of Age

Figure 7-51 *Maxillary incisors in a 6- to 8-week-old kitten. (A) Deciduous incisors, (B) deciduous canine tooth, and (C) unerupted permanent incisors.* **Figure 7-52** *Maxillary incisors in a 10- to 12-week-old kitten. (A) Erupting permanent incisors, (B) deciduous incisors, (C) unerupted incisors, (D) deciduous canine tooth, and (E) unerupted permanent canine tooth.* **Figure 7-53** *Maxillary incisors in a 12- to 14-week-old kitten. Compare this radiograph with the structures in Figure 7-52 to gain a better understanding of the growth that can occur within 2 weeks. (A) Erupted permanent incisors, (B) deciduous incisor, (C) unerupted incisors, (D) deciduous canine tooth, and (E) unerupted permanent canine tooth.* **Figure 7-54** *Maxillary incisors in a 4- to 5-month-old kitten. (A) Permanent incisors, (B) deciduous canine tooth, (C) permanent canine tooth, and (D) open apex.*

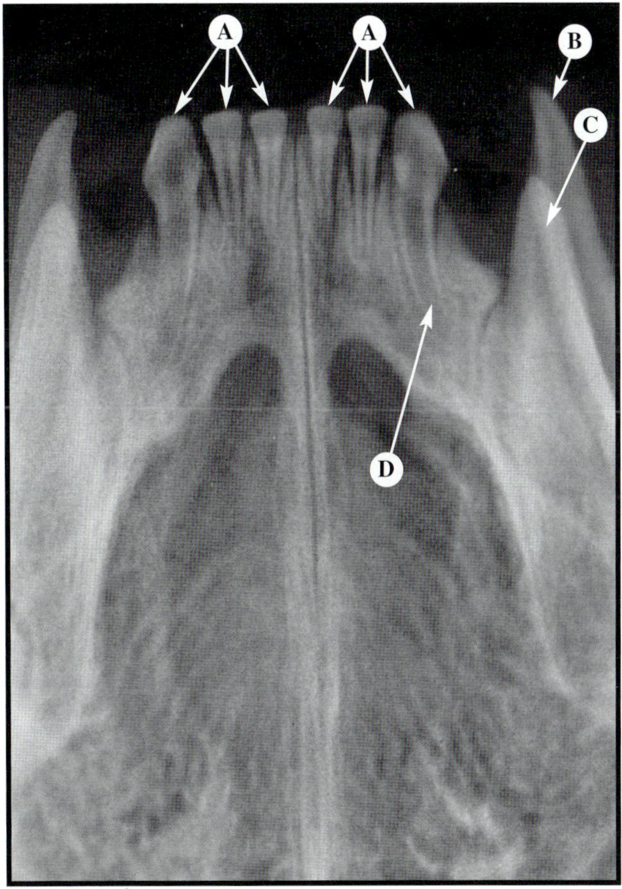

FIGURE 7-54 Structural Anatomy at 4 to 5 Months of Age

NORMAL ANATOMY OF THE MAXILLARY INCISORS—CATS *(continued)*

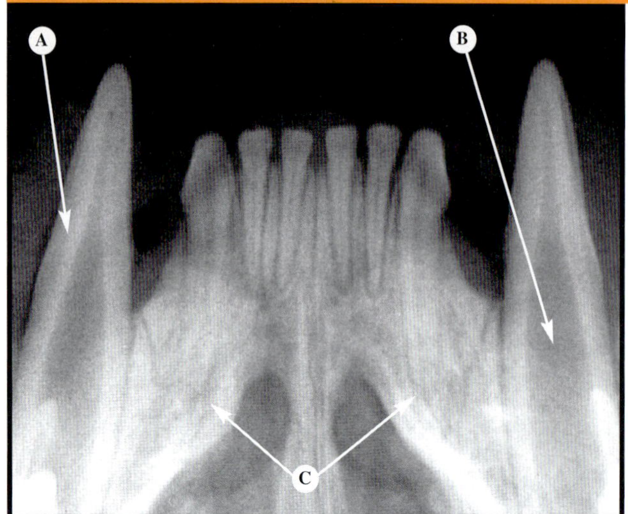

FIGURE 7-55 Structural Anatomy at 6 to 8 Months of Age

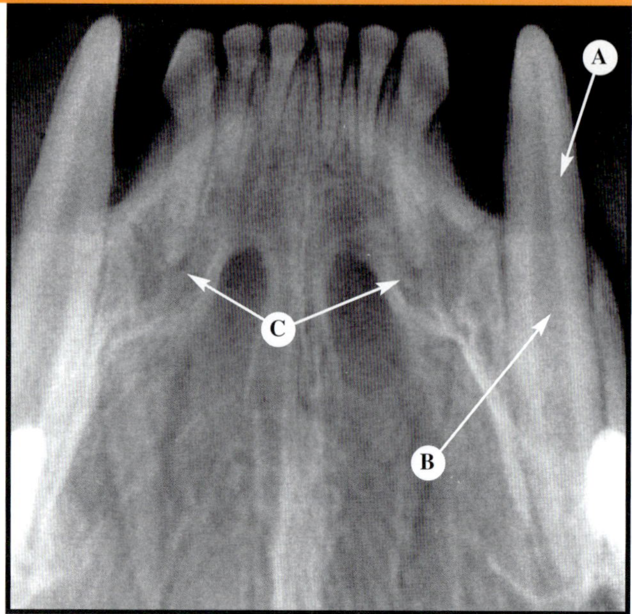

FIGURE 7-56 Structural Anatomy at 12 to 18 Months of Age

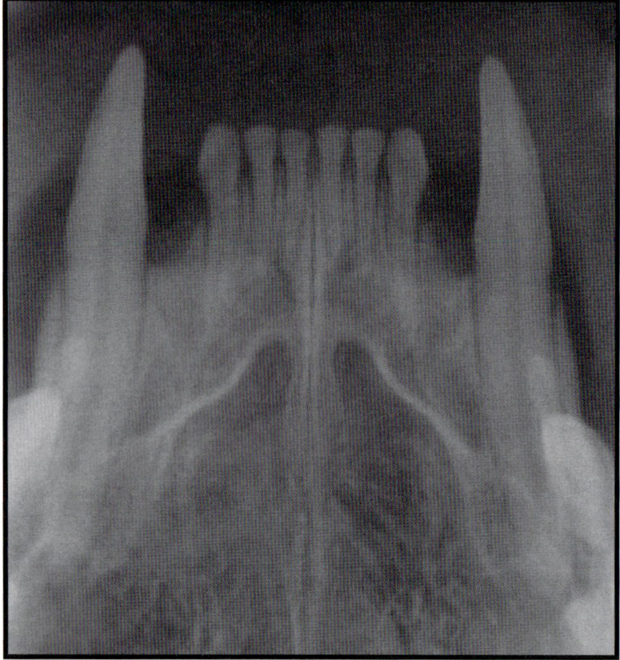

FIGURE 7-57 Structural Anatomy of a Young Adult Cat

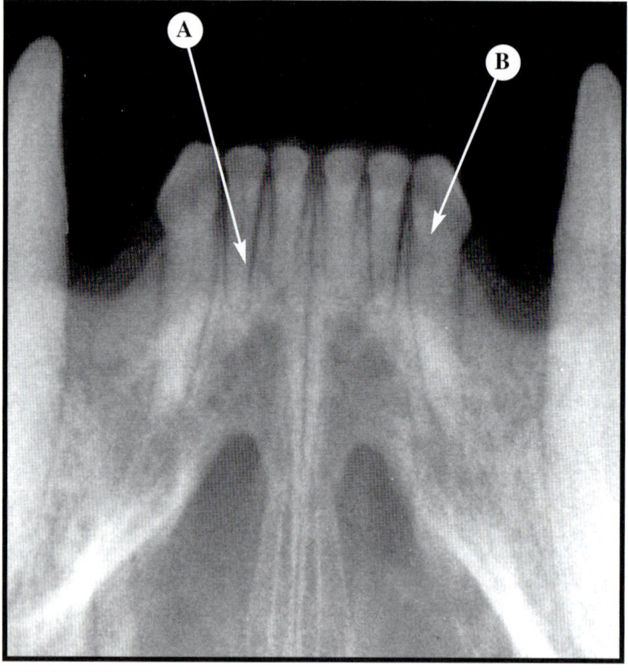

FIGURE 7-58 Structural Anatomy of a Geriatric Cat

Figure 7-55 *Permanent incisors in a 6- to 8-month-old cat. Note the* (A) *narrow dentinal walls,* (B) *wide canal, and* (C) *incomplete apical closure.* **Figure 7-56** *Permanent incisors in a 12- to 18-month-old cat. Note the* (A) *wider dentinal walls,* (B) *narrowing lumen, and* (C) *complete apical closure.* **Figure 7-57** *The permanent incisors in this cat look very similar to those in Figure 7-56; however, when comparing the appearance of the canine teeth, this cat is slightly older than the cat in Figure 7-56.* **Figure 7-58** *Maxillary incisors in a geriatric cat. Note the* (A) *wide dentinal walls and* (B) *constricted root canal.*

NORMAL ANATOMY OF THE MANDIBULAR CANINE TEETH—CATS

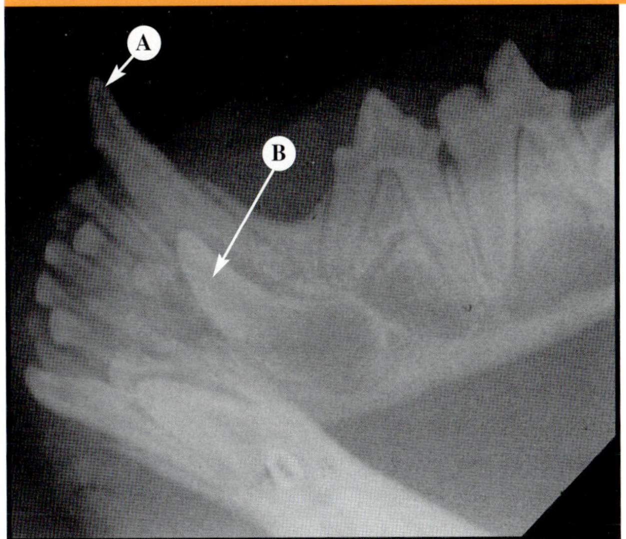

FIGURE 7-59 Structural Anatomy at 6 to 8 Weeks of Age

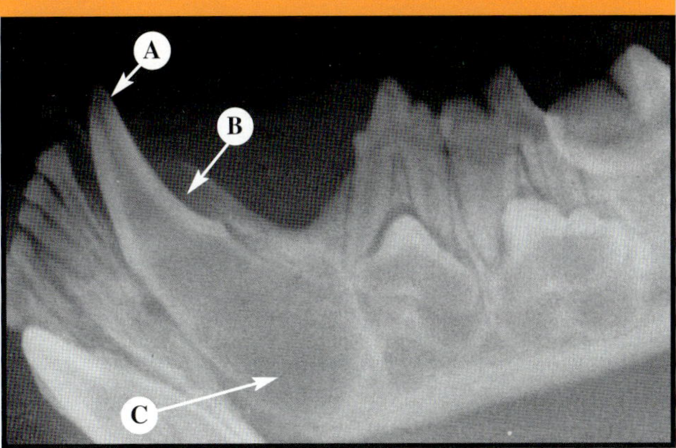

FIGURE 7-60 Structural Anatomy at 5 Months of Age

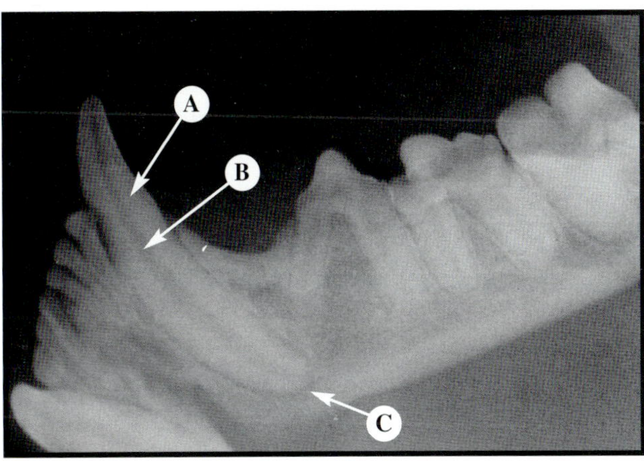

FIGURE 7-61 Structural Anatomy at 12 to 18 Months of Age

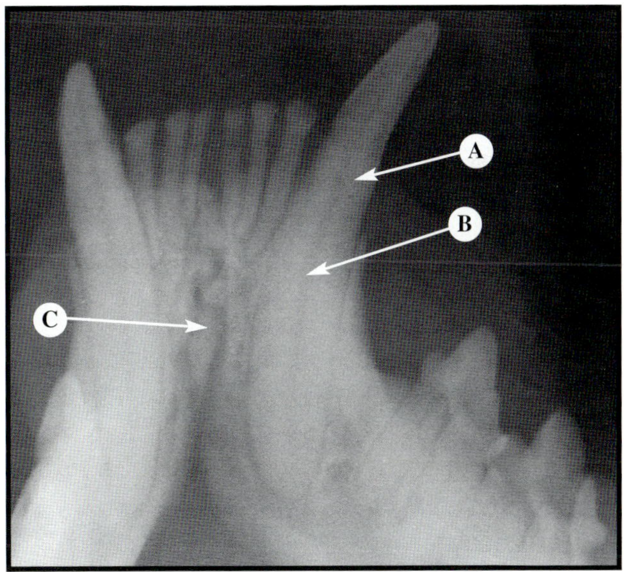

FIGURE 7-62 Structural Anatomy of a Geriatric Cat

Figure 7-59 *Mandibular canine teeth in a 6- to 8-week-old kitten.* (A) *Deciduous canine tooth and* (B) *unerupted permanent canine tooth.* **Figure 7-60** *Permanent canine teeth in a 5-month-old kitten.* (A) *Erupting permanent canine tooth,* (B) *wide periodontal space, and* (C) *incomplete length of the root, open apex, and wide lumen in the root canal.* **Figure 7-61** *Mandibular canine teeth in a young adult cat (12 to 18 months of age). Note the* (A) *thicker dentinal walls,* (B) *narrowing lumen, and* (C) *apical closure.* **Figure 7-62** *Permanent canine teeth in a geriatric cat. Note the* (A) *thick dentinal wall,* (B) *constricted root canal, and* (C) *mandibular symphysis.*

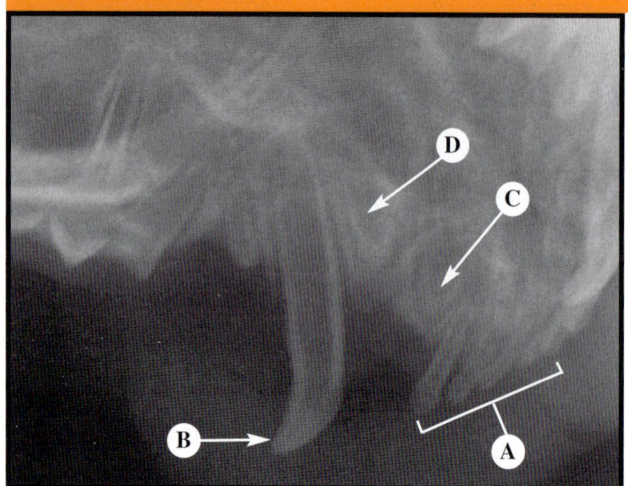

FIGURE 7-63 Structural Anatomy at 6 to 8 Weeks of Age

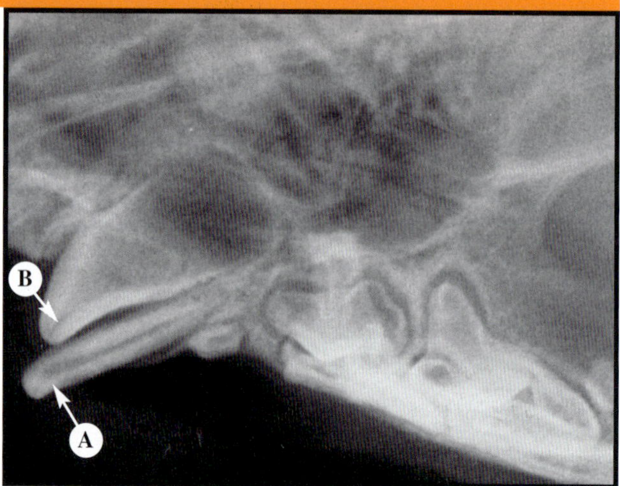

FIGURE 7-64 Structural Anatomy at 4 to 5 Months of Age

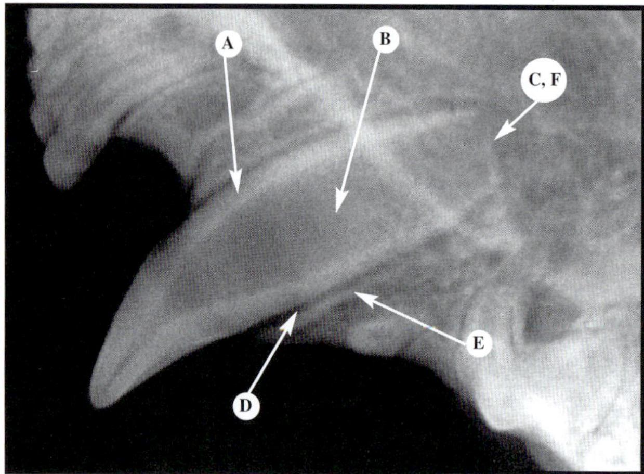

FIGURE 7-65 Structural Anatomy at 8 Months of Age

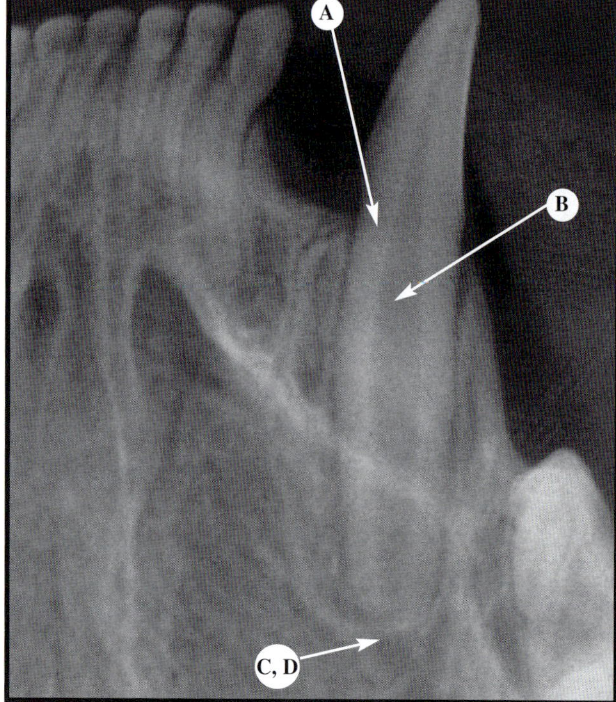

FIGURE 7-66 Structural Anatomy of a Young Adult Cat

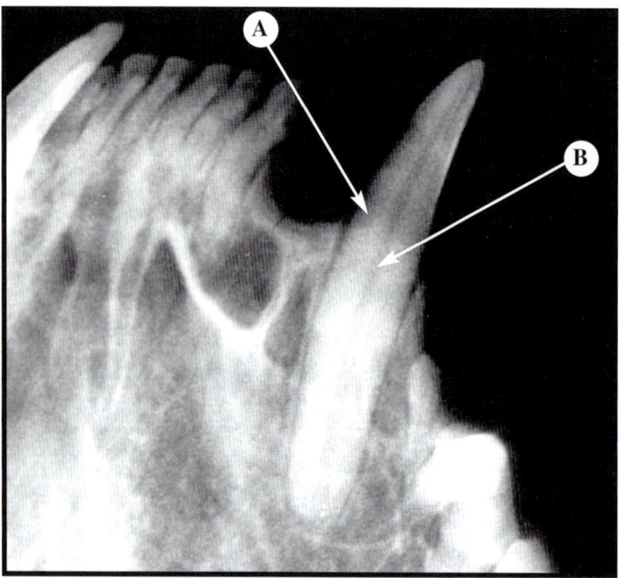

FIGURE 7-67 Structural Anatomy of a Mature Cat

Figure 7-63 *Maxillary canine teeth in a 6- to 8-week-old kitten. (A) Deciduous incisors, (B) deciduous canine tooth, (C) unerupted permanent incisor 3, and (D) unerupted canine tooth.* **Figure 7-64** *Permanent canine teeth in a 4- to 5-month-old kitten. (A) Deciduous canine tooth and (B) erupting permanent canine tooth.* **Figure 7-65** *Permanent canine teeth in an 8-month-old cat. Note the (A) narrow dentinal walls, (B) wide lumen, (C) incomplete apical closure, (D) wide periodontal space, (E) prominent lamina dura, and (F) incomplete length of the root.* **Figure 7-66** *Maxillary canine teeth in a young adult cat. Note the (A) wider dentinal walls, (B) narrowing lumen, (C) apex beginning to close, and (D) normal root length.* **Figure 7-67** *Maxillary canine tooth in a mature adult cat (8 to 10 years of age). Note the (A) thick dentinal walls and (B) narrow lumen.*

NORMAL ANATOMY OF THE MANDIBULAR PREMOLARS AND MOLAR—CATS

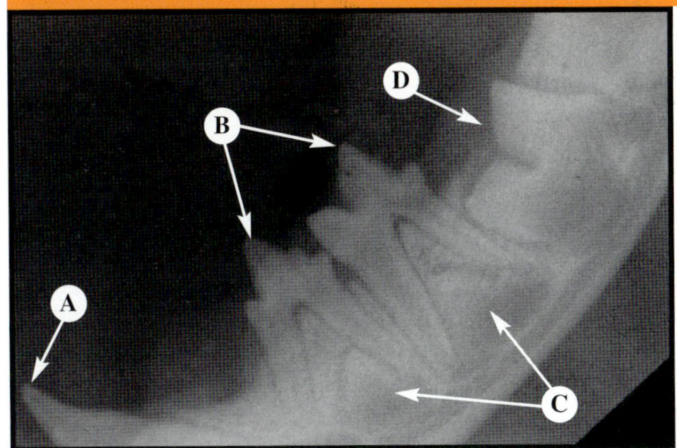

FIGURE 7-68 Structural Anatomy at 6 to 8 Weeks of Age

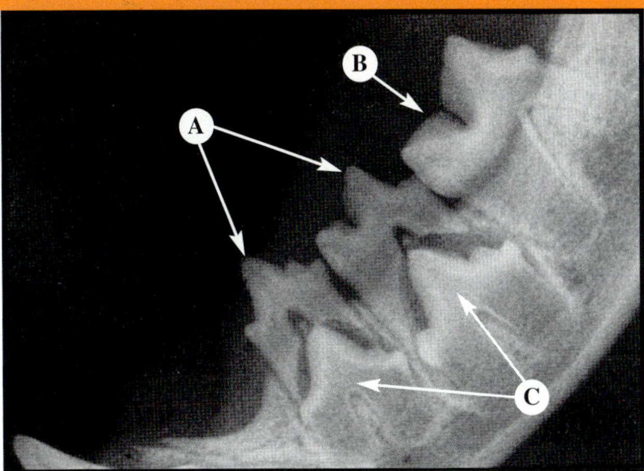

FIGURE 7-69 Structural Anatomy at 5 Months of Age

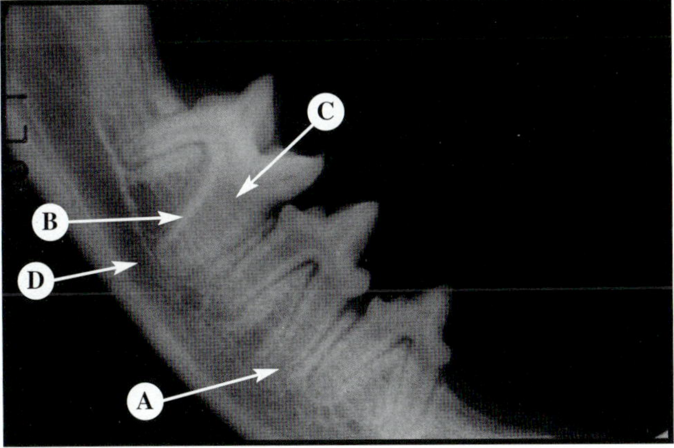

FIGURE 7-70 Structural Anatomy at 8 Months of Age

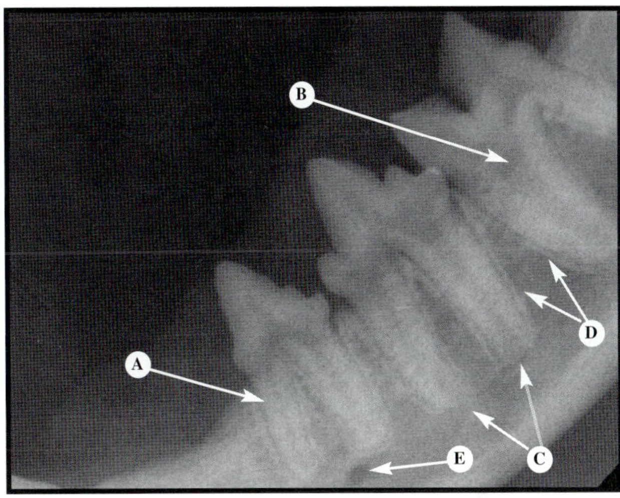

FIGURE 7-71 Structural Anatomy at 8 to 10 Months of Age

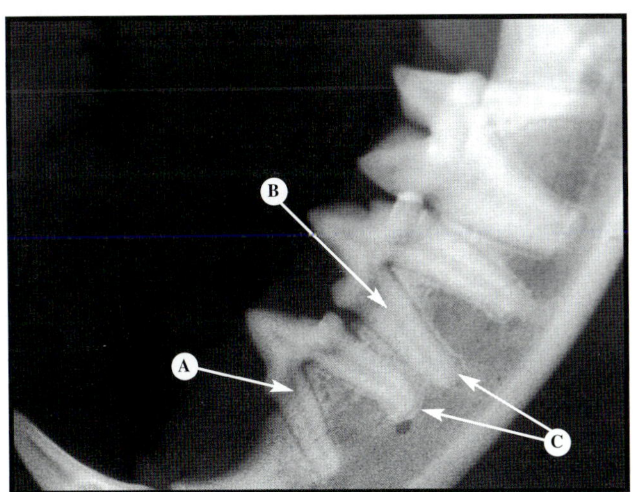

FIGURE 7-72 Structural Anatomy of a Young Adult Cat

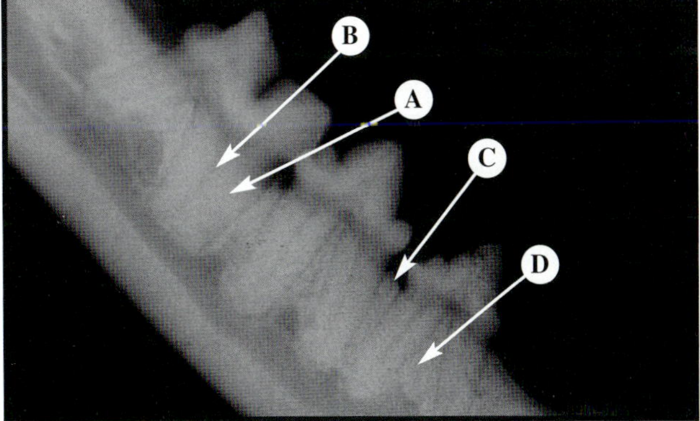

FIGURE 7-73 Structural Anatomy of a Geriatric Cat

Figure 7-68 *Mandibular premolars in a 6- to 8-week-old kitten. Note that the precursor for the fourth premolar has the same morphology as that of a molar. (A) Deciduous canine tooth, (B) deciduous premolars, (C) unerupted permanent premolars, and (D) unerupted molar.* **Figure 7-69** *Mandibular premolars in a 5-month-old kitten. (A) Deciduous premolars undergoing resorption, (B) molar, and (C) unerupted permanent premolars.* **Figure 7-70** *Permanent premolars in an 8-month-old cat. Note the (A) incomplete apex, (B) thin dentinal walls, (C) wide lumen, and (D) mandibular canal.* **Figure 7-71** *Permanent premolars in an 8- to 10-month-old cat. Note the (A) wider dentinal walls, (B) wide lumen, (C) partial apical closure, (D) prominent lamina dura, and (E) mental foramen.* **Figure 7-72** *Permanent premolars in a mature cat. Note the (A) wide dentinal walls, (B) narrow lumen, and (C) complete apical closure.* **Figure 7-73** *Mandibular premolars in a geriatric cat. Note the (A) thick dentinal walls, (B) narrow lumen, (C) mild crestal regression, and (D) indistinct lamina dura.*

NORMAL ANATOMY OF THE MAXILLARY PREMOLARS AND MOLAR—CATS

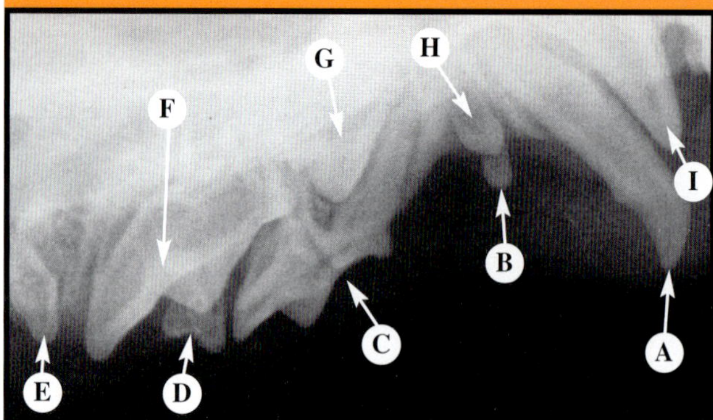

FIGURE 7-74 Structural Anatomy at 3 to 4 Months of Age

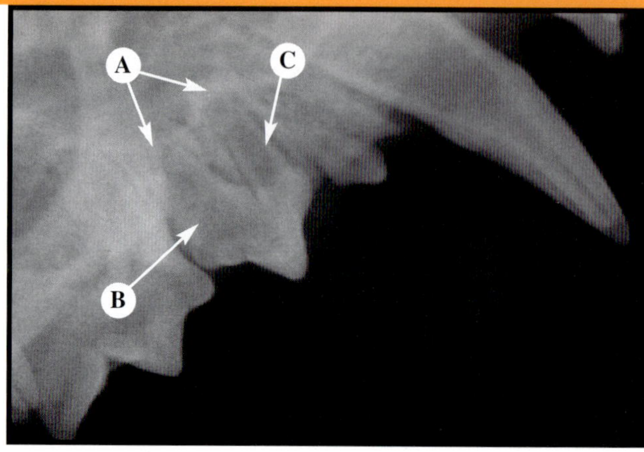

FIGURE 7-75 Structural Anatomy at 6 Months of Age

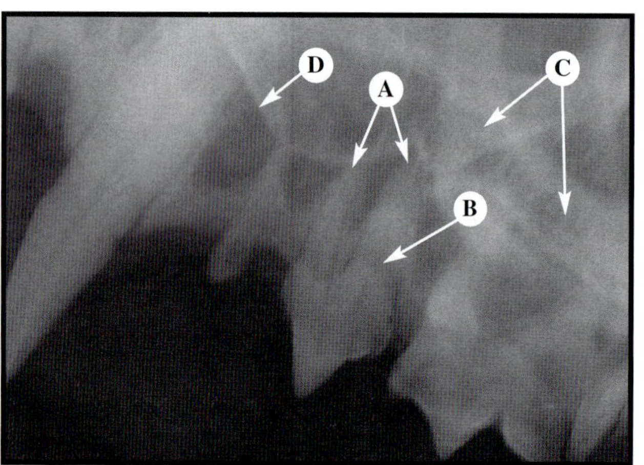

FIGURE 7-76 Structural Anatomy of an Adult Cat

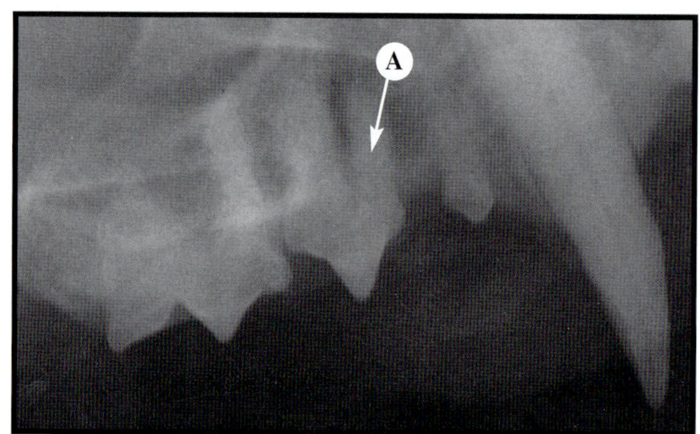

FIGURE 7-77 Structural Anatomy of a Geriatric Cat

Figure 7-74 *Maxillary premolars in a 3- to 4-month-old kitten. Note that the precursor for the fourth premolar has the same morphology as that of a molar, whereas the precursor for premolar 3 has morphology that resembles that of a permanent premolar 4. (A) Deciduous canine tooth, (B) deciduous precursor for premolar 2, (C) deciduous precursor for premolar 3, (D) deciduous precursor for premolar 4, (E) molar, (F) unerupted permanent premolar 4, (G) unerupted permanent premolar 3, (H) unerupted permanent premolar 2, and (I) unerupted permanent canine tooth.* **Figure 7-75** *Maxillary premolars in a 6-month-old kitten. Note the (A) open apices, (B) wide lumen, and (C) thin dentinal wall.* **Figure 7-76** *Maxillary premolars in an adult cat (3 years of age). Note the (A) closed apices, (B) thick dentinal walls, (C) zygomatic arch, and (D) confluence of the bones of the palate and maxilla.* **Figure 7-77** *Maxillary premolars in a geriatric cat. Note the (A) thick dentinal walls and constricted lumen.*

REFERENCES *(continued)*

   Small Animal Practice, vol 22, no. 6. Philadelphia, WB Saunders Co, 1992, pp 1265–1277.

4. Eisner E: Problems associated with veterinary dental radiography, in Manfra Marretta S (ed): *Problems in Veterinary Medicine,* vol 2, no. 1. Philadelphia, JB Lippincott Co, 1990, pp 46–84.

5. Evans HE, Christensen GS: *Miller's Anatomy of the Dog,* ed 2. Philadelphia, WB Saunders Co, 1979, pp 146–154..

6. Boyd JS, Paterson C, May AH: *Color Atlas of Clinical Anatomy of the Dog and Cat.* Philadelphia, Mosby Year Book 1991, pp 15–30.

7. Harvey CE, Flax BM: Feline oral-dental radiographic examination and interpretation, in Harvey CE (ed): *The Veterinary Clinics of North America. Small Animal Practice,* vol 22, no. 6. Philadelphia, WB Saunders Co, 1992, pp 1279–1296.

8. Schebitz H, Wilkens H: *Atlas of Radiographic Anatomy of the Dog and Cat,* ed 4. Philadelphia, WB Saunders Co, 1986, pp 12–34, 154–164.

9. Harvey CE, Dubielzig RR: Anatomy of the oral cavity in the dog and the cat, in Harvey CE (ed): *Veterinary Dentistry.* Philadelphia, WB Saunders Co, 1985, pp 11–22.

10. Evans HE, Christensen GS: *Miller's Anatomy of the Dog,* ed 2. Philadelphia, WB Saunders Co, 1979, pp 415–422.

11. Eisenmenger E, Zetner K: *Veterinary Dentistry.* Philadelphia, Lea & Febiger, 1985, pp 14–17.

12. Emily P: Intraoral radiology, in Frost P (ed): *The Veterinary Clinics of North America. Small Animal Practice,* vol 16, no. 5. Philadelphia, WB Saunders Co, 1986, pp 808–816.

13. Hooft J, Mattheeuws D, Van Bree P: Radiology of deciduous teeth resorption and definitive teeth eruption in the dog. *J Small Anim Pract* 20:175–180, 1979.

14. Cahill DR, Marks SC Jr: Chronology and histology of exfoliation and eruption of mandibular premolars in dogs. *J Morphol* 171:213–218, 1982.

15. Main JHP: A histological survey of the hammock ligament. *Arch Oral Biol* 10:343–351, 1965.

16. Morgan JP, Miyabayashi T: Dental radiology: Aging changes in permanent teeth of beagle dogs. *J Small Anim Pract* 32:11–18, 1991.

Chapter Eight

DEVELOPMENTAL PROBLEMS AND DENTAL ANOMALIES

Developmental problems and anomalous conditions of the oral cavity deserve radiographic evaluation because simple visual examination is inadequate. Many diseases and disorders have similar presentations that require different treatment measures and therefore must be differentiated by radiographic evaluation. Anatomic conditions, such as extra roots or dilacerations, can complicate extraction or endodontic procedures without the benefits of radiographic assessment.

DEVELOPMENTAL PROBLEMS

MISSING TEETH

Many breed standards for dogs require that a specified number of teeth be present to qualify for the conformation ring. Therefore, early detection of permanent teeth can be beneficial to breeders and show exhibitors who need to know which puppies have full dentition.[1,2] Permanent teeth can be evaluated on radiographs (Figures 8-1 through 8-8) when puppies (and kittens) are 12 weeks of age, even though the permanent teeth have not erupted. Although the permanent teeth are not fully formed at this age, the crowns have calcified and are visible on radiographs. Radiographs of the following regions are usually adequate to assess the dentition properly: upper and lower incisors and both sides of the upper and lower cheek teeth.

The presence or absence of permanent teeth can also be determined by evaluating radiographs of retained deciduous teeth (Figures 8-9 through 8-11). These radiographs are also of value before extraction of retained deciduous teeth. The amount of root structure remaining after resorption and the relation between the permanent tooth bud and surgical area can be determined.

DELAYED ERUPTION

Delayed eruption and tooth impactions have been associated with various conditions, including nutritional deficiency, endocrinopathy, genetic defects, crowded dentitions, retained deciduous teeth, trauma, odontomas, mesiodens, dentigerous cysts, and other anomalies (Figures 8-12 through 8-17).[3,4]

The structural makeup of mesiodens, odontomas, and dentigerous cysts can cause abnormal eruption. Mesiodens is a very small, often comma-shaped supernumerary tooth that occurs at the midline near the central incisors. Despite the size, a tiny crown and developing root can usually be discerned.

Odontomas are presently classified as mixed tumors that originate in odontogenic tissue. The radiographic presentation and biologic behavior of odontomas can vary considerably, depending on the degree of histopathologic differentiation and stage of development. An odontoma is usually discovered during radiographic examination of a young animal that has missing teeth or delayed eruption. Swelling of the jaw may be evident. The tumor is comprised of fluid and varying amounts of differentiating tissue that contain mineralized elements of enamel, dentin, and cementum. If tissue differentiation does not occur, less-mineralized ameloblastoma-like epithelium can develop. Compound odontomas often contain well-formed, recognizable elements of teeth; these elements become more mineralized with time, as do normal teeth. Complex odontomas contain a nondescript mass of radiopaque tissue that can become large. All odontomas are space-occupying lesions, although some are more aggressive and invasive than others. In general, they are well-demarcated lesions that are sometimes encased in a radiolucent cystic region. Odontomas are usually benign. Histopathologic examination is, however, always advised to determine the exact nature and biologic activity of the lesion.

Dentigerous cysts are fluid-filled follicular structures that initially surround the crown of an unerupted developing tooth. If expansion of the fluid is slow and the area around the

(continues on page 97)

DEVELOPMENTAL PROBLEMS—MISSING TEETH

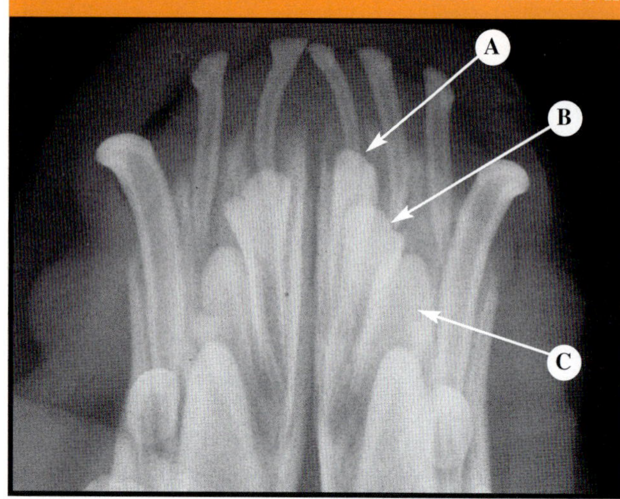

FIGURE 8-1 Missing Deciduous and Permanent Incisors

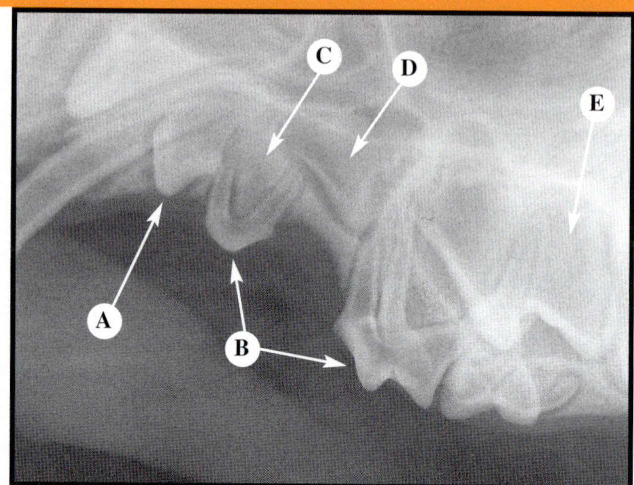

FIGURE 8-2 Missing Upper Second Premolar

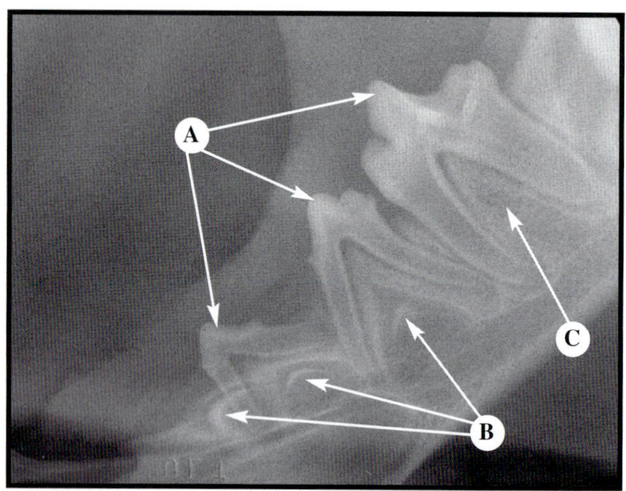

FIGURE 8-3 Missing Lower Fourth Premolar

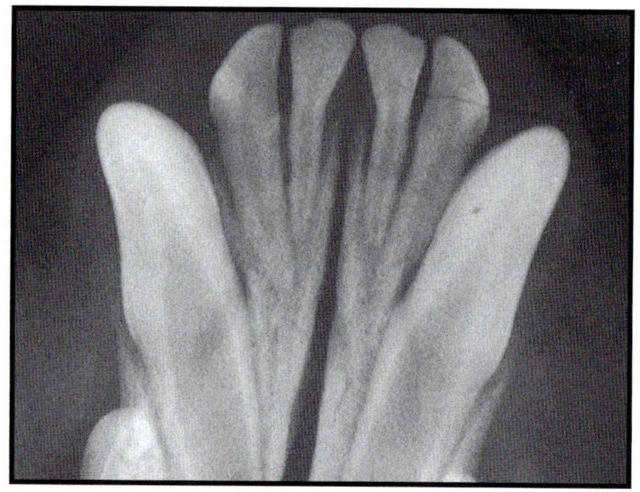

FIGURE 8-4 Missing Central Incisors

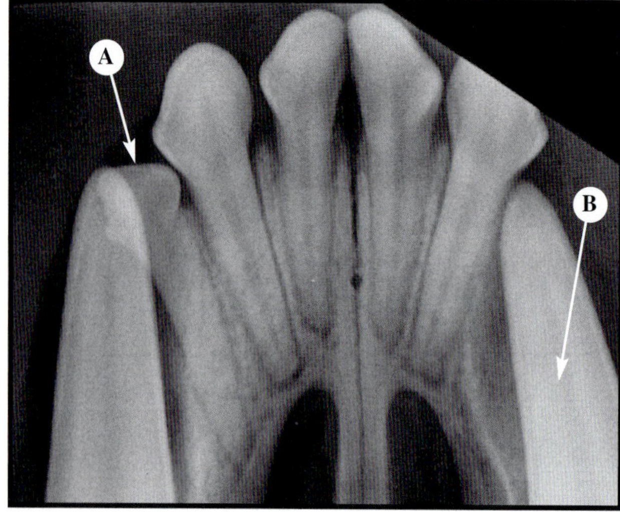

FIGURE 8-5 Missing Corner Incisor

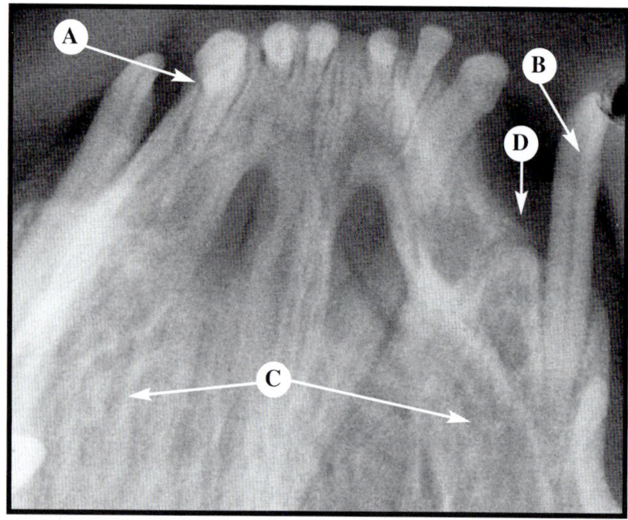

FIGURE 8-6 Missing Permanent Canine Teeth in a Cat

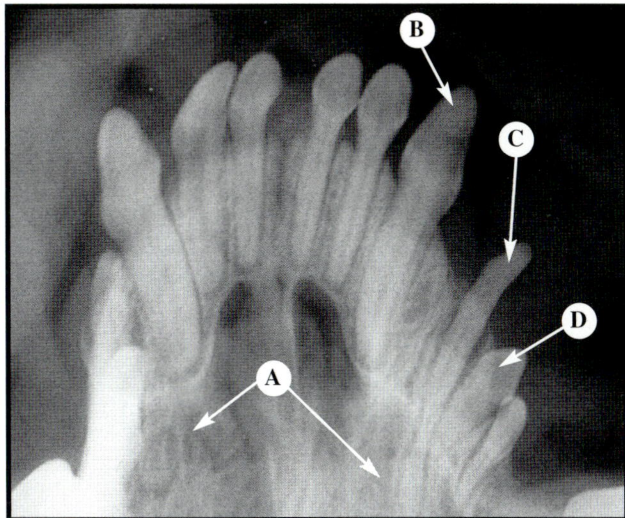

FIGURE 8-7 Missing Permanent Canine Teeth in a Dog

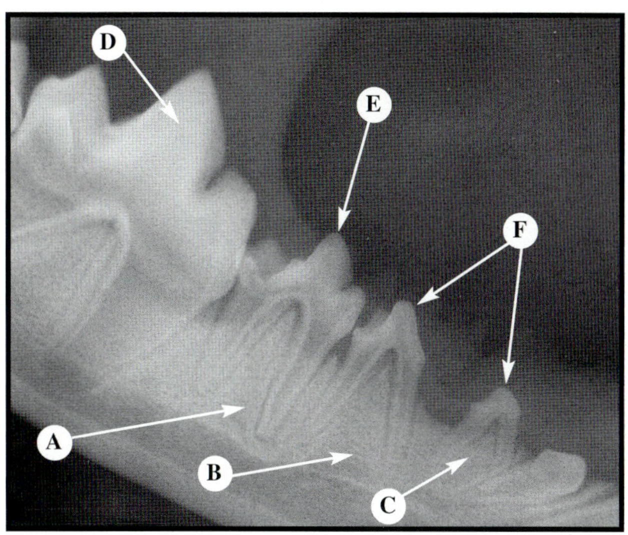

FIGURE 8-8 Missing Lower Permanent Premolars

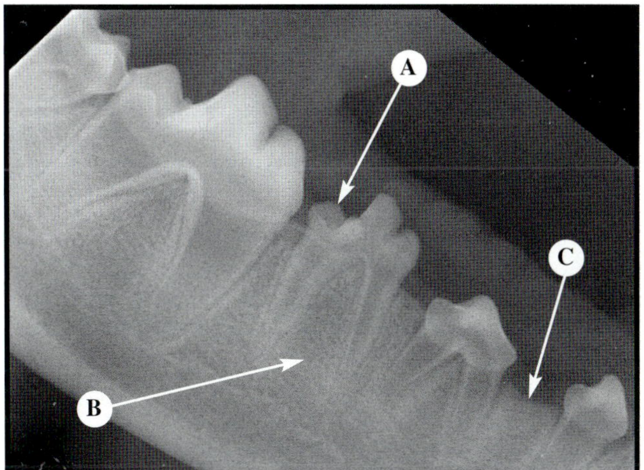

FIGURE 8-9 Retained Deciduous Premolar and Missing Premolar

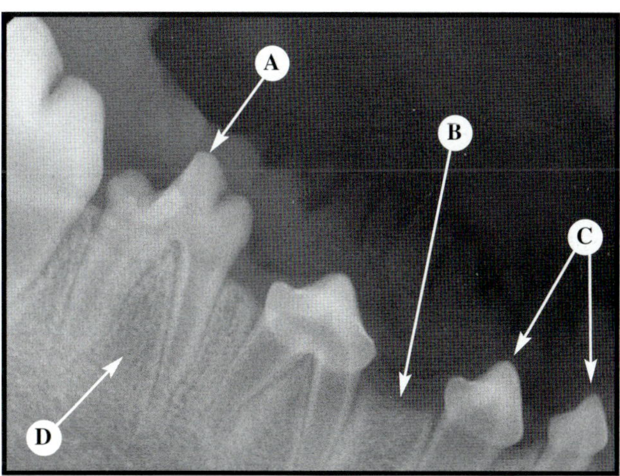

FIGURE 8-10 Missing and Supernumerary Teeth

Figure 8-1 *Both the deciduous and permanent central incisors on the left side of the radiograph appear to be missing, but another radiographic view should be taken to rule out superimposition. A deciduous tooth must be present for the corresponding permanent tooth to form. (A) Normal permanent central incisor, (B) normal permanent intermediate incisor, and (C) normal permanent corner incisor.* **Figure 8-2** *Missing premolars might be a genetic predisposition in some breeds of dogs. (A) First premolar, (B) deciduous premolars, (C) missing second permanent premolar, (D) third permanent premolar, and (E) fourth permanent premolar.* **Figure 8-3** *Missing lower premolars might also be a genetic predisposition in some breeds of dogs. (A) Deciduous premolars, (B) permanent premolars 1, 2, and 3, and (C) missing permanent fourth premolar.* **Figure 8-4** *The cause of missing incisors is unknown, although the bilateral symmetry in this radiograph suggests a genetic condition. In addition to evaluating radiographs, practitioners must obtain a complete patient history to determine whether trauma or disease may be present.* **Figure 8-5** *Only five incisors are apparent in this radiograph. The narrow skeletal structure and rostroversion of the canine teeth are familial in some breeds of dogs, especially Shelties. (A) A microdont in malocclusion with a canine tooth is visible (see the sections on Delayed Eruption and Supernumerary Teeth for more information on microdonts) and (B) a rostroversion of the canine tooth.* **Figure 8-6** *Deciduous teeth eventually undergo resorption and exfoliation and are not permanently retained. (A) Permanent incisors, (B) deciduous canine tooth, (C) missing permanent canine teeth, and (D) alveolar bone in a 9-month-old cat.* **Figure 8-7** *The deciduous canine tooth is undergoing resorption and eventual exfoliation. (A) Missing permanent canine teeth. (B) permanent incisor, (C) retained deciduous canine tooth, and (D) first premolar.* **Figure 8-8** *Missing teeth between the first premolar and molar, as shown in this radiograph, is familial in some breeds of dogs. (A) Missing fourth permanent premolar, (B) missing third permanent premolar, (C) missing second permanent premolar, (D) first molar, and (E and F) deciduous premolars.* **Figure 8-9** *Retention can be associated with missing permanent teeth. The deciduous tooth tends to dissolve at about 14 to 18 months of age, even if a permanent tooth is not present. (A) Retained lower deciduous premolar, (B) missing lower permanent fourth premolar, and (C) missing lower permanent second premolar.* **Figure 8-10** *This radiograph is an example of missing, retained, and extra teeth in the same animal. (A) Retained deciduous premolar, (B) missing second premolar, (C) two first premolars, and (D) missing fourth premolar.*

DEVELOPMENTAL PROBLEMS—MISSING TEETH *(continued)*

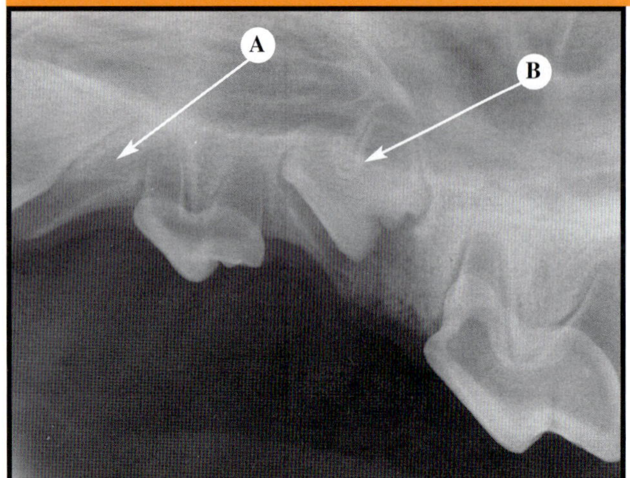

Figure 8-11 *This radiograph is an example of* (A) *missing first premolar* (B) *unerupted third premolar in the same arcade. Surgical intervention for impacted teeth would be indicated if root formation were to proceed or near completion without eruptive movement. We recommend taking a second radiographic view before performing surgery. Ectopic teeth often undergo delayed eruption and impaction, and a tooth may therefore be in an odd position.*

FIGURE 8-11 Missing Premolar and Delayed Eruption

DEVELOPMENTAL PROBLEMS—IMPACTED OR RETAINED TEETH AND DELAYED ERUPTION

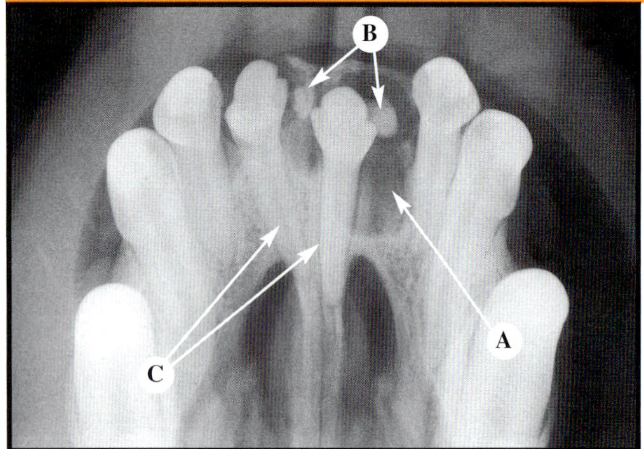

FIGURE 8-12 Odontoma

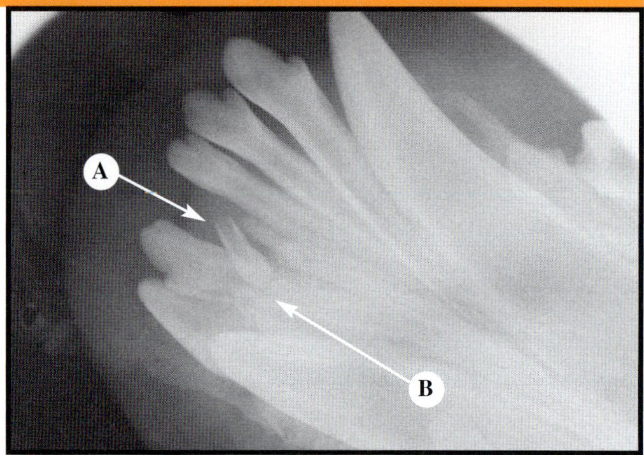

FIGURE 8-13 Mesiodens

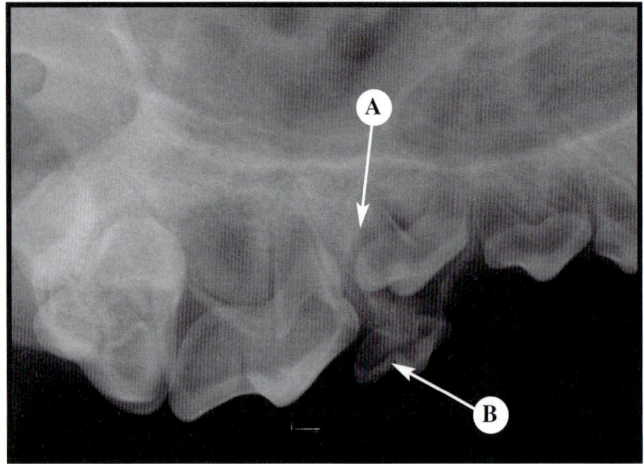

FIGURE 8-14 Retained Deciduous Premolar

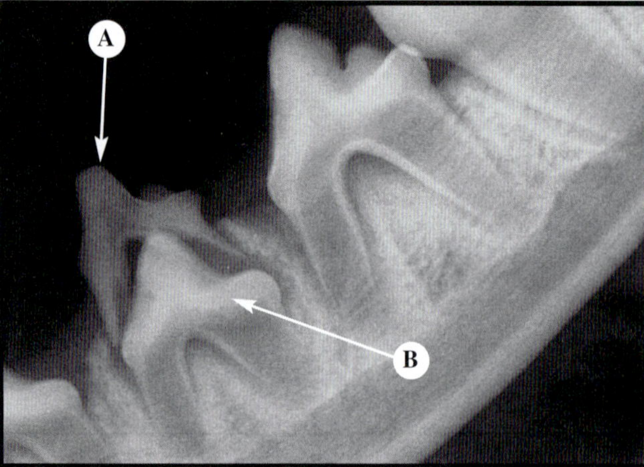

FIGURE 8-15 Retained Deciduous Premolar

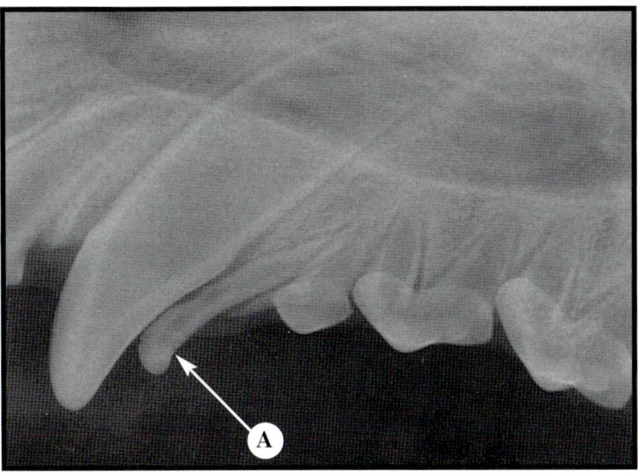

FIGURE 8-16 Retained Deciduous Canine Tooth

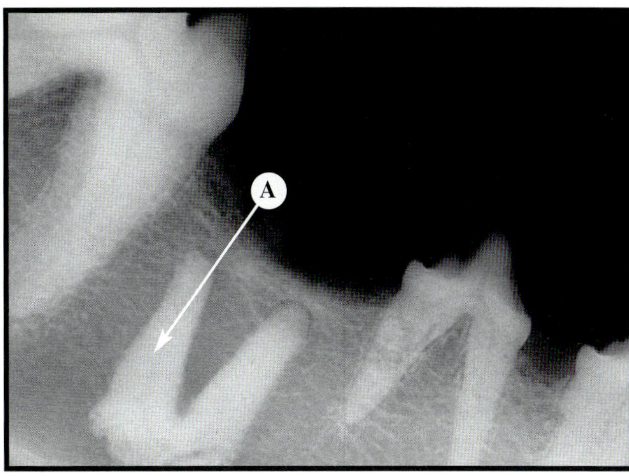

FIGURE 8-17 Inverted Embedded Premolar

Figure 8-12 *Odontomas, which can occur in young dogs and cats, can prevent eruption of adjacent teeth. (A) Radiolucent structure, (B) denticles (also referred to as toothlets), and (C) unerupted permanent incisors.* **Figure 8-13** *Mesiodens often affect eruption of adjacent teeth. Because more than one mesiodens may be present, multiple radiographs capturing different views should be taken before and after removal. (A) Mesiodens and (B) unerupted central incisor.* **Figure 8-14** *The retained deciduous premolar is preventing eruption of the permanent third premolar and should be removed. (A) Unerupted permanent third premolar and (B) retained deciduous premolar.* **Figure 8-15** *Another example of a retained deciduous premolar that requires surgical removal. (A) Retained deciduous premolar and (B) unerupted permanent third premolar.* **Figure 8-16** *The retained deciduous canine tooth is blocking the normal eruption pathway, thereby rostrally displacing the permanent canine tooth. Such crowding can accelerate periodontal disease. The problem is common in toy breeds of dogs. (A) Retained deciduous canine tooth.* **Figure 8-17** *Unerupted teeth are sometimes subject to cystic degeneration and therefore should be monitored on a regular basis. (A) Embedded lower third premolar.*

DEVELOPMENTAL PROBLEMS—DELAYED ERUPTION OF GENETIC ORIGIN

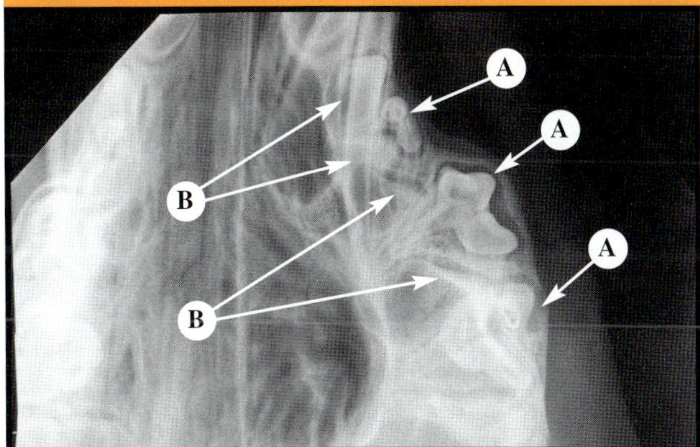

FIGURE 8-18 Delayed Eruption in a 4-Month-Old Tibetan Terrier

Figure 8-18 *Delayed eruption of deciduous teeth can be self-limiting to the animal but should resolve without surgical intervention before the animal is 6 months of age. The problem is familial and therefore should be monitored. (A) Unerupted fully formed deciduous premolars and (B) unerupted developing permanent premolars.* **Figure 8-19** *Retained deciduous teeth, delayed eruption, and impaction of permanent teeth are familial in Tibetan terriers, Shih Tzus, and Lhasa apsos. (A) Retained deciduous incisors, (B) absence of bone resorption, and (C) fully formed roots of unerupted permanent incisors.*

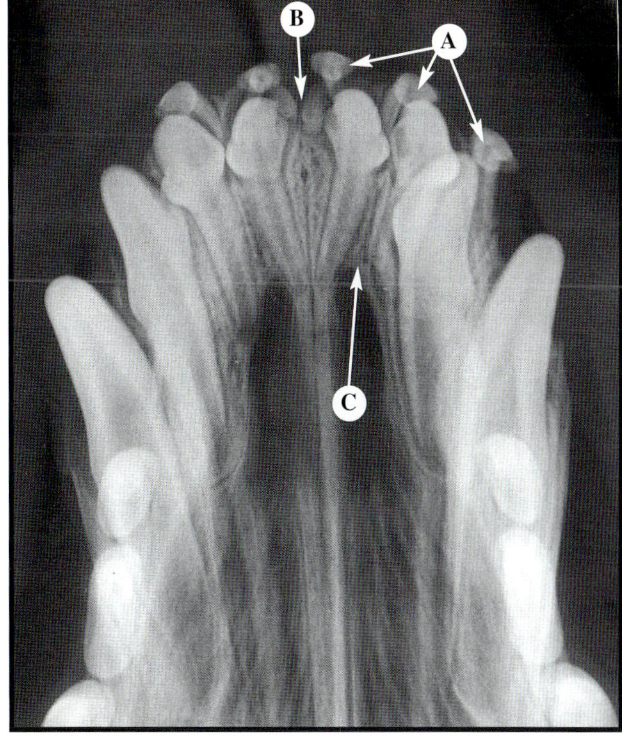

FIGURE 8-19 Delayed Eruption in a 6-Month-Old Tibetan Terrier

DEVELOPMENTAL PROBLEMS—DELAYED ERUPTION OF GENETIC ORIGIN *(continued)*

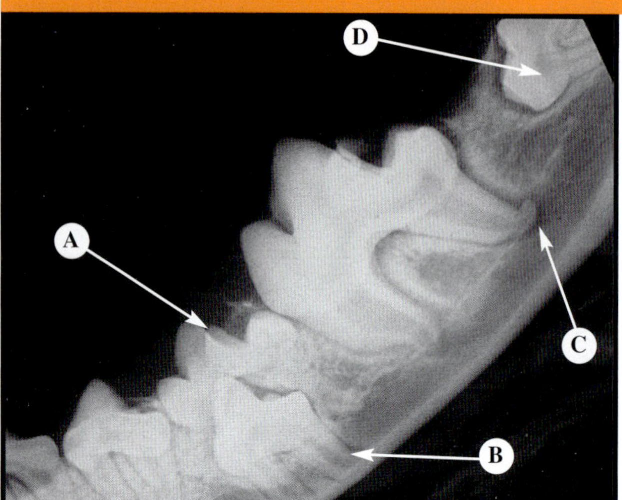

Figure 8-20 *Teeth that are affected by delayed eruption or impaction often develop dilaceration (see section on Anomalous Tooth Formation). The problem is familial in Tibetan terriers. (A) Incompletely erupted deciduous premolar, (B) impacted fourth premolar, (C) J-hook apex, and (D) impacted and malpositioned molar.*

FIGURE 8-20 Delayed Eruption in a 17-Month-Old Tibetan Terrier

DEVELOPMENTAL PROBLEMS—MALOCCLUSIONS AND ECTOPIC ERUPTION

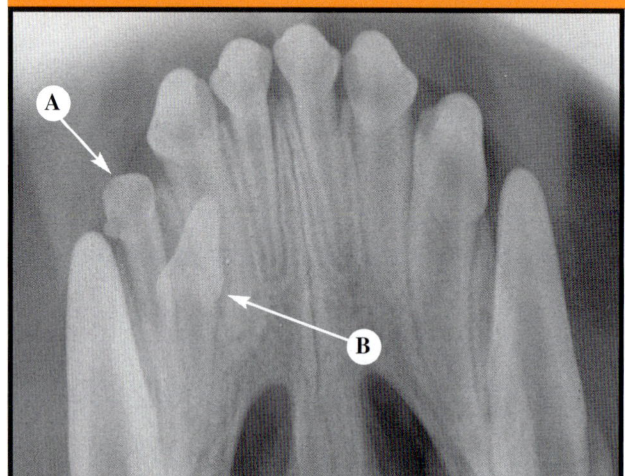

FIGURE 8-21 Ectopic Eruption of Incisor

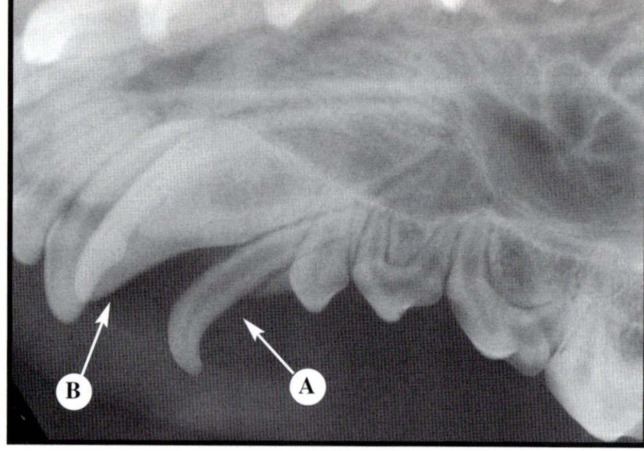

FIGURE 8-22 Rostroversion of the Upper Canine Tooth

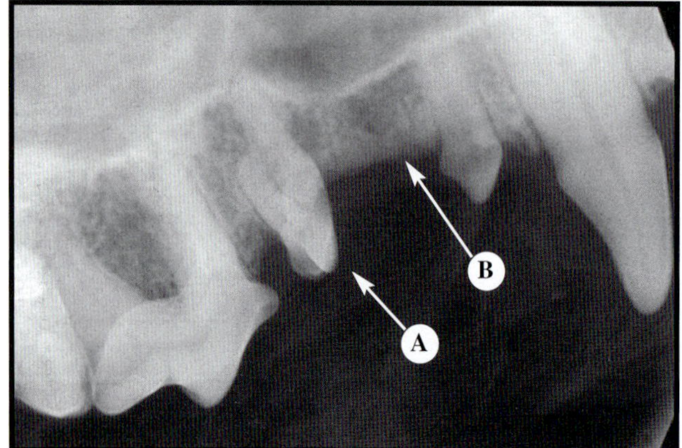

FIGURE 8-23 Rotated Premolar

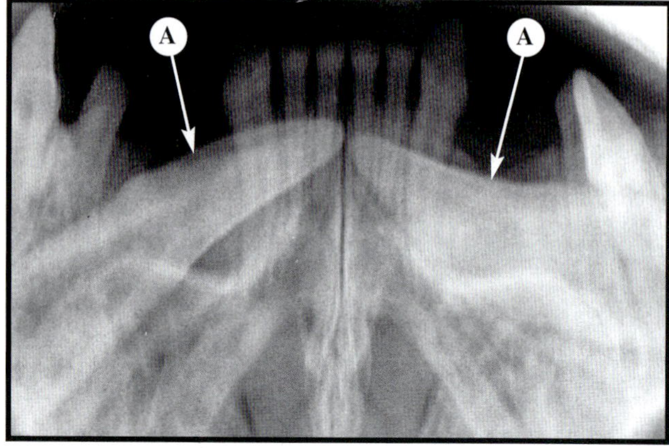

FIGURE 8-24 Ectopic Eruption and Malpositioned Canine Teeth in a Cat

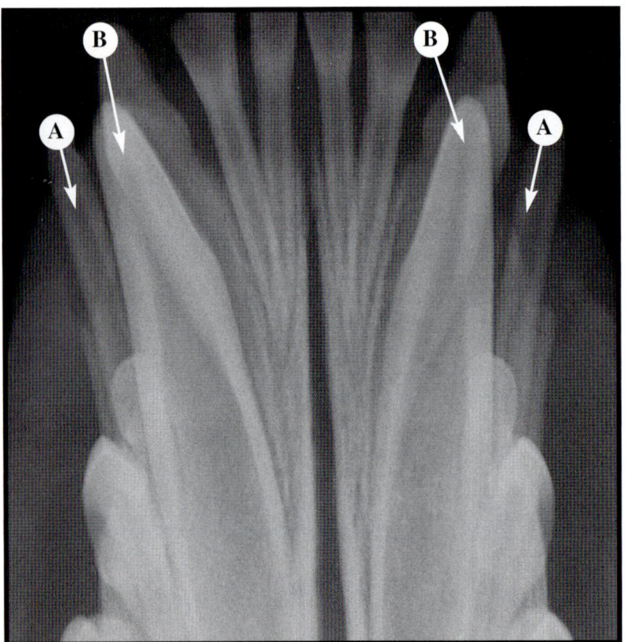

FIGURE 8-25 Retained Deciduous Canine Teeth

Figure 8-21 Note the impaction of the unerupted permanent incisor. (A) Retained deciduous incisor and (B) ectopic pathway preventing eruption of permanent incisor. **Figure 8-22** *The large gap between the deciduous and permanent canine teeth suggests that retention of the deciduous tooth was not the primary cause for displacement of the permanent tooth. (A) Deciduous canine tooth and (B) permanent canine tooth.* **Figure 8-23** *Rotated premolars are a common finding in brachycephalic breeds and, although not in the patient depicted in the radiograph, can be associated with tooth crowding. Malpositioned teeth are vulnerable to periodontal disease. (A) Rotated premolar and (B) missing premolar.* **Figure 8-24** *In this radiograph, the ectopic pathway preventing eruption of the canine teeth is directed toward the midline. This problem occurs in brachycephalic breeds of cats. (A) Palatally directed canine teeth.* **Figure 8-25** *Retained deciduous canine teeth in the mandible are associated with lingually displaced canine teeth, which is one of the most important and serious malocclusions that occur in dogs. (A) Retained deciduous canine teeth and (B) permanent canine teeth that are lingually displaced.*

DEVELOPMENTAL PROBLEMS—SUPERNUMERARY TEETH

FIGURE 8-26 Eight Upper Incisors

FIGURE 8-27 Extra Lower First Premolar in a Puppy 10 to 12 Weeks of Age

Figure 8-26 These overcrowded teeth are misaligned to accommodate the extra teeth. (A) Deciduous tooth remnant and (B) supernumerary teeth. **Figure 8-27** *The supernumerary premolars shown in this radiograph could eventually cause crowding. This problem is familial in some breeds. (A) Two first premolars and (B) permanent premolars 2, 3, and 4.*

(continued from page 91)

missing tooth is otherwise healthy, a dentigerous cyst may not be discovered until after the lesion is large enough to be detected on clinical examination.[5,6] (See Chapter 13 for additional examples of these lesions.)

Delayed eruption has been reported in Tibetan terriers and apparently is genetic in origin (Figures 8-18 through 8-20). The deciduous teeth, and eventually the permanent teeth, may have soft tissue or bony impaction. Eruption, resorption, and exfoliation of deciduous teeth may be delayed for several weeks to possibly several months, with eventual displacement of the permanent teeth and resultant malocclusion. Eruption of the permanent teeth may also be delayed for months and may occur in an abnormal sequence. Although delayed eruption does not seem to be painful and is often self-corrected by

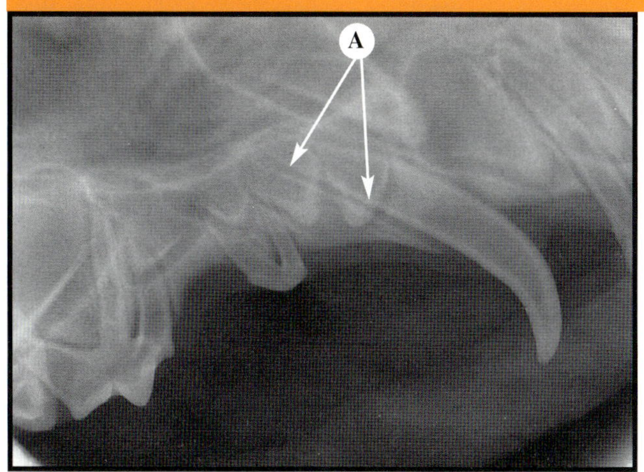

FIGURE 8-28 Extra Upper First Premolar in a Puppy 10 to 12 Weeks of Age

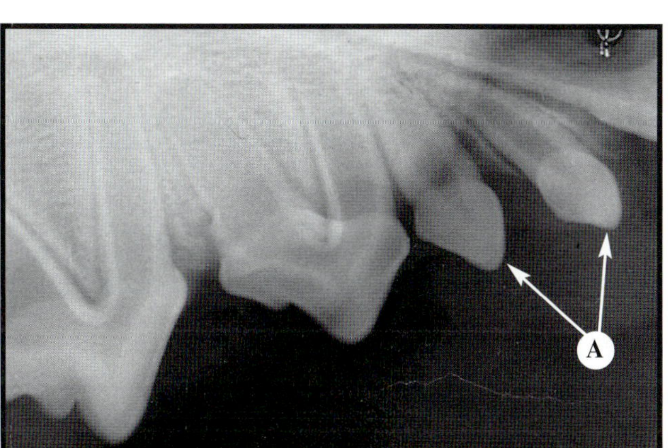

FIGURE 8-30 Supernumerary Upper First Premolar

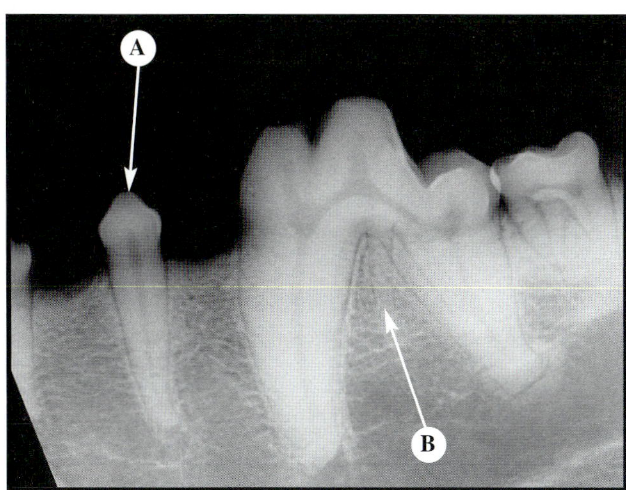

FIGURE 8-32 Microdont

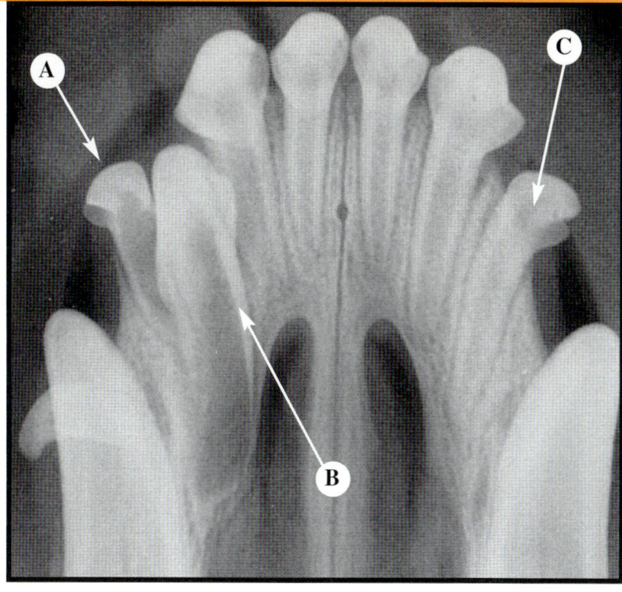

FIGURE 8-29 Supernumerary Incisor

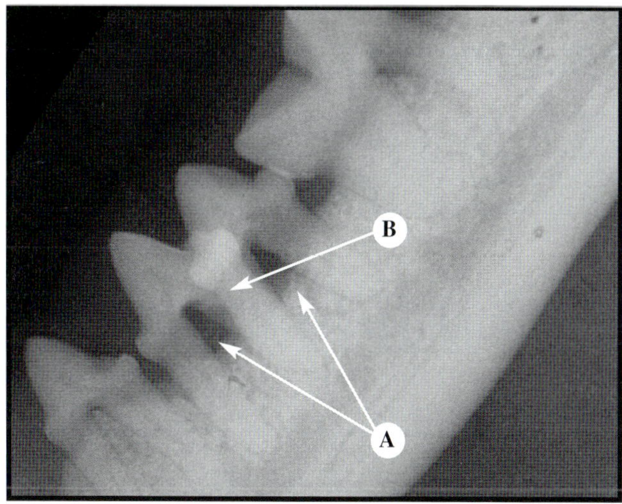

FIGURE 8-31 Supernumerary Lower Premolar in a Cat

Figure 8-28 *Another case of two first premolars; however, crowding should be anticipated. This problem also is familial in some breeds. (A) Two first premolars.* **Figure 8-29** *Supernumerary teeth are often smaller in size than normal teeth and may be conical or peg shaped. These supernumerary teeth often affect the eruption of other teeth. (A) Supernumerary peg-shaped tooth, (B) permanent incisor, and (C) microdont.* **Figure 8-30** *Mild crowding can occur when extra teeth are present. This problem is familial in some breeds. (A) Two first premolars.* **Figure 8-31** *Crowded teeth, such as those shown in this radiograph of a cat's lower premolars, can be vulnerable to periodontal disease. (A) Periodontal bone recession that is localized to the crowded premolars and (B) summation image caused by superimposition.* **Figure 8-32** *In this radiograph, the conical (peg-shaped) tooth resembles the first premolar, suggesting a differential diagnosis of transposition of the first premolar to the position of the fourth premolar. (A) Conical single-rooted tooth and (B) supernumerary root of a molar.*

TABLE 8–1
DIFFERENTIATION OF DEFECTIVE TOOTH FORMATION[4,11]

Problem Area	Cause	Clinical Appearance	Radiographic Signs
Enamel	*Hypoplasia*	*Pitted crowns; thin enamel in focal or generalized areas*	*Enamel that is present has normal radiodensity; defects will appear radiolucent*
	Hypocalcification	*Normal thickness but enamel is soft; easy to remove with an instrument*	*Enamel not as radiodense as dentin; presence of caries*
	Hypomaturation	*Normal thickness but enamel is soft; easy to remove with an instrument; presence of white opaque flecks or snow caps*	*Enamel usually has same radiodensity as dentin*
Dentin	*Dentinogenesis imperfecta*	*Discolored or stained teeth; absence of enamel support; severe attrition*	*Thin roots with wide root canals; constricted cementoenamel junction; pulp chamber obliterated*
	Dentin dysplasia	*Crowns may appear normal*	*Stubby roots with points or rootless teeth; pulp obliterated; periapical lucency*
Enamel and Dentin	*Odontodysplasia*	*Affects a region in a quadrant; yellow-brown discoloration; deformed teeth*	*Delayed or failed eruption; presence of ghost teeth; decreased radiodensity (often indistinct)*

one year of age, monitoring the condition can be helpful in determining when and if surgical intervention is required to remove retained deciduous teeth or to eliminate impactions.[3]

MALOCCLUSIONS AND ECTOPIC ERUPTION

Malocclusions occur when the teeth erupt into an abnormal position and can be caused by a retained deciduous tooth, ectopic pathways that prevent tooth eruption, rotation of tooth buds, or crowding (Figures 8-21 through 8-25). Retained deciduous teeth are common in toy breeds of dogs. The small skeletal size of toy breeds and some brachycephalic breeds in relation to the size of the teeth often results in crowding and rotation of the teeth. Crowded or malpositioned teeth are more prone to plaque accumulation and plaque-induced disease.[7,8]

Anecdotes on breed predisposition to many orthodontic conditions have been reported, but partially impacted and mesially tipped canine teeth is a consistent finding in the Shetland sheepdog. Deciduous canine teeth are often retained, but whether retention is truly a factor in the development of this particular type of malocclusion remains questionable.[8,9]

SUPERNUMERARY TEETH

Although supernumerary teeth are undesirable, they may not be a clinical problem unless they cause crowding and malocclusion, form dentigerous cysts, or interrupt the eruption of permanent teeth. Radiographs can help to locate and identify supernumerary teeth (Figures 8-26 through 8-31), especially if extraction is necessary. Some supernumerary teeth are smaller than normal and are then referred to as microdonts. Microdonts may not have the same anatomic structure as their normal-sized counterparts and often are peg shaped or conical (Figure 8-32). Small supernumerary teeth that are located at or near the midline between the central incisors are mesiodens. The supernumerary teeth that erupt in ectopic pathways in either a buccal or lingual–palatal aspect to the normal arch are known as peridens.[4,5]

DENTAL ANOMALIES

ANOMALOUS TOOTH FORMATION

Problems with tooth formation may result in defective formation of enamel or dentin or unusual anatomic shapes. Radiographs can help to identify aberrant tooth anatomy (Table 8-1) that might interfere with extraction, complicate endodontic treatment, or influence restorative options.

Dilaceration is a disturbance in tooth formation that results in a bend or crease in the tooth. The condition may be caused by trauma to a developing tooth, impaction of a developing tooth, or a developmental anomaly. Distortion can occur anywhere in the crown or root and, if severe enough, may prevent complete eruption of a tooth or expose the pulp. Root dilacerations can present a challenge when extraction or

DENTAL ANOMALIES—ANOMALOUS TOOTH FORMATION

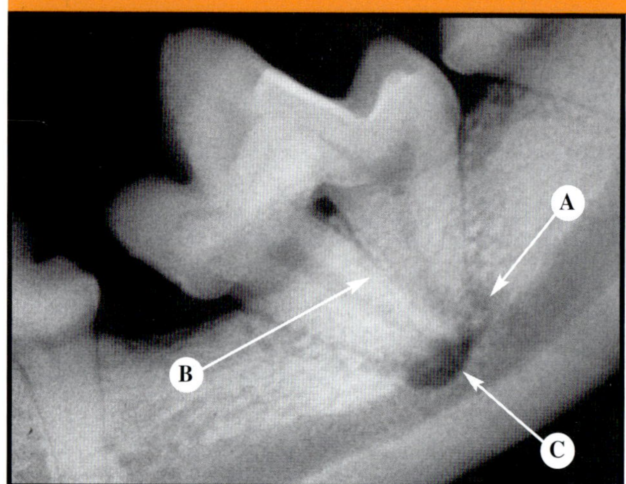

FIGURE 8-33 Dilaceration Involving V-Root Malformation

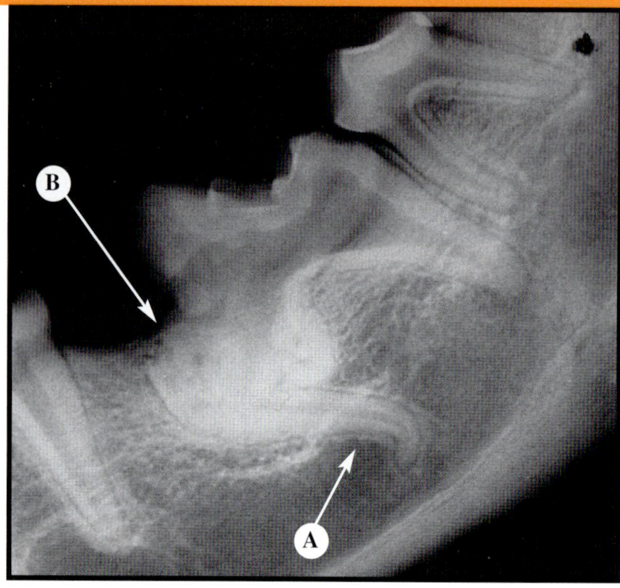

FIGURE 8-34 Dilaceration Involving Malformation of the Root and Crown

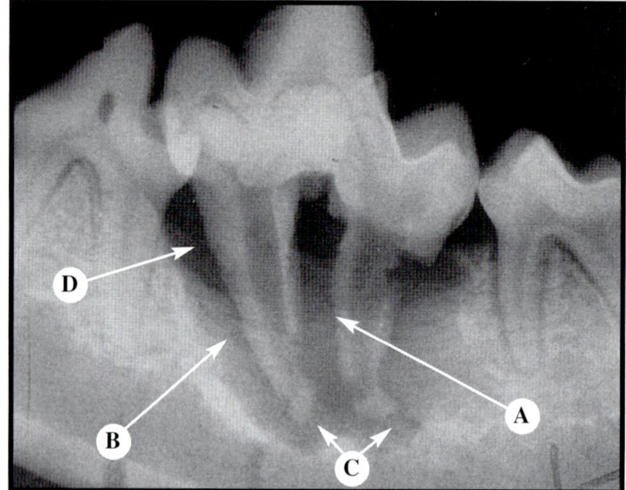

FIGURE 8-35 Dilaceration Involving a Combination Lesion

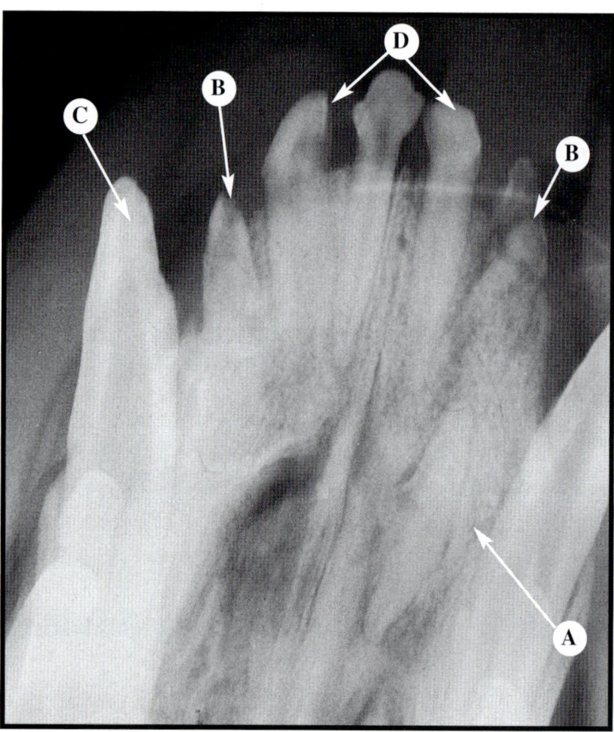

FIGURE 8-36 Odontodysplasia

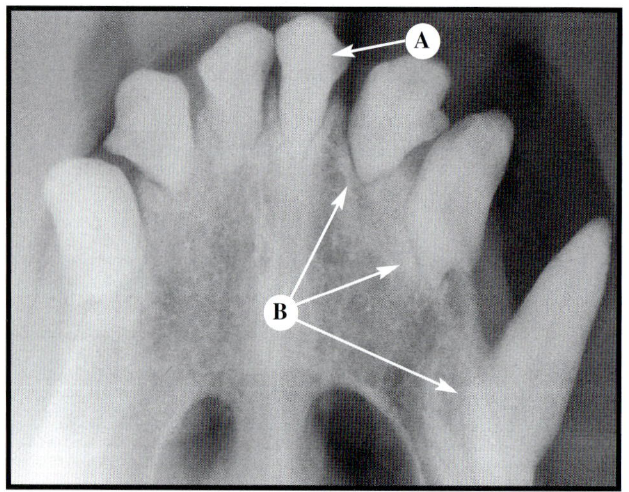

FIGURE 8-37 Dentin Dysplasia in a 2-Year-Old Dog

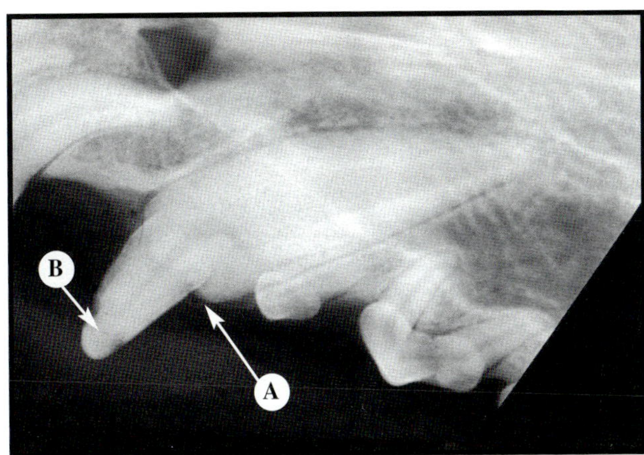

FIGURE 8-38 Crown Dilaceration

Figure 8-33 *Malformed teeth can cause endodontic-related disease. (A) Malformed apex, (B) roots that are joined at the apex diverge with the crown, and (C) periapical radiolucency.* **Figure 8-34** *Root dilaceration, such as that shown here, can complicate tooth extraction. Trauma is the typical cause. A malformed tooth is vulnerable to both endodontic and periodontal disease. (A) Malformed mesial root (J-hook) and (B) malformed crown showing folded and gnarled enamel.* **Figure 8-35** *An endodontic–periodontal lesion of malformed teeth, such as that involving the tooth shown here, can result in necrosis. (A) Dilacerated V-root system undergoing resorption, (B) wide periodontal space, (C) periapical radiolucency, and (D) periodontal bone loss.* **Figure 8-36** *Malformed teeth, such as those shown here, can result in attrition. The embedded tooth should be monitored on a regular basis. (A) Malformed embedded tooth, (B) partially erupted malformed teeth that are dysplastic and hypocalcified, (C) defects on the enamel, and (D) malformed teeth.* **Figure 8-37** *When short roots exist, a patient is at an increased risk for spontaneous exfoliation. The anomaly is possibly genetic in origin. (A) Obliterated pulp and (B) short roots.* **Figure 8-38** *Trauma or inflammation of a developing tooth bud is often indicated as causing dilaceration of the crown. (A) Folded or creased enamel and (B) hypoplasia of the enamel.*

DENTAL ANOMALIES—ROOT ABNORMALITIES

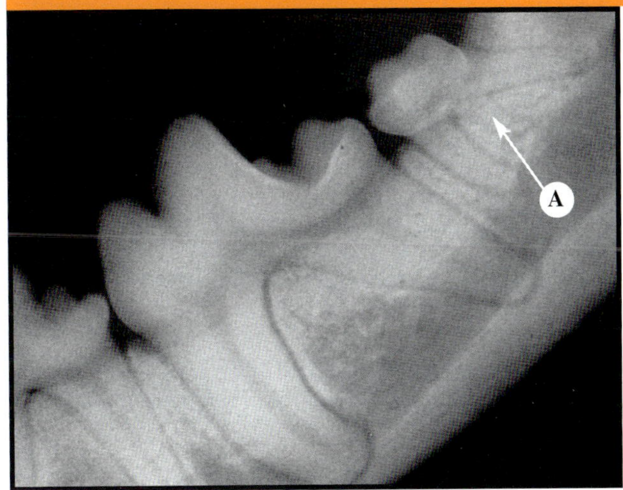

FIGURE 8-39 Extra Root in a Tooth

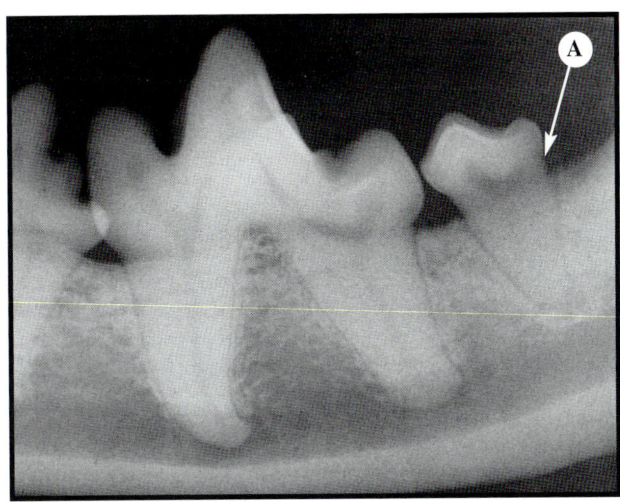

FIGURE 8-41 Single-Rooted Molar

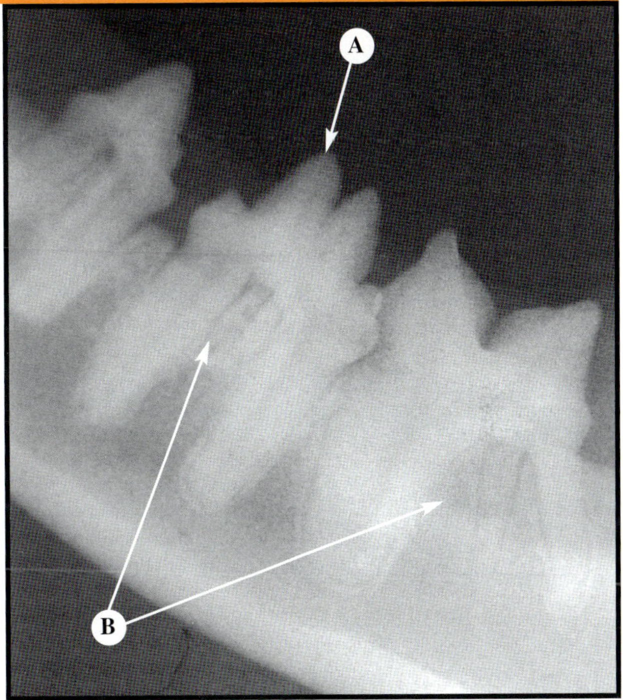

FIGURE 8-40 Supernumerary Roots and Bifid Crown

Figure 8-39 *Extra dental roots can occasionally occur but usually have no pathologic significance; however, they can cause complications if the tooth needs to be extracted or if endodontic treatment is required. (A) Lower second molar with an extra root.* **Figure 8-40** *Supernumerary roots usually have no pathologic significance. In addition, partial gemination of a premolar (see section on Fusion and Gemination) probably has no pathologic significance unless periodontal disease develops at the furcation or if the tooth needs to be extracted. (A) Bifid crown and (B) supernumerary roots.* **Figure 8-41** *A molar with a single root has no pathologic significance. (A) Lower second molar with only one root.*

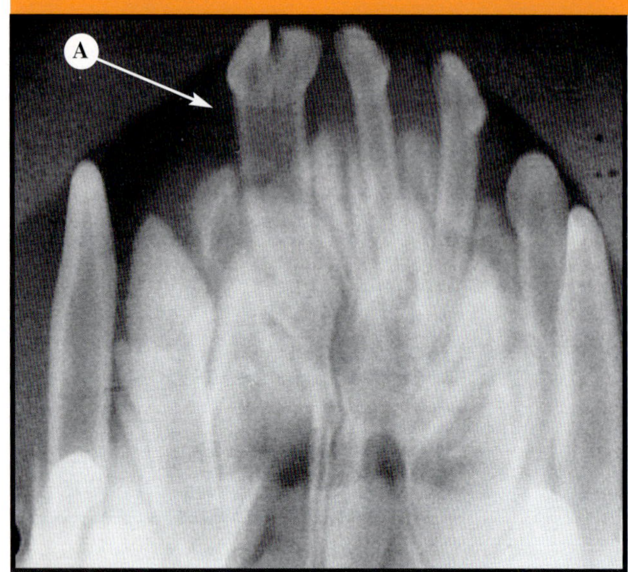

FIGURE 8-42 Fusion of Deciduous Incisors

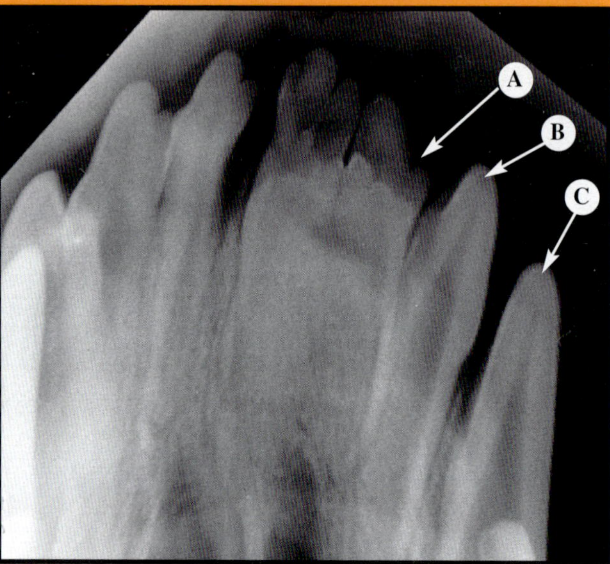

FIGURE 8-43 Fusion of Permanent Incisors

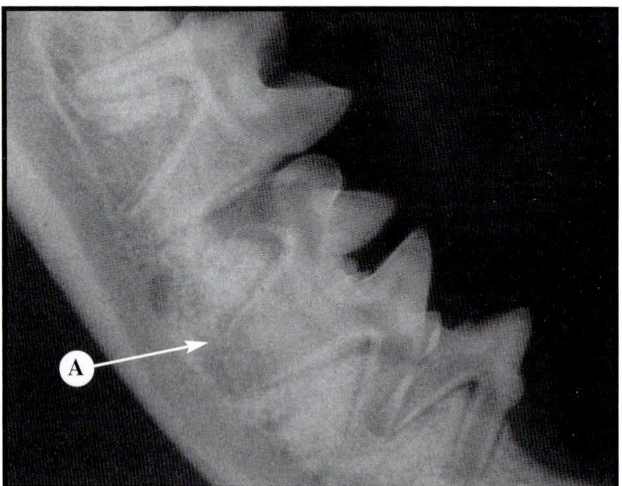

FIGURE 8-44 Gemination of Lower Premolar in a Cat

Figure 8-42 *Fusion of deciduous incisors can affect the corresponding permanent teeth.* (A) *Fusion of the deciduous central and intermediate incisors.* **Figure 8-43** *Fused teeth share a common root canal system.* (A) *Fusion of the permanent central and intermediate incisors,* (B) *corner incisor, and* (C) *canine tooth.* **Figure 8-44** *Clinical examination of the patient depicted in this radiograph would indicate the presence of two separate teeth; however, the root system is shared, which would prevent separate extraction of the teeth. (Figure 8-44 courtesy of Daniel Carmichael, DVM, Diplomate, AVDC, Ardsley, New York.)*

ROOT ABNORMALITIES

Supernumerary roots are rarely a problem but are clinically relevant if teeth with supernumerary roots require extraction or endodontic treatment. Radiographs can assist the practitioner in identifying these areas (Figures 8-39 through 8-41).

FUSION AND GEMINATION

Fusion occurs when two teeth are joined or fuse together during development (Figures 8-42 and 8-43). There is usually a reduced number of teeth in the arch. The fused teeth may appear as a bifid crown or two recognizable teeth joined by dentin or enamel. The pulp chamber, root canal, or crown may present an unusual configuration. Teeth that are fused or joined by cementum only are actually affected by concrescence and not by fusion. Hypercementosis can also eventually evolve into concrescence[4,5] (see Chapter 9).

Gemination (twinning) occurs when a tooth bud attempts to divide (Figure 8-44). The pulp chamber may be enlarged or partially divided. The crown may be partially cleft or bifid or completely bifurcated and therefore appear to be a normal tooth with a supernumerary tooth next to it. A normal num-

endodontic treatment is necessary[4,5] (Figures 8-33 through 8-35).

Disturbances in formation, mineralization, or maturation of the enamel or dentin can result in tooth defects and problems with wear and caries (Figures 8-36 through 8-38). Defects can result from developmental anomalies, trauma, infection, inflammation, genetic predisposition, nutritional deficiency, or metabolic conditions that occur when the tooth is being formed. Defects that exhibit chronologic banding of the teeth are likely the result of a compromising event (e.g., systemic illness) that occurred in the animal's life while the tooth was developing[4,5,10–12] (Table 8-1).

SKELETAL ANOMALIES—CLEFT PALATE AND GROWTH IRREGULARITIES

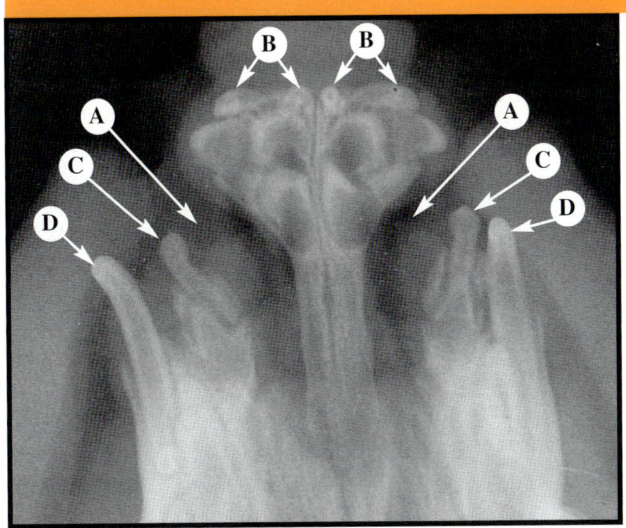

FIGURE 8-45 Cleft Palate

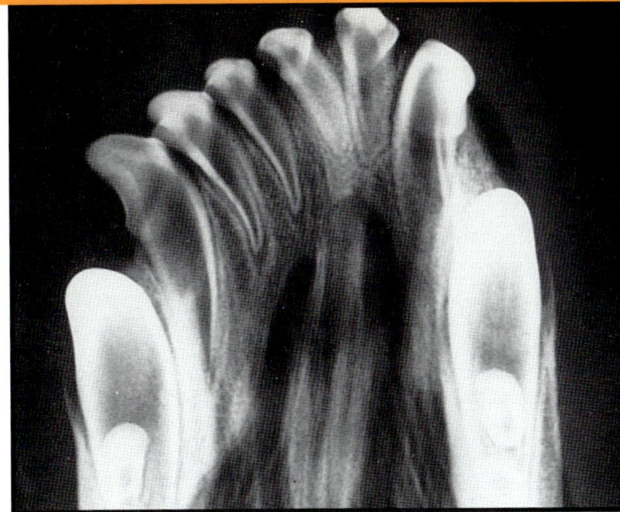

FIGURE 8-46 Ventrodorsal View of Wry Bite

Figure 8-45 *Congenital oronasal fistula, such as shown in this radiograph, would require surgical correction. The condition is considered to be a genetic problem.* (A) *Bilateral clefts,* (B) *deciduous central and intermediate incisors,* (C) *deciduous corner incisors, and* (D) *deciduous canine teeth.* **Figure 8-46** *Wry bite is believed to be genetic in origin and can be associated with facial trauma during growth or secondary malocclusion. Note that the curvature of the maxilla to the left was caused by disparate growth rates of the left and right maxillary bones. (Figure 8-45 courtesy of Johnathon R. [Bert] Dodd, BS, DVM, Fellow, Academy of Veterinary Dentistry, Austin, Texas.)*

ber of teeth are usually present if the twinned tooth is counted as one tooth.[5]

SKELETAL ANOMALIES

Radiographic examination of a skeletal abnormality can determine the severity and extent of the condition and assist in treatment measures. Cleft palate can result from an insult during fetal growth (e.g., drug exposure), be acquired through injury or disease, or be inherited (Figure 8-45). The reported incidence of cleft palate in brachycephalic breeds of dogs and in Siamese cats has increased. Clefts can involve the lip, primary palate, or secondary palate, including the soft palate. Disparity in jaw length or growth pattern can also be inherited, acquired by an adverse dental interlock (resulting from poor synchronization of growth in the mandible and maxilla or tooth eruption), or caused by trauma during growth[13–15] (Figure 8-46).

REFERENCES

1. Emily P: Intraoral radiology, in Frost P (ed): *The Veterinary Clinics of North America. Small Animal Practice*, vol 16, no. 5. Philadelphia, WB Saunders Co, 1986, pp 801–816.

2. Harvey CE, Dubielzig RR: Anatomy of the oral cavity in the dog and cat, in Harvey CE (ed): *Veterinary Dentistry*. Philadelphia, WB Saunders Co, 1985, p 22.

3. Aller MS: Retained deciduous teeth and delayed development of dentition of Tibetan terriers, in *Proceedings of Veterinary Dental Forum '90*. Las Vegas, NV, American Veterinary Dental College and Academy of Veterinary Dentistry, 1990, pp 75–78.

4. Sapp JP, Eversole LR, Wysocki GP: *Contemporary Oral and Maxillofacial Pathology*. Philadelphia, CV Mosby Co, 1997, pp 2–37, 147–149.

5. Goaz PW, White SC: *Oral Radiology Principles and Interpretation*. Philadelphia, CV Mosby Co, 1994, pp 340–368, 442–449.

6. Hale FA, Wilcock BP: Compound odontoma in a dog. *J Vet Dent* 13(3):93–95, 1996.

7. Colmery B, Frost P: Periodontal disease etiology and pathogenesis, in Frost P (ed): *The Veterinary Clinics of North America. Small Animal Practice*, vol 16, no. 5. Philadelphia, WB Saunders Co, 1986, pp 817–833.

8. Hennet P, Harvey CE: Diagnostic approach to malocclusions in dogs. *J Vet Dent* 9(2):23–26, 1992.

9. Saidla J: Inherited dental problems in the dog and the cat, in *Proceedings of Veterinary Dentistry '89*. New Orleans, LA, American Veterinary Dental College and Academy of Veterinary Dentistry, 1989, p 66, addendum.

10. Rossman LE, Garber DA, Harvey CE: Disorders of teeth, in *Veterinary Dentistry*. Philadelphia, WB Saunders Co, 1985, pp 79–83.

11. Bittegeko SB, Arnberg J, Nkya R, Tevik A: Multiple dental developmental abnormalities following canine distemper infection. *JAAHA* 31(1):42–5, 1995.

12. Miles DA, Van Dis M, Kaugars GE, Lovas JG: *Oral and Maxillofacial Radiology: Radiologic/Pathologic Correlations*. Philadelphia, WB Saunders Co, 1991, 306–318.

13. Hennet PR, Harvey CE: Craniofacial development and growth in the dog. *J Vet Dent* 9(2):11–18, 1992.

14. Ross DL: Orthodontics for the dog, in Frost P (ed): *The Veterinary Clinics of North America. Small Animal Practice*, vol 16, no. 5. Philadelphia, WB Saunders Co, 1986, pp 939–954.

15. Hoskins JD: *Veterinary Pediatrics: Dogs and Cats from Birth to Six Months*. Philadelphia, WB Saunders Co, 1990, pp 142–143.

Chapter Nine

INTERPRETATION OF PERIODONTAL DISEASE

The supporting structures of the dentition, or periodontium, consist of the cementum, periodontal ligament, alveolar bone, and gingiva. Periodontal disease involves both hard and soft tissues that surround each tooth and is one of the most common oral conditions that affect domestic cats and dogs. The most common scenario occurs when bacterial plaque-induced inflammation of the gingiva extends into the bone and causes bone resorption. The gingiva may recede to the bone, form pockets around the resultant bone defects, or undergo hyperplasia. The expression and degree of gingivitis and periodontitis and the age of onset are not completely predictable because the conditions are influenced by many genetic and environmental factors. The ultimate result, however, is loss of tooth support.[1-4]

Radiography is a valuable tool in the diagnosis and treatment of periodontal disease and should be considered a partner to the results of clinical examination. Because periodontal lesions often demonstrate periods of dormancy and exacerbation, clinical evaluation of soft tissue can vary from one appointment to another. Although radiographs basically provide no information about the present status of soft tissue, they do reveal changes to the condition of mineralized tissue. Therefore, complete clinical assessment of periodontal tissue depends on clinical examination and radiography[5-7] (Figure 9-1).

RADIOGRAPHIC SIGNS OF PERIODONTAL DISEASE

The radiographic appearance of periodontal disease varies according to the type, location, and the extent of bone resorption around a tooth (Table 9-1). The distribution of bone loss within the oral cavity may be an important clue in determining the factors (e.g., occlusal trauma, crowded dentition, or restoration overhangs) that are contributing to increased plaque accumulation and bone destruction. Initially, the crestal bone exhibits a jagged or eroded surface, and a break or lack of definition in the continuity of the lamina dura occurs at the mesial or distal aspect of the crest. The angle of the crest to the neck of the tooth widens to form a wedge-shaped radiolucent defect. Resorption patterns of bone then vary.

Horizontal bone loss occurs when bone resorption occurs at about the same level in a particular region or group of teeth; the crestal bone remains essentially parallel to the occlusal plane (Figure 9-2). Angular (or vertical) bone loss occurs when

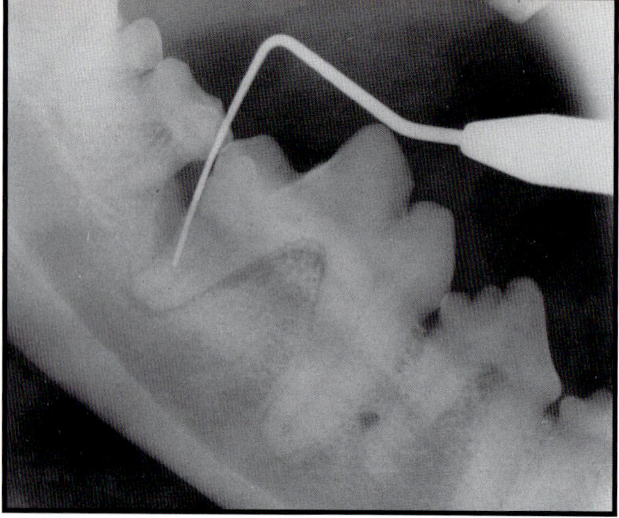

EVALUATION OF THE PERIODONTIUM— CLINICAL EVALUATION

FIGURE 9-1 *Thorough evaluation of the periodontium requires clinical examination (using a probe) and radiography. The probe was radiographed for emphasis.*

TABLE 9 – 1 **RADIOGRAPHIC SIGNS OF PERIODONTAL DISEASE[4,7–9]**
Resorption of the Alveolar Crest
Loss of Integrity in the Lamina Dura
Increased Periodontal Space
Bone Resorption: Horizontal, Angular, or at the Furcation

resorption occurs in an apical direction along a specific root surface with limited effect on surrounding bone. A bony pocket that is shaped by angular bone loss or formation of a crater may have one or more walls of bone lining its boundaries (Figure 9-3). Bone loss at the furcation occurs in the space between the roots of multirooted teeth (Figure 9-4). If any one of these resorptive processes reaches a lateral canal or the apex of a tooth, the blood and nerve supplies to the pulp become involved and pulpitis subsequently follows. When a periodontal lesion is extensive enough to involve the pulp, a combined periodontal–endodontic lesion can develop. Often, a widened periodontal space or a localized region of radiolucency in the periradicular bone is present at the site of the involved lateral or apical canals. In addition, if bone loss is severe enough, the remaining adjacent bone will spontaneously fracture under the stress of normal activities. If resorption and pocketing are extensive in the maxilla, erosion into the nasal sinus is possible. The result can be an oronasal fistula in which the oral cavity communicates with the nasal cavity through the periodontal tract. Reliable diagnosis of oronasal fistulas is made by clinical examination because often the fistulas are not evident on radiographs because of superimposition of the root over the fistulous site. On occasion, the loss of integrity of the radiopaque line formed by the confluence of the maxillary and palatal bones (the so-called white line) may be associated with an oronasal fistula. The eventual loss of bony support in periodontal disease also causes increased mobility of teeth, which can affect occlusion, most often in the incisors of dogs and canine teeth of cats. Unstable teeth may not stay in occlusion, but their position may shift and even undergo supereruption.

EVALUATION OF THE PERIODONTIUM—COMPARISON OF BONE HEIGHT

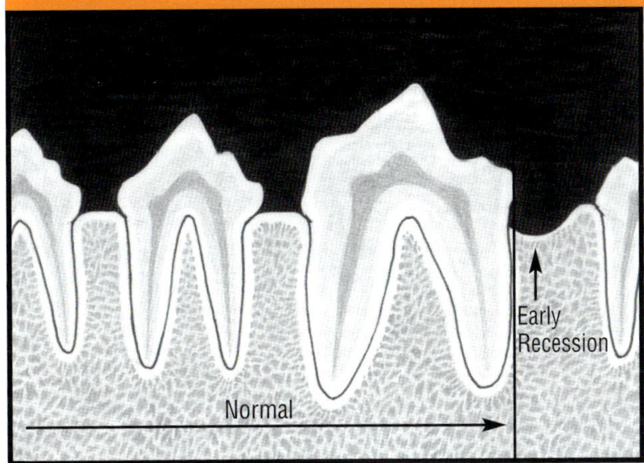

FIGURE 9-2A Early Bone Changes

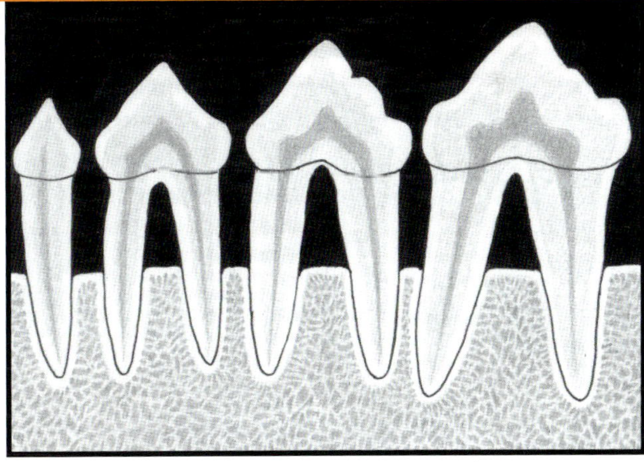

FIGURE 9-2B Horizontal Loss

Figure 9-2A *This illustration shows early crestal bone change. Erosion of the alveolar crest distal to the molar is evident, whereas other regions are normal.*
Figure 9-2B *This illustration shows generalized horizontal bone loss.*

EVALUATION OF THE PERIODONTIUM—INFRABONY POCKETS

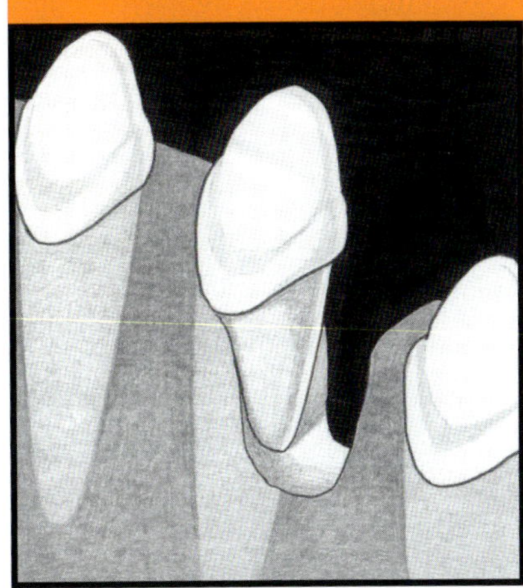

FIGURE 9-3A Single-Wall Defect

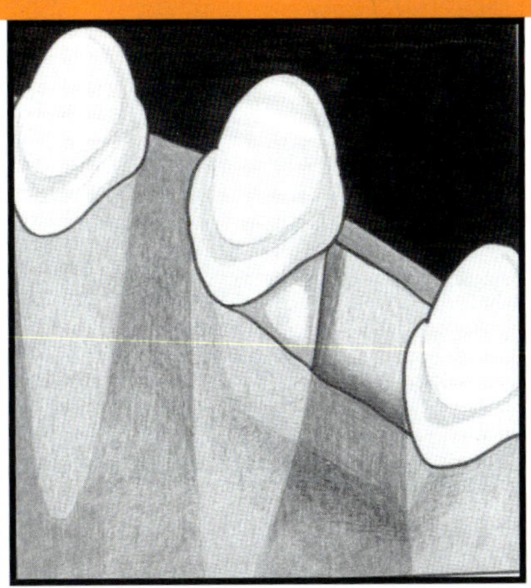

FIGURE 9-3B Two-Wall Defect

Figures 9-3A through 9-3E *Infrabony defects are classified according to the number of bony walls that line the defect and face the diseased root.*

(continues on next page)

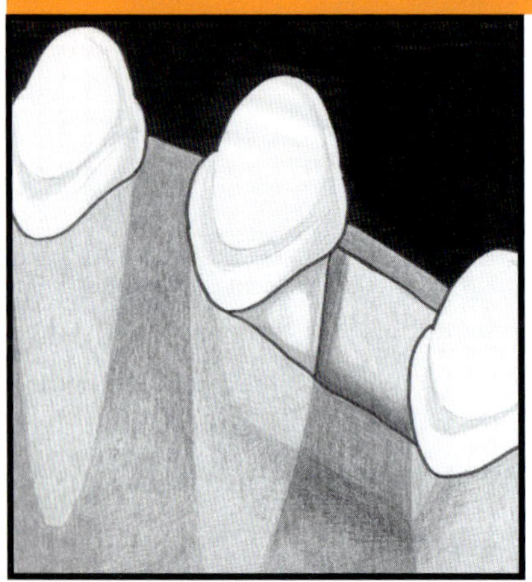

FIGURE 9-3C Two-Wall Crater Defect

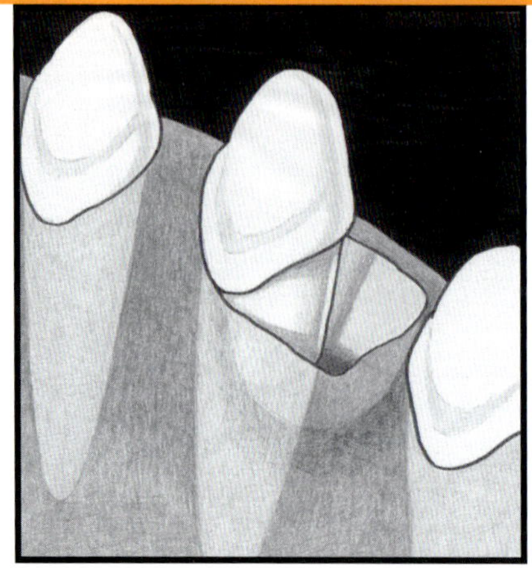

FIGURE 9-3D Three-Wall Defect

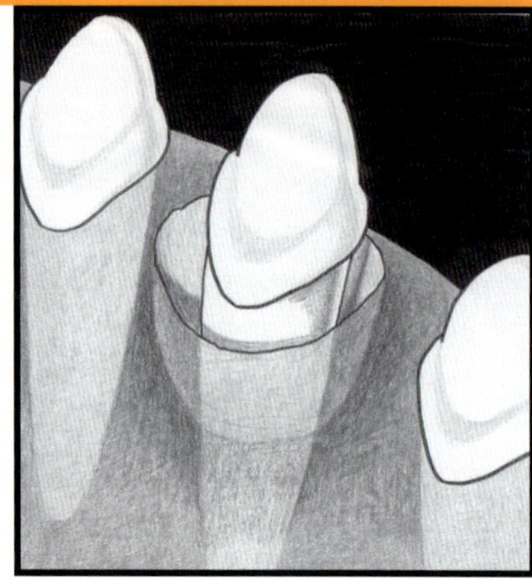

FIGURE 9-3E Four-Wall Cup Defect

(continued from previous page)

Figures 9-3A through 9-3E *Infrabony defects are classified according to the number of bony walls that line the defect and face the diseased root.*

Because bone recession exposes root surfaces to the oral environment, the teeth are more vulnerable to root caries. In cats, root-resorptive lesions can be associated with gingivitis and periodontal disease. Both root caries and root resorption are demonstrated on radiographs by radiolucent defects in the root.[4,5,7-12] Chapter 10 also discusses combination lesions; whereas Chapter 11 addresses caries, resorptive lesions, and the pathology of fractures.

Consistent and successful treatment of periodontal disease relies extensively on being able to identify the various permutations and expressions of clinical signs, many of which can be identified on radiographs. Radiographic detection of periodontal bone disease has some limitations, however. Bone loss cannot be detected on radiographs until 30% to 50% of bone mineral has been lost. Therefore, in general, actual bone loss is greater than the images presented on radiographs. In addition, changes to the buccal and lingual plates of bone can be obscured by the root structure. Accurate identification of bony pockets can be obscured by the buccal or lingual plates of bone that form the boundaries of the defect. The use of gutta percha points, dental probes, or contrast material can help to delineate the boundaries of the bony pocket on radiographs (Figure 9-1). The apparent height of the crestal bone, which is an important feature that can define the extent of horizontal bone loss, can vary according to the angle of the x-ray beam. Bone loss at the furcation can easily be missed because the area is readily obscured by minor changes to the angle of the beam. Therefore, a small change in the radiodensity of the furcation and surrounding area or a widened peri-odontal space at the furcation deserves further scrutiny of the furcation. When marked bone loss occurs around only a single root of a multirooted tooth, the furcation is also likely to be involved and must be considered when planning treatment measures, even if bone loss is not obvious at the furcation.[5,7,13]

CHANGES AFFECTED BY THE AGING PROCESS

Sometimes the normal acceptable signs of aging of periodontal tissue and significant early or mild periodontal disease are difficult to distinguish. With aging, the density of supporting bone normally increases and the lamina dura is less discernible. An indistinct lamina dura that is affected by the aging process could be misconstrued as loss of continuity or integrity in the lamina dura and interpreted as a sign of periodontal disease. An increase in the coarseness and density of trabecular bone could be misinterpreted as sclerosis or a response to chronic inflammation of the bone. Some mild regression of the alveolar crest is considered to be normal during the aging process; however, if regression causes root exposure, the regression would be considered pathologic (Figure 9-5). Differentiation between inconsequential changes attributed to aging and early periodontal disease must be based on clinical evaluation and radiographic interpretation. [4,8,14,15]

On occasion, hypercementosis is an incidental finding that presents as clinically normal; however, the condition is considered to be part of a compensatory or repair process. Excessive deposits of cementum can cause the affected tooth to be bulbous or club shaped. The problem often affects teeth in the apical region, which can present a challenge if extraction is

EVALUATION OF THE PERIODONTIUM—THREE-DIMENSIONAL ANALYSIS OF BONE LOSS

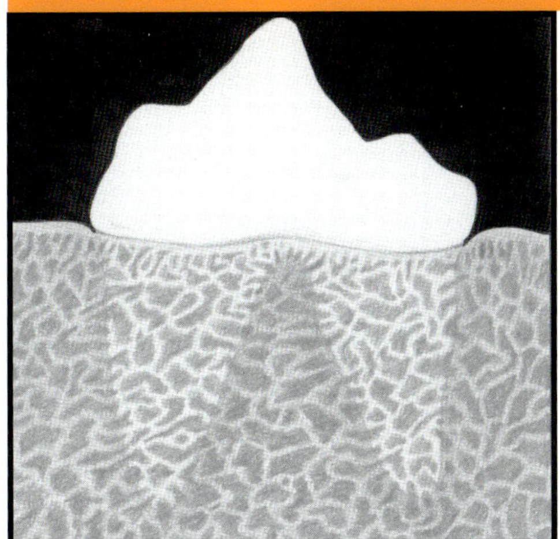

FIGURE 9-4A Lateral View of Normal Bone Height at the Lingual Aspect

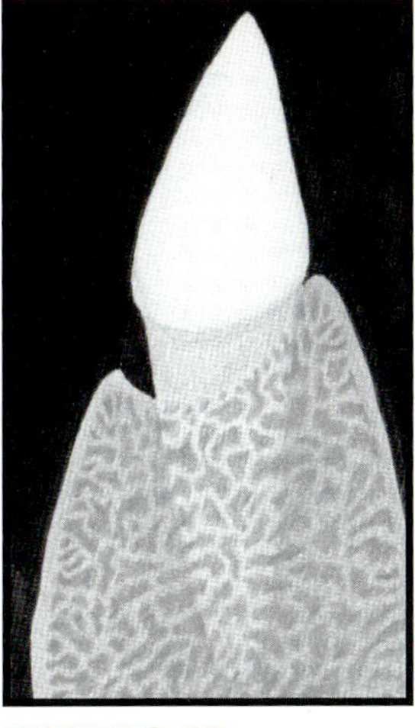

Figures 9-4A, 4B, and 4C *These illustrations depict bone height from three views. When viewed from a lateral aspect, one side of the tooth may be affected whereas the other side is not.* **Figure 9-4D** *This radiograph shows bone loss at the furcation.* (A) *Bone loss at the furcation and visible lamina dura of an intact bone plate,* (B) *buccal bone height, and* (C) *lingual bone height.*

FIGURE 9-4B Mesiodistal View of Bone Height

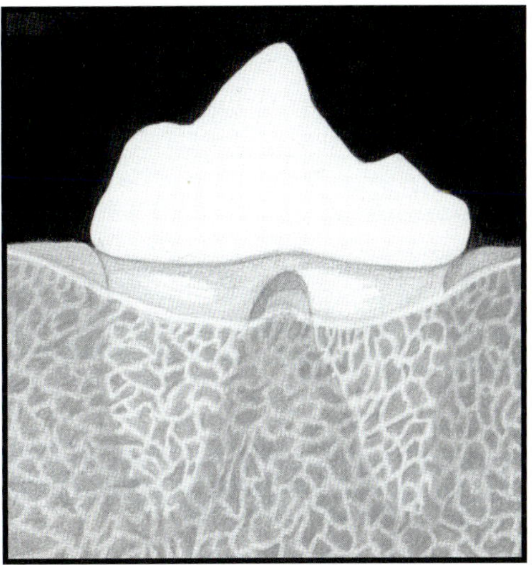

FIGURE 9-4C Lateral View of Decreased Bone Height at the Buccal Aspect

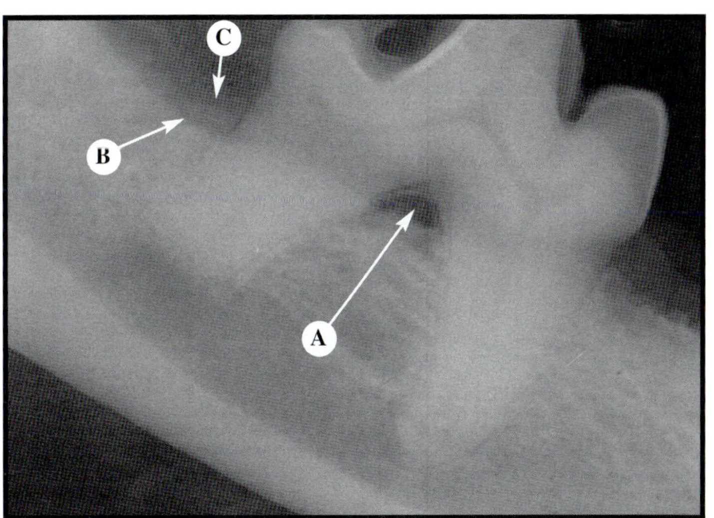

FIGURE 9-4D Bone Loss at the Furcation

necessary. The affected tooth is usually surrounded by a normal periodontal ligament space and lamina dura. Adjacent teeth that are also affected by hypercementosis, however, may produce sufficient cementum to cause eventual concrescence. Hypercementosis is associated with an increase or decrease in occlusal forces, chronic inflammation, resorptive lesions of the root (cats), aging, hyperpituitarism, and dental anomalies[4,7,10,14,16] (Figure 9-6)

EVALUATION OF TREATMENT MEASURES

The criteria for successful treatment of periodontal disease are based on (1) clinical improvement to the gingival attachments, presence of inflammation, degree of mobility, and depth of pockets and (2) radiographic improvement to the quality and height of bone with normalization of the periodontal ligament space. Radiographic demonstration of successful periodontal treatment depends on whether standardized techniques are followed when positioning the primary x-ray beam, the patient, and the film and when exposing and processing the film (see Chapters 1, 2, 3, and 4). Changes to bone structure can then be attributed to a clinical response rather than technical differences in radiographs.

Foreshortening and elongation are two common errors that can affect dimensional measurements (see Chapters 2 and 4). The effects of these errors can be minimized by calculating the radiographic crown-to-root ratio or percent of root length to compare the changes in the height of bone shown in two radiographs. The crown-to-root ratio compares the portion of the

EVALUATION OF THE PERIODONTIUM— THE AGING PROCESS

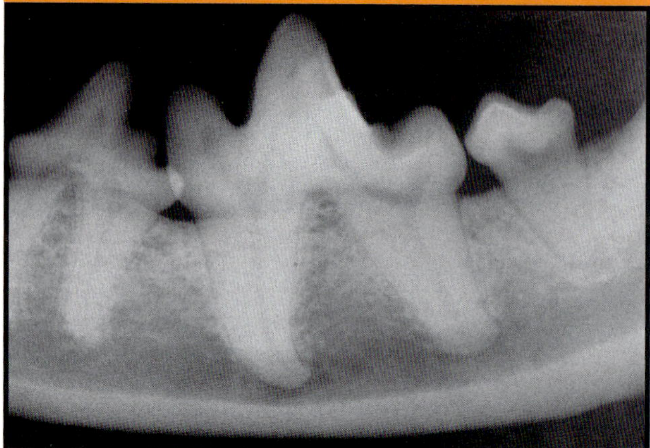

FIGURE 9-5A Normal Aging Periodontium

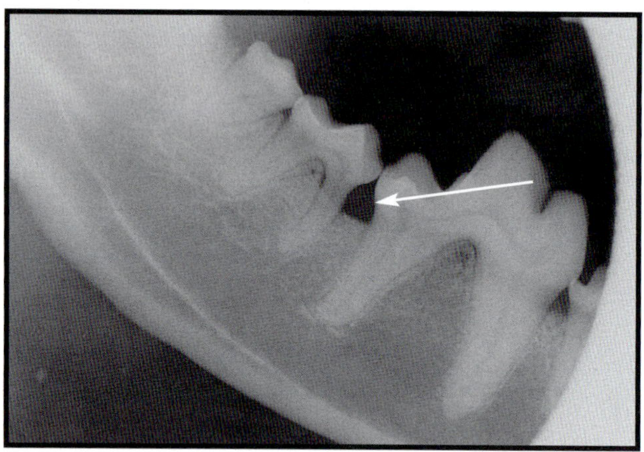

FIGURE 9-5B Periodontal Disease

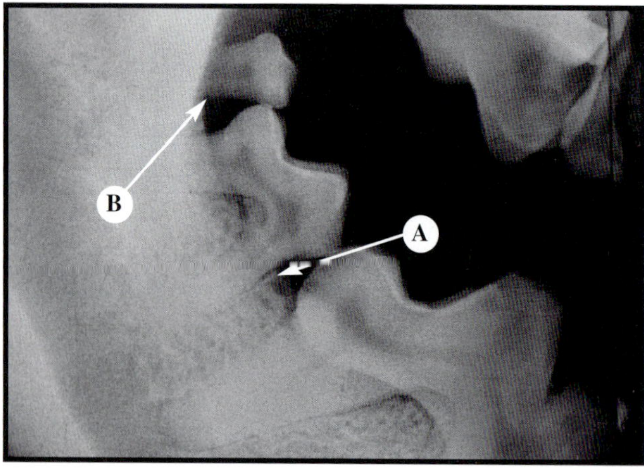

FIGURE 9-5C Changes to the Periodontium

Figure 9-5A *Normal periodontium affected by the aging process shows slight regression and flattening of the alveolar crest.* **Figure 9-5B** *Mild horizontal recession of bone and root exposure* (arrow). **Figure 9-5C** (A) *A jagged alveolar crest and* (B) *supereruption and root exposure are substantial changes that occur with periodontal disease.*

EVALUATION OF THE PERIODONTIUM— HYPERCEMENTOSIS

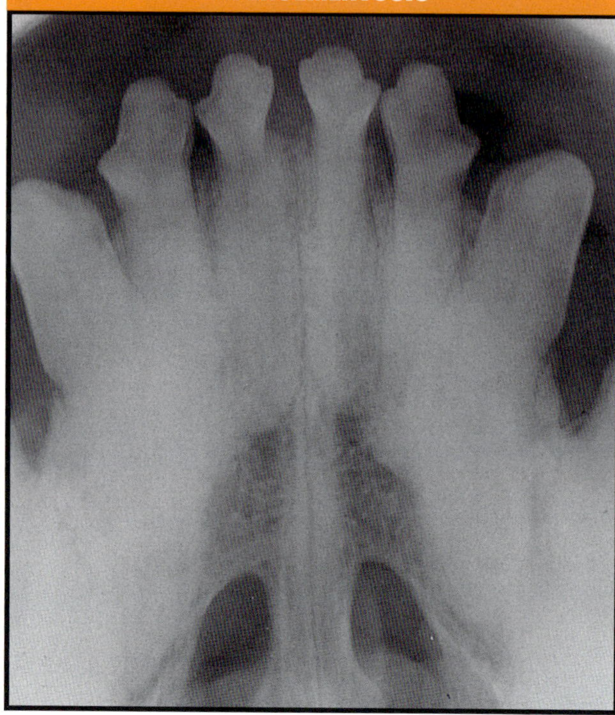

FIGURE 9-6 *This radiograph shows the bulbous and club-shaped roots associated with hypercementosis.*

entire tooth coronal to the crestal bone level with the portion of the tooth apical to the bone level (Figure 9-7). An increasing crown-to-root ratio indicates a worsening prognosis. Bone loss from the root can also be expressed as a percent of root length. The measurement of the distance from the cemento-enamel junction to the crestal bone divided by the distance measured from the cementoenamel junction to the root apex calculates the percent of bone loss in decimals[7,10,11,13,17] (Figure 9-7).

The radiographs presented in Figures 9-9 through 9-14 address some of the basic structural changes to the periodontium that can be detected on radiographs. Various phases of bone loss in both dogs and cats are depicted in Figures 9-15 through 9-38. In addition, various types of pathologic conditions and fractures that can accompany bone loss are shown in Figure 9-39 through 9-60.

REFERENCES

1. Colmery B, Frost P: Periodontal disease etiology and pathogenesis, in Frost P (ed): *The Veterinary Clinics of North America. Small Animal Practice*, vol 16, no. 5. Philadelphia, WB Saunders Co, 1986, pp 817–833.
2. Grove TK: Problems associated with the management of periodontal disease in clinical practice, in Manfra Marrietta S (ed):

(continues on page 123)

EVALUATION OF THE PERIODONTIUM—CALCULATING RATIOS

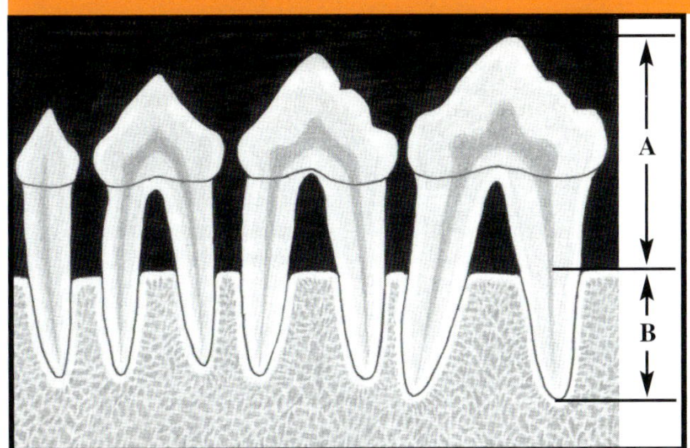

FIGURE 9-7A Crown-to-Root Ratio

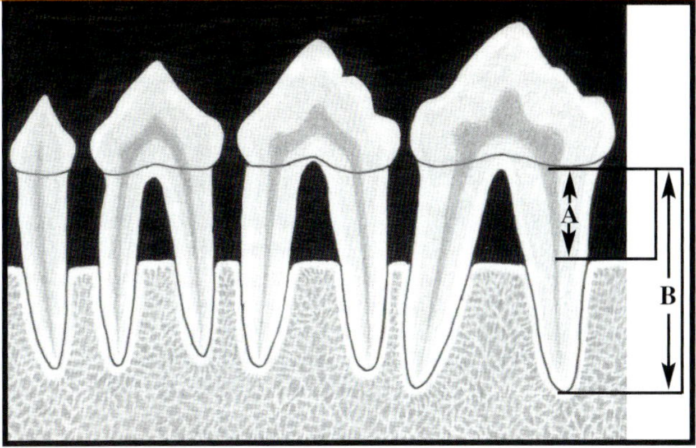

FIGURE 9-7B Percent of Root Length

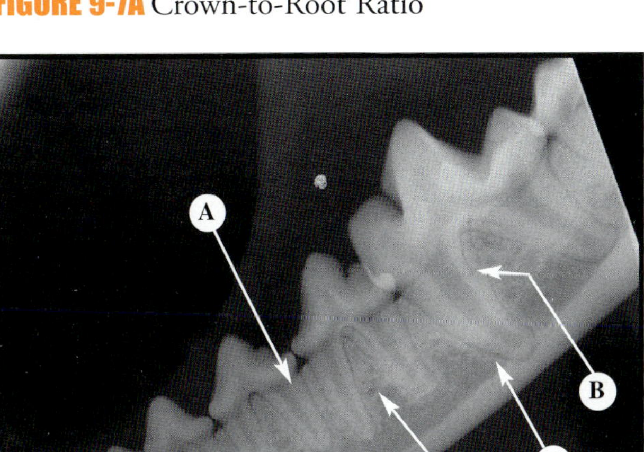

FIGURE 9-8 Normal Periodontium

Figure 9-7A This illustration shows how the crown-to-root ratio measures the length of the structure of the tooth that is (A) coronal to the level of crestal bone (clinical crown) and (B) apical to the level of crestal bone (clinical root). Figure 9-7B This illustration shows how the percent of root length compares (A) the length of the root above the bone line with (B) that of the entire root length. Figure 9-8 This radiograph shows a normal periodontium. (A) Crestal height at the cementoenamel junction, (B) normal width of the periodontal space, (C) normal lamina dura, and (D) trabecular pattern of the bone.

BASIC RADIOGRAPHIC SIGNS OF PERIODONTAL DISEASE

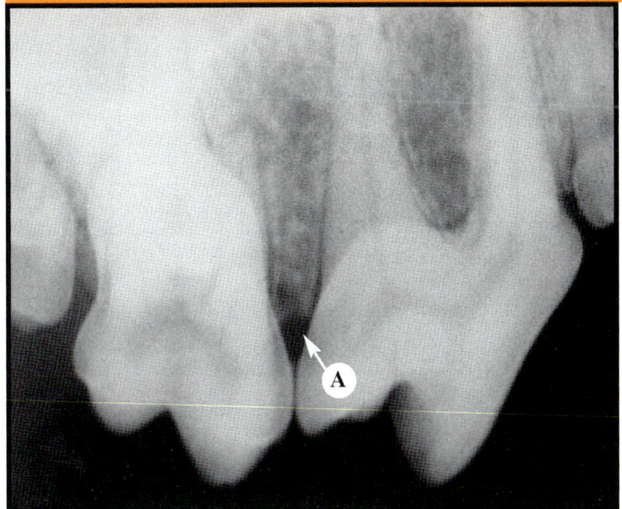

FIGURE 9-9A Jagged Crestal Bone in Dogs

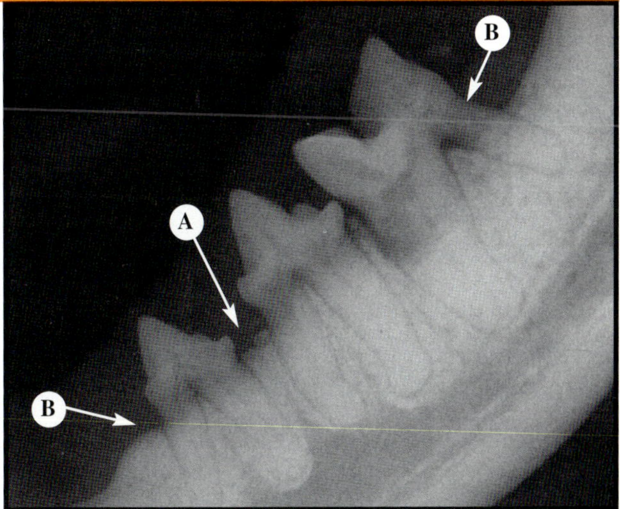

FIGURE 9-9B Jagged Crestal Bone in Cats

Figures 9-9A and 9-9B The (A) jagged crestal bone and loss of integrity of the lamina dura are evident in these radiographs of a dog and cat. In Figure 9-9A, the crestal bone is undergoing recession. Note the (B) horizontal bone loss in Figure 9-9B. The contour of the crestal bone is being remodeled because of changes in bone affected by periodontal disease.

BASIC RADIOGRAPHIC SIGNS OF PERIODONTAL DISEASE (continued)

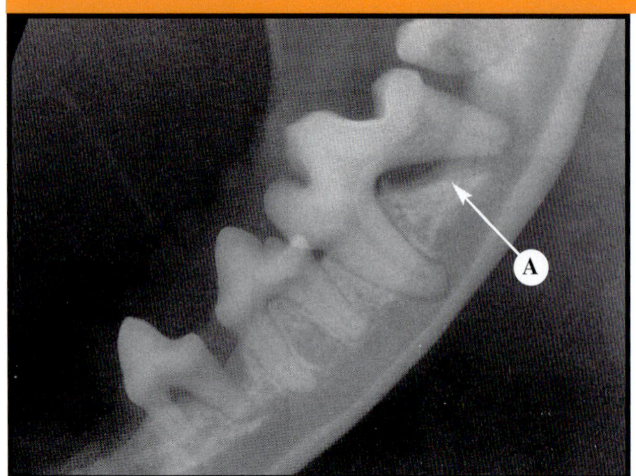

FIGURE 9-10 Widened Periodontal Space

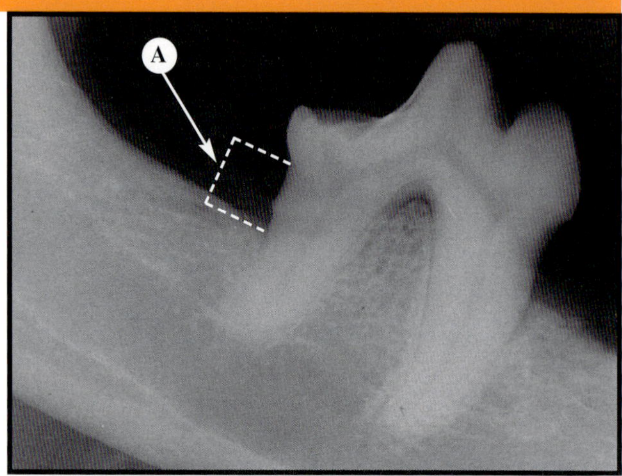

FIGURE 9-11 Bone Loss

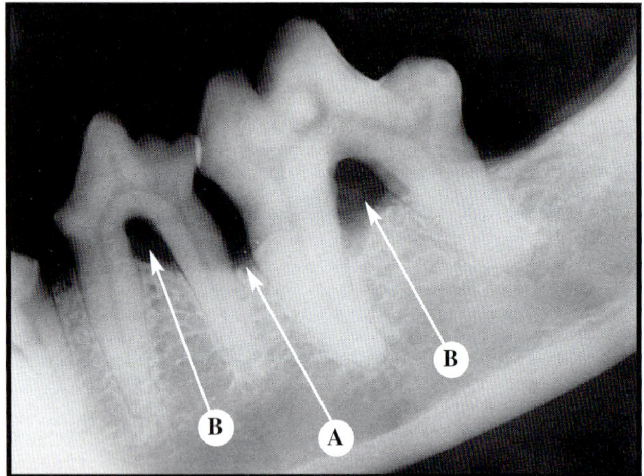

FIGURE 9-12 Recession at the Furcation

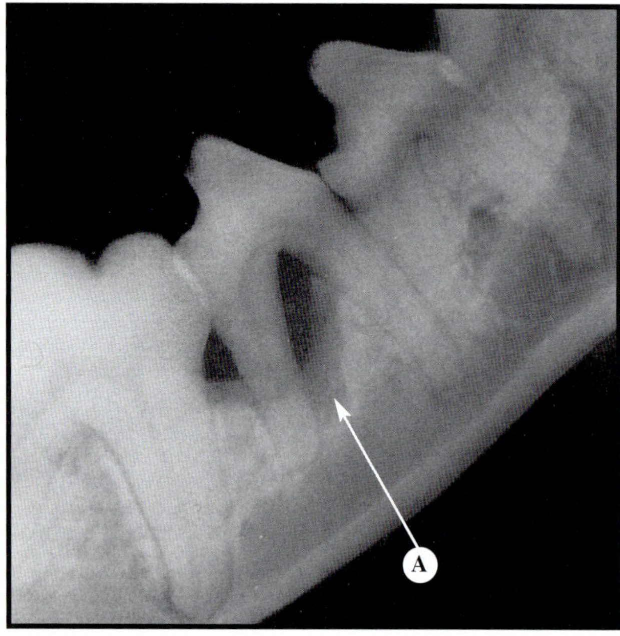

FIGURE 9-13 Vertical Recession (Angular Defects)

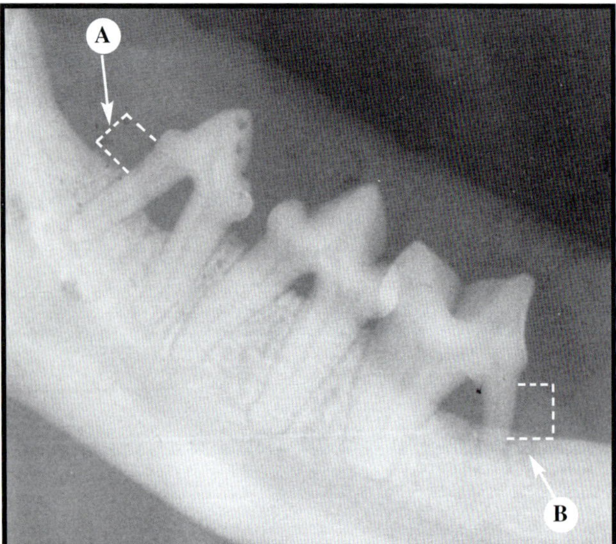

FIGURE 9-14 Horizontal Recession

Figure 9-10 *Inflammation in the periodontal space can cause resorption of the adjoining alveolar wall. Such resorption causes* (A) *widening of the space and loss of integrity in the lamina dura. When the apex is involved, the lesion would be a combination periodontal–endodontic lesion.* **Figure 9-11** *Bone loss occurs in the area from the cementoenamel line to the level of crestal bone.* (A) *Note the bone recession.* **Figure 9-12** *When one tooth is affected by recession, adjacent teeth often are affected too.* (A) *Interproximal recession and* (B) *recession at the furcation.* **Figure 9-13** *Vertical recession is apical loss of bone support down one root or one tooth with little or no effect on adjoining structures.* (A) *Vertical bone loss.* **Figure 9-14** *Severe horizontal loss of bone height and exposure at the furcation.* (A) *50% loss of bone and* (B) *horizontal loss of bone and a crater defect.*

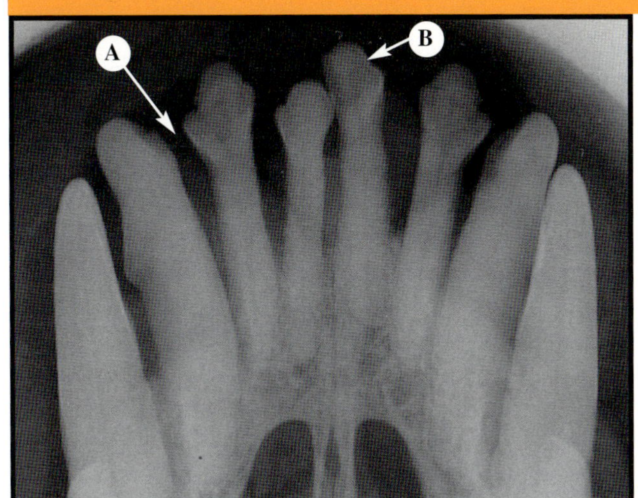

FIGURE 9-15 Advanced Bone Recession of the Maxillary Incisors

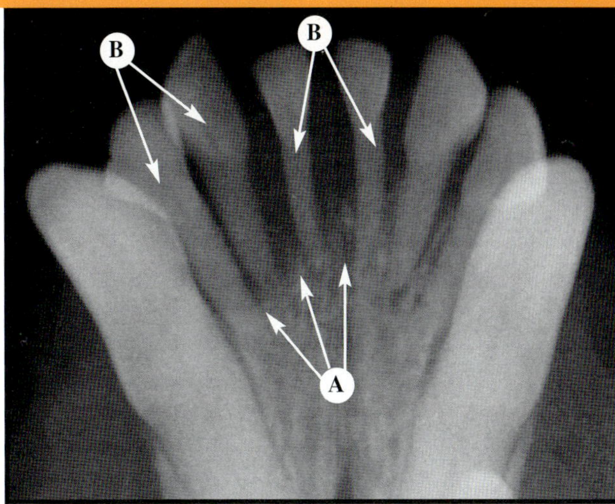

FIGURE 9-16 Advanced Bone Recession of the Mandibular Incisors

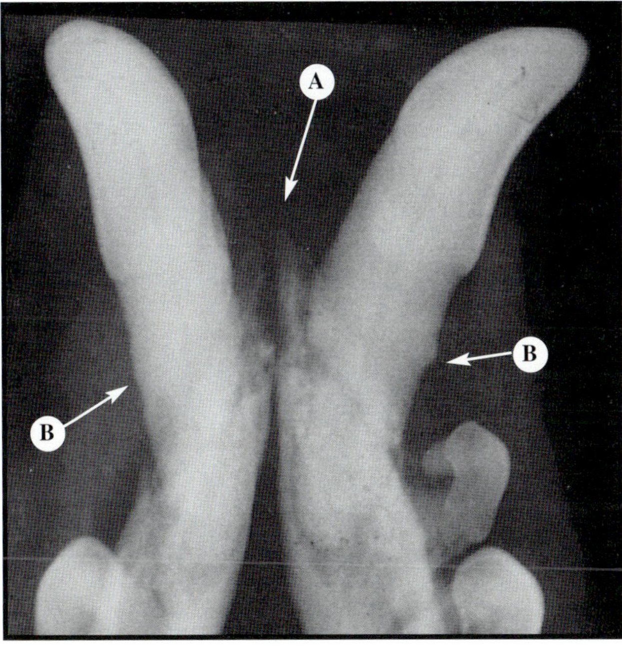

FIGURE 9-17 Advanced Bone Loss of the Mandible

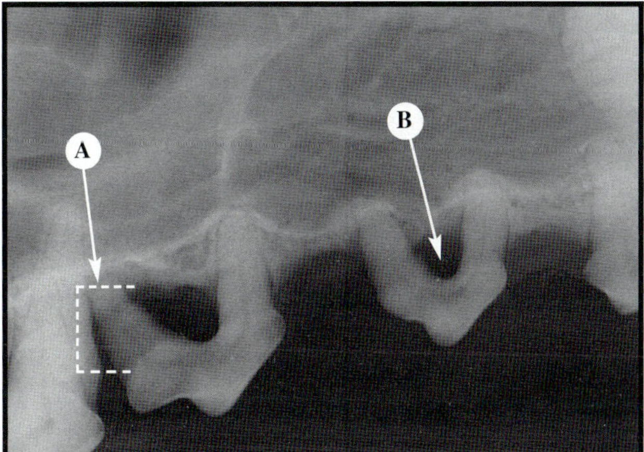

FIGURE 9-18 Advanced Bone Recession of the Maxillary Premolars

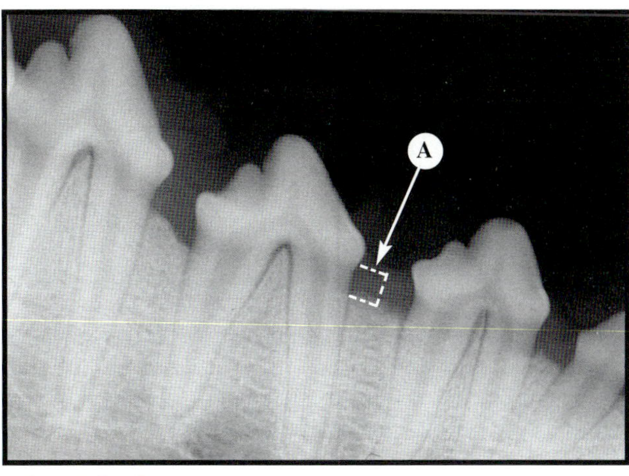

FIGURE 9-19 Moderate Bone Recession of the Mandibular Premolars

Figure 9-15 *Supereruption of the incisor further compromises root support.* (A) *Bone loss and* (B) *supereruption of the incisors.* **Figure 9-16** *Almost all bony support is lost in this radiograph of the mandibular incisors.* (A) *Level of crestal bone and* (B) *exposed root.* **Figure 9-17** *An example of advanced mandibular bone loss. This dog would be at risk for mandibular fracture and tongue prolapse if the canine teeth were extracted.* (A) *Loss of mandibular bone and incisors and* (B) *loss of bone around the canine teeth.* **Figure 9-18** *This radiograph shows a common pattern of bone loss in dogs. Exposure at the furcations allows increased accumulation of plaque.* (A) *Nearly complete bone loss and* (B) *bone loss at the furcation.* **Figure 9-19** *The furcation is at risk for being exposed if the bone loss progresses.* (A) *Crestal bone loss.*

HORIZONTAL BONE RECESSION—DOGS *(continued)*

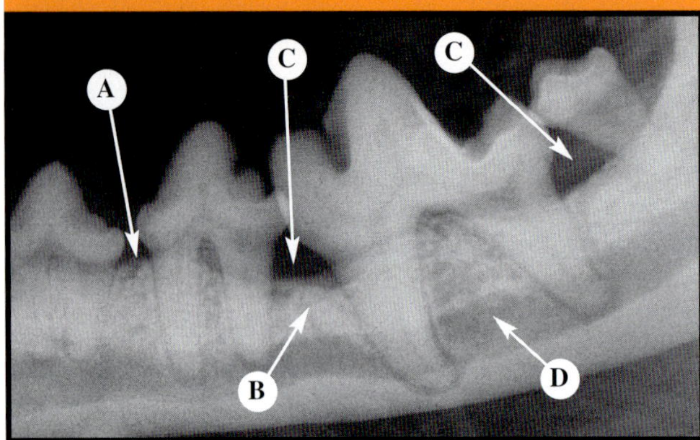

Figure 9-20 *Pockets between the teeth tend to allow greater accumulation of plaque than occurs in other regions of the mouth. (A) Jagged crestal bone, (B) wedge-shaped radiolucent defect, (C) bone recession, and (D) mandibular canal. Note the decreased radiopacity at the furcations.*

FIGURE 9-20 Advanced Bone Recession of the Mandibular Premolars and Molars

HORIZONTAL BONE RECESSION—CATS

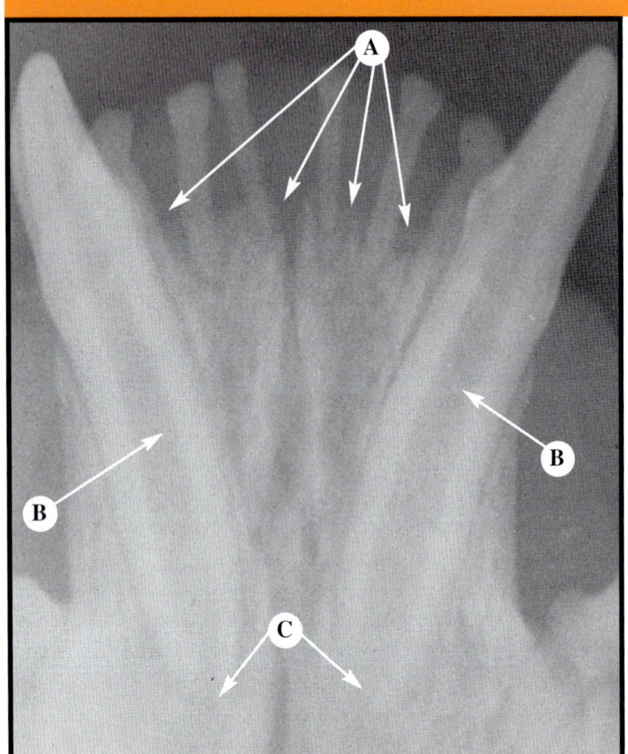

FIGURE 9-21 Advanced Bone Recession of the Mandibular Incisors

Figure 9-21 *Although the cat shown is this radiograph is young, the bone recession is advanced because of juvenile-onset gingivitis and periodontitis. (A) Loss of bone height, (B) wide root canal, and (C) incomplete formation of the apex.* **Figure 9-22** *Advanced bone recession. Hypercementosis may complicate extraction. (A) Crestal bone loss and (B) club-shaped roots.* **Figure 9-23** *Radiograph of a mature cat with advanced periodontal recession. Note that the incisors have little remaining bony support. (A) Loss of bone height and (B) narrow root canal and closed apex.*

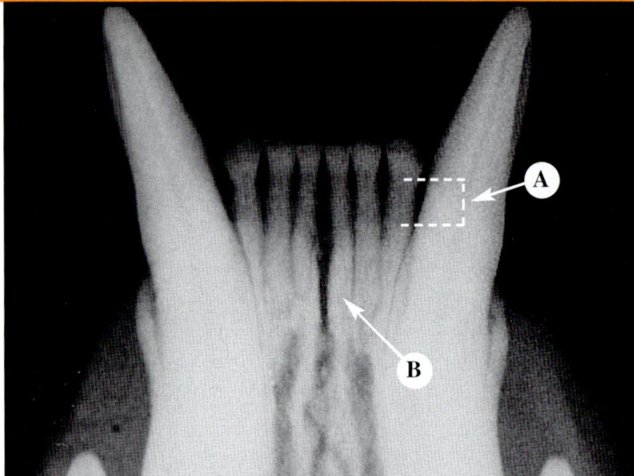

FIGURE 9-22 Advanced Bone Recession of the Mandibular Incisors

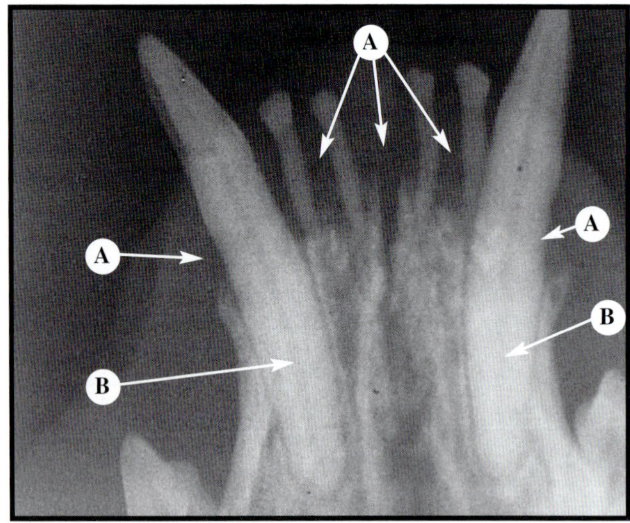

FIGURE 9-23 Advanced Bone Recession of the Mandibular Incisors

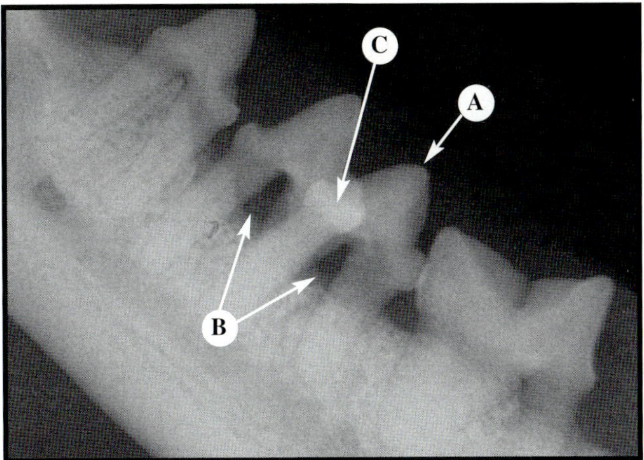

FIGURE 9-24 Advanced Bone Recession of the Mandibular Premolars

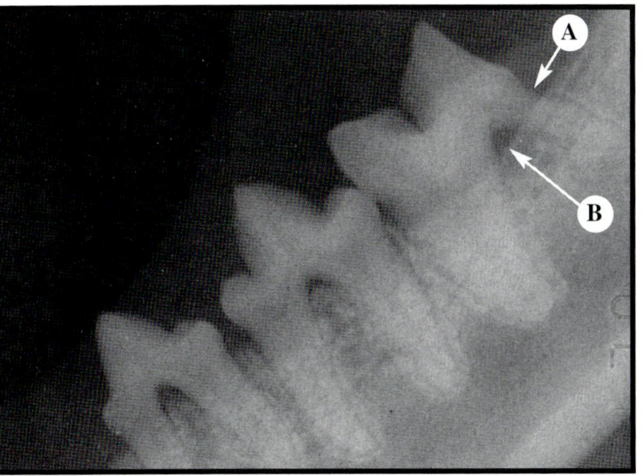

FIGURE 9-25 Mild Bone Recession of the Mandibular Premolars and Molars

Figure 9-24 *Advanced bone loss. The crowded dentition allows increased accumulation of plaque. (A) Supernumerary premolar, (B) horizontal bone loss and exposure at the furcation, and (C) superimposition of premolars (summation image).* **Figure 9-25** *Mild to moderate horizontal recession of bone. (A) Loss of crestal bone and (B) bone loss at the furcation.* **Figure 9-26** *Advanced generalized periodontal disease. Note the bone loss at the (A) furcation, (B) roots of the premolar, and (C) single bone plate over the molar.*

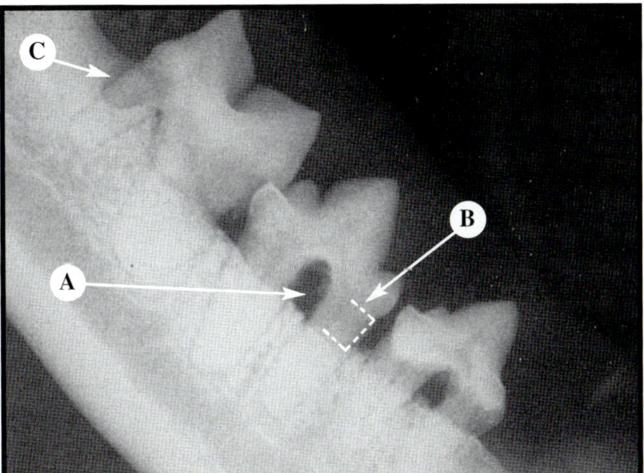

FIGURE 9-26 Advanced Bone Recession of the Mandibular Premolars and Molars

ANGULAR BONE RECESSION—DOGS

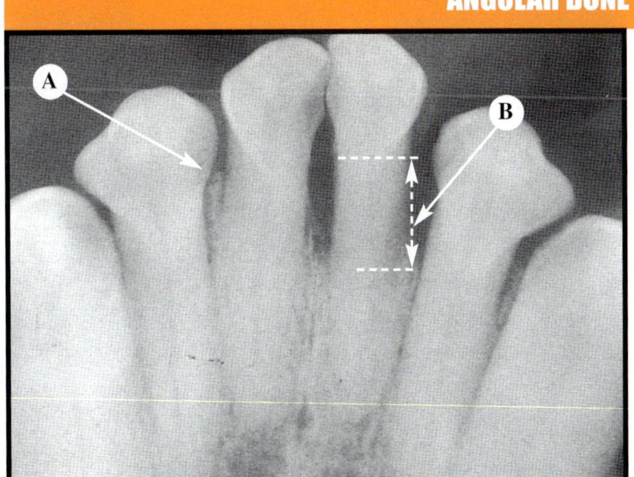

FIGURE 9-27 Advanced Bone Recession of the Incisor

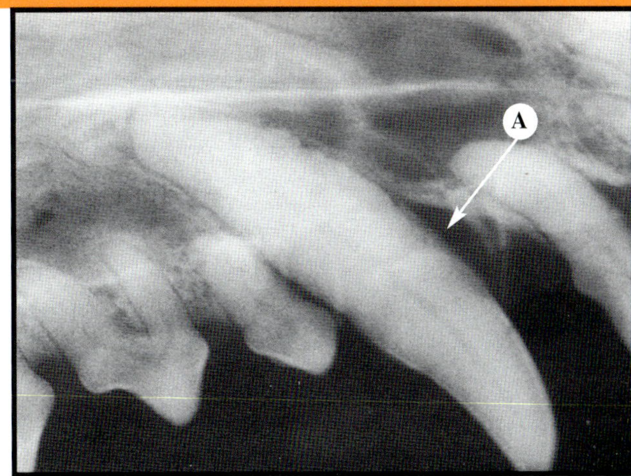

FIGURE 9-28 Moderate Bone Recession of the Canine Tooth

Figure 9-27 *One half of the root attachment is lost. Note the (A) normal height of crestal bone and (B) bone loss.* **Figure 9-28** *The bone recession in this canine tooth increases the risk for developing an oronasal fistula. (A) Bone loss producing a cleft in the bone.*

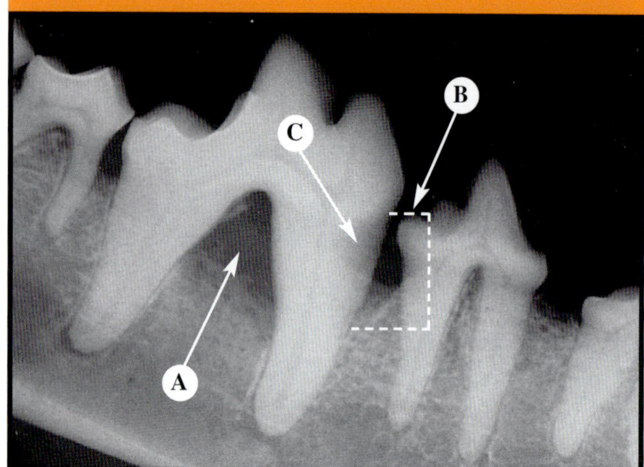

FIGURE 9-29 Advanced Bone Recession of the Mandibular Molar

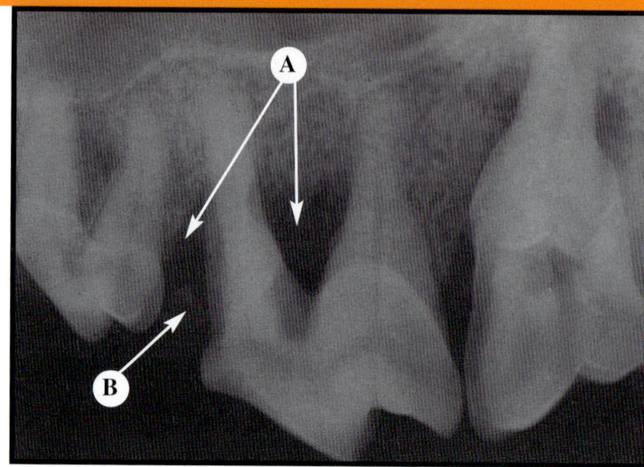

FIGURE 9-30 Vertical Bone Loss of the Maxillary Premolar

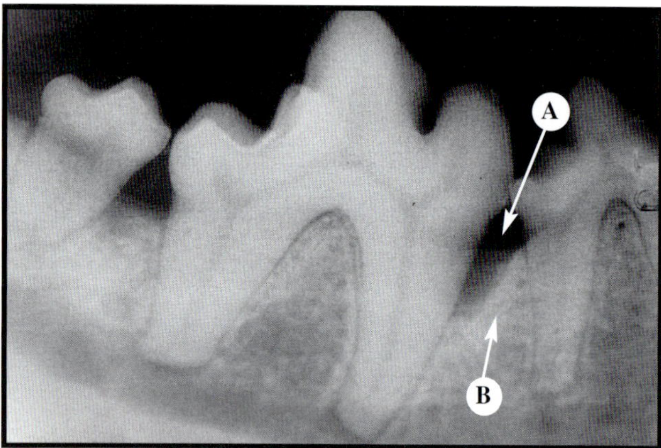

FIGURE 9-31 Bone Loss and Sclerosis

Figure 9-29 *Angular defects are often hidden clinically by gingiva and therefore are undetected unless radiographed or discovered by probing. (A) Bone loss at the furcation, (B) vertical bone loss, and (C) radiolucent defect or cervical burnout (see Chapter 11).* **Figure 9-30** *Vertical bone loss is often hidden by gingival tissue, which can be inflamed and become proliferative. Proliferative gingiva should always be biopsied. (A) Vertical bone loss involving the mesial roots and furcation and (B) region of an epulis (see Chapter 13).* **Figure 9-31** *The bone loss shown in this radiograph is possibly traumatic in origin. Sclerosis may be associated with low-grade inflammation or may indicate that the bone attempted to heal, thereby limiting disease progression. (A) Wedge-shaped radiolucent defect and (B) sclerotic bone margins.* **Figure 9-32** *This defect appears to be close to the apex. (A) Vertical bone defect of the incisor and canine tooth.*

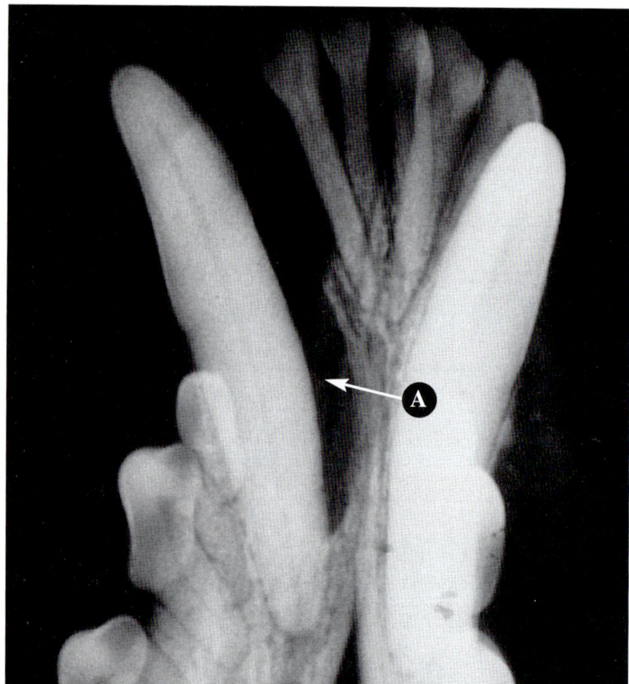

FIGURE 9-32 Advanced Vertical Bone Loss of the Mandibular Canine Tooth and Incisor

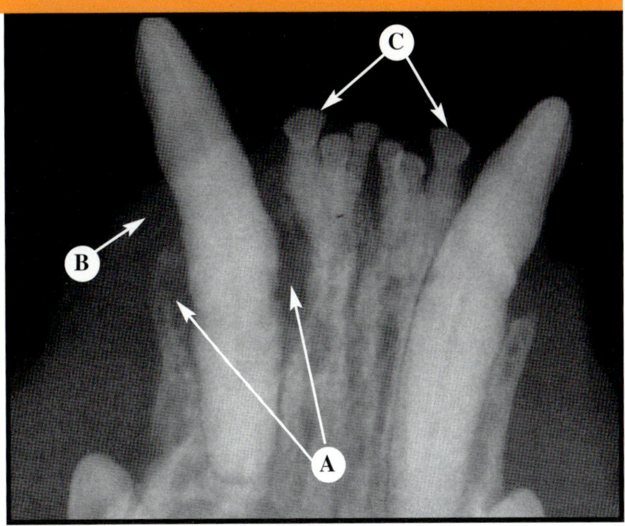

FIGURE 9-33 Vertical Bone Recession of the Maxillary Canine Tooth

FIGURE 9-34 Vertical Bone Defect of the Mandibular Canine Tooth

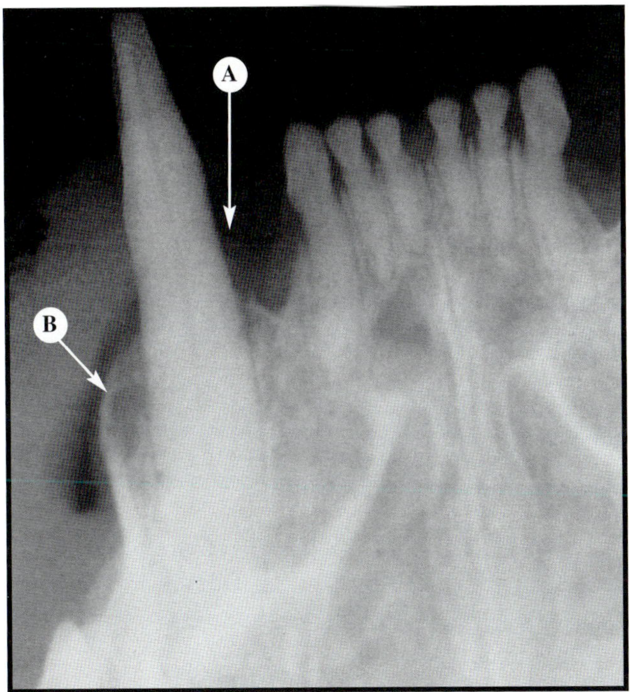

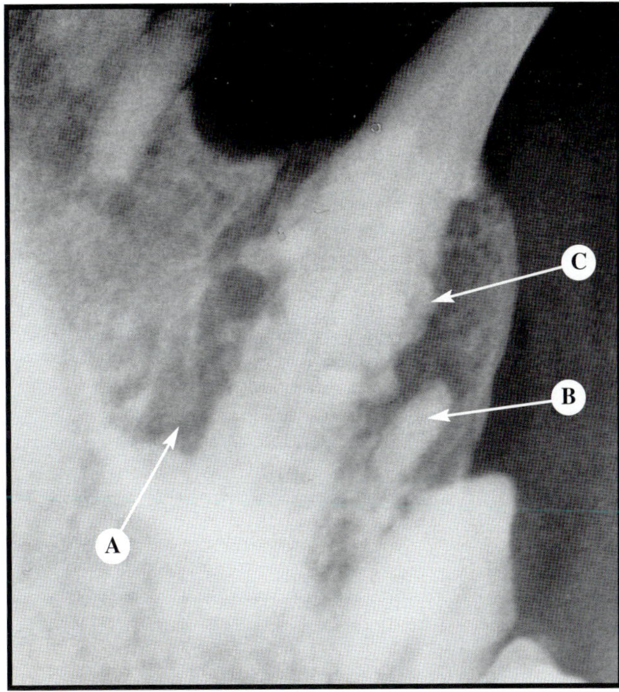

FIGURE 9-35 Bone Recession of the Maxillary Canine Tooth

FIGURE 9-36 Advanced Bone Recession of the Maxillary Canine Tooth

Figure 9-33 *The bone recession shown here is a common pattern in cats. The bone overlying the defect becomes very prominent. (A) Normal height of crestal bone and (B) infrabony pocket.* **Figure 9-34** *Plaque accumulation is often more extensive in vertical bone defects than in horizontal bone defects because the plaque is hidden from view and not diagnosed for treatment. Note the hypercementosis of the incisors. (A) Vertical bone loss, (B) horizontal bone loss, and (C) incisors with club-shaped roots.* **Figure 9-35** *The circumferential loss of bone shown in this radiograph is typical around the canine teeth of cats. (A) This arrow depicts what would be the normal height of the crestal bone (cementoenamel junction) and (B) vertical bone loss creates an infrabony pocket around the root.* **Figure 9-36** *The cuplike pattern to the bone loss is typical around the canine teeth of cats. The prognosis would be grave with the amount of loss shown here. In addition, the cat would be at risk for developing an oronasal fistula. (A) 75% to 90% loss of bone, (B) second premolar, and (C) calculus.*

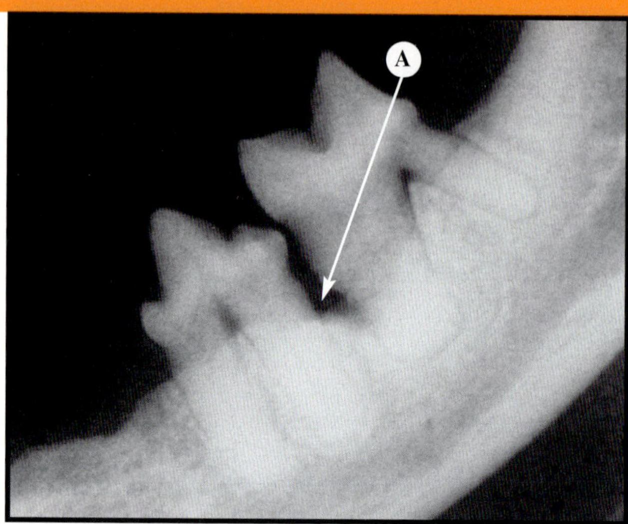

ANGULAR BONE RECESSION—CATS *(continued)*

FIGURE 9-37 Moderate Bone Recession of the Mandibular Premolar

FIGURE 9-38 Vertical Bone Recession of the Mandibular Molar

Figure 9-37 *Bone loss that initially affects only one tooth may be angular and eventually affect adjacent teeth, thereby creating horizontal loss. (A) Loss of bone height, (B) bone loss at the furcation, and (C) normal height of crestal bone.* **Figure 9-38** *Vertical bone recession. The distal root of the premolar is also affected by the same bony defect. (A) Angular bony defect of the mesial root of the molar.*

COMBINED PATTERNS OF BONE LOSS— MULTIPLE BONE RECESSION

FIGURE 9-39 *Various patterns of bone loss of the mandibular premolars and molars. These are often found in the same region. (A) Vertical bone loss, (B) angular bone loss and bone loss at the furcation, and (C) horizontal bone loss.*

COMBINED PATTERNS OF BONE LOSS— BONE LOSS AND CRATER DEFECTS

FIGURE 9-40 *Multiple patterns of bone loss of the mandibular premolars and molar. The crater defects distal to the mandibular molar of cats are often covered by soft tissue and are missed because their location is less accessible. (A) Vertical bone loss of the distal molar root, (B) crater defect, (C) bone loss at the furcation, and (D) mandibular canal.*

COMBINED PATHOLOGIC CONDITIONS—PERIODONTAL–ENDODONTIC LESIONS

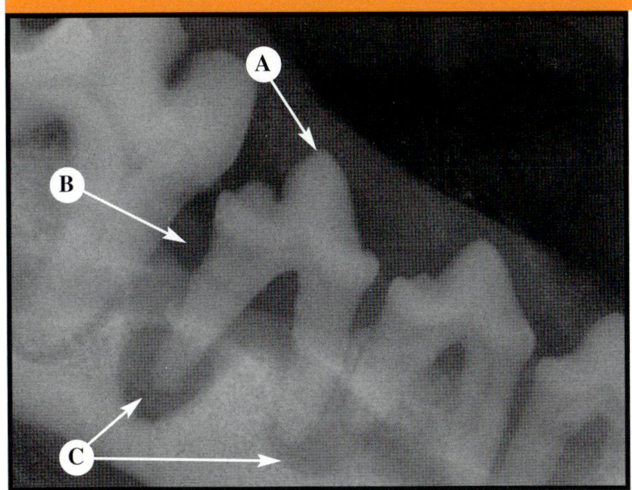

FIGURE 9-41

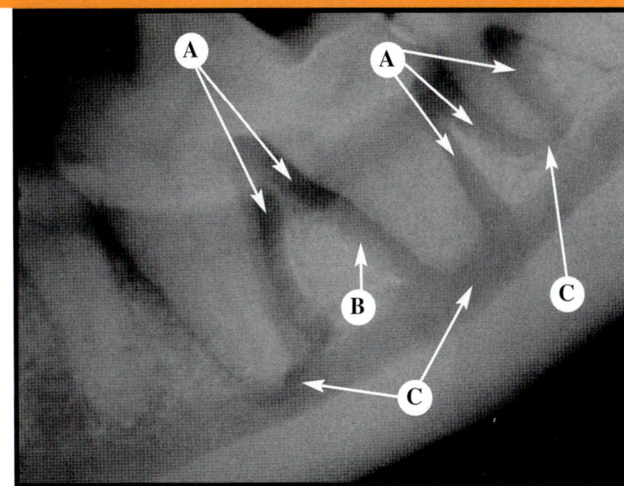

FIGURE 9-42

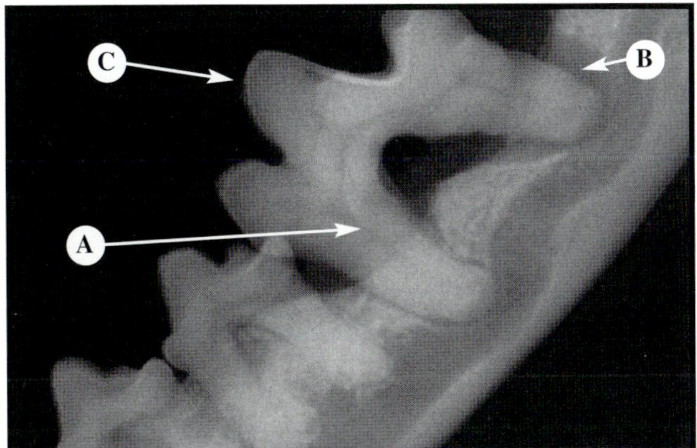

FIGURE 9-43

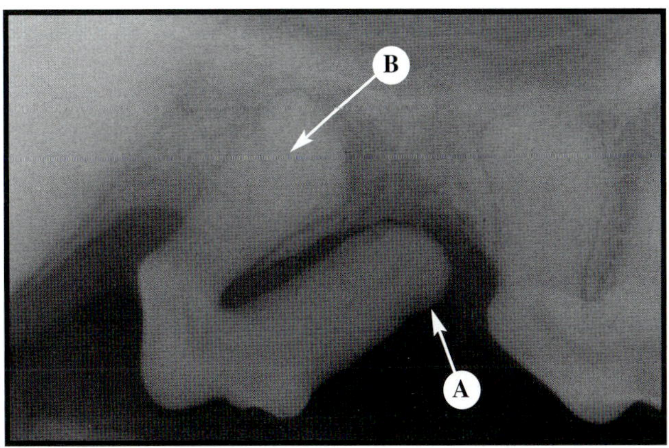

FIGURE 9-44

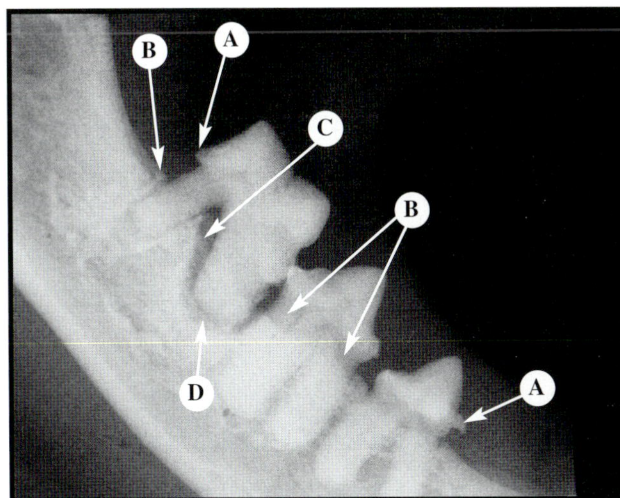

FIGURE 9-45

Figure 9-41 *The endodontic involvement shown in this radiograph is an extension of periodontal disease. (A) No apparent damage to the crown, (B) bone loss, and (C) two areas of periapical radiolucency.* **Figure 9-42** *The periodontal–endodontic lesion shown in this radiograph was caused by advanced periodontitis and multiple vertical defects of the bone. (A) Wide periodontal space, (B) loss of integrity in the lamina dura, and (C) periapical radiolucency.* **Figure 9-43** *This periodontal–endodontic lesion is an extension of vertical bone loss at the distal root. (A) A moderate amount of vertical bone loss at the mesial root, (B) loss of integrity in the lamina dura and wide periodontal space involving the apex of the distal root, and (C) intact crown.* **Figure 9-44** *In this radiograph, the mesial root is the only support for the tooth (with possible compensatory hypercementosis). The lesion is the result of vertical bone loss at the distal root. (A) Complete loss of bone at the distal root and (B) a club-shaped root.* **Figure 9-45** *This lesion is caused by vertical bone loss at the mesial root, although horizontal bone loss also is present to some degree. (A) Calculus overhangs, (B) horizontal bone loss, (C) vertical bone loss, and (D) wide periapical periodontal space.*

COMBINED PATHOLOGIC CONDITIONS—PERIODONTAL–ENDODONTIC LESIONS *(continued)*

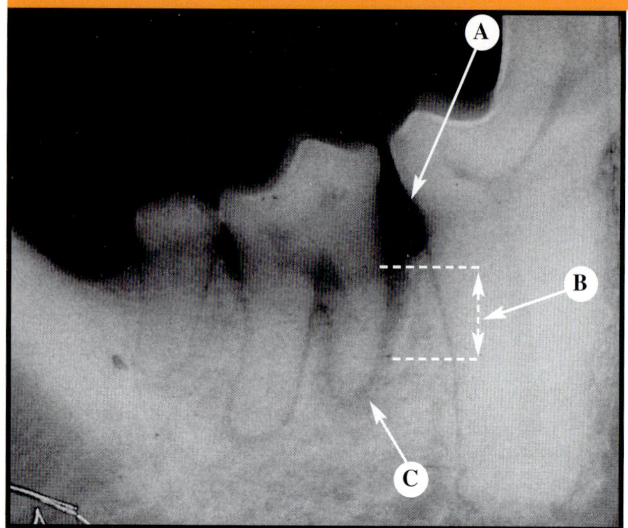

FIGURE 9-46

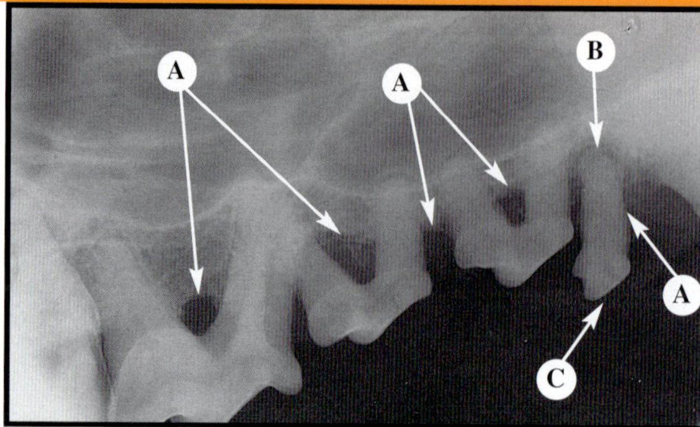

FIGURE 9-47

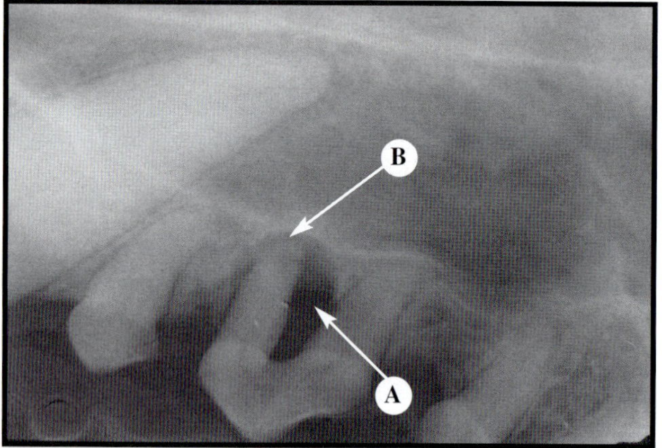

FIGURE 9-48

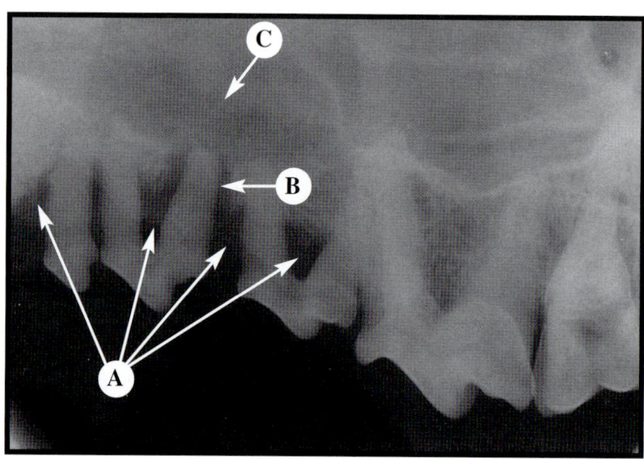

FIGURE 9-49

Figure 9-46 *Vertical bone loss and apical periodontitis of the mesial root of the mandibular molar. (A) Former normal height of crestal bone, (B) decreased height of crestal bone and wide periodontal space, and (C) wide periapical periodontal space.* **Figure 9-47** *A periodontal–endodontic lesion and supereruption of the first premolar secondary to horizontal bone loss. (A) Horizontal bone loss, (B) wide periodontal space, and (C) supereruption.* **Figure 9-48** *This periodontal–endodontic lesion of the premolar was caused by bone loss at the furcation. (A) Vertical bone loss at the furcation and (B) wide periodontal space involving the apex.* **Figure 9-49** *This lesion is secondary to periodontal bone loss. The animal is possibly at risk for developing an oronasal fistula. (A) Horizontal bone loss, (B) vertical bone loss, and (C) wide periodontal space and loss of integrity in the radiopaque line that corresponds to the confluence of the maxillary and palatal bones (also referred to as the white line).*

COMBINED PATHOLOGIC CONDITIONS—ORONASAL FISTULAS

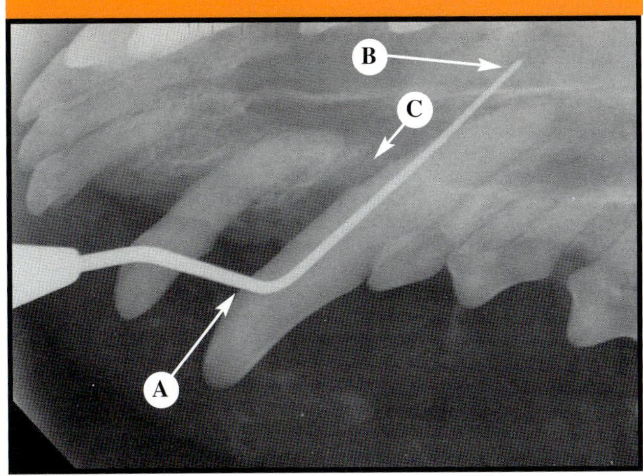

FIGURE 9-50

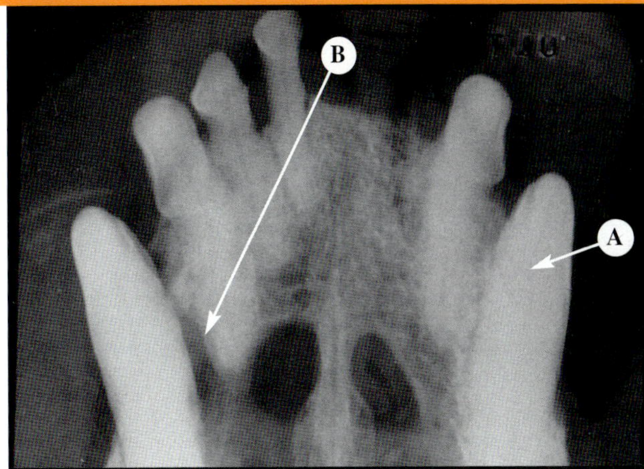

FIGURE 9-51

Figure 9-50 *An oronasal fistula secondary to periodontal pocketing and bone loss. Note the* (A) *position of the periodontal probe,* (B) *the probe penetrating beyond the tooth apex, and* (C) *the loss of the white line as a landmark.* **Figure 9-51** *This oronasal fistula is secondary to periodontal pocketing and vertical bone loss. Note the* (A) *normal side and* (B) *wide periodontal space and loss of integrity in the lamina dura.*

COMBINED PATHOLOGIC CONDITIONS—BONE LOSS AND ROOT RESORPTION

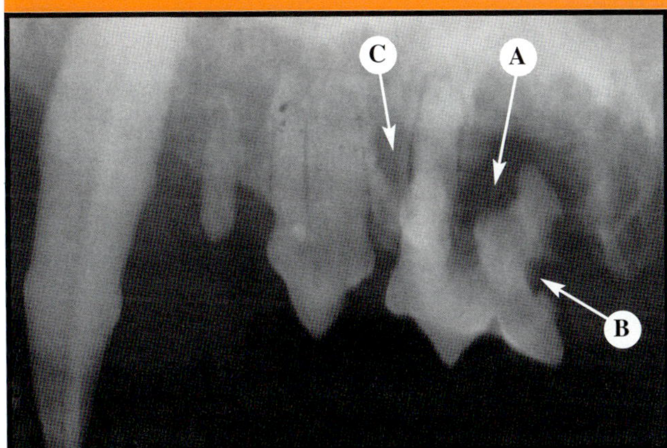

FIGURE 9-52

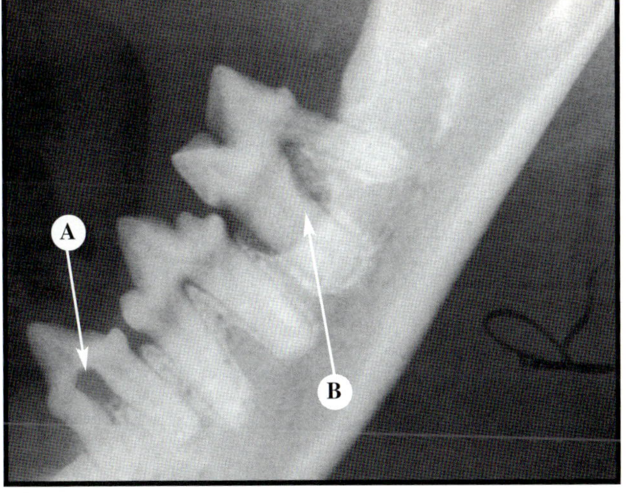

FIGURE 9-53

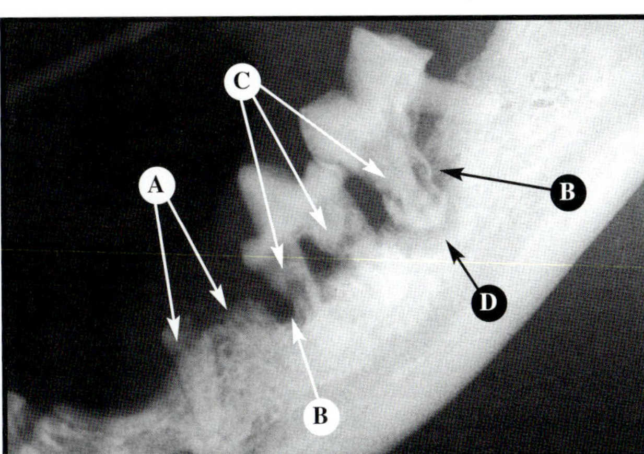

FIGURE 9-54

Figure 9-52 *This periodontal–endodontic lesion with root resorption is secondary to periodontitis and bone loss.* (A) *Wide periodontal space and bone loss,* (B) *root defects, and* (C) *angular bone loss and root defect.* **Figure 9-53** *The root resorption shown in this radiograph is associated with periodontal disease.* (A) *Bone loss at the furcation and a crestal defect in the tooth and* (B) *angular bone loss and root defect.* **Figure 9-54** *A periodontal–endodontic lesion of the molar with root resorption and periodontal bone loss and also root resorption of the premolar. Retained root fragments open to the oral cavity serve as a nidus of inflammation.* (A) *Retained root fragments,* (B) *angular bone loss,* (C) *irregular root contour and radiolucent defects, and* (D) *wide periodontal space.*

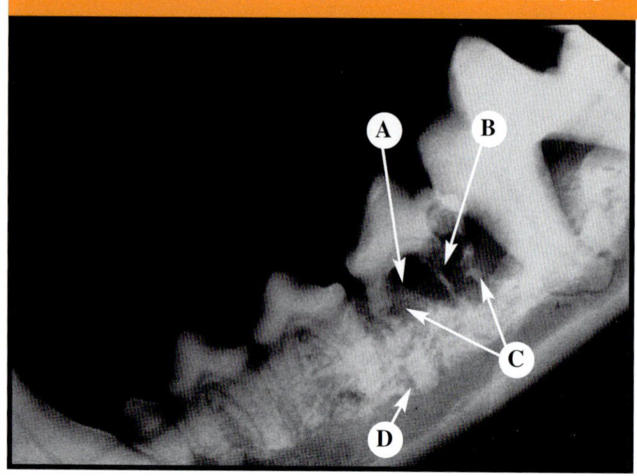

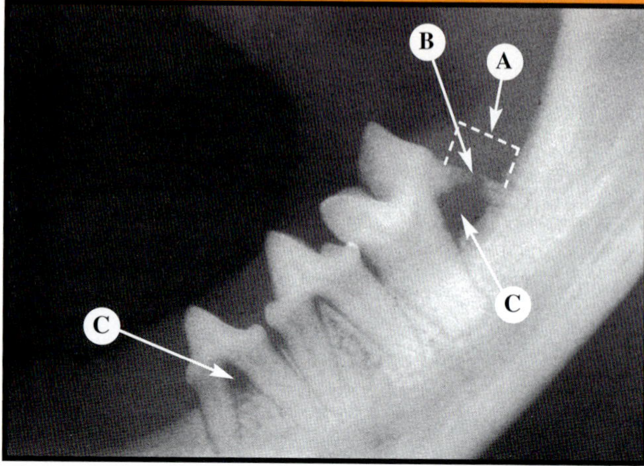

FIGURE 9-55 Angular Bone Loss

FIGURE 9-56 Endodontic Involvement

Figure 9-55 *Root resorption associated with angular bone loss. The apices are likely to fracture during extraction and may complicate healing if retained. (A) Angular bone loss, (B) radiolucent defect, (C) irregular root contour, and (D) apex.* **Figure 9-56** *Various pathologic conditions are present in this radiograph, including a periodontal–endodontic lesion. Because of the severity of the defect and the fracture line of the exposed distal root, endodontic involvement is assumed, even with the absence of a periapical lesion. The fracture line is a lateral opening to the pulp. (A) Bone loss at the distal root, (B) root defects and fracture line, and (C) bone loss at the furcation.*

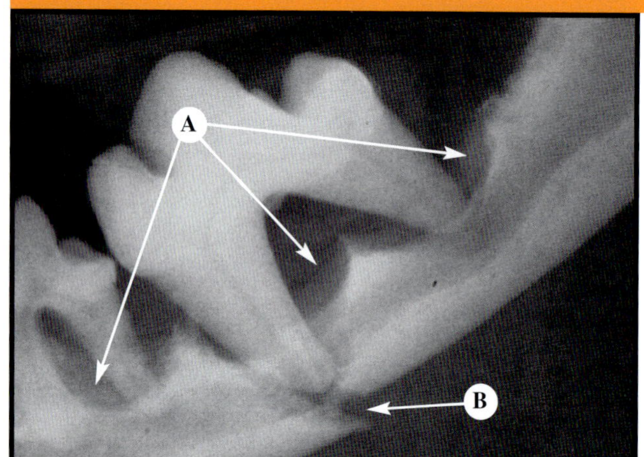

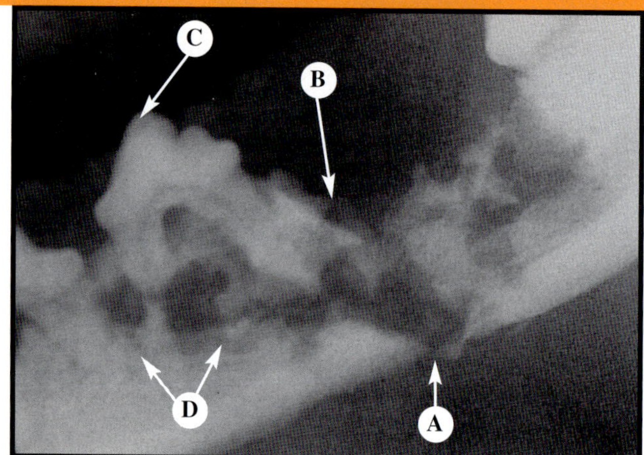

FIGURE 9-57 Mandibular Fracture Secondary to Periodontal Bone Loss

FIGURE 9-58 Complications of Fracture Healing

Figure 9-57 *The fracture shown in this radiograph is secondary to periodontal bone loss of the mandible. (A) An arc-like pattern to bone loss and (B) the fracture line.* **Figure 9-58** *The healing of this fracture is being complicated by bone loss and tooth fragments with bony sequestra in the fracture line. (A) Fracture line at previous molar site, (B) bone loss, (C) sequestered bone with premolar fragment, and (D) apical root fragments of the premolar.*

COMBINED PATHOLOGIC CONDITIONS—TOOTH EXTRACTION

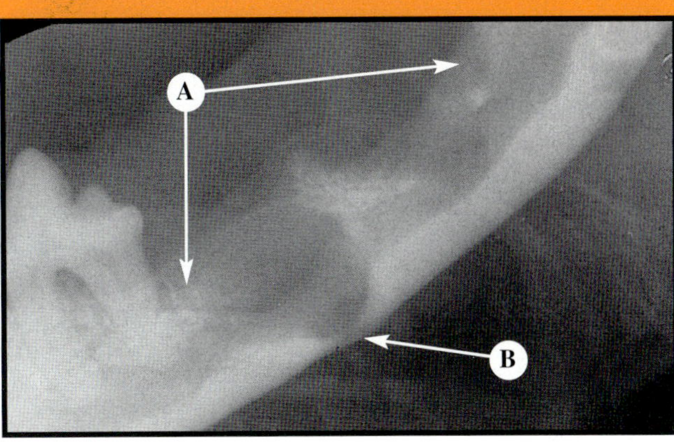

FIGURE 9-59A Before Extraction

FIGURE 9-59B After Extraction

Figure 9-59A *A periodontal–endodontic lesion before extraction. Care must be taken during extraction to avoid causing iatrogenic fractures. (A) Angular bone loss and (B) wide periodontal space and decreased density in the cortex.* **Figure 9-59B** *The tooth was successfully extracted; however, the thin mandible must be protected from activities that might cause pathologic fractures during healing. (A) Bone loss and (B) intact but very thin cortical plate.*

OSSEOUS INDUCTION

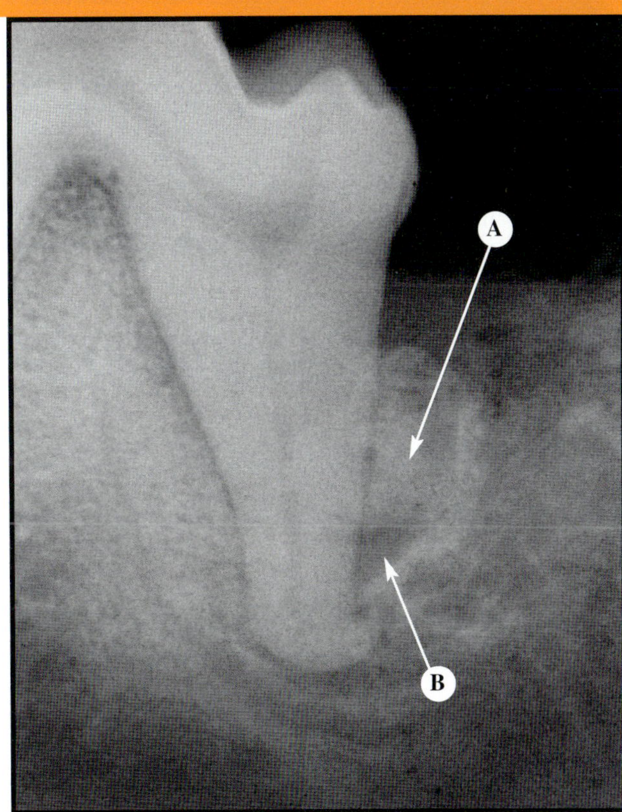

FIGURE 9-60A Before Treatment

FIGURE 9-60B Immediately After Treatment

Figure 9-60A *The bone defect in this radiograph extends almost to the apex, thereby placing the tooth at risk for developing an endodontic lesion. (A) Infrabony pocket and (B) sclerotic margin.* **Figure 9-60B** *This radiograph was taken immediately after treatment had been completed. The induction material may not be thoroughly packed into the defect. (A) Demineralized bone applied into the bone defect and (B) base of the infrabony pocket.*

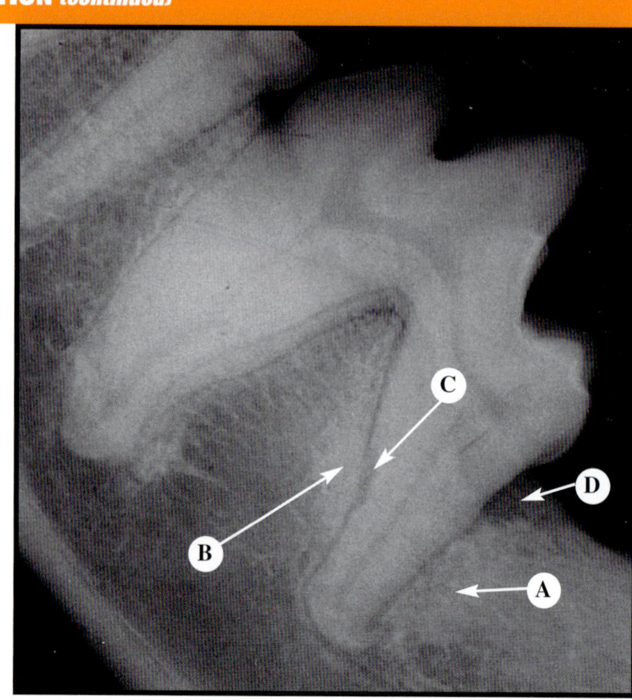

FIGURE 9-60C Three Months After Treatment

FIGURE 9-60D One Year After Treatment

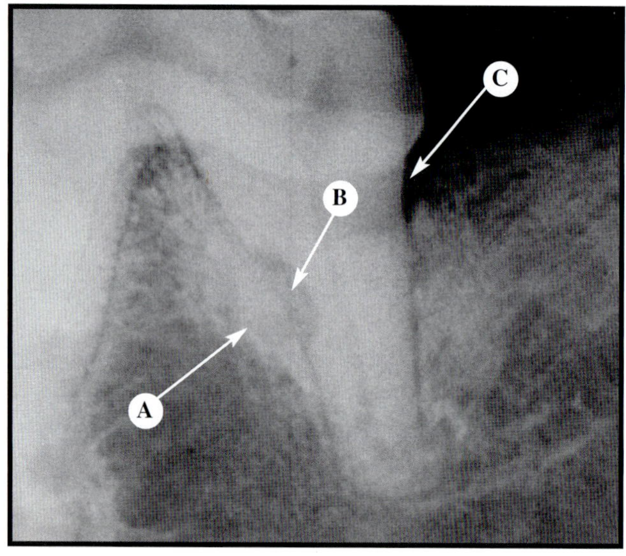

FIGURE 9-60E Two Years After Treatment

FIGURE 9-60F Four Years After Treatment

Figure 9-60C *The radiolucency previously associated with the bony defect does not appear to be worsening, and no rarefaction is apparent around the induction material. (A) Induction material and (B) film artifacts.* **Figure 9-60D** *A periodontal lesion continues to be present at the coronal aspect of the distal root. Because the induction material seems to be replaced by bone, however, the defect has been partially corrected. Root resorption and sclerosis may be secondary to periodontitis or the result of long-standing pulpitis at a lateral canal site. (A) Site of induction material, (B) sclerosis, (C) root defect, and (D) wedge-shaped periodontal radiolucency.* **Figure 9-60E** *Root resorption and sclerosis have progressed, thereby suggesting persistent inflammation. Although smaller, the periodontal lesion persists. (A) sclerosis, (B) root defect, and (C) periodontal bone loss.* **Figure 9-60F** *Four years after the original defect was detected on radiographs, root resorption continues to progress. (A) Bone sclerosis and root defects, (B) vertical bone defect, and (C) radiolucent border around the radiopaque region. The lesion roughly corresponds to the original bone defect and induction material, thereby suggesting that reformation of the defect may be occurring.*

REFERENCES *(continued from page 108)*

Problems in Veterinary Medicine Dentistry, vol 2, no. 1. Philadelphia, JB Lippincott Co, 1990, pp 110–136.

3. Page RC, Schroeder HE: *Periodontitis in Man and Other Animals. A Comparative Review.* New York, Karger, 1982, pp 127–158.

4. Reichart PA, Durr UM, Triadan H, Vekendey G: Periodontal disease in the domestic cat. A histopathologic study. *J Periodon Res* 19:67–95, 1984.

5. Carranza FA: *Glickman's Clinical Periodontology,* ed 7. Philadelphia, WB Saunders Co, 1990, pp 501–523.

6. Smith MM, Zontine WJ, Willits NH: A correlative study of the clinical and radiographic signs of periodontal disease in dogs. *JAVMA* 186(12):1286–1290, 1985.

7. Grove TK: Periodontal disease, in Harvey CE (ed): *Veterinary Dentistry.* Philadelphia, WB Saunders Co, 1985, pp 59–77.

8. Zontine WJ: Dental radiographic technique and interpretation, in Suter PF (ed): *The Veterinary Clinics of North America*, vol 4, no. 4. Philadelphia, WB Saunders Co, 1974, pp 741–761.

9. Harvey CE, Flax BM: Feline oral-dental radiographic examination and interpretation, in Harvey CE (ed): *The Veterinary Clinics of North America*, vol 22, no. 6. Philadelphia, WB Saunders Co, 1992, pp 1279–1295.

10. Goaz PW, White SC: *Oral Radiology: Principles and Interpretation,* ed 3. Philadelphia, CV Mosby Co, 1994, pp 327–339, 376–378, 384–385.

11. Razmus TF: Caries, periodontal disease, and periapical changes, in Miles DA, Van Dis ML (eds): *The Dental Clinics of North America*, vol 38, no. 1. Philadelphia, WB Saunders Co, 1994, pp 13–27.

12. Lyon KF: Subgingival odontoclastic resorptive lesions classification, treatment, and results in 58 cats, in Harvey CE (ed): *The Veterinary Clinics of North America. Small Animal Practice*, vol 22, no. 6. Philadelphia, WB Saunders Co, 1992, pp 1417–1432.

13. Jeffcoat MK, Reddy MS: A comparison of probing and radiographic methods for detection of periodontal disease progression. *Curr Opin Dentistry* 1:45–51, 1991.

14. Morgan JP, Miyabayashi T: Dental radiology: Aging changes in permanent teeth of beagle dogs. *J Small Anim Pract* 21:11–18, 1991.

15. Morgan JP, Miyabayashi T, Anderson J, Kling B: Periodontal bone loss in the aging beagle dog: A radiographic study. *J Clin Periodont* 17:630–635, 1990.

16. Sapp JP, Eversole LR, Wysocki GP: *Contemporary Oral and Maxillofacial Pathology.* Philadelphia, CV Mosby Co, 1997, pp 11–12.

17. Lynch SE: Methods for evaluation of regenerative procedures. *J Periodontol* 63:1085–1092, 1992.

Chapter Ten

INTERPRETATION OF ENDODONTIC DISEASE

Because the endodontic system and periapical tissue are completely hidden from clinical view, radiography is the primary means for evaluating these structures. Radiographs can reveal the nature and severity of endodontic lesions and can help practitioners to assess the morphologic condition of the canal and maturity of the teeth. Radiography is essential when determining the correct treatment measures and when monitoring the status of the pulp or root canal during and after surgery.

Radiographic diagnosis of endodontic lesions must, however, be kept in perspective. In many cases, endodontic lesions do not develop pathologic signs that are visible on radiographs until changes in the anatomy of either the tooth structure itself or periradicular tissue can be detected. Early radiographs of a tooth that has irreversible pulpitis or necrotic pulp can appear normal. Pulpal inflammation can be present before acute clinical signs (e.g., swelling, sinusitis, abscessation, gingival inflammation, cellulitis, or fistulation) are apparent. An astute pet owner may remember a history of minor trauma and notice a fractured crown or discolored tooth. On the other hand, the pet may initially present no overt physical signs except changes in eating or chewing habits, an episode of unexplained drooling, transient depression, or other subtle behavioral changes that might be indicative of an occult endodontic lesion. Eventually, detectable changes in the endodontic pathology will affect the tooth and periradicular tissue but often only after a chronic long-standing lesion has developed. Therefore, a tooth that is endodontically compromised should not be ignored in the absence of radiographic signs. Likewise, a suspicious clinical sign should not be ignored if a tooth grossly appears normal. A definitive diagnosis should always be made on the basis of clinical signs and radiographic findings (Table 10-1).[1-5]

ENDODONTIC ANATOMY

Immature teeth have wide pulp chambers and relatively wide apical foramina that allow toxins and inflammatory mediators to access periapical tissue when pulpitis and infection occur. Young animals form periapical lesions more rapidly than adults do because immature teeth have thin dentinal walls and an open apex; therefore, in immature teeth, the pulp is not cloistered from periapical tissue as well as occurs in mature teeth. Percolation of higher concentrations of toxins, inflammatory mediators, and other irritants more likely occurs through an open apex. The vital pulp of an immature

TABLE 10 – 1 **CLINICAL SIGNS ASSOCIATED WITH ENDODONTIC DISEASE**[1-5]
Physical Findings
Worn, cracked, or fractured teeth *Caries* *Cervical resorption* *Gingivitis, periodontitis, or periodontal pockets* *Discolored tooth* *Swelling* *Increased calculus on the affected side* *Formation of abscess* *Fistulation* *Cellulitis* *Sinusitis; unilateral nasal discharge or conjunctivitis* *Tooth mobility* *Fever* *Pulp polyp* *Deep restorations*
Behavioral Findings
Drooling *Depression* *Difficulty chewing or eating* *Chewing on one side* *Avoids using mouth* *Decreased use of chew toys* *Decreased intake of hard food* *Less forceful bite* *Sneezing* *Withdraws on palpation or percussion of tooth* *Excessive licking motions* *Drops food* *Rubbing or pawing at face* *Wincing or jerking head while eating*

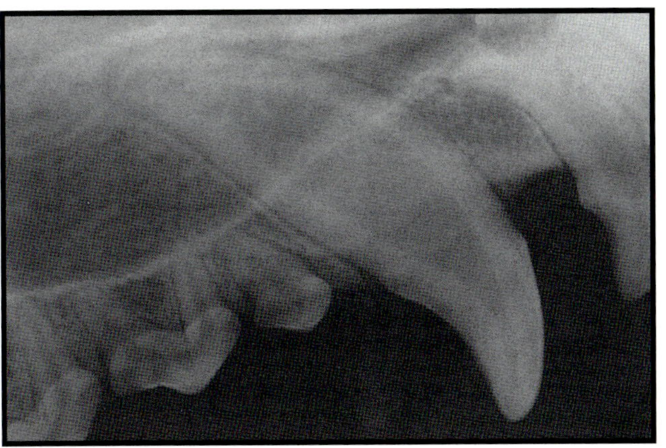

FIGURE 10-1A Lateral View

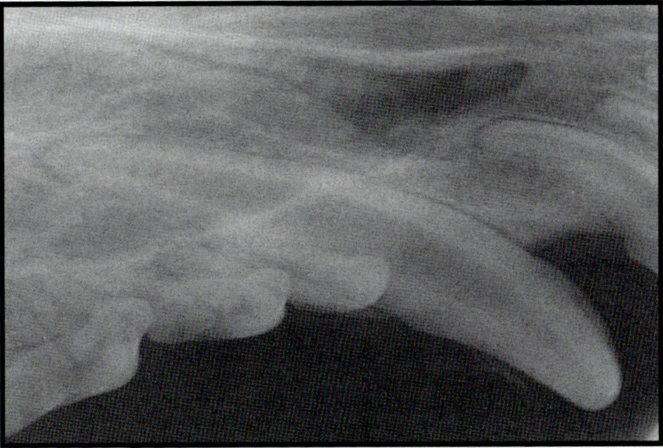

FIGURE 10-1B Oblique View

Figure 10-1A and 1B *The different horizontal angles of the primary x-ray beam as depicted in these figures demonstrate that the shape of the root canal is not a perfect cylindrical tube.*

tooth does, however, have greater healing capabilities, partially because of its larger blood supply.[1,6]

In most dogs and cats, mature teeth have multiple apical foramina instead of one opening to accommodate the pulpal vessels and nerves. On radiographs, the apex appears solidly closed and the root canal seems to end inside the tooth (a small distance from the apex). These numerous small openings and canals, often called the apical delta, are not usually visible on radiographs except when delineated by radiopaque-obturating material or studied in a research environment. Because the canals in the delta are so small, they cannot be easily accessed by instruments and canal obturation is essentially impossible.[7-10]

Other variables that affect the anatomy of root canals are lateral or accessory canals, supernumerary roots, and anomalous root formations. Radiographs can be very helpful in identifying root anatomy and therefore can forewarn practitioners of existent anomalies or the presence of extra structures. For example, if a radiograph shows that the root canal is not located in the center of the root, more than one root canal may be present. Lateral canals exit at the lateral aspect of the root or in furcations rather than through the apical delta. These canals have a significant role in the pathology of periradicular lesions that are located away from the apical region.[4,11,12]

Root canals are not perfect cylindrical spaces within the teeth. In general, the contour of root canals follows the curvature of the root. Often, however, the root canal is flattened in one dimension, thereby creating an oval shape on cross section. In addition, the diameter of the root canal is not uniform throughout the tooth. Therefore, the same root canal can have a completely different shape and diameter in different radiographic views, depending on the horizontal or vertical angle of the primary x-ray beam (Figure 10-1).

RADIOGRAPHIC SIGNS OF ENDODONTIC DISEASE

Radiographic signs of endodontic disease can involve changes in the bone, morphologic structure of the canal, and tooth structure. Inflammatory mediators, toxins, and other irritants that are released when pulpitis is present elicit a host response. If the pulpitis does not resolve, various lesions and conditions (e.g., bone rarefaction, osteosclerosis, canal calcification, and tooth resorption) can be induced[2,4,5,13] (Table 10-2).

Radiolucent bone lesions that are caused by endodontic disease have several distinguishing features. Endodontic lesions are most often found at the apex. The accessory canals, which exit the root other than at the apex, are also a source of similar periradicular lesions; these endodontic lesions can mimic periodontal lesions. The periodontal ligament space can be widened at the apex, or a larger periapical lucent defect with a round, oval, or hanging-drop appearance may be present. The integrity of the lamina dura is lost in these regions. The radiolucent defect

TABLE 10 – 2
RADIOGRAPHIC SIGNS OF ENDODONTIC DISEASE[2,5,6,17,19]

Bone

Wide periodontal space
Periradicular radiolucent defects
Periradicular radiopacities
Discontinuous lamina dura
Thick lamina dura

Tooth

Arrested development and maturation
Pulp calcification
Internal resorption
External resorption

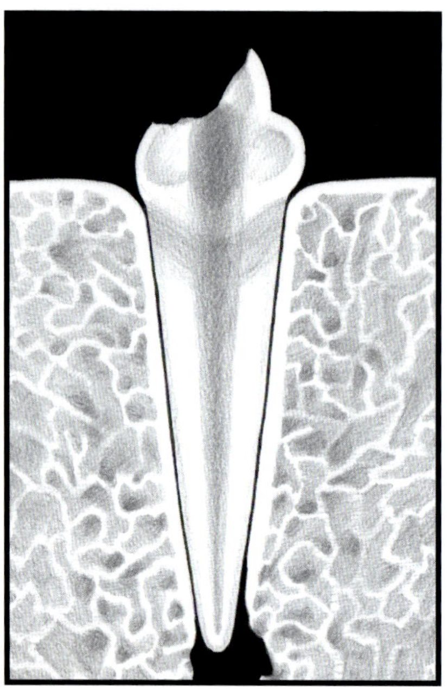

FIGURE 10-2

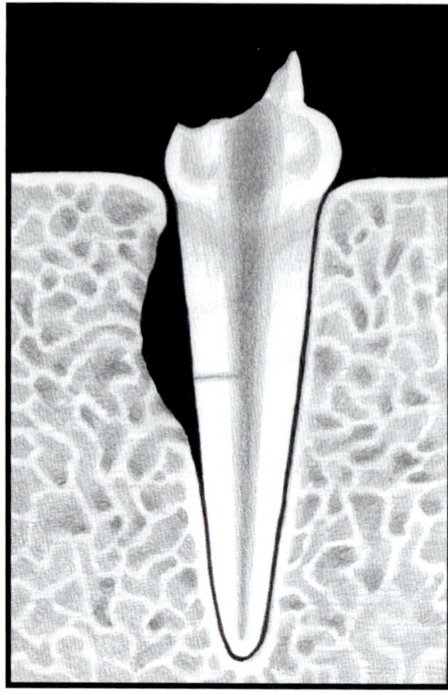

FIGURE 10-3

Figure 10-2 *The loss of integrity in the lamina dura and periapical lucency are depicted in this illustration.* **Figure 10-3** *This illustration demonstrates abnormal widening of the periodontal space.*

will remain associated with the root despite changes in the angle of the x-ray beam or different radiographic views. Because about 30% to 60% of the bone in the region must be destroyed before that destruction can be detected on radiographs, the leading edge of the lesion may be smaller in the radiographs than the actual lesion. In general, the pulp of a tooth with periapical involvement is considered to be partially or completely nonvital[2,13,14] (Figures 10-2 and 10-3).

Radiopaque bone lesions caused by endodontic disease are diffusely opaque with an indistinct border that varies in shape and extent but is somewhat concentrically arranged around the apex. Regional thickening of the lamina dura may occur or, if the opacity extends further into the trabecular bone, condensing osteitis may be present. A wide periodontal space may be concurrently present. Increasing radiopacity of the trabecular bone and the resultant indistinct lamina dura must be differentiated from harmless conditions, such as idiopathic osteosclerosis, and from normal patterns that occur with aging. Radiopaque changes that occur because of endodontic disease are usually associated with low-grade inflammation, where the irritants are somewhat diluted.[13,15,16]

The presence or absence of changes in the morphologic condition of the canal is valuable information when assessing the vitality and activity of the pulp. Obtaining radiographs of comparable teeth can be helpful for evaluation. If a tooth shows delayed development or maturation compared with the other teeth, the pulp should be considered nonvital.

Odontoblasts that line the root canal and pulp chamber stop depositing dentin if they are nonvital. The vital teeth continue to mature and develop thicker dentinal walls and more narrow root canals, whereas the morphologic structure of a nonvital tooth remains unchanged. The disparity in the diameters of root canals is readily apparent on radiographs with time, especially in a young animal when growth is normally rapid. On the other hand, if maturation of the tooth is accelerated, the root canal should appear more narrow than the canals of the other teeth. Pulp calcification can be a normal change that is associated with the aging process; however, if calcification is limited to a given tooth only, the pulp has probably sustained injury. When calcification is diffuse, the canal eventually becomes atretic. When calcification is discrete, pulpal stones reside within the chamber and partially obstruct or occlude the patency of the root canal[6,17,18] (Figures 10-4 to 10-6).

Both internal and external root resorption are expressions of endodontic disease because inflammation of the pulp can initiate resorption of the tooth structure. Resorptive lesions are generally radiolucent defects unless reparative bone or sclerosis replaces and obscures the defect. Internal resorption enlarges and alters the contour of the root canal, excavating

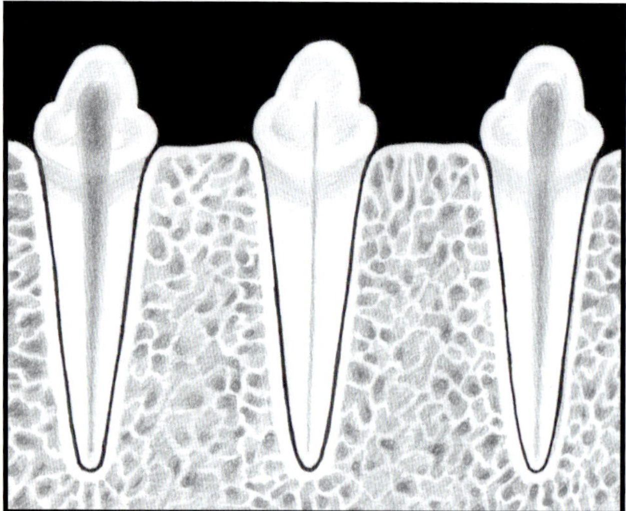

FIGURE 10-4 Accelerated Endodontic Calcification

Figure 10-4 *This illustration depicts accelerated endodontic calcification in the central incisor. Note that the diameters of the other roots are larger.*

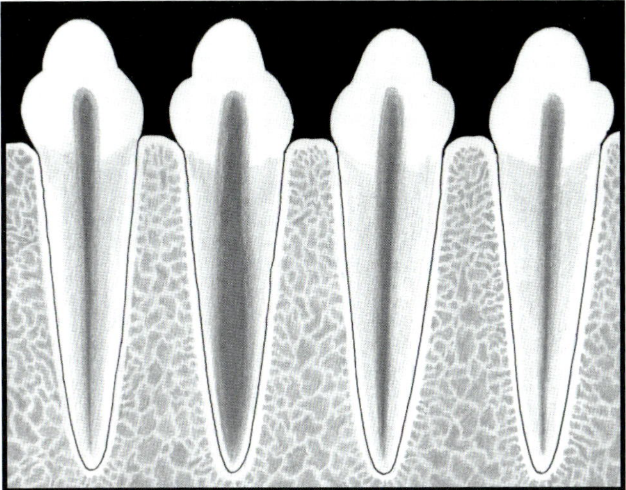

FIGURE 10-5 Incisor with Necrotic Pulp

Figure 10-5 *Necrotic pulp of an incisor is depicted in this illustration. Note the arrested development of the central left incisor.* **Figure 10-6** *This illustration shows a necrotic pulp of the canine tooth on the right. Note the asymmetry of the two canine teeth.*

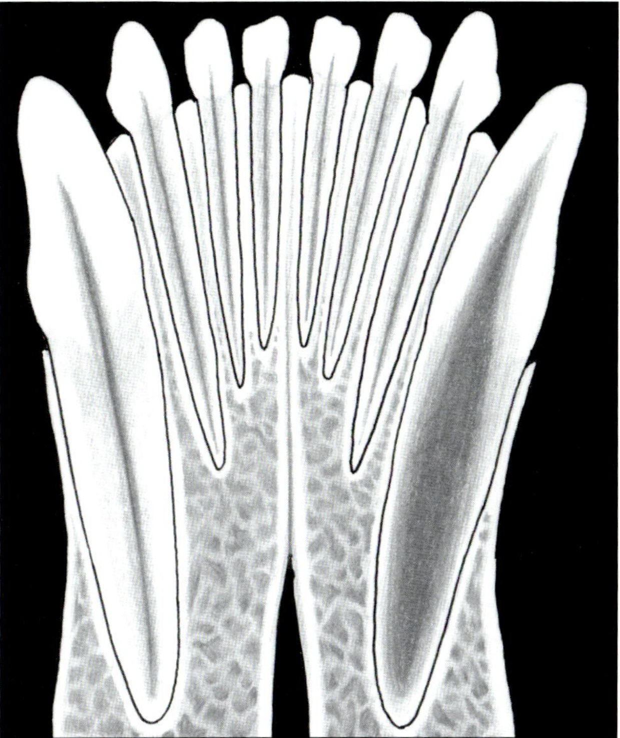

FIGURE 10-6 Canine Tooth with Necrotic Pulp

from the center of the canal to the peripheral wall as it progresses. External resorption occurs externally on the root surface, laterally, or at the apex and advances from the outside wall toward the root canal. When external resorptive lesions are superimposed over the radiographic image of the pulp, the outline of the root canal is still discernible. Changes to the angle of the primary x-ray beam can reveal whether the position of an external lesion has shifted with respect to the root canal, whereas the position of an internal lesion remains in close association with the root canal regardless of the radiographic view[5,17,19] (Figures 10-7 and 10-8).

CAUSES OF ENDODONTIC PATHOSIS

Pulpitis is ultimately the source of endodontic disease, regardless of whether the precipitative cause is trauma, thermal injury, infection, caries, deep restorations, malocclusions, periodontal disease, or anomalous tooth formation.[20-22]

DIRECT PULP EXPOSURE

Pulp exposure of the crown is a common presenting sign. The most prevalent circumstances for overt pulpal exposure in dogs and cats are fractures of the crown; in addition, resorptive lesions are also prevalent in cats. Frank pulp exposure allows oral bacteria and other irritants to contaminate the interior of a tooth, which eventually results in pulpitis.[1-3,5]

INDIRECT PULP EXPOSURE

Even without direct exposure, the pulp can become inflamed if it is indirectly exposed to the oral environment. Any condition that exposes dentin can create indirect pulp exposure, even though the pulp seems to be safely encased within the tooth. Dentin is a solid but porous material that is honey-combed with countless microscopic tubules into which the cells that line the pulp chamber project cytoplasmic extensions. Toxins, irritants, and bacteria can reach the pulp through these tubules, thereby causing pulpitis to develop.[6]

ROOT FRACTURES

Root fractures that occur apical to the level of periodontal attachment may show no clinical signs of endodontic involvement other than increased tooth mobility. The more coronal the fracture line is, the more that tooth mobility increases. The fracture line may not be detectable on initial radiographs unless the injury has caused a separation of the fractured pieces. When root fracture is suspected, several radiographic views should be taken. Follow-up radiographs often reveal a more distinct or even a previously undiagnosed fracture line because separation of the segments can occur with edema, mastication, granulation tissue, connnective tissue healing, or resorption. Root fractures can heal, especially if the fracture line is sheltered within an intact alveolus. Teeth with root fractures tend to maintain vital pulps, unless injury to the pulp is severe enough to cause necrosis. On occasion,

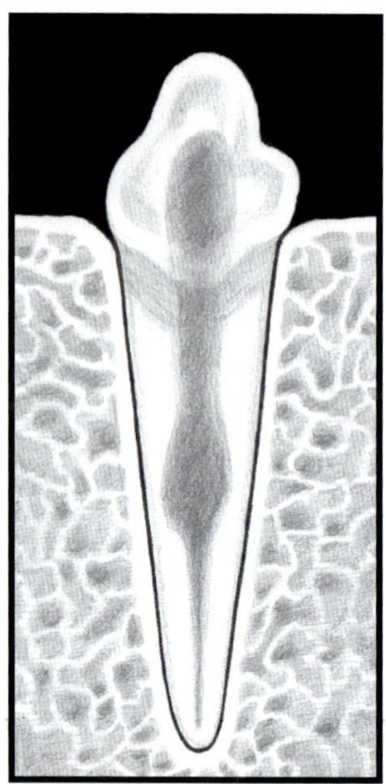

FIGURE 10-7A Internal Resorption

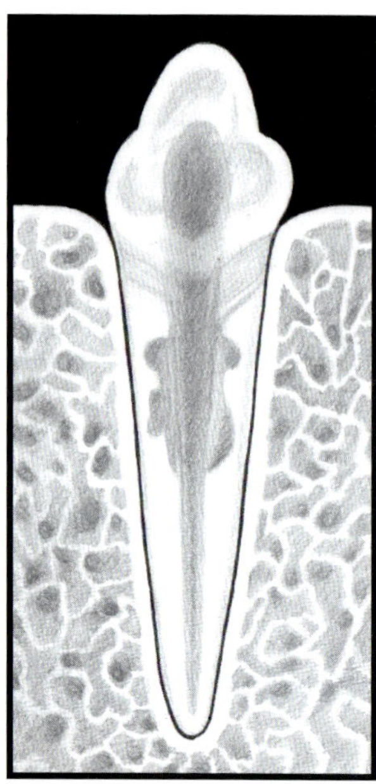

FIGURE 10-7B External Resorption

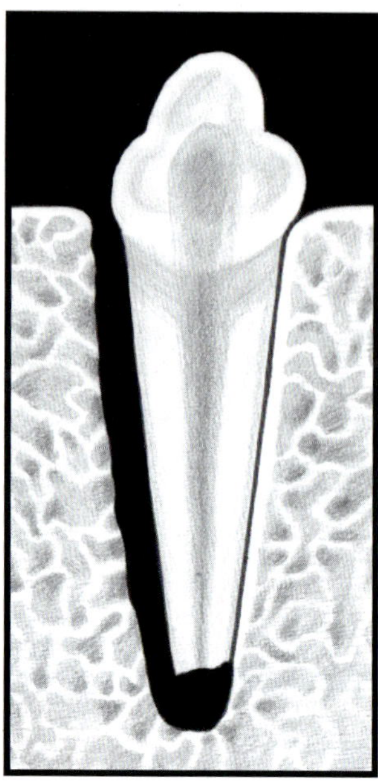

FIGURE 8 Resorption of the Tooth Apex

Figure 10-7A *This illustration depicts internal resorption. Note the lack of definition to the walls of the root canal.* **Figure 10-7B** *In comparison, note what occurs with external resorption. The walls are still visible.* **Figure 10-8** *Resorption at the apex is depicted in this illustration.*

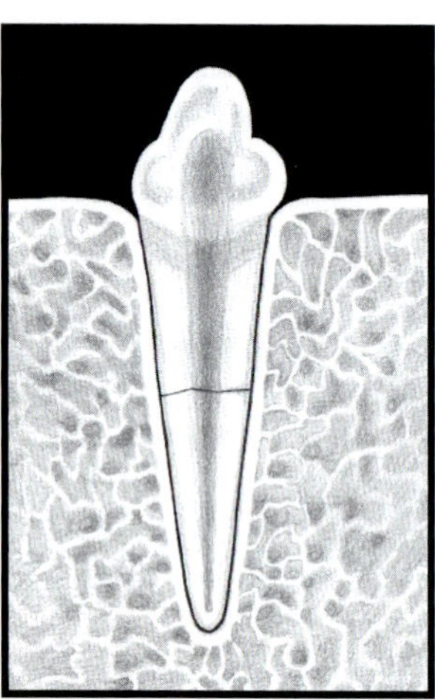

FIGURE 10-9A

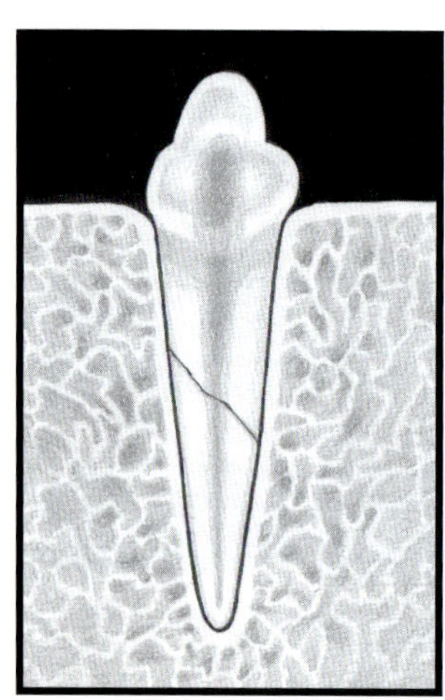

FIGURE 10-9B

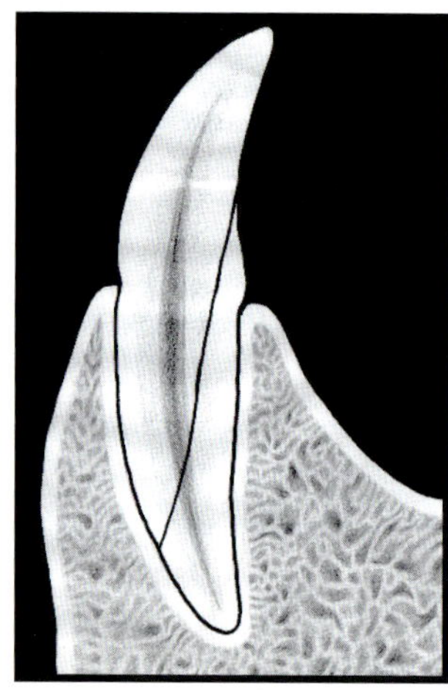

FIGURE 10-9C

Figures 10-9A, 9B, and 9C *These three illustrations depict fractures of the root. Figure 10-9A shows a horizontal fracture, 10-9B an oblique fracture, and 10-9C a vertical fracture. Note the communication between the endodontic system and oral cavity evident in the illustration of the vertical fracture.*

only the apical segment remains vital and the coronal segment does not. If the fracture line is coronal to the level of periodontal attachment, oral fluids and bacteria can easily access the pulp and pulpitis will develop. When follow-up radiographs indicate that bone in or adjacent to the fracture line has been destroyed, pulpal necrosis is likely. The prognosis for vertical root fractures is very poor because healing is impaired by contamination from the oral cavity and by instability of the fragments[23] (Figure 10-9).

ENDODONTIC–PERIODONTAL LESIONS

Communication between the pulp and periodontium occurs through the apical foramina, lateral canals, and denti-nal tubules, thereby allowing irritants to move from one system to another. As a result, both endodontic and periodontal lesions can develop in the same tooth. These lesions are referred to as combined lesions and are classified according to the lesion of origin. Differential diagnosis of the three types of combination lesions depends on clinical parameters and radiographic findings.[24-26]

Class I lesions, or endodontic–periodontal lesions, are endodontic in origin. They begin in the pulp and progress into the periodontal tissue. The periodontal defect is usually very narrow and extends to the apex or a lateral canal. Class I lesions often have a typical J-shaped radiolucent lesion around one root and its apex. Evidence of either direct or indirect

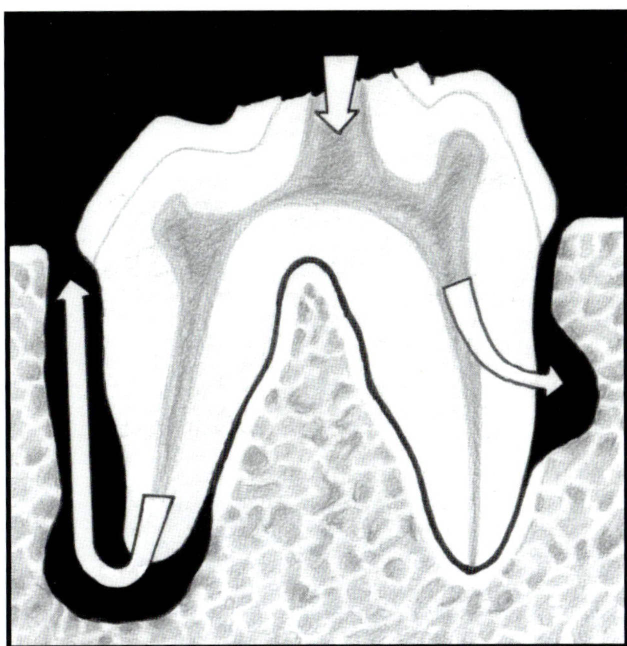

FIGURE 10-10 Class I Lesion

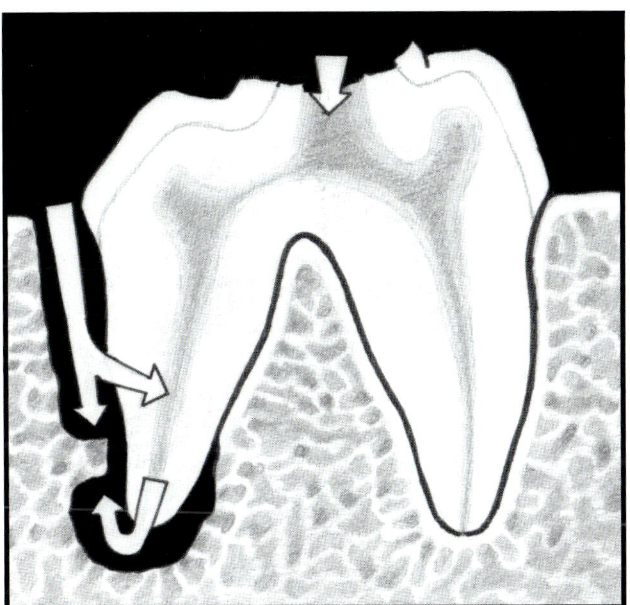

FIGURE 10-12 Class III Lesion

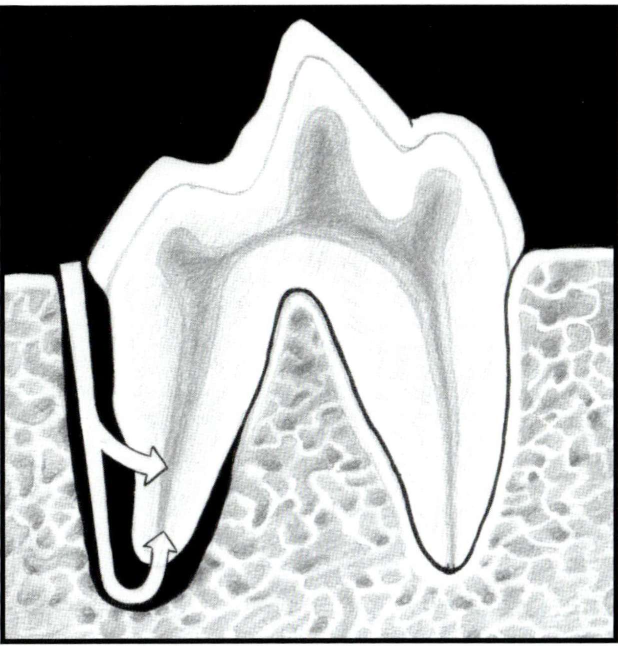

FIGURE 10-11 Class II Lesion

Figure 10-10 *This illustration depicts a class I lesion, which is referred to as an endodontic–periodontal lesion. The* arrows *demonstrate how the lesion begins in the pulp and progresses into the periodontal tissue.* **Figure 10-11** *In contrast to class I lesions, class II lesions (which are referred to as periodontal–endodontic lesions) begin as a wide periodontal pocket that deepens to the apex or a lateral canal as shown by the flow of the* arrows. **Figure 10-12** *Class III lesions are true endodontic–periodontal lesions. The* arrows *demonstrate how the development of the lesions occurs simultaneously but independently. (Figures 10-10 through 10-12 from Marretta SM, Schloss AJ, Klippert LS: Classification and prognostic factors of endodontic–periodontal lesions in the dog. J Vet Dent 9(2):27–30, 1992. Adapted with permission.)*

Disease also includes radiographs of periodontal–endodontic lesions.

Class III lesions, or true combined endodontic–periodontal lesions, present when pathologic evidence of endodontic and periodontal lesions affects a single tooth concurrently, even though the two lesions originated independently. The affected tooth often shows signs of direct pulp exposure with an unrelated wide periodontal defect. The periapical and periodontal lesions may be distinctly separate on radiographs or may be coalesced into a single, large, extensive radiolucent defect[24-26] (Figure 10-12).

DENTAL ANOMALIES

Malformed crowns occasionally have creases, invaginations, or defects in the enamel and dentin that can open communication between the pulp and oral cavity, thereby contaminating the pulp with oral fluids and bacteria. The communication may not be obvious on clinical examination; however, with time, evidence of an endodontic lesion will become

pulp exposure is frequently apparent on the crown; the pulp is usually necrotic[24-26] (Figure 10-10).

Class II lesions, or periodontal–endodontic lesions, are periodontal in origin. They begin as a wide periodontal pocket that deepens to the apex or a lateral canal. The crown is usually intact, and the pulp may be vital. Angular V-shaped or horizontal bone loss is demonstrated on radiographs.[24-26] For comparative value, Figure 10-11 has been included in this chapter; however, Chapter 9 on Interpretation of Periodontal

apparent on radiographs. Therefore, we recommend that anomalous teeth be radiographically evaluated on a regular basis.

RADIOGRAPHIC EVALUATION OF TREATMENT MEASURES

Appraisal of endodontic treatment involves both clinical and radiographic examination. If a tooth required treatment and the periradicular tissue appeared normal on initial radiographs, then follow-up radiographs should continue to reflect the same findings. If periradicular radiolucent defects existed when treatment was initiated, subsequent radiographs would ideally demonstrate resolution of those defects. Complete regeneration of periradicular tissue, however, is not always evident on radiographs, even if the treatment was successful. Therefore, at minimum, clinical signs should not be present and new lesions should not have developed[27-29] (Table 10-3).

Follow-up examinations should continue for several years after treatment has ended. If a dog or cat had root canal treatment, a three-dimensional obturation of the canal space should be evident with no voids in the apical third of the canal or overextension of filling materials beyond the apex. Follow-up visits should confirm that clinical signs have abated and that signs of endodontic lesions are not evident on radiographs. On occasion, a periapical radiolucency might not resolve but instead have partially reduced or remained stable. If the lesion were actually an unresolved chronic inflammatory lesion, acute exacerbation of clinical signs would occur when a response to inflammation is mounted; this situation is referred to as a phoenix abscess or flare-up. Success with pulp-capping techniques is judged by the absence of adverse clinical signs, absence of lesions on radiographs, formation of a dentinal bridge, and continued root development and maturation. If pulp necrosis occurs, pulpal contamination and infection are still possible through anachoresis, even when no contamination occurs through the oral cavity. Anachoresis is the attraction of blood-borne microorganisms to inflamed or necrotic tissue. Therefore, bacteremias from home care, eating or chewing activities, prophylaxis, extractions, or other scenarios are potential sources of infection at any time.[21, 22, 27-30]

RADIOGRAPHIC EVALUATION

Recognizing the anatomic or pathologic differences between normal oral conditions and those associated with endodontic disease is essential to evaluate oral structures on radiographs. The following radiographs were selected to assist practitioners in differentiating the various endodontic defects and lesions that can affect the teeth in dogs and cats. Figures 10-13 to 10-30 provide a means of comparing the various signs of endodontic conditions that are visible on radiographs. Because the age of a cat or dog can affect the appearance of the teeth in radiographs, Figures 10-31 to 10-36 show some distinguishing characteristics of tooth anatomy during maturation and aging. Although positioning techniques are discussed at length in Chapters 1 though 4, some considerations that affect the monitoring of treatment protocols are unique to endodontics and therefore are addressed in Figures 10-37 to 10-39. These radiographs are followed by some figures that can assist practitioners in properly monitoring endodontic conditions (Figures 10-40 to 10-48). Finally, evaluation after surgery (Figures 10-49 to 10-87), complications (Figures 10-88 to 10-90), and long-term follow up (Figures 10-90 to 10-93) are addressed.

TABLE 10 – 3
RADIOGRAPHIC EVALUATION OF TREATMENT MEASURES[27-29]

Region	*Satisfactory Result*	*Questionable Outcome*	*Unsatisfactory Result*
Periodontal Ligament Space	*Normal*	*Slightly increased*	*Markedly increased*
Periradicular Radiolucent Defect or Rarefaction	*Osseous repair*	*Similarity in size or slightly smaller*	*Same size or larger; presence of new radiolucent defects*
Lamina Dura	*Normal*	*Thickened, sclerotic, or irregular*	*Lamina dura unchanged; no repair*
Tooth	*No resorption*	*Slight resorption*	*Progressive resorption*

REFERENCES

1. Mulligan TW: Endodontics, in Kirk RW (ed): *Current Veterinary Therapy. X. Small Animal Practice.* Philadelphia, WB Saunders Co, 1989, pp 954–959.

2. Emily P: Problems associated with the diagnosis and treatment of endodontic disease, in Manfra Marretta S (ed): *Problems in Veterinary Medicine: Dentistry*, vol 2, no. 1. Philadelphia, JB Lippincott Co, 1990, pp 152–182.

3. Williams CA: Endodontics, in Frost P (ed): *The Veterinary Clinics of North America. Small Animal Practice*, vol 16, no. 5. Philadelphia, WB Saunders Co, 1986, pp 875–893.

4. Dumsha TC, Hovland EJ: Problems in radiographic technique and interpretation, in Gutmann JL, Dumsha TC, Lovdahl PE, Hovland EJ (eds): *Problem Solving in Endodontics: Prevention,*

(continues on page 150)

RADIOGRAPHIC SIGNS OF ENDODONTIC CONDITIONS—FRACTURES

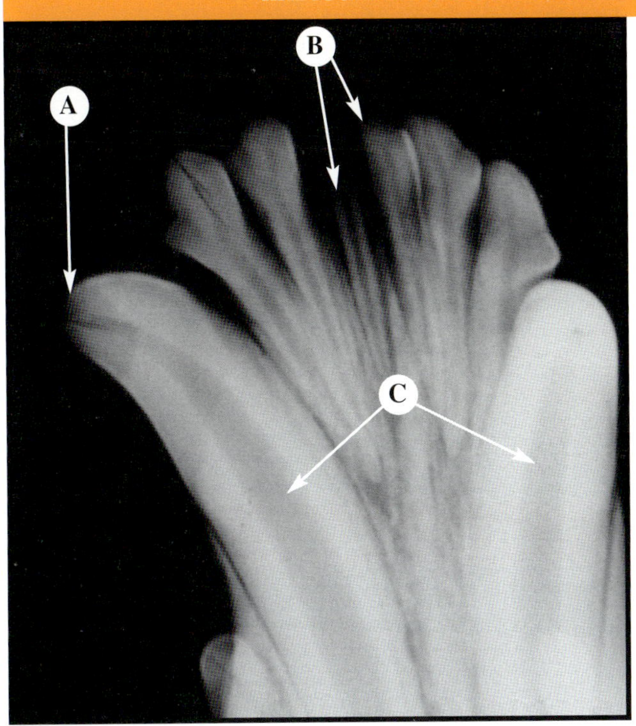

FIGURE 10-13 Superficial Fracture of the Crown

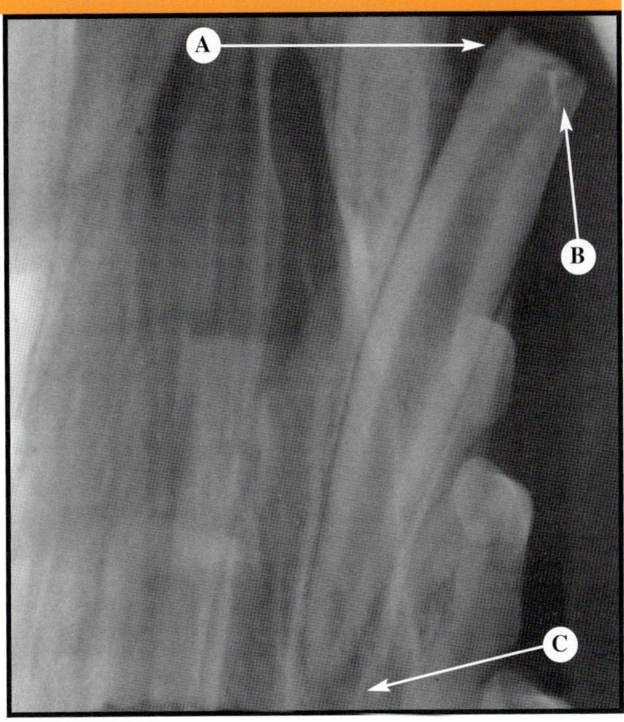

FIGURE 10-14 Deep Fracture of the Crown

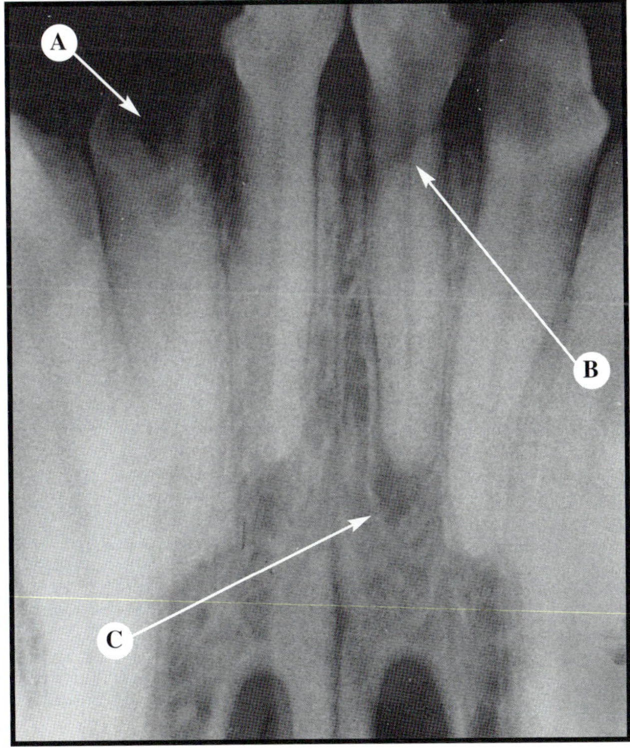

FIGURE 10-15 Fracture of the Root

Figure 10-13 *In this radiograph of a superficial fracture of the canine tooth, a tiny opening that was clinically evident is not representative of the actual size of the diameter of the root canal. (A) Coronal fracture exposing the pulp, (B) deep and superficial fractures of the incisor, and (C) comparative discrepancies in the diameter of the lumen, which are likely the result of the angle of the x-ray beam.* **Figure 10-14** *This fracture is apical enough to expose the pulp directly. The periapical changes confirm a diagnosis of endodontic disease. (A) Fracture of the canine tooth, (B) exposed pulp chamber with debris, and (C) periapical radiolucency and widening of the periodontal space.* **Figure 10-15** *Two teeth are affected. (A) Fractured crown, (B) fractured root, and (C) periapical radiolucency.*

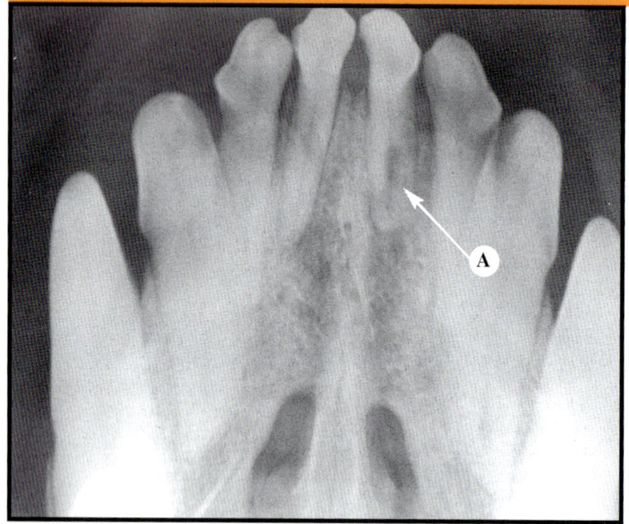

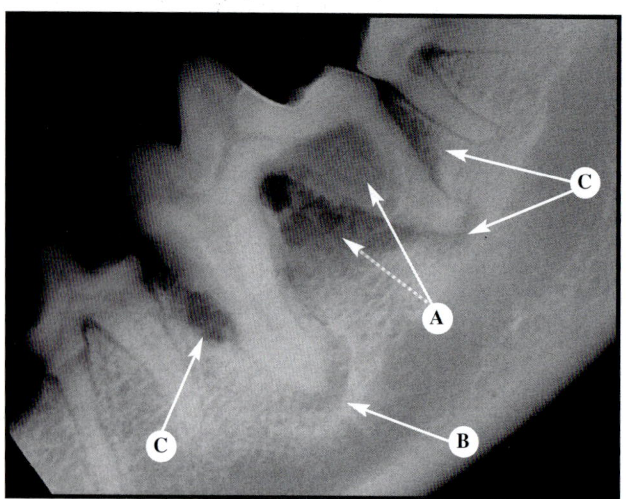

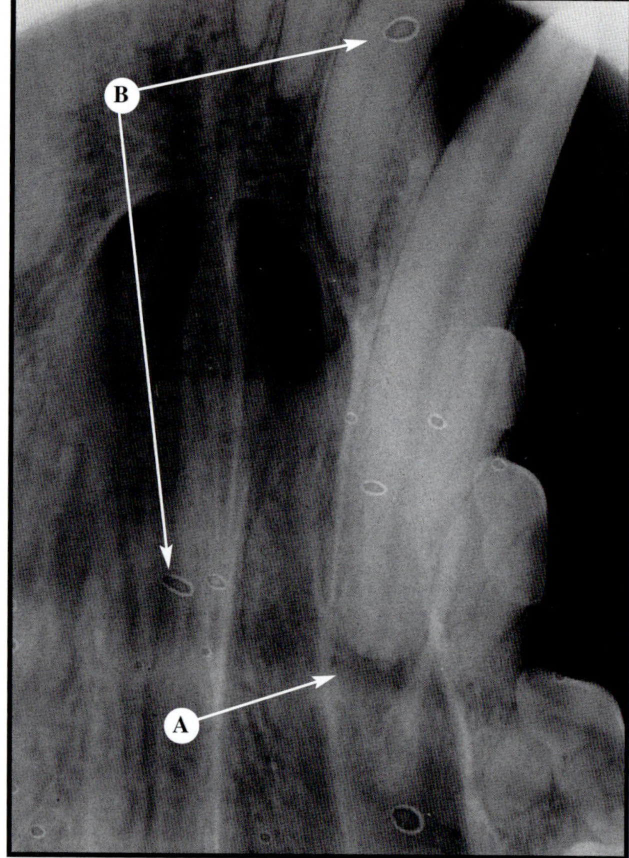

RADIOGRAPHIC SIGNS OF ENDODONTIC CONDITIONS—CARIES

FIGURE 10-16 Caries

Note the advanced pathologic involvement of the periapical and apical structures. (A) Deep caries and (B) apical resorption and periapical radiolucency.

RADIOGRAPHIC SIGNS OF ENDODONTIC CONDITIONS—RESORPTION

FIGURE 10-17 Internal Resorption in a Dog

FIGURE 10-18 External Resorption in a Combined Endodontic–Periodontal Lesion

FIGURE 10-19 External Resorption of the Apex in a Dog

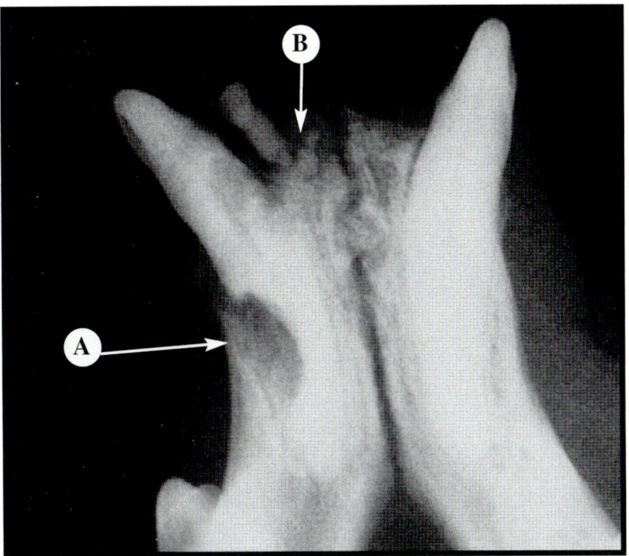

FIGURE 10-20 External Resorption in a Cat

FIGURE 10-21 External and Internal Resorption

Figure 10-17 *Internal resorption is a radiolucent defect seen in the root canal of the incisor. The apex is blunted and may also have undergone resorption. Another radiographic view is recommended to confirm that the resorption is actually internal. (A) Radiolucent defect within the root canal.* **Figure 10-18** *Example of deep external resorption caused by an endodontic–periodontal lesion. (A) External resorption and infrabony pocket, (B) periapical radiolucency, and (C) widened periodontal space.* **Figure 10-19** *Note the advanced pathologic involvement of the apex. The depth of the open apex must be accurately measured to prevent overdebridement and extrusion of obturation materials. (A) Apical resorption and (B) processing artifacts.* **Figure 10-20** *External resorption associated with an infrabony pocket has caused endodontic involvement of the canine tooth. (A) External resorption with infrabony pocket and (B) complete bone loss of the incisor.* **Figure 10-21** *Resorption caused by a class I endodontic–periodontal lesion. The pulp is probably necrotic. (A) Coronal fracture, (B) root resorption, (C) bone loss, and (D) widened periodontal space and external resorption.*

RADIOGRAPHIC SIGNS OF ENDODONTIC CONDITIONS—CALCIFICATION

FIGURE 10-22

Calcification of the endodontic system, which is a reaction to chronic irritation, can present an obstruction during root canal treatment. (A) Pulp calcification, (B) periapical radiolucency, (C) periradicular radiolucency and root resorption, and (D) fractured crown.

RADIOGRAPHIC SIGNS OF ENDODONTIC CONDITIONS—DIAMETER OF ROOT CANALS

FIGURE 10-23

The wider diameter of the canal shown in this radiograph is evidence that tooth maturation had stopped and suggests that pulpal death occurred earlier. When comparing the diameters of root canals, the primary x-ray beam should be positioned at the same angle. (A) Wide root canal diameter and (B) narrow root canal diameter.

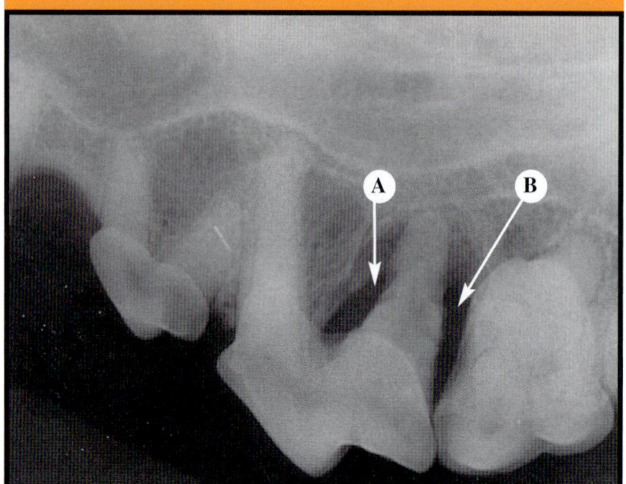

FIGURE 10-24

When a widened periodontal space extends close to the apical region, it is suggestive of a class II combined lesion. The pulp may still be vital. (A) Abnormally wide periodontal space and (B) V-shaped angular bone loss extending to the apex.

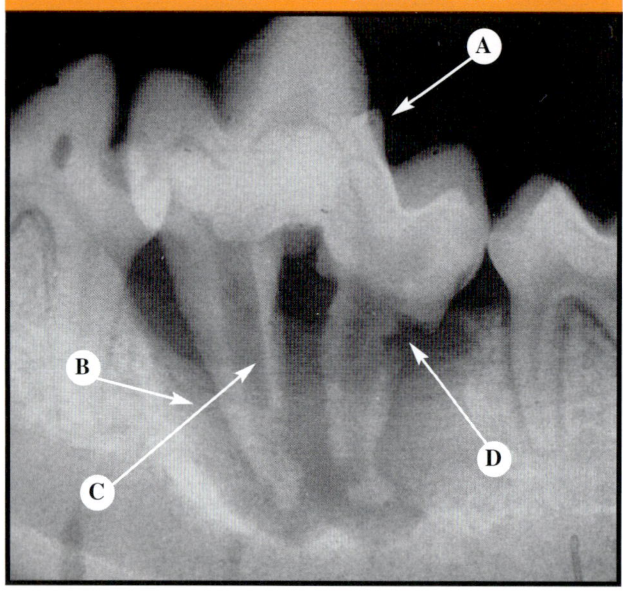

FIGURE 10-27

Dilacerated crown and root. Deformities often involve endodontic or periodontal structures and may develop a class III endodontic–periodontal lesion. (A) Deformed crown with enamel invaginations, (B) generalized widening of the periodontal space, (C) large diameter of the root canal, and (D) external resorption. Note the abnormal angle of the root; the roots converge rather than diverge.

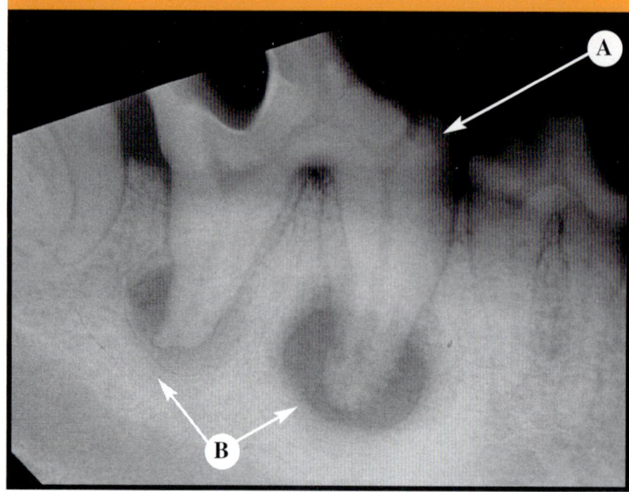

FIGURE 10-25 Periapical Radiolucency

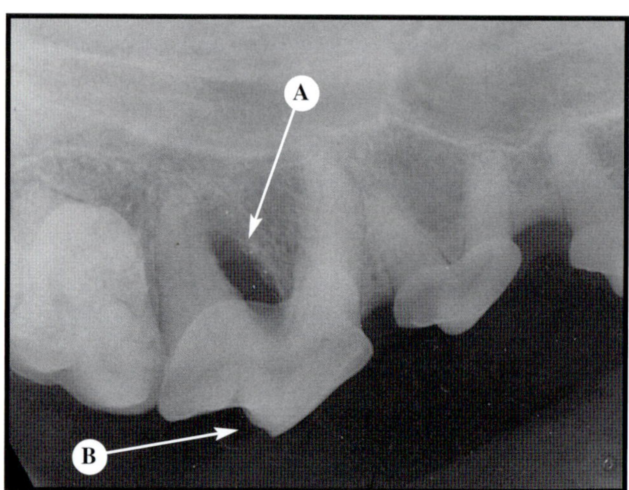

FIGURE 10-26 Periradicular Radiolucency

Figure 10-25 *Pathologic involvement of endodontic structures. (A) Fractured crown with direct pulpal exposure (through the pulp horns) and (B) periapical rarefaction.* **Figure 10-26** *A lateral canal associated with this periradicular space may be the endodontic source of the periodontal lesion. (A) Periradicular radiolucency and (B) coronal damage.*

RADIOGRAPHIC SIGNS OF ENDODONTIC CONDITIONS—COMBINED LESIONS

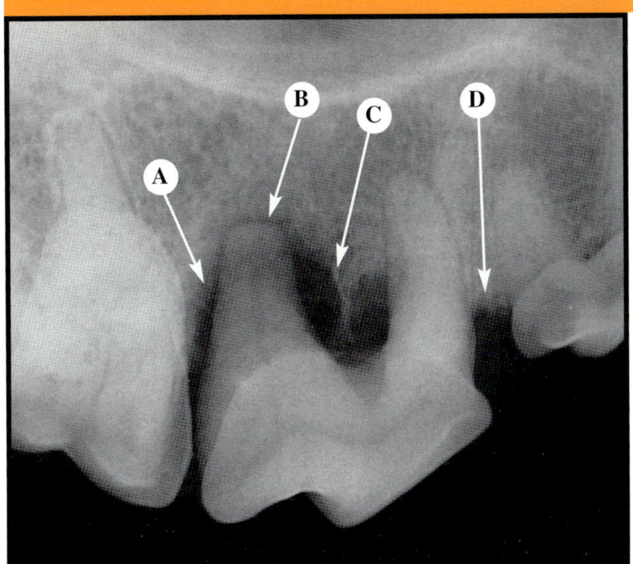

FIGURE 10-28 Class II Periodontal–Endodontic Lesion

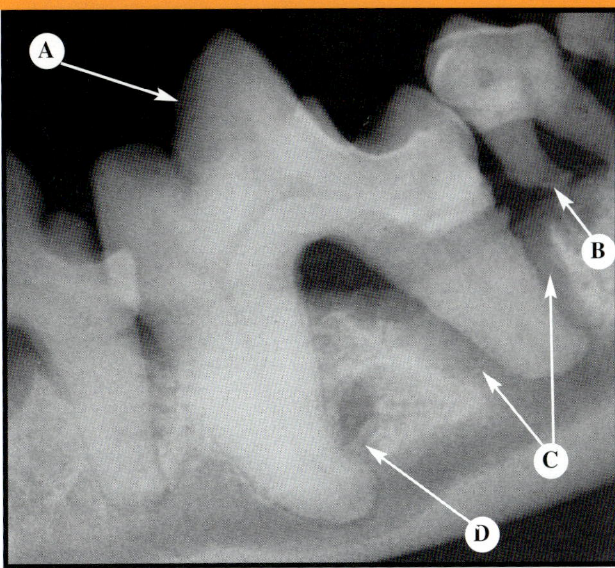

FIGURE 10-29 Class II Periodontal–Endodontic Lesion

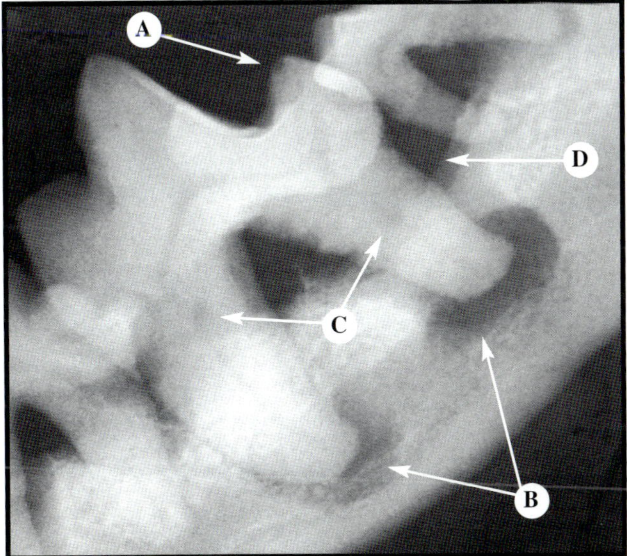

FIGURE 10-30 Class III Endodontic–Periodontal Lesion

RADIOGRAPHIC ATTRIBUTES AFFECTED BY AGE

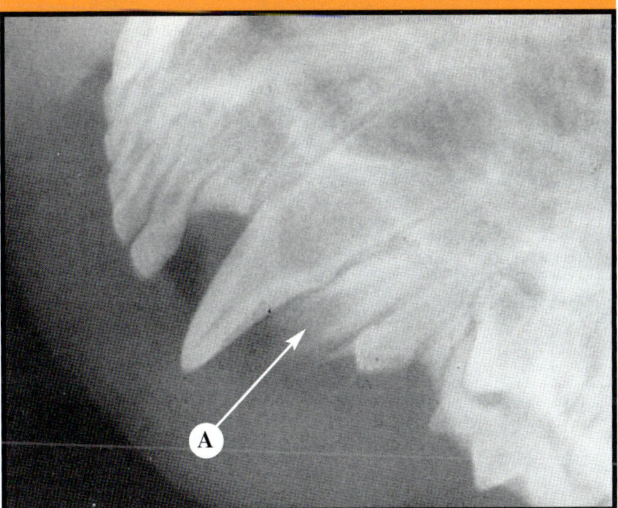

FIGURE 10-31 Deciduous Teeth

Figure 10-28 *An example of primary periodontal disease with secondary endodontic involvement. (A) Moderately widened periodontal space, (B) slightly widened periodontal space, (C) angular bone loss at the furcation, and (D) crestal bone loss.* **Figure 10-29** *Presence of a periodontal–endodontic lesion. Note the impending exfoliation of the second molar. (A) No apparent damage to the crown, (B) supereruption and horizontal bone loss extending beyond the apex, (C) very wide periodontal space, and (D) periradicular rarefaction.* **Figure 10-30** *Class III combined endodontic–periodontal lesion of the first molar. The second molar has a class II combined lesion. (A) Fracture of the coronal cusp, (B) periapical rarefaction, (C) internal resorption, and (D) horizontal bone loss.*

Figure 10-31 *In this radiograph of mixed dentition in a cat, no endodontic treatment would be recommended for the fractured deciduous canine tooth because the permanent tooth has erupted. The practitioner should elect to extract the root fragment. (A) Fractured deciduous canine tooth.*

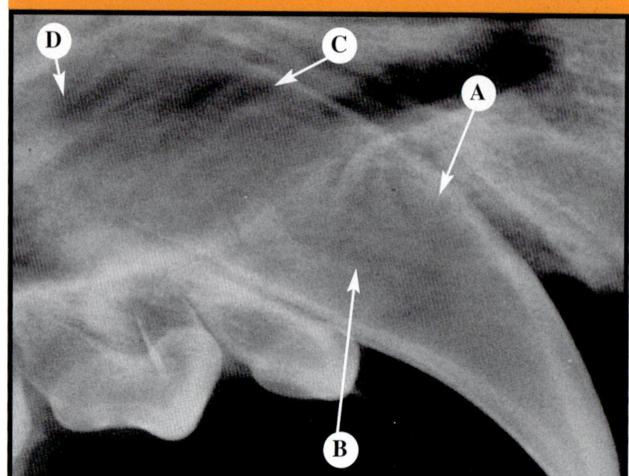

FIGURE 10-32 Erupting Permanent Tooth

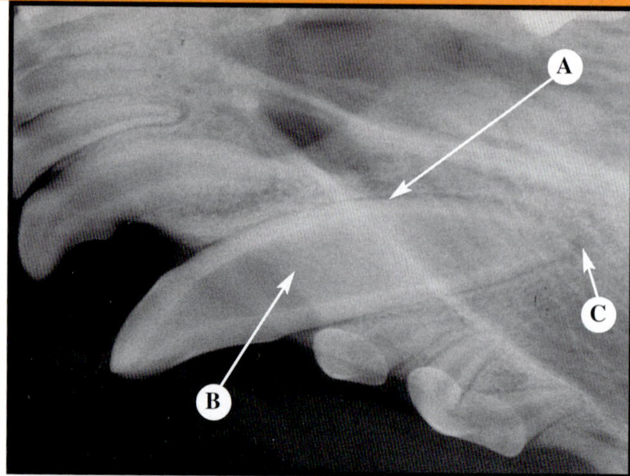

FIGURE 10-33 Immature Permanent Tooth

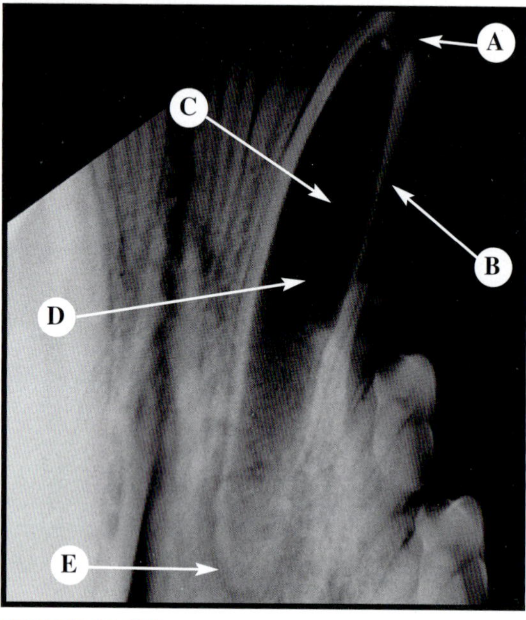

FIGURE 10-34 Immature Permanent Tooth
with Necrosis

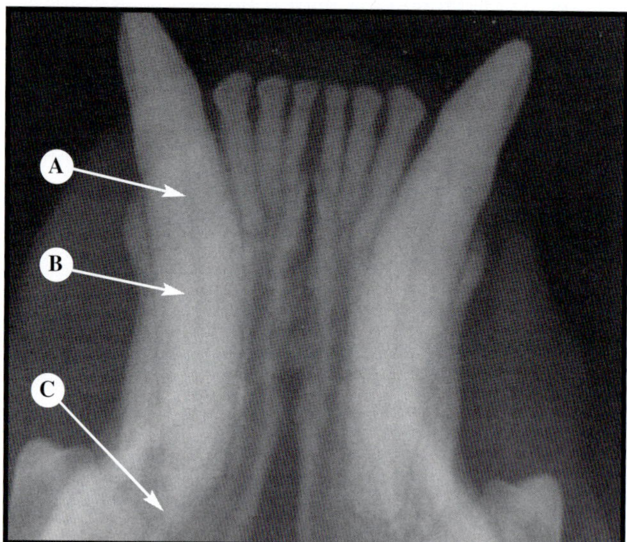

FIGURE 10-35 Mature Permanent Tooth in a Cat

Figure 10-32 *Conventional root canal therapy would be contraindicated in immature teeth. Such options as apexification and apexigenesis could be considered.* (A) *Thin dentinal walls,* (B) *wide root canal diameter,* (C) *incomplete length to the root, and* (D) *open apex.* **Figure 10-33** *Conventional root canal therapy would still be questionable because the tooth remains immature.* (A) *Narrow dentinal walls,* (B) *wide canal diameter, and* (C) *complete root length and closing apex.* **Figure 10-34** *Pulp exposure and necrosis occurred at 8 to 10 months of age.* (A) *Fractured crown,* (B) *narrow dentinal walls,* (C) *radiolucent necrotic pulp,* (D) *wide canal diameter, and* (E) *periapical radiolucency and wide periodontal space.* **Figure 10-35** *The tooth shown here is mature enough to accept conventional therapy if necessary.* (A) *Patent root canal,* (B) *thickened dentinal walls, and* (C) *completely formed apex.* **Figure 10-36** *Conventional use of instruments would be difficult in a geriatric patient because of atretic canals.* (A) *Thick dentinal walls,* (B) *narrow root canal,* (C) *fracture of the crown, and* (D) *periapical radiolucency.*

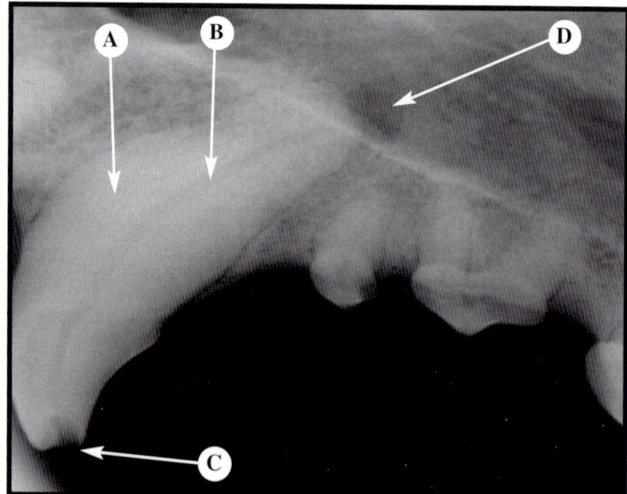

FIGURE 10-36 Mature Permanent Tooth in a
Geriatric Dog

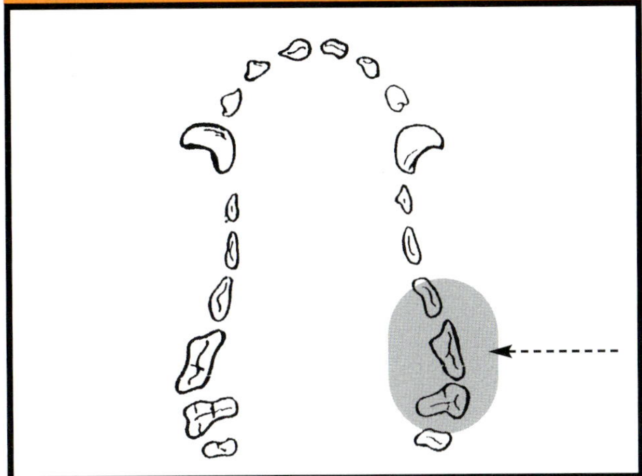

FIGURE 10-37A Image Field: Lateral View

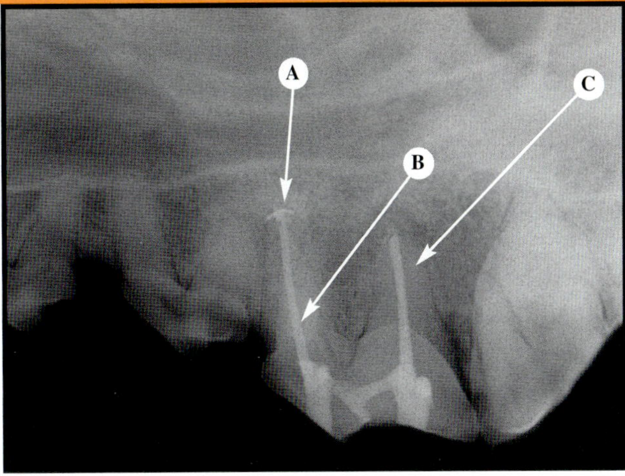

FIGURE 10-37B Radiographic Image: Lateral View

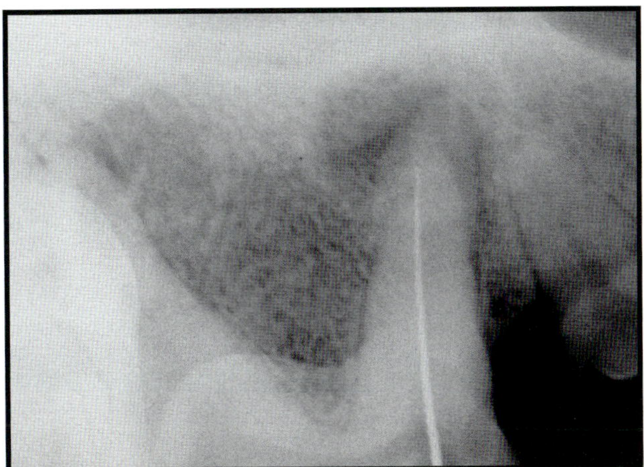

FIGURE 10-37C Position of the File: Lateral View

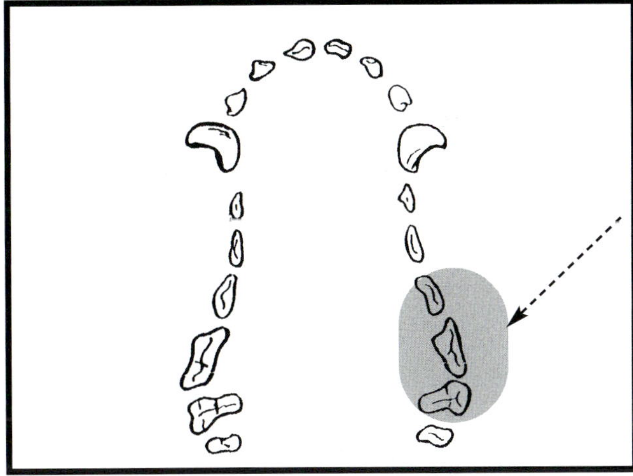

FIGURE 10-38A Image Field: Mesiolateral Oblique View

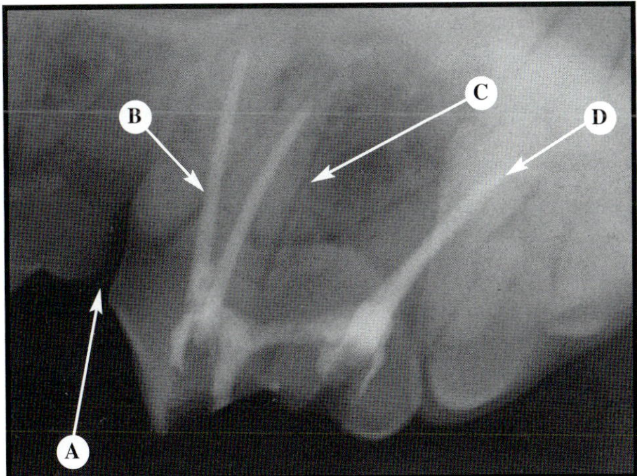

FIGURE 10-38B Radiographic Image: Mesiolateral Oblique View

Two views are used for evaluating the three root apices of the maxillary fourth premolar: mesiolateral oblique and distolateral oblique. **Figure 10-37A** *Image field of lateral view (bisecting angle).* **Figure 10-37B** *If the x-ray beam is positioned at a horizontal angle of 90° to the interproximal spaces, the image of the mesiobuccal and palatal roots will be superimposed. Such superimposition makes evaluation of specific apices very difficult, if not impossible. In this radiograph, the two mesial roots are superimposed. (A) Apical area of the two mesial roots, (B) superimposed root canal filling of the two mesial roots, and (C) distal root.* **Figure 10-37C** *The position of the file cannot be determined from this view.* **Figures 10-38A and 10-39A** *Image fields of the mesiolateral oblique and distolateral oblique views.* **Figures 10-38B and 10-39B** *The tubehead should be positioned at the usual bisecting angle laterally directed at the fourth premolar. The tubehead is then turned in the horizontal plane so that the cone is directed at the fourth premolar from either a slight mesial or distal aspect. Such positioning will shift the two mesial roots so that they are no longer superimposed and can be evaluated individually. In Figure 10-38B, (A) is the interdental space, (B) the palatal root, (C) the mesiobuccal root, and (D) the distal root superimposed over the first molar. In Figure 10-39B, (A) is the mesiobuccal root superimposed over the third premolar, (B) the palatal root, and (C) the distal root.* **Figures 10-38C and 39C** *The file can be seen in the mesiobuccal root.*

(figures continue on next page)

POSITIONING TECHNIQUES FOR MONITORING TREATMENT MEASURES *(continued)*

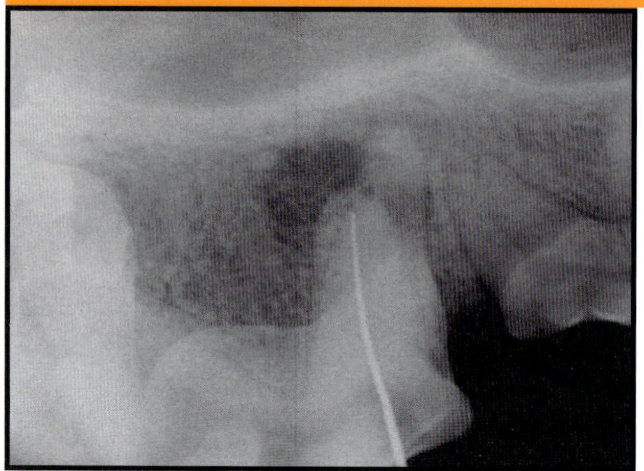

FIGURE 10-38C Position of the File: Mesiolateral Oblique View

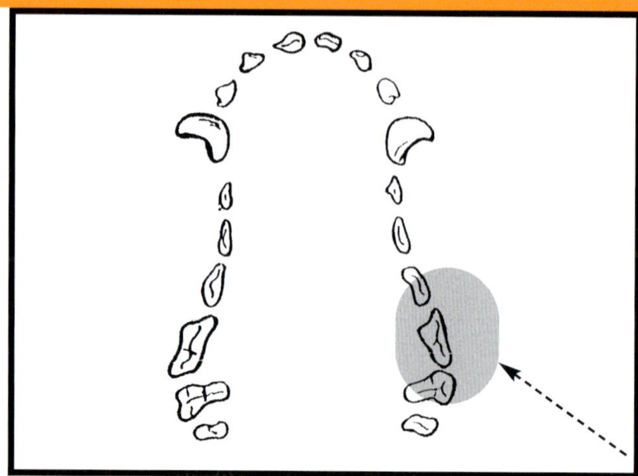

FIGURE 10-39A Image Field: Distolateral Oblique View

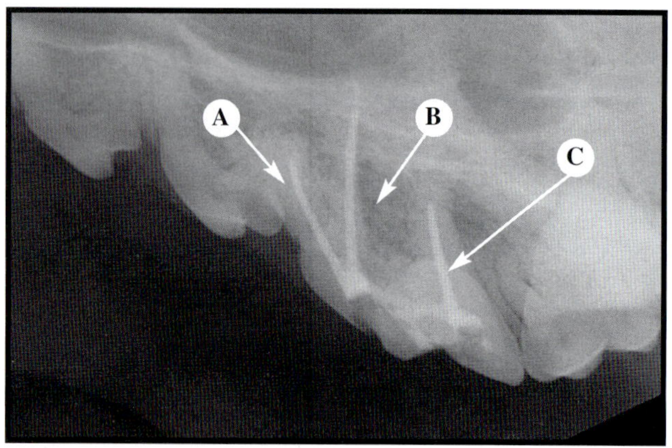

FIGURE 10-39B Radiographic Image: Distolateral Oblique View

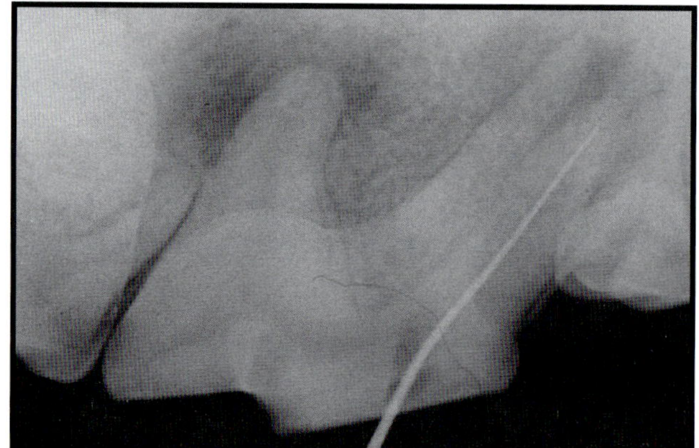

FIGURE 10-39C Position of the File: Distolateral Oblique View

MONITORING TREATMENT MEASURES—USE OF INSTRUMENTS

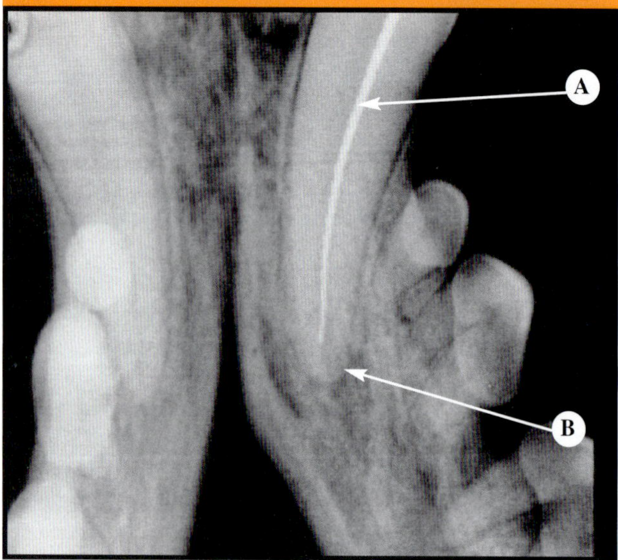

FIGURE 10-40 Monitoring the Length of the Canal

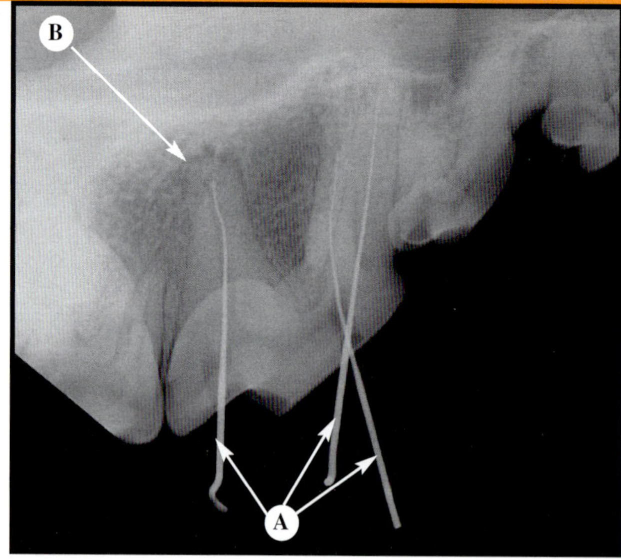

FIGURE 10-41 Monitoring Placement of the File

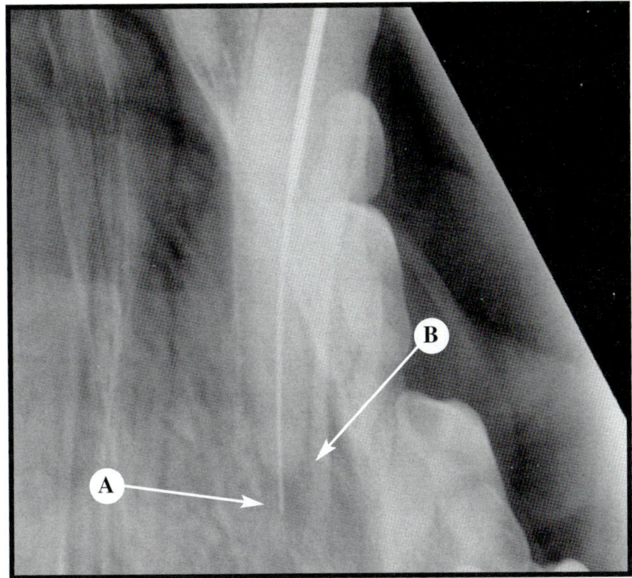

FIGURE 10-42 Improper Use of Instruments

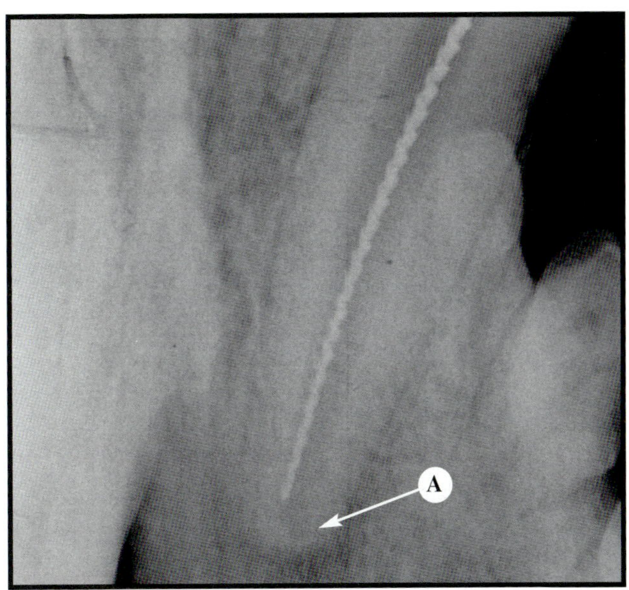

FIGURE 10-43 Improper Use of Instruments

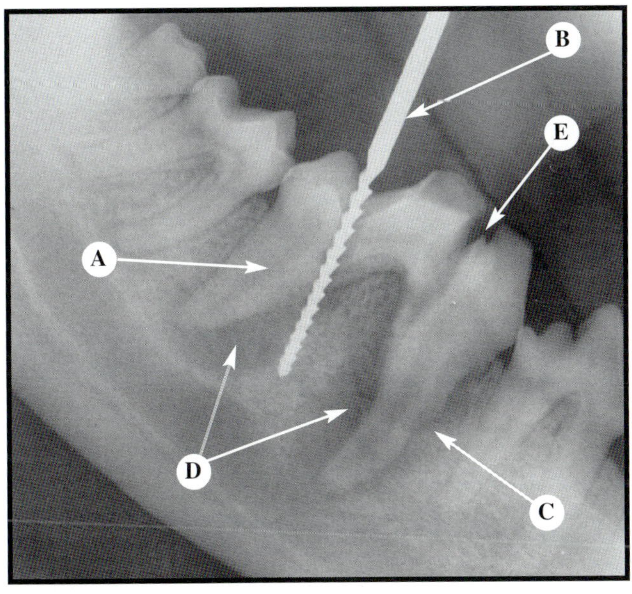

FIGURE 10-44 Lateral Wall Perforation

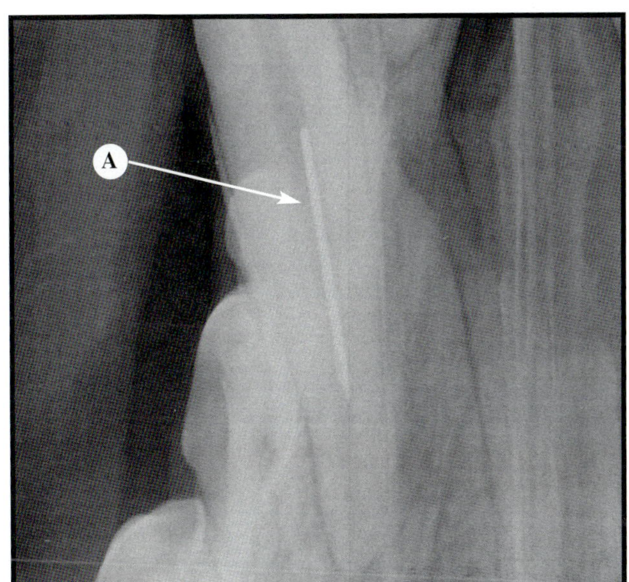

FIGURE 10-45 Broken Gates Glidden Drill

Figure 10-40 *The length of the canal can be determined by measuring a file that has been inserted into the canal. Radiographs show the relation of the file to the apical canal. (A) File inserted into the canal and (B) apical delta region.* **Figure 10-41** *The file should be inserted into the end of the canal and radiographs taken for confirmation. (A) Files inserted into the canals and (B) apical delta region.* **Figure 10-42** *In this radiograph, (A) the file is perforating the apex; (B) shows the apex of the tooth. The significance of this radiograph is that the apex is open and the canal measures deeper than it actually is. Periapical trauma is evident, with extrusion of necrotic tissue into the periapical region. Overly aggressive use of instruments must be avoided.* **Figure 10-43** *The size of the file and its insertion are incorrect. The file is "ledging" into the lateral wall 2 mm short of the apical extent of the canal. Ledging can lead to perforation and incomplete debridement. (A) Apical canal.* **Figure 10-44** *A misdirected file can cause perforation. Such an error in procedure would require special endodontic treatment. Note the class I endodontic–periodontal lesion. (A) Root canal, (B) misdirected endodontic file, (C) external resorption, (D) wide periodontal spaces, and (E) misdirected filing site.* **Figure 10-45** *The broken Gates Glidden drill may obstruct conventional access to the apex. Surgical treatment may be required. (A) Broken Gates Glidden drill.*

MONITORING TREATMENT MEASURES—OBTURATION

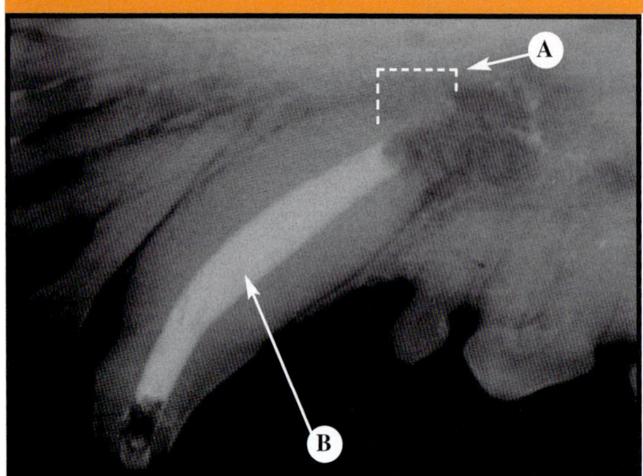

FIGURE 10-46 Improper Retrograde Preparation

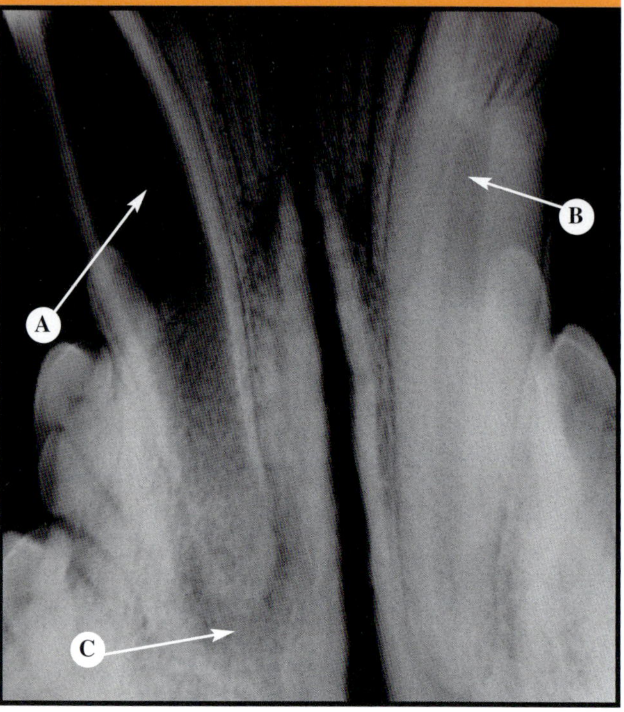

FIGURE 10-47 Indication for Apexification

Figure 10-46 *Apicoectomy should be complete to the level of the fill. This is an example of poor endodontic technique if not corrected. Radiographs should be taken before closure. (A) Incomplete apicoectomy and (B) conventional obturation.* **Figure 10-47** *Apexification is an option. Differences in the size of the lumen and apical development between the two canine teeth indicate early pulpal death of the tooth on the left. (A) Necrotic immature canine tooth, (B) normal vital canine tooth, and (C) periapical radiolucency.* **Figure 10-48** *The fill to the apex has been adequately placed. (A) Obturation using calcium hydroxide.*

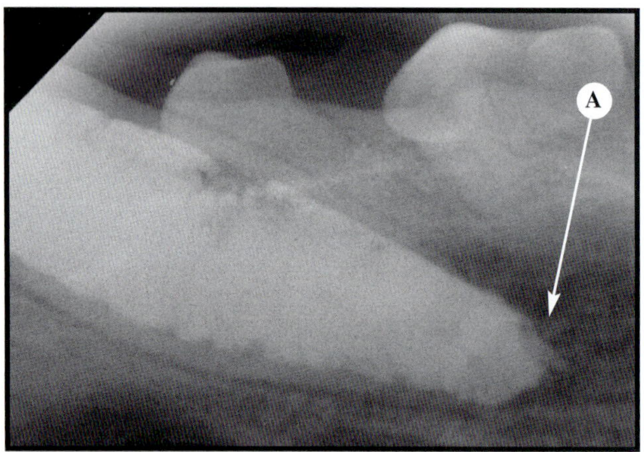

FIGURE 10-48 Apexification Obturation

EVALUATION AFTER SURGERY—FOLLOW-UP FOR VITAL PULPOTOMY

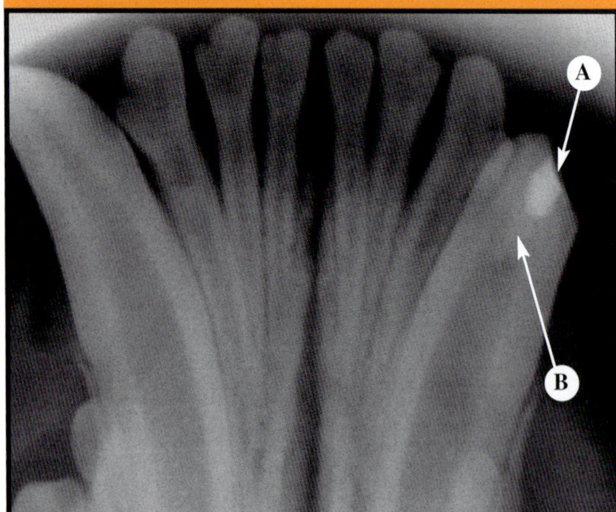

FIGURE 10-49A Standard X-ray Taken Immediately After Surgery

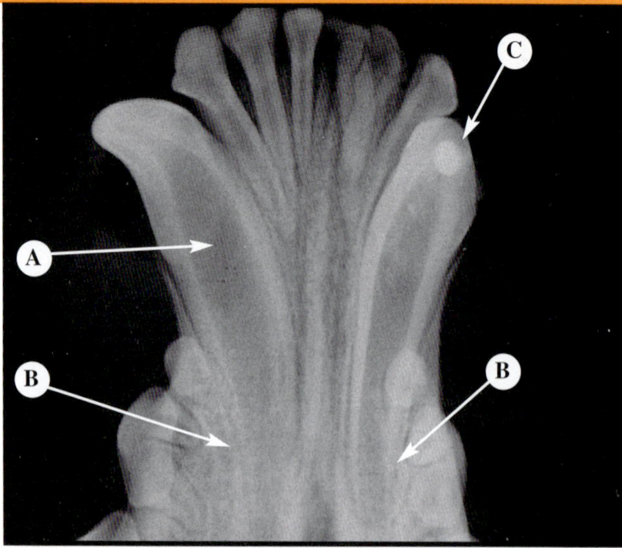

FIGURE 10-49B 4-Month Follow-up Radiograph

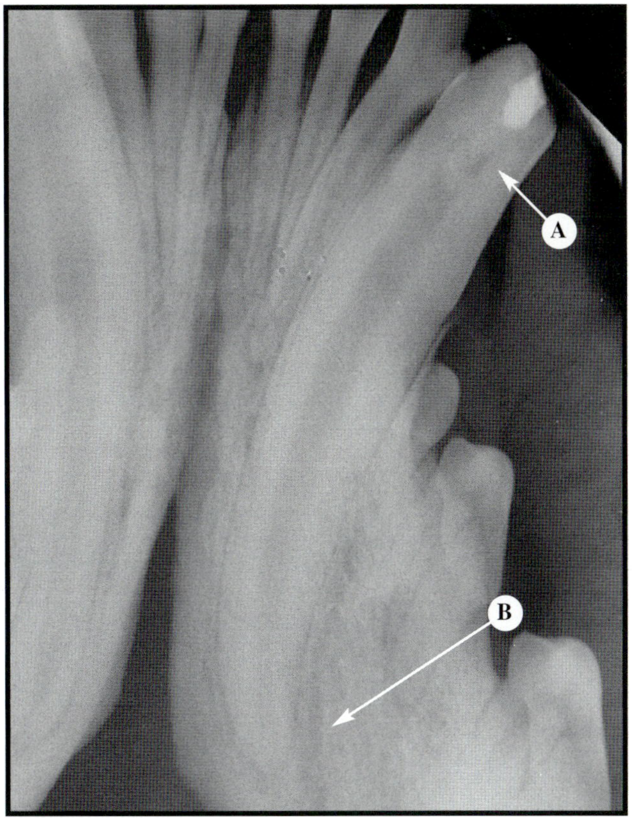

FIGURE 10-49C 12-Month Follow-up Radiograph

Figure 10-49A *This radiograph demonstrates the current recommended method of performing vital pulpotomy of immature teeth.* (A) *Composite and intermediate layers and* (B) *calcium hydroxide layer.* **Figure 10-49B** *Because this view is oblique, there is an apparent disparity in the sizes of the canal. The symmetry of the apices suggests successful therapy.* (A) *The diameter of the canal appears to be wider in the normal canine tooth,* (B) *development of the apex seems symmetric, and* (C) *site of pulpotomy and restoration. Another radiographic view would be recommended.* **Figure 10-49C** *The canal lumen of the two canine teeth are symmetric. The untreated tooth can be used to determine the successful treatment of contralateral teeth. The treated tooth is maturing normally.* (A) *Medication and dentinal bridge and* (B) *complete lengthening of the root and formation of the apex.*

EVALUATION AFTER VITAL PULPOTOMY—MONITORING ENDODONTIC COMPLICATIONS

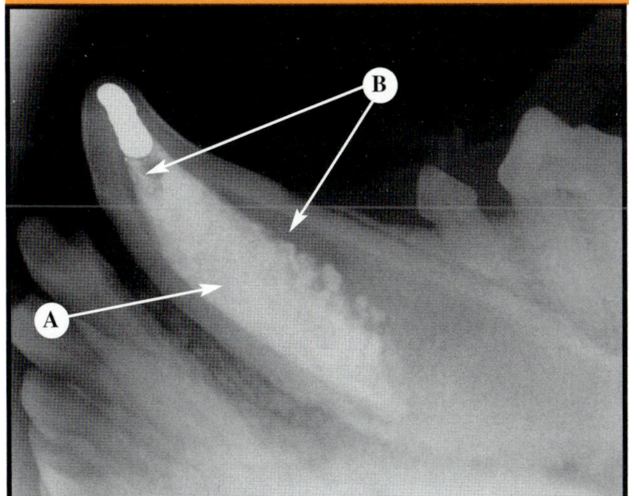

FIGURE 10-50 Overmedication

Figure 10-50 *Excessive treatment can compromise the vitality of the pulp. The technique followed here was poor.* (A) *Calcium hydroxide powder and* (B) *filling voids.* **Figure 10-51** *The treatment of the left tooth in this radiograph was successful; however, treatment of the right tooth failed.* (A) *Normal maturing tooth,* (B) *apical resorption,* (C) *periapical radiolucency, and* (D) *wider canal diameter.*

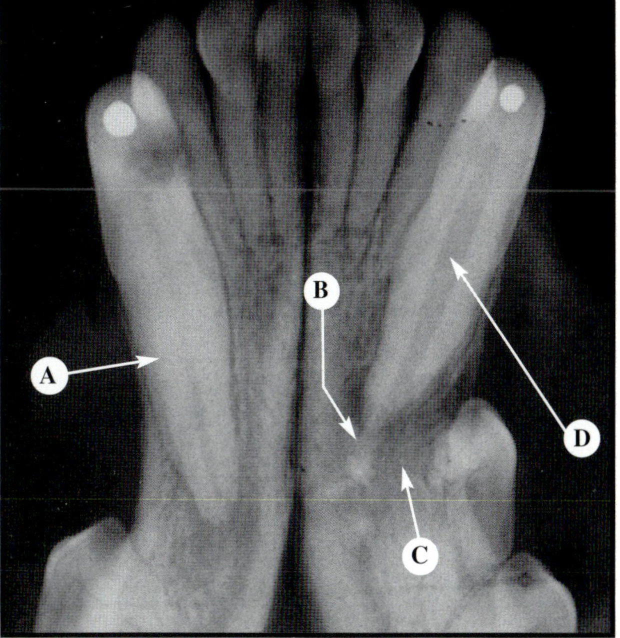

FIGURE 10-51 Treatment Failure in Follow-up Radiograph

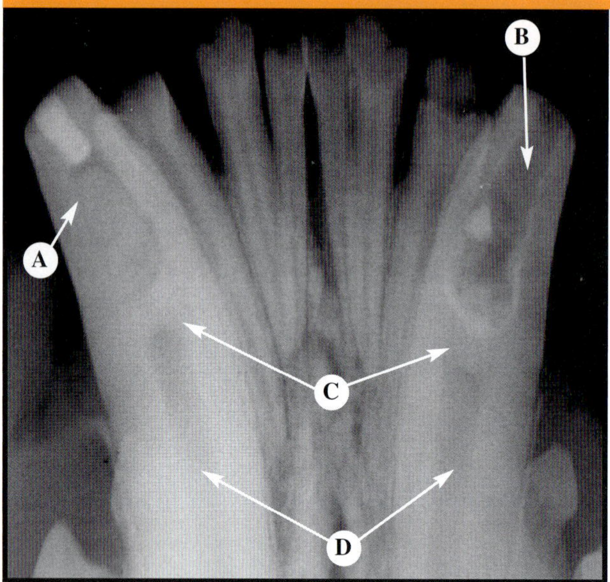

FIGURE 10-52 Treatment Success in Follow-up Radiograph

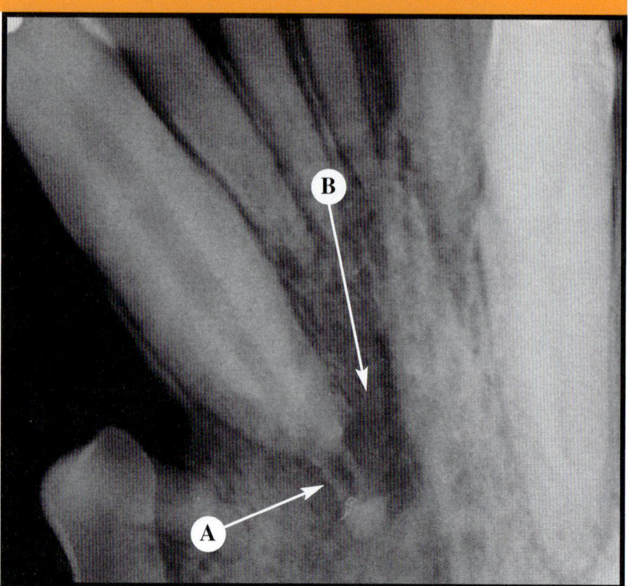

FIGURE 10-53 External Apical Resorption

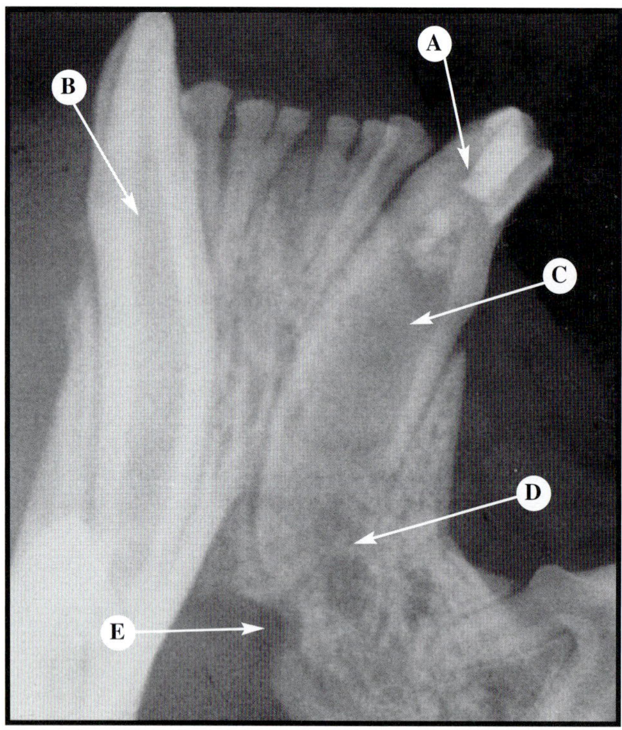

FIGURE 10-54A Treatment Failure: 6-Month Follow-up

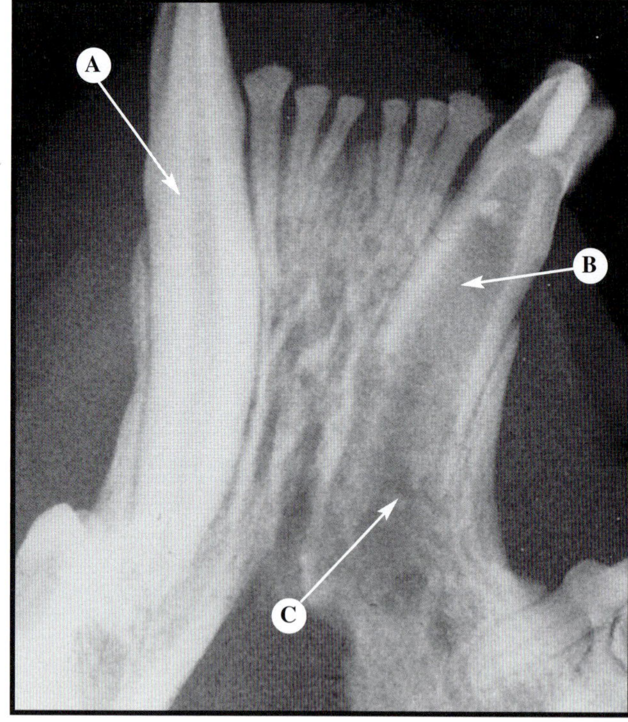

FIGURE 10-54B Treatment Failure: 2-Year Follow-up

Figure 10-52 *In this radiograph, the treatment appears to have been successful despite the loss of preparation on one side. Repair and continued periodic monitoring would be indicated. (A) Retention of initial treatment and restoration, (B) loss of restoration, (C) secondary dentinal bridges, and (D) symmetric diameters of the lumen.* **Figure 10-53** *Treatment of the tooth has failed, and extraction may leave apical fragments. (A) External apical resorption and fragments and (B) periapical bone radiolucency.* **Figure 10-54A** *In this 6-month follow-up radiograph, treatment seems to have failed. Severe inflammation and fistulation are present. (A) Treatment modality, (B) normal tooth, (C) wide canal diameter, (D) apical resorption, and (E) bony lysis.* **Figure 10-54B** *In the 2-year follow-up radiographs to Figure 10-54A, the procedure definitely failed and resorption has progressed. (A) Normal maturation, (B) wide canal diameter, and (C) apical resorption and periapical radiolucency.*

EVALUATION AFTER ROOT CANAL THERAPY—SUCCESSFUL OBTURATION

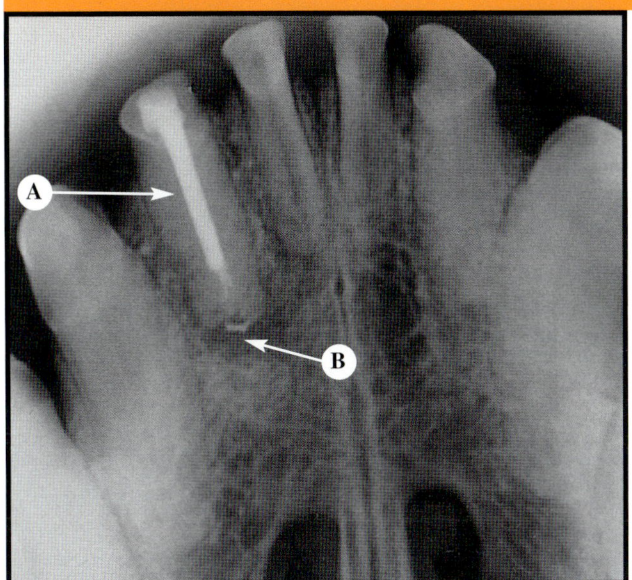

FIGURE 10-55 Incisor Tooth

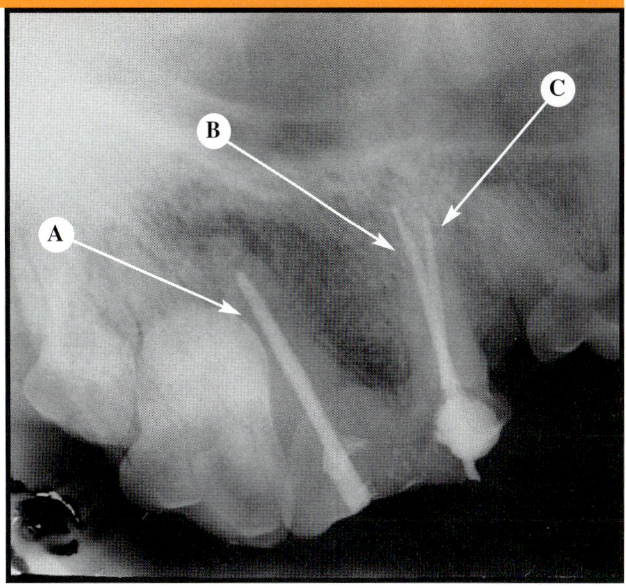

FIGURE 10-56 Maxillary Fourth Premolar

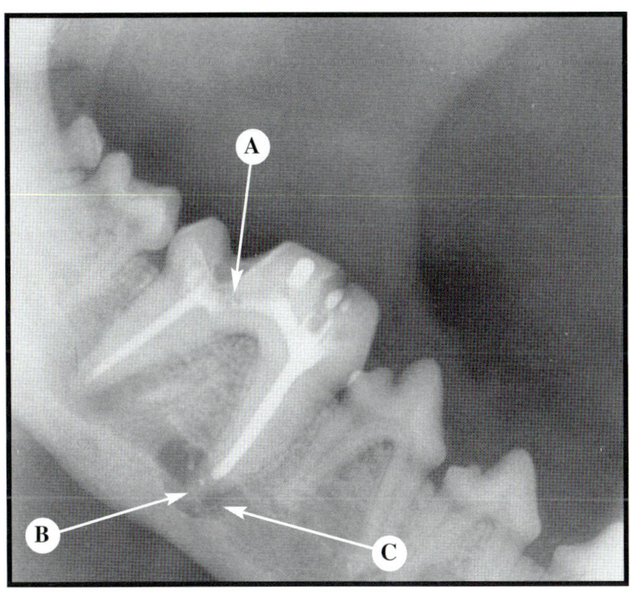

FIGURE 10-57 Mandibular First Molar

Figure 10-55 *Slight extrusion of cement, as shown here, is acceptable. (A) Obturated canal and (B) extrusion of cement through the apex.* **Figure 10-56** *This mesiolateral oblique view shows that all the canals are obturated. The unfilled pulp chamber is acceptable as long as the canals have been properly sealed. (A) Distal root, (B) mesiobuccal root, and (C) palatal root.* **Figure 10-57** *Both canals are obturated. The slight extrusion of cement is acceptable. Note the presence of an endodontic–periodontal lesion. (A) Filling defect in the pulp chamber, (B) slight cement extrusion, and (C) periapical radiolucency.* **Figure 10-58** *This radiograph shows a technique that uses plasticized gutta percha with a carrier file. (A) Endodontic file and (B) obturated canal in a geriatric patient.*

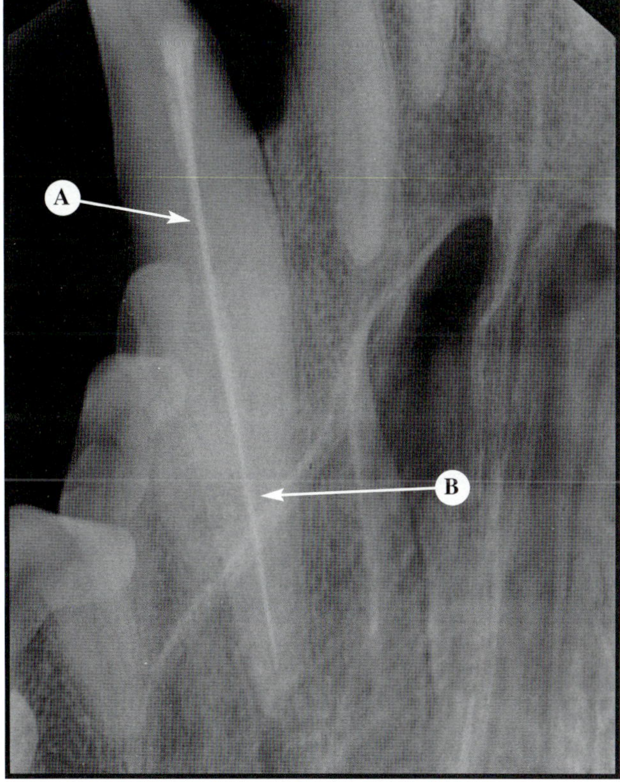

FIGURE 10-58 Plasticized Gutta Percha in Canine Tooth

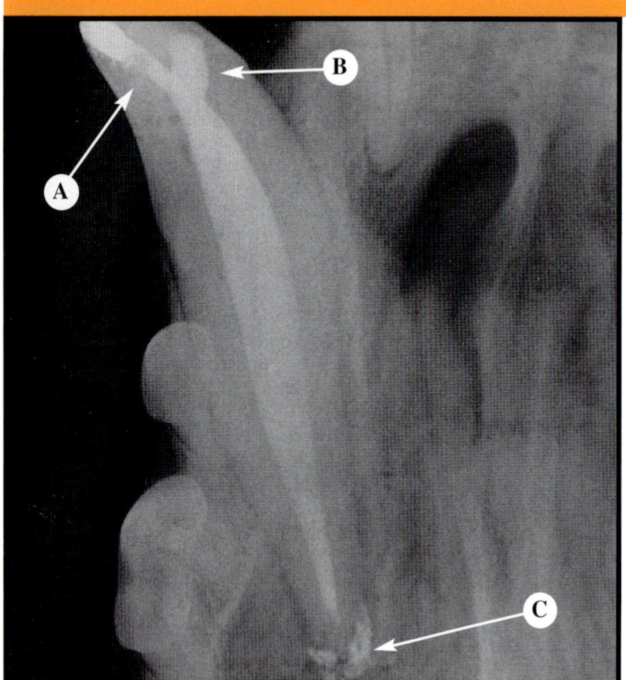

FIGURE 10-59 Cement Extrusion in Canine Tooth

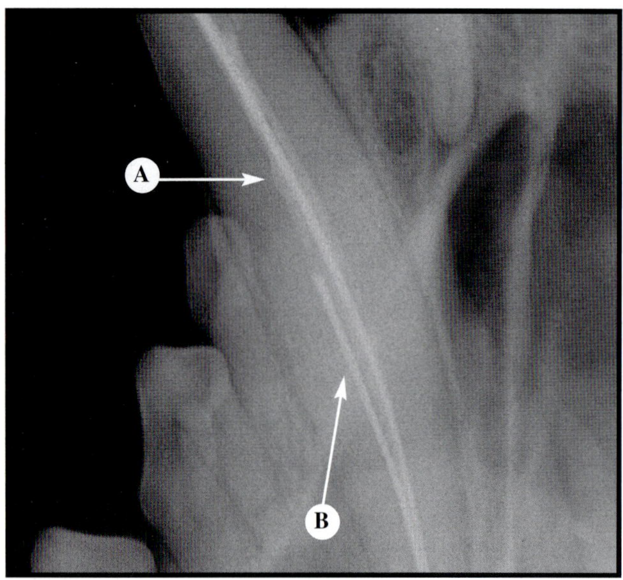

FIGURE 10-61 Bypassing a Broken File

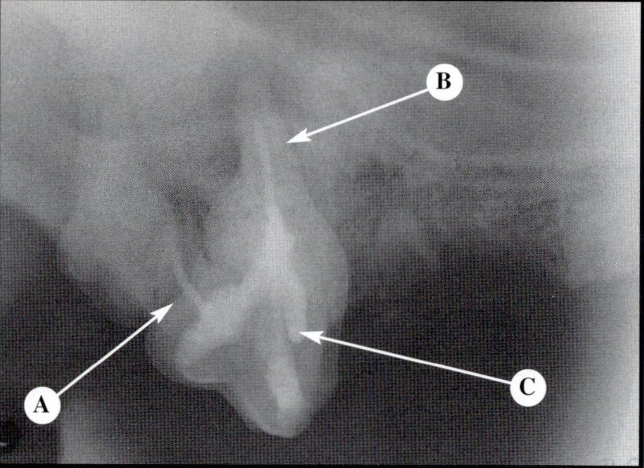

FIGURE 10-60 Maxillary First Molar

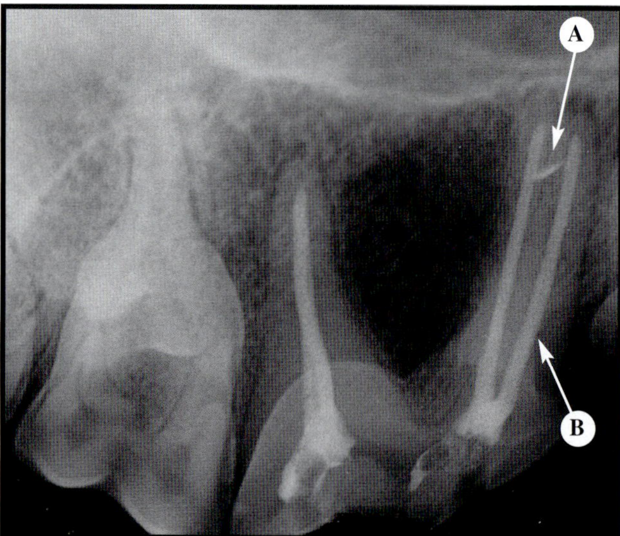

FIGURE 10-62 Lateral Canal of the Maxillary Fourth Premolar

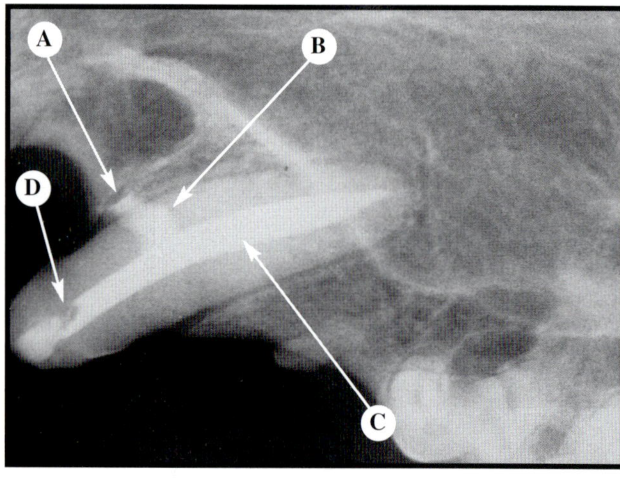

FIGURE 10-63 Lateral Canal of the Canine Tooth

Figure 10-59 *The cement extrusion suggests satisfactory apical seal, although excessive amounts would be undesirable.* (A) *Filled pulp chamber,* (B) *access site, and* (C) *cement extrusion.* **Figure 10-60** *Successful obturation of the upper first molar is difficult without taking several radiographic views. Distortion is inherent in the bisecting angle technique.* (A) *Distobuccal root,* (B) *palatal root, and* (C) *mesiobuccal root.* **Figure 10-61** *Apical obturation is possible around the obstruction.* (A) *Plasticized gutta percha carrier and* (B) *fractured implement.* **Figure 10-62** *The lateral canal is filled with cement.* (A) *Palatal root with the lateral canal and* (B) *mesiobuccal root.* **Figure 10-63** *The lateral canal appears to be filled sufficiently. The apical and periapical regions should be closely monitored.* (A) *Cement extrusion,* (B) *lateral canal,* (C) *obturated primary canal, and* (D) *filling void.*

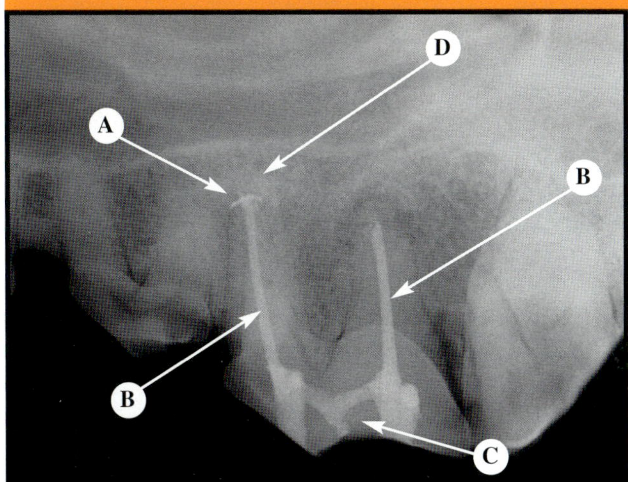

FIGURE 10-64 Questionably Successful Obturation

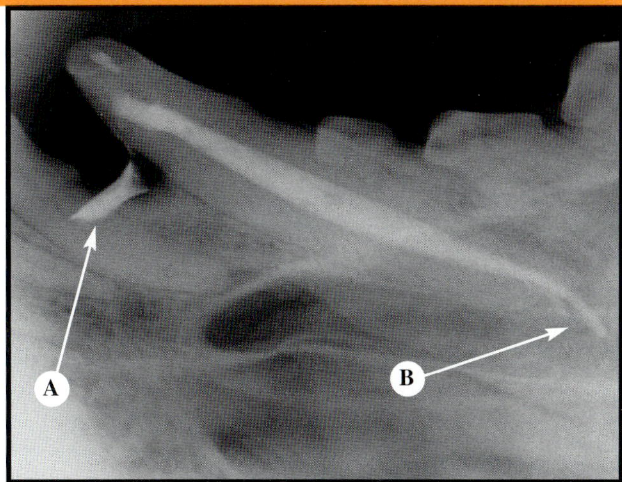

FIGURE 10-65 Overobturation

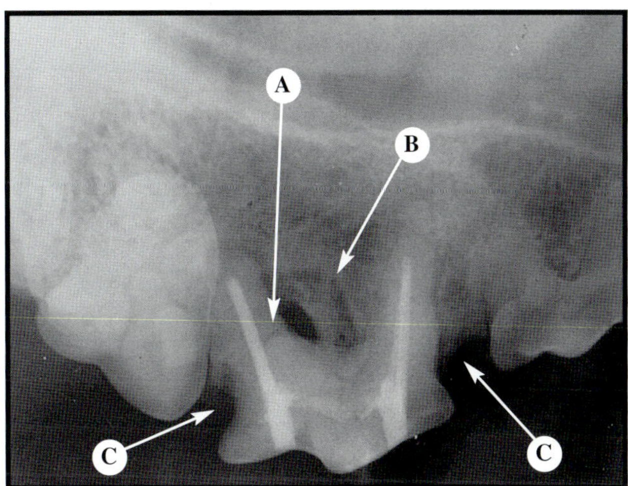

FIGURE 10-66 Insufficient Obturation of Lateral Canal

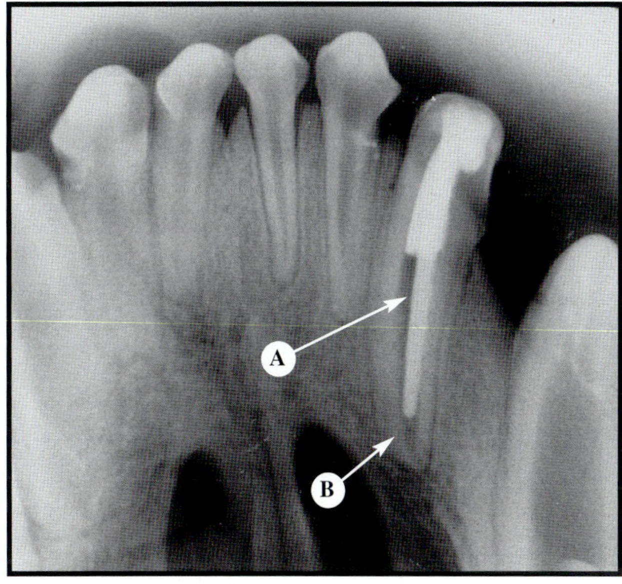

FIGURE 10-67 Insufficient Obturation of an Incisor

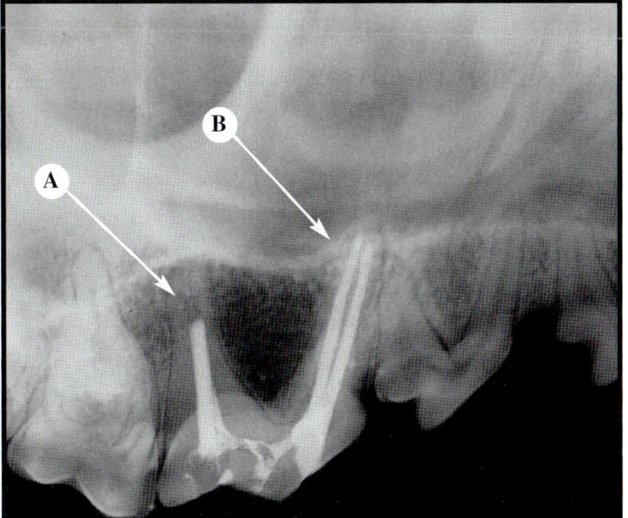

FIGURE 10-68 Insufficient Obturation of a Premolar

Figure 10-64 *All root canals should be filled in multirooted teeth. Because only two of the three roots appear to be filled, another radiographic view should be taken for verification. (A) Slight cement extrusion, (B) fully obturated root canals, (C) well-filled pulp chamber, and (D) palatal root apex.* **Figure 10-65** *Foreign-body reaction is likely to occur if the gutta percha remains in place. Cone penetration suggests an open apex. (A) Debris and (B) penetration of the gutta percha cone into the periapical space.* **Figure 10-66** *The prognosis would be guarded without adequate sealing of the lateral canal. A diagnostic differential for the root canal would be a fracture. (A) No endodontic cement in lateral canal of the distal root, (B) periradicular radiolucency, and (C) periodontal recession.* **Figure 10-67** *The distal one third of the canal must be completely sealed to ensure a good prognosis. Recapitulation and reobturation are needed. (A) No endodontic cement in the apical one half of the canal and (B) the gutta percha cone does not extend to the apex.* **Figure 10-68** *Recapitulation and reobturation of the distal root are indicated. (A) The obturation does not reach the apical extent of the distal root and (B) the mesial roots are obturated.*

EVALUATION AFTER ROOT CANAL THERAPY—UNSUCCESSFUL OBTURATION *(continued)*

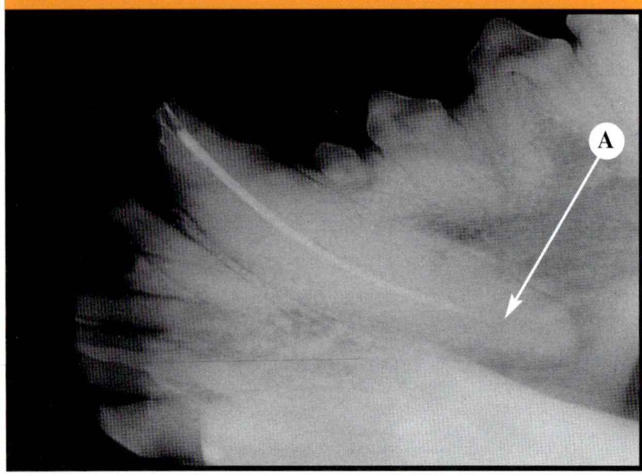

FIGURE 10-69 Insufficient Filling in a Canine Tooth

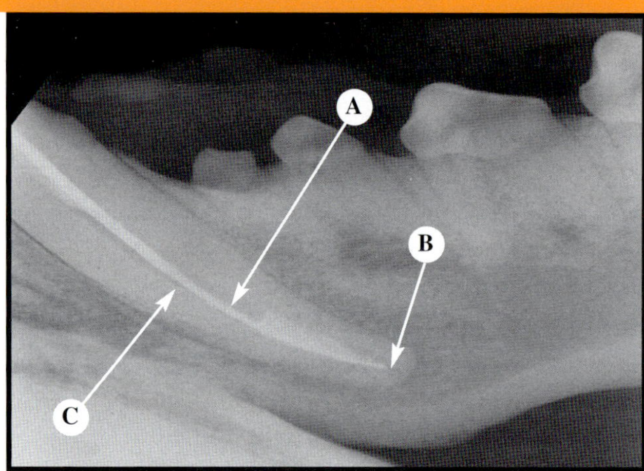

FIGURE 10-70 Insufficient Obturation Using Plasticized Gutta Percha

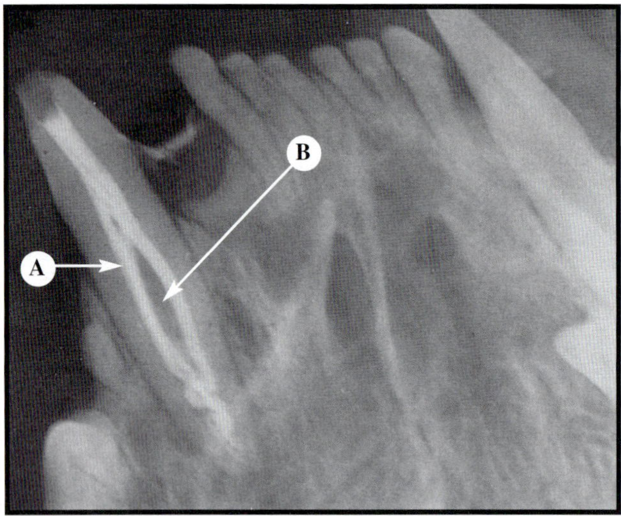

FIGURE 10-71 Insufficient Lateral Condensation

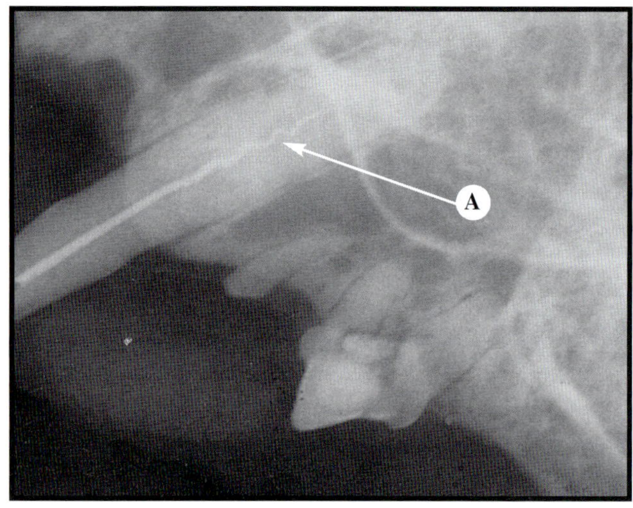

FIGURE 10-72 Incorrect Diameter of the Gutta Percha Cone

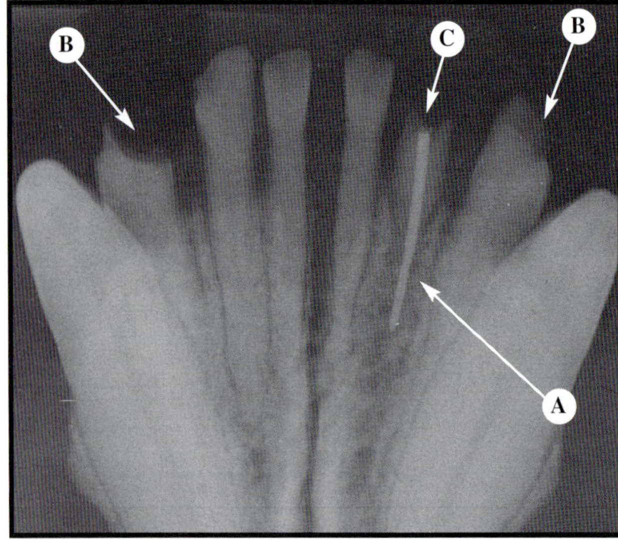

FIGURE 10-73 Root Resorption of an Incisor

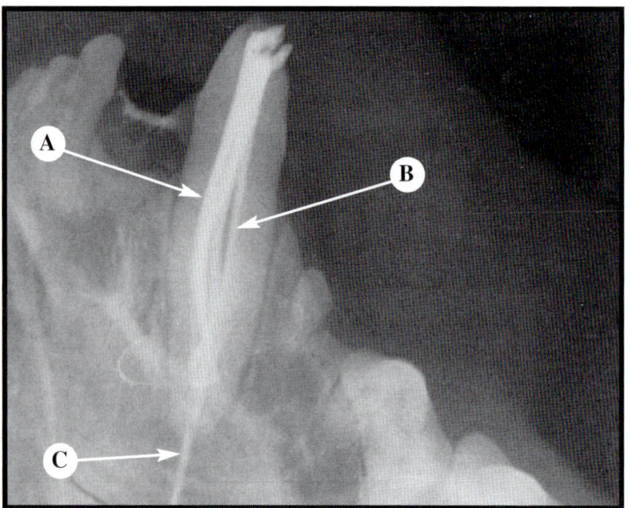

FIGURE 10-74 Overobturation of a Canine Tooth

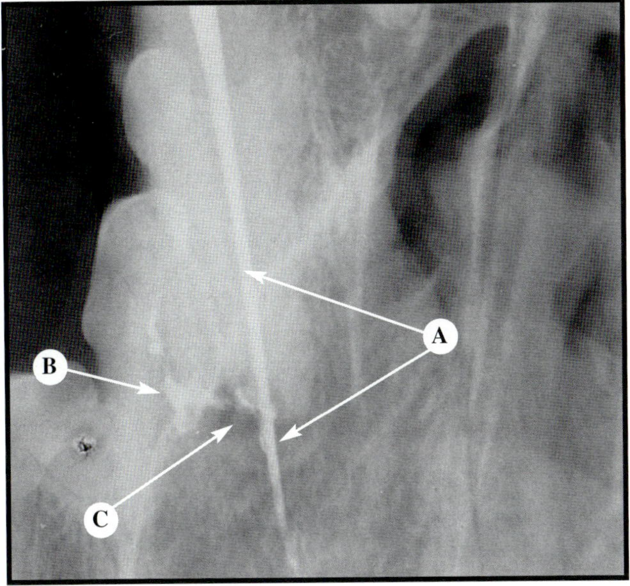

FIGURE 10-75 Overobturation of a Canine Tooth

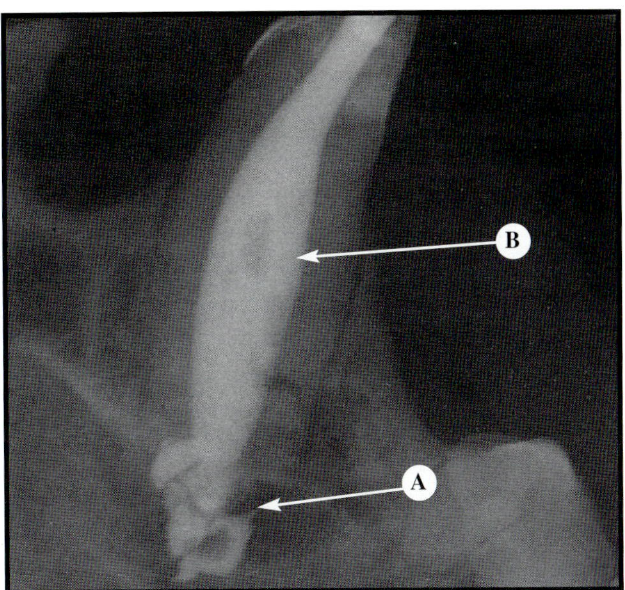

FIGURE 10-76 Extrusion of Cement

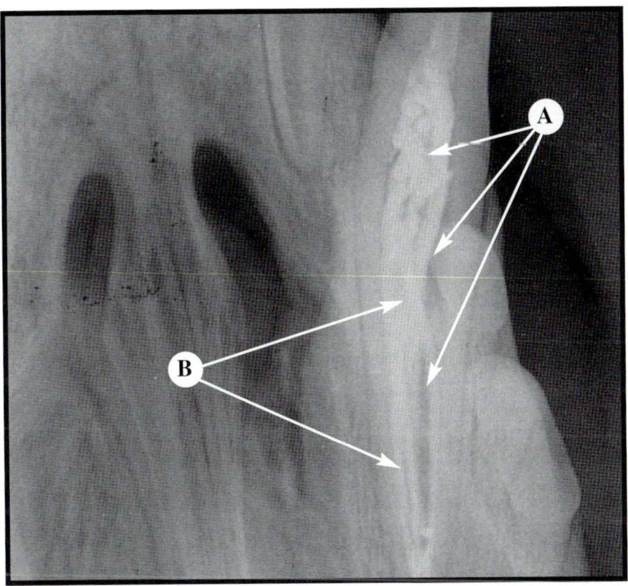

FIGURE 10-77 Insufficient Lateral Condensation

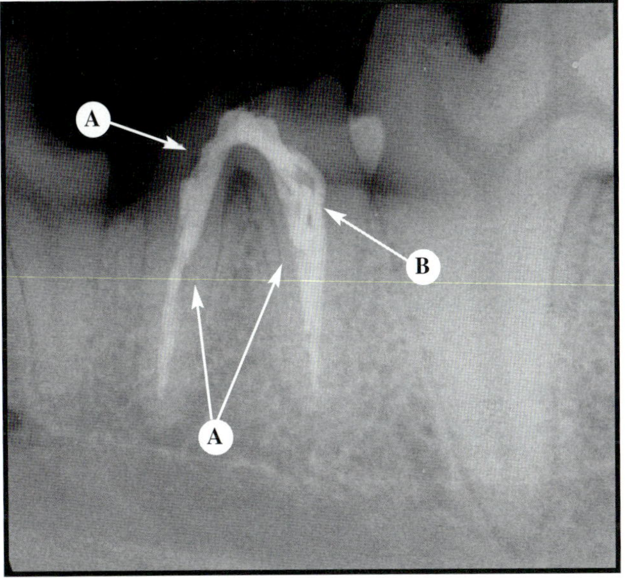

FIGURE 10-78 Incorrect Use of Gutta Percha

Figure 10-69 *In geriatric patients, root canals are often too constricted to allow conventional debridement. (A) Obturation material does not extend to the apical root canal.* **Figure 10-70** *This radiograph shows multiple filling voids. Recapitulation is indicated. (A) Absence of gutta percha, (B) insufficient gutta percha at the apex, and (C) carrier file.* **Figure 10-71** *This radiograph shows that poor technique was followed. There is inadequate lateral and vertical condensation. Recapitulation is indicated. (A) Gutta percha and (B) insufficient gutta percha cement.* **Figure 10-72** *The diameter of the gutta percha cone is too small for the apical canal. The cone can wrinkle when subjected to vertical condensation. A larger cone should be selected to match the diameter of the apical canal. The coronal portion would have to be reshaped. (A) Accordion appearance to gutta percha.* **Figure 10-73** *Treatment obviously failed on the basis of this 24-month follow-up radiograph. Root resorption has occurred compared with the length of the contralateral incisor. (A) Gutta percha cone, (B) additional fractured crowns, and (C) lost restoration.* **Figure 10-74** *This radiograph demonstrates poor technique. The apex is open. Recapitulation and reobturation are indicated. (A) Multiple points of gutta percha, (B) insufficient cement, and (C) gutta percha protruding through the apex.* **Figure 10-75** *The prognosis here would be guarded. Advanced pathologic involvement of the apex and fistulation are present. Surgery may be necessary. (A) Gutta percha cone, (B) endodontic cement following the direction of the fistula, and (C) periapical radiolucency and misshapened apex.* **Figure 10-76** *This immature tooth shows massive cement extrusion, and the apex is open. Apexification could be considered. (A) Excessive cement extrusion and (B) filled immature canal.* **Figure 10-77** *This radiograph demonstrates poor technique. An insufficient amount of cement was used. Reobturation would be indicated. (A) Multiple filling defects and (B) multiple gutta percha cones.* **Figure 10-78** *Poor canal preparation can be noted from the irregular, roughened walls. An insufficient amount of cement is present, and the use of many small cones creates voids. Recapitulation and reobturation would be indicated. (A) Multiple filling defects and (B) multiple small gutta percha cones.*

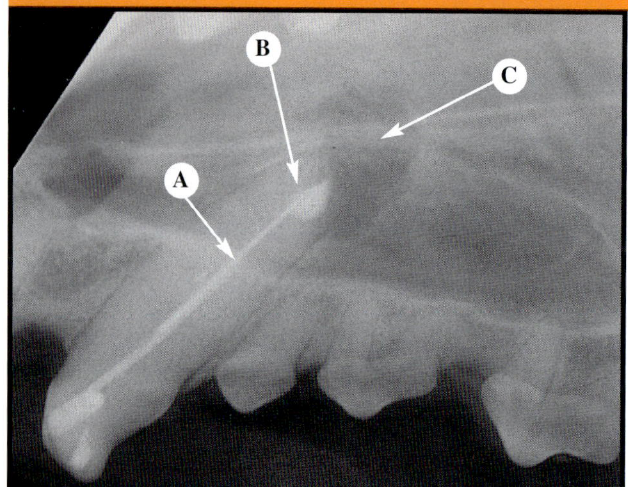

FIGURE 10-79 Canine Tooth

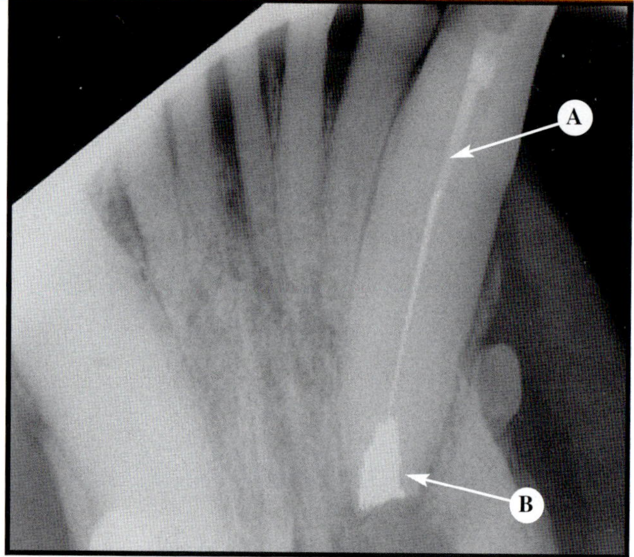

FIGURE 10-80 Canine Tooth

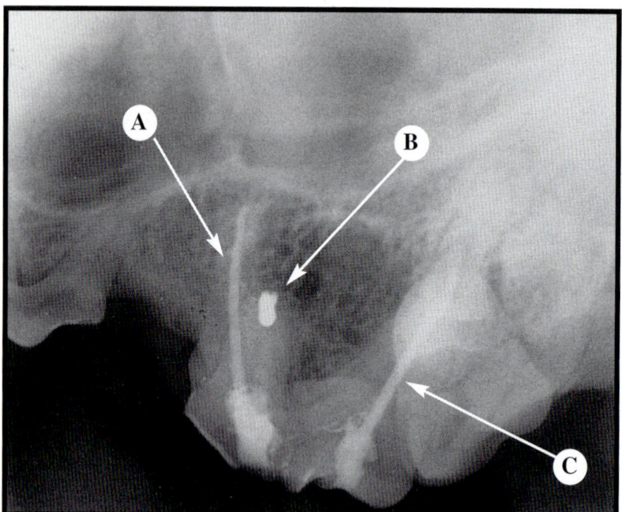

FIGURE 10-81 Maxillary Fourth Premolar

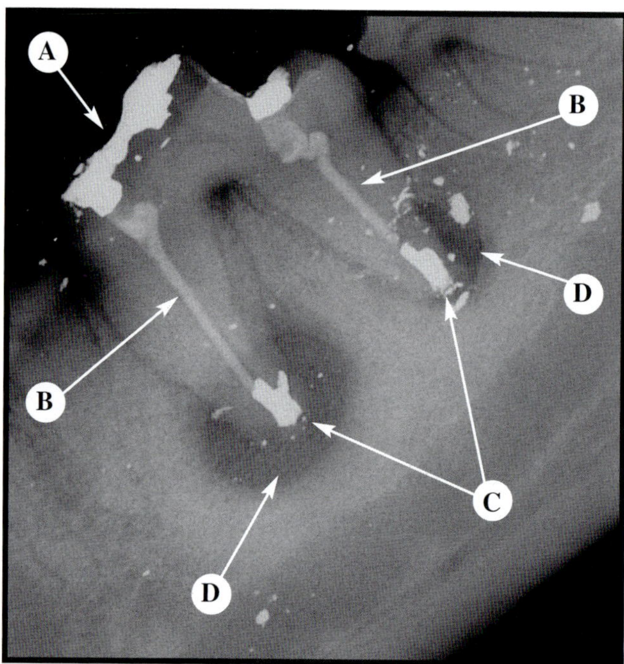

FIGURE 10-82 Mandibular First Molar

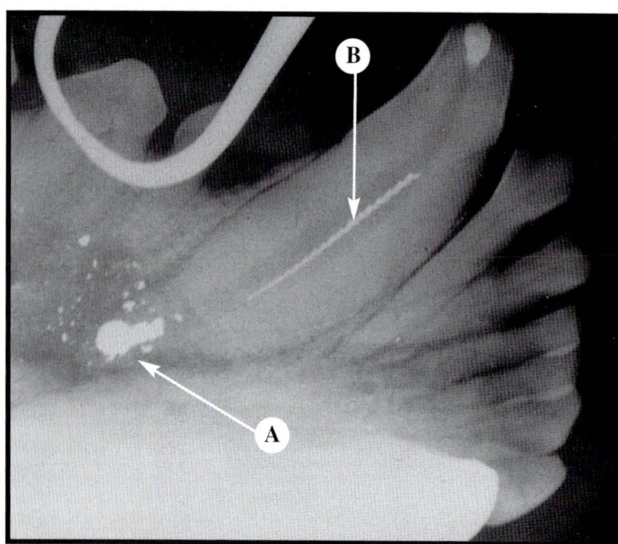

FIGURE 10-83 Fractured Implement

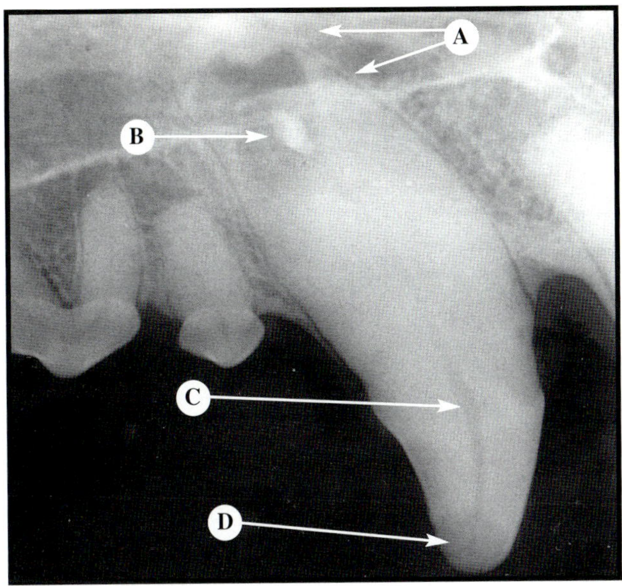

FIGURE 10-84 Incomplete Apicoectomy

Figure 10-79 *Proper technique seems to have been followed here, but two radiographic views would be required for accurate evaluation. The radiolucency that is apparent is the site of surgical access and apicoectomy. (A) Conventional root canal therapy, (B) retrograde filling, and (C) periapical radiolucency.* **Figure 10-80** *Proper technique also seems to have been followed here, but again two radiographic views need to be taken. (A) Conventional root therapy and (B) retrograde filling.* **Figure 10-81** *The mesiobuccal root canal was too constricted to be treated with conventional therapy. (A) Palatal root, (B) retrograde filling at the mesiobuccal root, and (C) the distal root superimposed over the first molar.* **Figure 10-82** *Tiny specks of amalgam debris are easier to notice on radiographs than when an area is grossly examined. (A) Restorative material, (B) conventional treatment, (C) retrograde fillings, and (D) periapical radiolucency.* **Figure 10-83** *A retrograde technique was followed to seal the apex. Completion of conventional therapy, however, is always recommended before electing to perform surgery. The canal seems wide enough to allow circumvention of a broken file. (A) Retrograde filling and (B) broken file.* **Figure 10-84** *The entire apex must be removed to avoid possible endodontic leakage and abscessation. Poor technique was followed here. (A) Incomplete apicoectomy, (B) retrograde filling, (C) no conventional obturation, and (D) missing restoration.*

EVALUATION AFTER SURGERY—HEMISECTED TEETH

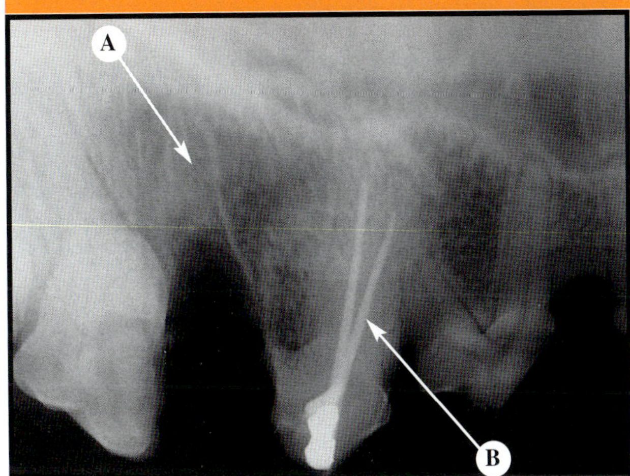

FIGURE 10-85 Mesial Roots of the Maxillary Fourth Premolar

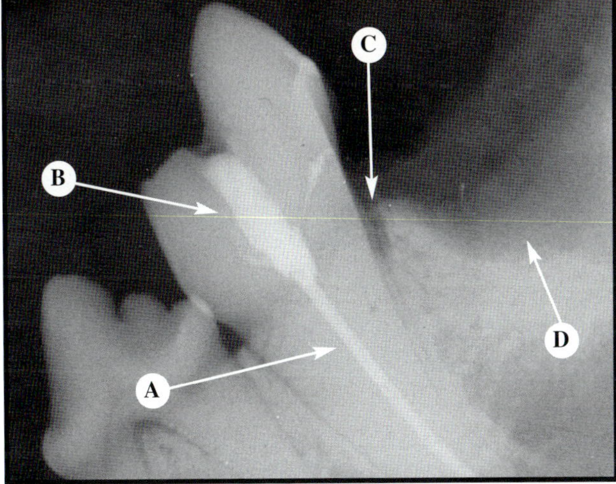

FIGURE 10-86 Mesial Root of the Mandibular First Molar

Figure 10-85 *The salvage technique demonstrated in this radiograph would be appropriate for multirooted teeth only. (A) Extraction site at the alveolus of the distal root and (B) obturation of the canals of the mesial roots.* **Figure 10-86** *The mesial root and the crown were retained for function. (A) Obturation of the canal of the mesial root, (B) obturation of the pulp chamber, (C) wide periodontal space, and (D) extraction site at the distal root.* **Figure 10-87** *If a tooth is vital, pulpotomy is an alternative to root canal treatment or complete extraction. (A) Restoration, (B) pulpotomy site and closure, and (C) original extraction site of the mesial root.*

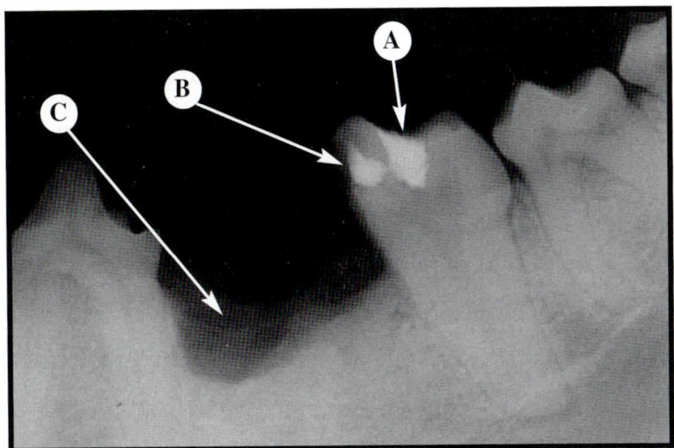

FIGURE 10-87 Distal Root of the Maxillary First Molar After Pulpotomy

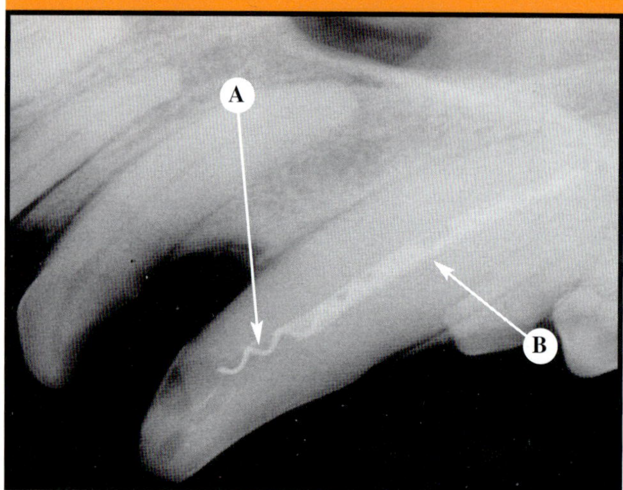

FIGURE 10-88 Broken Implement

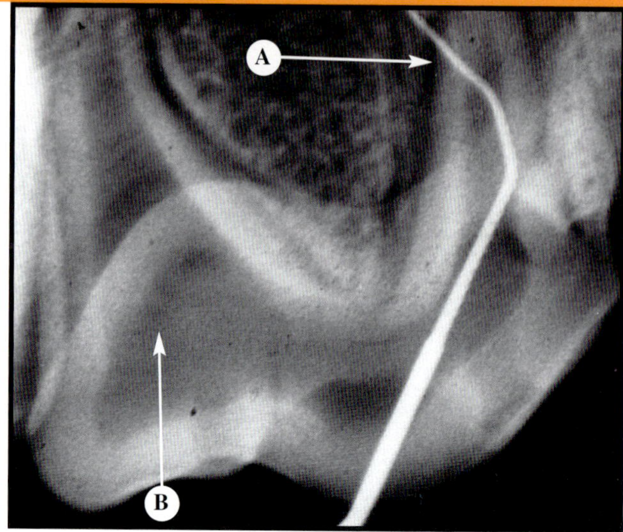

FIGURE 10-89 Lateral Wall Perforation

Figure 10-88 *Fractured implements can interfere with proper sealing of the apex. Inadequate obturation is evident. (A) Broken lentula and (B) filling voids.* **Figure 10-89** *Radiographs should be taken before surgery to assess the tooth structure and thereby avoid inserting instruments improperly. (A) Perforation of the lateral wall and (B) an immature tooth.* **Figure 10-90** *The gutta percha is exposed secondary to root resorption. Note the decreased root support around the crown. (A) Gutta percha and (B) periapical radiolucency and root resorption.*

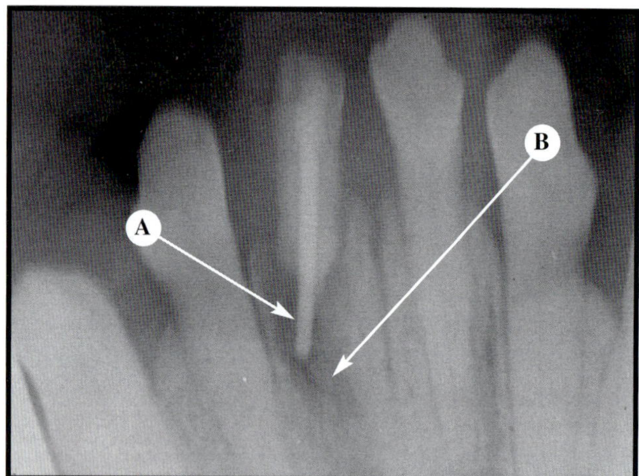

FIGURE 10-90 Root Resorption

REFERENCES *(continued from page 130)*

Identification, and Management, ed 2. Philadelphia, Mosby Year Book, 1992, pp 12–30.

5. Holmstrom SE: Feline endodontics. in Harvey CE (ed): *The Veterinary Clinics of North America. Small Animal Practice,* vol 22, no. 6. Philadelphia, WB Saunders Co, 1992, pp 1433–1451.

6. Trowbridge HO, Kim A: Pulp development, structure, and function, in Cohen S, Burns RC (eds): *Pathways of the Pulp.* Philadelphia, CV Mosby Co, 1994, pp 296–331.

7. Hennet PR, Harvey DV, Harvey CE: Apical root canal anatomy of canine teeth in cats. *Am J Vet Res* 57(11):1545–1548, 1996.

8. Gamm DJ, Howard PE, Harmeet W, Nencka DJ: Prevalence and morphologic features of apical deltas in the canine teeth of dogs. *JAVMA* 202(1):63–70, 1993.

9. Harvey CE, Dubielzig RR: Anatomy of the oral cavity in the dog and cat, in Harvey CE (ed): *Veterinary Dentistry.* Philadelphia, WB Saunders Co, 1985, pp 11–22.

10. Orsini P, Hennet P: Anatomy of the mouth and teeth of the cat, in Harvey CE (ed): *The Veterinary Clinics of North America. Small Animal Practice,* vol 22, no. 6. Philadelphia, WB Saunders Co, 1992, pp 1265–1277.

11. Wiggs RB: Problem solving in veterinary endodontics, in Spodnick GJ (ed): *Seminars in Veterinary Medicine and Surgery (Small Animal): Veterinary Dentistry.* Philadelphia, WB Saunders Co, 1993, pp 165–178.

12. Simon JHS, Werksman LA: Endodontic–periodontal relations, in Cohen S, Burns RC (eds): *Pathways of the Pulp,* ed 6. Philadelphia, CV Mosby Co, 1994, pp 513–530.

13. Walton RE: Endodontic radiography, in Walton RE, Torabinejad M (eds): *Principles and Practice of Endodontics.* Philadelphia, WB Saunders Co, 1989, pp 125–144.

14. Wood NK, Goaz PW, Jacobs MC: Periapical radiolucencies, in Wood NK, Goaz PW (eds): *Differential Diagnosis of Oral Lesions,* cd 4. Philadelphia, Mosby Year Book, 1991, pp 308–331.

15. Wood NK, Goaz PW, Lehnert JF: Periapical radiopacities, in Wood NK, Goaz PW (eds): *Differential Diagnosis of Oral Lesions,* ed 4. Philadelphia, Mosby Year Book, 1991, pp 549–564.

16. Goaz PW, White SC: *Oral Radiology: Principles and Interpretation,* ed 3. Philadelphia, CV Mosby Co, 1994, pp 381–397.

17. Torabinejad M, Walton RE: Pulp and periapical pathosis, in Walton RE, Torabinejad M (eds): *Principles and Practice of Endodontics.* Philadelphia, WB Saunders Co, 1989, pp 38–39.

18. Eisner E: Problems associated with veterinary dental radiogra-

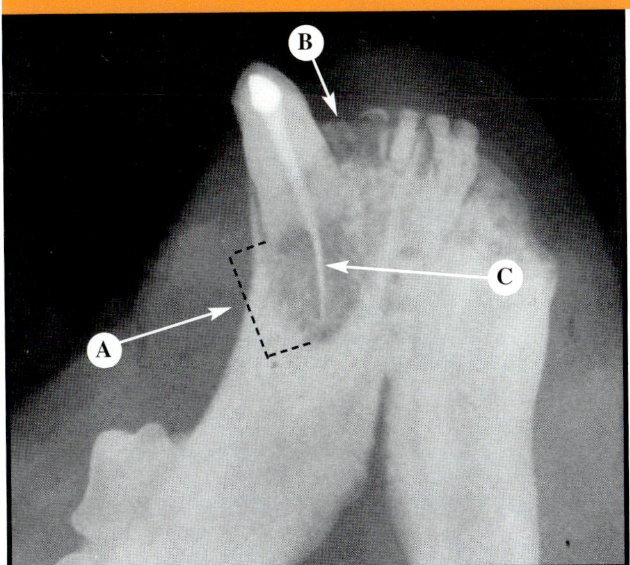

FIGURE 10-91 Failed Root Canal Treatment

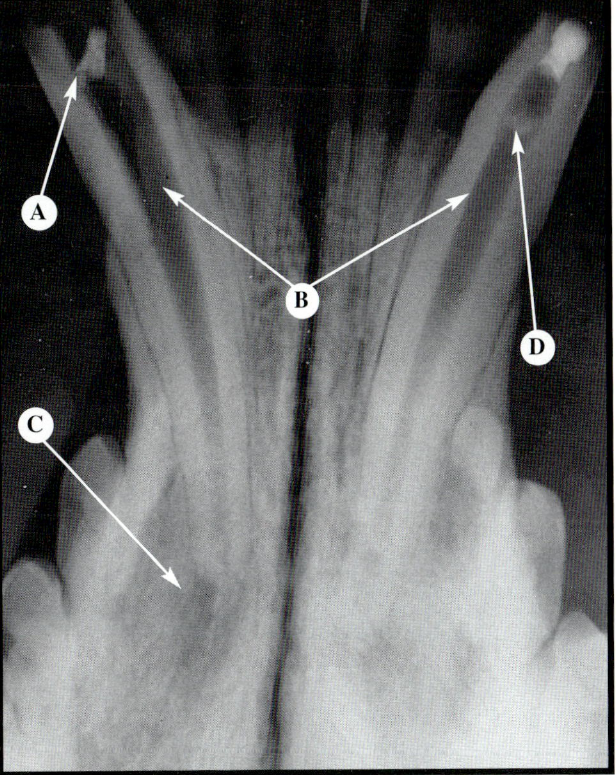

FIGURE 10-92 Failed Pulpotomy

Figure 10-91 *Treatment obviously failed here. The probability of osteomyelitis developing exists. Extraction would be indicated. (A) Resorption of the distal one half of the length of the root, (B) moth-eaten pattern to the bone affecting the adjacent incisors, and (C) gutta percha.*
Figure 10-92 *The pulp of the left tooth is necrotic, and the vitality of the right canine tooth is questionable. Comparison with original radiographs taken before treatment is recommended. (A) Lost preparation material, (B) similar diameters in the lumen, (C) periapical radiolucency and root resorption, and (D) dentinal bridge or original pulp dressing.* **Figure 10-93** *Note the class I endodontic–periodontal lesion. (A) Periapical radiolucency and (B) external resorption with widened periodontal space and wide lumen in the canal.*

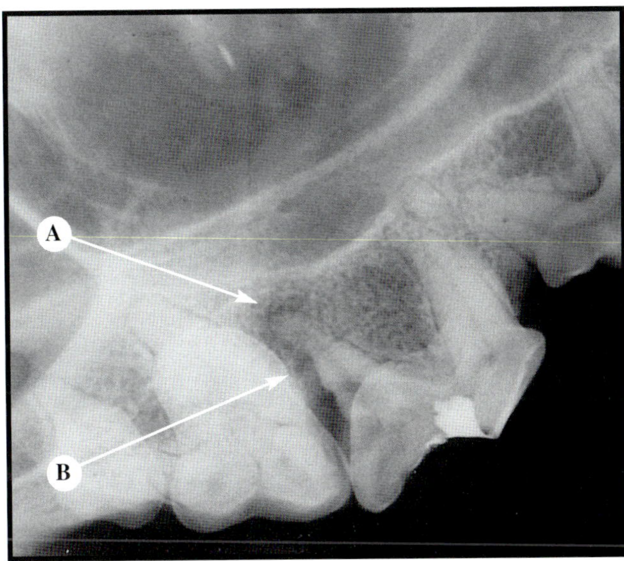

FIGURE 10-93 Failed Pulpotomy and Deep Restoration

phy, in Manfra Marretta S (ed): *Problems in Veterinary Medicine: Dentistry.* Philadelphia, JB Lippincott Co, 1990, pp 46–84.

19. Trope M, Chivian N: Root resorption, in Cohen S, Burns RC (eds): *Pathways of the Pulp,* ed 6. Philadelphia, CV Mosby Co, 1994, pp 486–512.

20. Lobprise HB: Principles of endodontic therapy, in Spodnick GJ (ed): *Seminars in Veterinary Medicine and Surgery (Small Animal): Veterinary Dentistry.* Philadelphia, WB Saunders Co, 1993, pp 155–163 .

21. Simon JHS: Periapical pathology, in Cohen S, Burns RC (eds): *Pathways of the Pulp,* ed 6. Philadelphia, CV Mosby Co, 1994, pp 337–362.

22. Kettering JD, Torabinejad M: Microbiology and immunology, in Cohen S, Burns RC (eds): *Pathways of the Pulp,* ed 6. Philadelphia, CV Mosby Co, 1994, pp 363–376.

23. Fountain SB, Camp JH: Traumatic injuries, in Cohen S, Burns RC (eds): *Pathway of the Pulp,* ed 6. Philadelphia, CV Mosby Co, 1994, pp 436–485.

24. Marretta SM, Schloss AJ, Klippert LS: Classification and prognostic factors of endodontic-periodontic lesions in the dog. *J Vet Dent* 9(2):27–30, 1992.

25. Tourbinejad M: Endodontic/periodontic interrelationships, in Walton RE, Torabinejad M (eds): *Principles and Practice of Endodontics.* Philadelphia, WB Saunders Co, 1989, pp 433–445.

26. Simon JHS, Werksman LA: Endodontic-periodontal relations, in Cohen S, Burns, RC (eds): *Pathways of the Pulp,* ed 6. Philadelphia, CV Mosby Co, 1994, pp 513–530.

27. Gutmann JL, Ptt Ford TR: Problems in the assessment of succcss

and failure, in Gutmann JL, Dumsha TC, Lovdahl PE, Hovland EJ (eds): *Problem Solving in Endodontics: Prevention, Identification and Management,* ed 2. Philadelphia, Mosby Year Book, 1988, pp 1–11.

28. Stabholz A, Friedman S, Tamse A: Endodontic failures and retreatment, in Cohen S, Burns RC (eds): *Pathways of the Pulp,* ed 6. Philadelphia, CV Mosby Co, 1994, pp 690–728.

29. Valle G: Evaluation of success and failure, in Walton RE, Torabinejad M (eds): *Principles and Practice of Endodontics.* Philadelphia, WB Saunders Co, 1989, pp 311–320.

30. Camp JH: Pediatric endodontic treatment, in Cohen S, Burns RC (eds): *Pathways of the Pulp,* ed 6. Philadelphia, CV Mosby Co, 1994, pp 633–671.

Chapter Eleven

ACQUIRED DEFECTS: CARIES AND REGRESSIVE CHANGES

Defects in the structure of teeth can be acquired through carious processes, resorption, trauma, erosion, abrasion, or attrition. The standard system for describing defects of the crown is the G. V. Black method of classification, which is based on the anatomic location of the lesion and not on the severity or cause.[1] Table 11-1 outlines a modified version of this classification system, and Figures 11-1 to 11-6 illustrate examples of the various classes. Radiographs that capture dental abnormalities of the crown are presented later in this chapter.

CARIES

When bacterial activity causes demineralization of enamel or cementum, the resultant defect is called a caries. Clinically, carious lesions are soft and sticky. When tooth destruction is extensive, the pulp may be exposed and endodontic disease can follow. The demineralization process usually originates in pits and fissures on the enamel, spreads to the dentinoenamel junction, and continues to penetrate the tooth along the dentinoenamel junction. Such a lesion is visible on radiographs as a thin radiolucent line between the enamel and dentin; eventually the defect extends toward the pulp in a spherical pattern. As the lesion spreads through the dentin, it undermines the enamel; the result is cavitation of the occlusal surface. On radiographs, a carious lesion appears as a radiolucent defect with diffuse margins. Some of the figures presented later in this chapter show the radiographic images of caries. More than 40% of a tooth is affected by demineralization before the lesion can be detected on a radiograph. Therefore, the margins are not an accurate depiction of the advancing edge of the lesion. Caries rarely affect the teeth of animals; however, if a caries does develop, the most common location is the occlusal surfaces of the molars in dogs. Carious lesions should be treated.[3-5]

RESORPTION

The classification of resorptive lesions on the basis of depth and location has been proposed.[6-8] Mixed lesions result when a lesion extends from one region to another or when several lesions coalesce. Mixed lesions constitute end-stage pathology and are often so advanced that determining their origin is difficult. The depth of a lesion can be best determined by clinical examination using an explorer or probe. Table 11-2 lists the four stages that classify the depth of lesions, and Figures 11-7 to 11-10 illustrate each stage. The region (Table 11-3) is best determined on radiographs, examples of which are provided in Figures 11-11 to 11-13.

Root resorption is associated with chronic inflammation, periodontal disease, endodontic disease, endocrinopathy, chronic vomiting, dietary risk factors, trauma, and idiopathic conditions. Resorption of tooth substances can occur internally, externally, or both. External tooth resorption occurs more often in cats than in dogs, and histopathologic origin has been linked to osteoclastic activity of surrounding tissue rather than to a carious process. In cats, resorptive lesions have

TABLE 11–1
MODIFIED G. V. BLACK CLASSIFICATION SYSTEM FOR CROWN DEFECTS[1,2]

Class I	Pits and fissures on the occlusal surfaces of molars
Class II	Defect on the interproximal surfaces of premolars and molars
Class III	Defect on the interproximal surfaces of incisors; incisal edges are not involved
Class IV	Defect on the interproximal surfaces and incisal edges of incisors
Class V	Defect on the gingival third of labial, buccal, lingual, or palatal surfaces of the teeth
Class VI	Defect on the incisal edges of anterior teeth and cusp tips of posterior teeth

CLASSIFICATION OF THE LOCATION OF DEFECTS

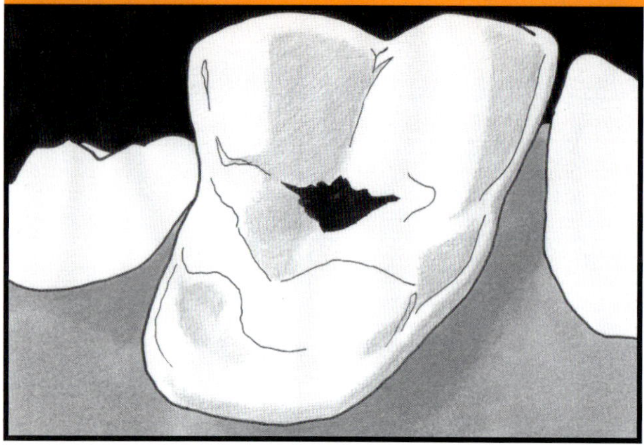

FIGURE 11-1A

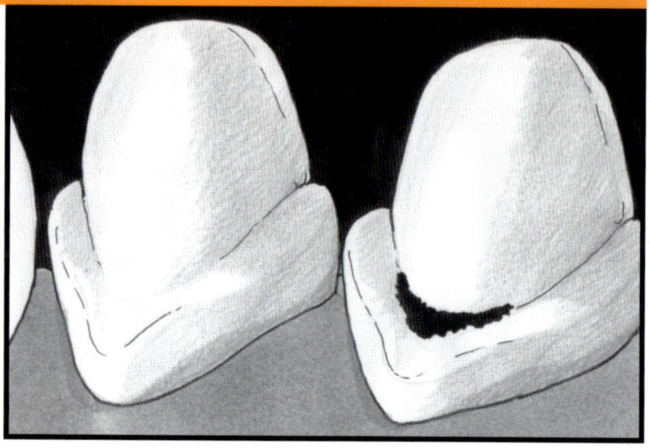

FIGURE 11-1B

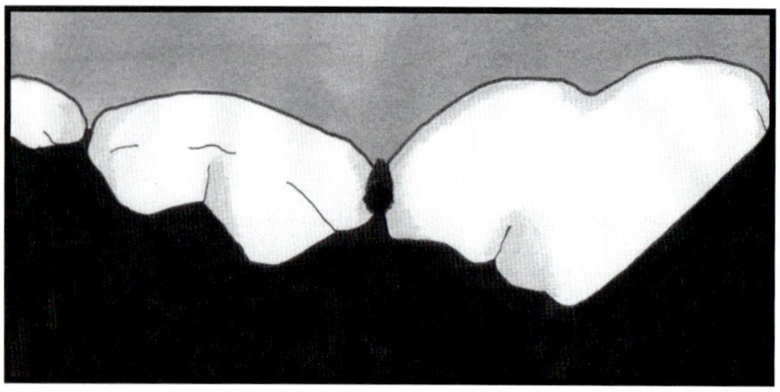

FIGURE 11-2

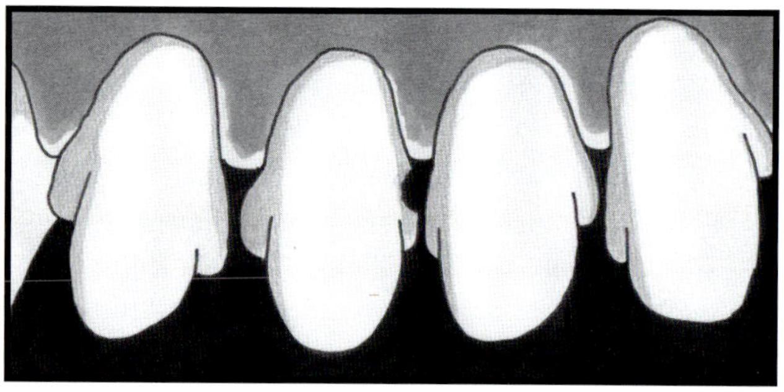

FIGURE 11-3

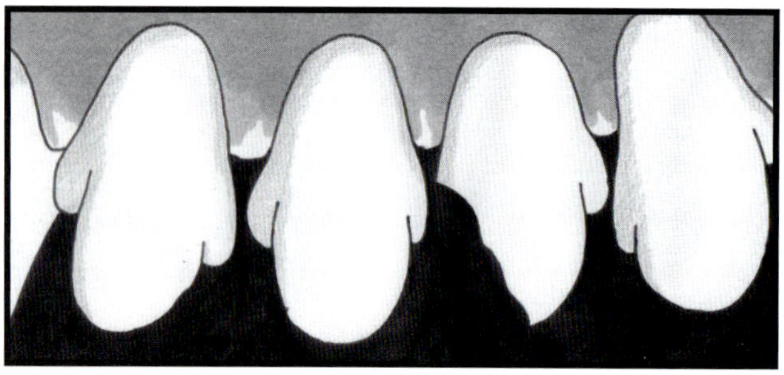

FIGURE 11-4

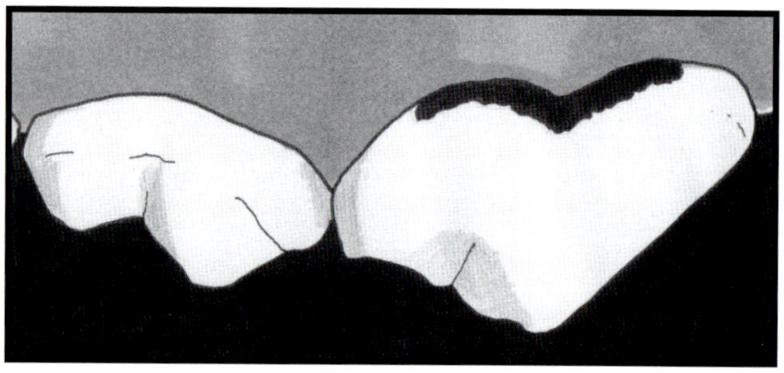

FIGURE 11-5A

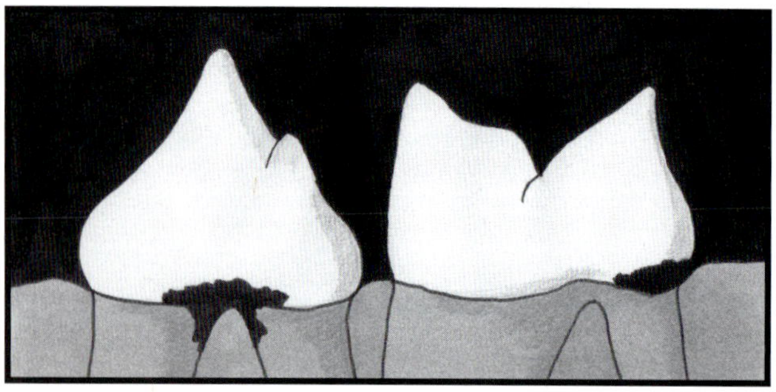

FIGURE 11-5B

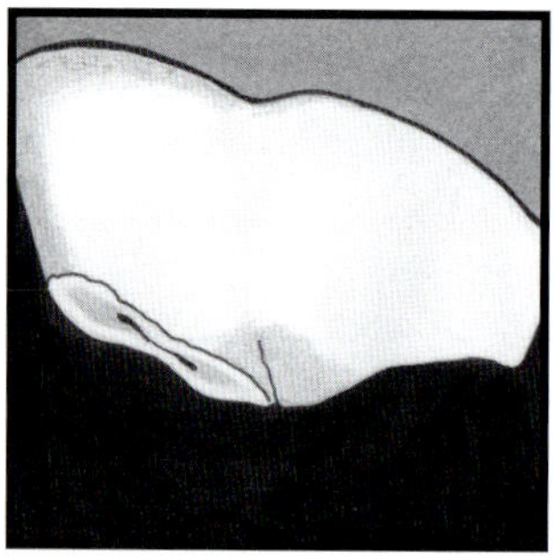

FIGURE 11-6A

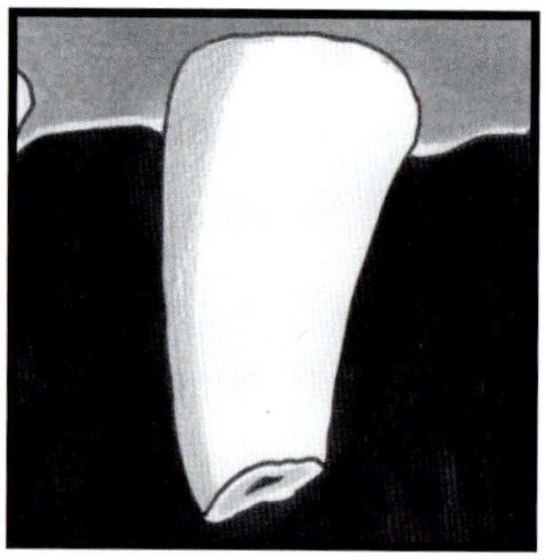

FIGURE 11-6B

Figure 11-1A *Class I defect (pits and fissures) is a common defect of the first molar in dogs.* **Figure 11-1B** *Class I defect located in the cingulum of an incisor.* **Figure 11-2** *Class II interproximal defect.* **Figure 11-3** *Class III interproximal defect.* **Figure 11-4** *Class IV interproximal defect.* **Figures 11-5A and 11-5B** *Class V cervical line defects. Figure 11-5A shows a typical lesion that occurs in dogs, whereas 11-5B shows the common locations for defects in cats.* **Figures 11-6A and 11-6B** *Class VI defects. Figure 11-6A shows a cusp fracture of the maxillary fourth premolar, and 11-6B shows a typical fracture of the canine tooth.*

been referred to as cervical line erosions, neck lesions, external odontoclastic resorption, and feline cavities. These lesions are usually detected at the neck of the tooth and exhibit external resorption; however, internal resorption can occur. The neck, crown, root, or all three structures may be involved. The lesions feel hard when touched with a dental probe or explorer and contain sharp edges, even though they may be filled with soft granulation tissue or hyperplastic gingival tissue. Deep lesions expose the pulp. Although internal resorption cannot be clinically detected, the crown will acquire a pink hue as the tooth wall becomes thinner.

On radiographs, resorptive lesions appear radiolucent, although the patterns of radiolucency can vary. Localized resorptive lesions are discrete and well demarcated and have sharp or scalloped margins. Invasive resorptive lesions are more diffuse and give the tooth a moth-eaten or striated appearance. Roots may undergo replacement resorption, ankylosis, or

TABLE 11–2
STAGES THAT DESCRIBE THE DEPTH OF DEFECTS[6]

Stage 1	Shallow defects of the enamel that do not enter the dentin
Stage 2	Defects that enter the dentin through enamel or cementum; the pulp is not exposed
Stage 3	Defects that enter the pulp chamber; some loss of crown structure occurs
Stage 4	Loss of crown structure; loss or ankylosis of root structure may occur

sequestration. When replacement resorption occurs, the tooth structure is replaced by a bony structure; and the contour of the root becomes irregular and notched. Ankylosis occurs when the periodontal ligament is replaced with bone, thereby fusing the

STAGES THAT IDENTIFY THE DEPTH OF A DEFECT

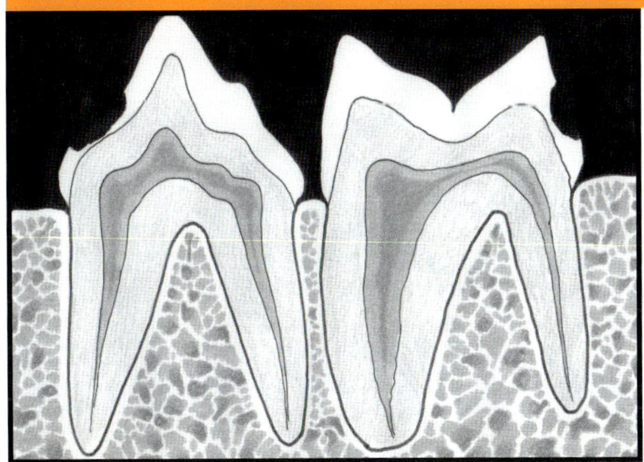

FIGURE 11-7 Stage 1

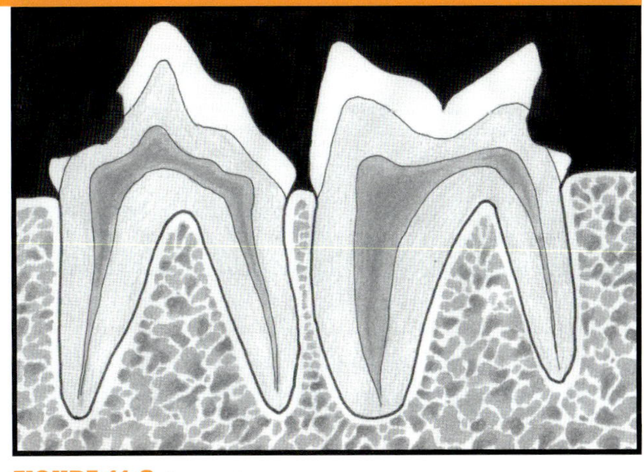

FIGURE 11-8 Stage 2

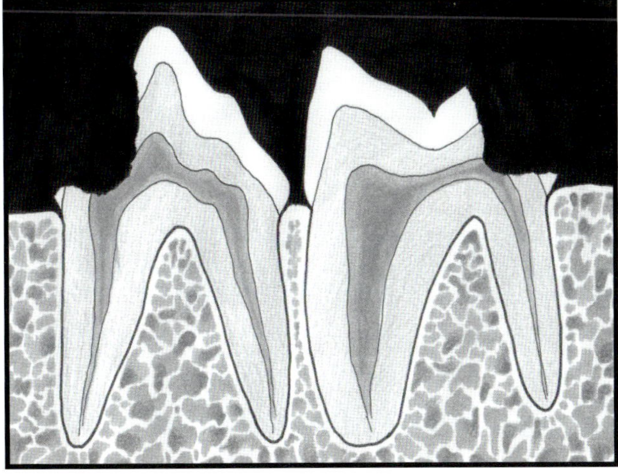

FIGURE 11-9 Stage 3

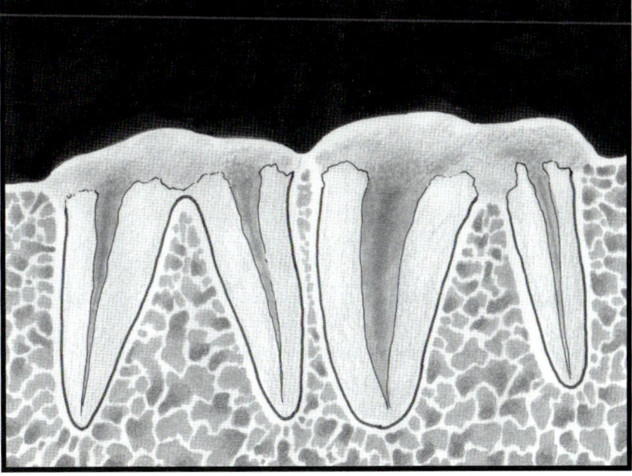

FIGURE 11-10 Stage 4

Figures 11-7 through 11-10 *These illustrations depict the four stages that describe the depth of the defect (see Table 11-2).*

TYPES OF DEFECTS BY REGION

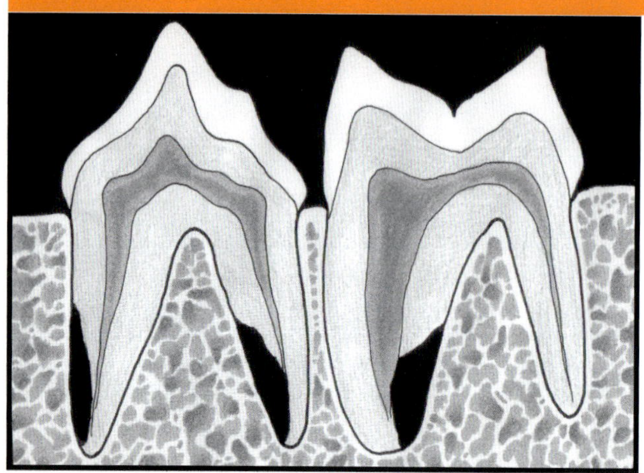

FIGURE 11-11 Intraosseous

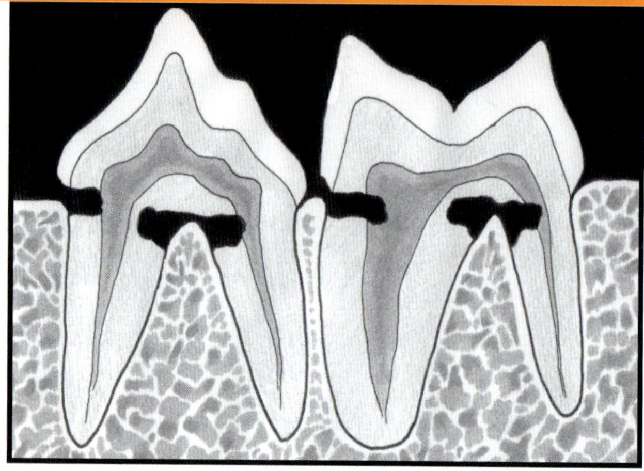

FIGURE 11-12 Crestal

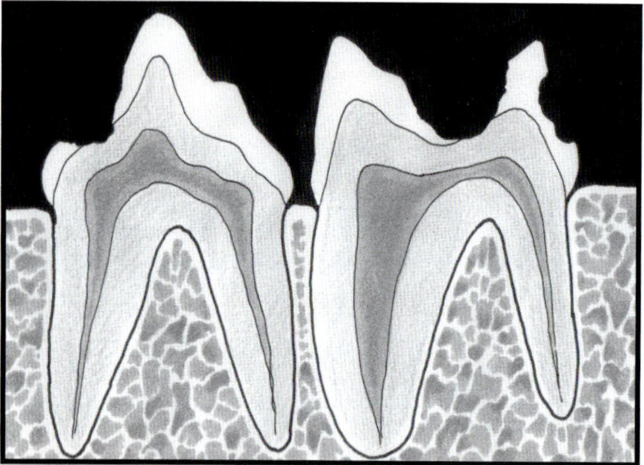

FIGURE 11-13 Supraosseous

Figure 11-11 *The origin of intraosseous lesions (or resorptions) is often unknown; however, they can sometimes be diagnosed with endodontic disease.* **Figure 11-12** *Crestal defects are a common site for external resorptions in cats. These resorptions continue to progress into the crown and root. Slab fractures and root caries are also causes of defects in this region.* **Figure 11-13** *Supraosseous defects are most often associated with resorptive lesions in cats, carious processes in dogs, and crown fractures in cats and dogs.*

DIFFERENTIATING CERVICAL BURNOUT AND RESORPTION

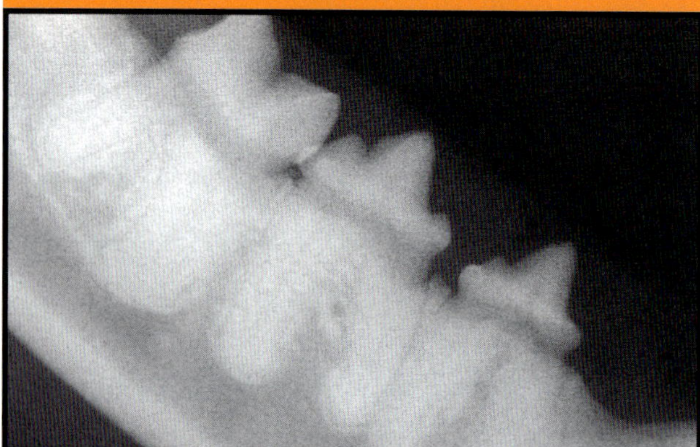

FIGURE 11-14A Cervical Burnout

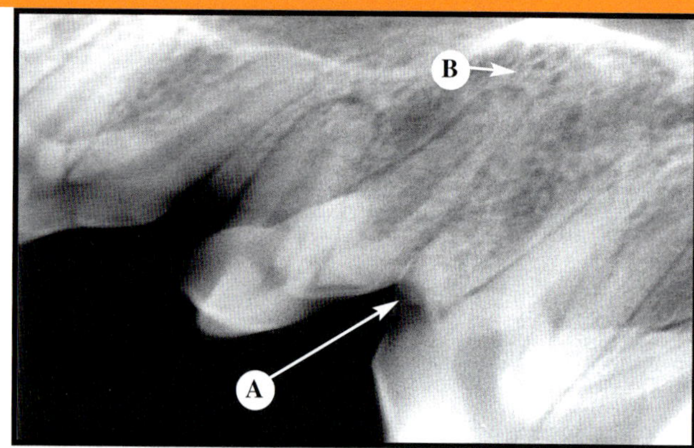

FIGURE 11-14B Cervical Burnout

Figure 11-14A *The cervical area has a more radiolucent band compared with the crown and root. Horizontal bone loss accentuates the artifact of cervical burnout by further decreasing the mass in the cervical area. True crestal defects are not present on these teeth.* **Figure 11-14B** *Cervical burnout and indistinct roots often occur when the horizontal angle of the x-ray beam is excessive. Such positioning of the beam is often done purposely to prevent superimposition of root images in radiographs of the fourth upper premolar and first molar. To rule out a lesion of the enamel, new radiographs should be taken using a different angle and the tooth should be reinspected closely.* (A) *Cervical burnout and* (B) *indistinct root.*

TABLE 11–3
LOCATION OF DEFECTS BY REGION[8]

Intraosseous	No direct communication with the oral cavity by way of periodontal lesion; resorption originates in the bone; root initially affected
Crestal	Direct or indirect communication to the oral cavity by way of periodontal lesion; alveolar crestal bone is involved; neck and furcations initially affected
Supraosseous	Direct exposure to the oral cavity; crown primarily involved

root to the bone and making extraction by conventional means very challenging. If sclerosis is present, ankylosis can be difficult to discern because assessment of the periodontal ligament space is obscured. Radiography is essential in determining the appropriate treatment for resorptive lesions. The extent of lesions is more accurately and reliably evaluated on radiographs because they are often more extensive and serious than is apparent during clinical examination; the latter can determine the depth and extent of lesions only in accessible areas of the tooth (i.e., the crown and neck).[5,9-12]

When viewing radiographs, artifacts from cervical burnout should not be confused with or interpreted as a resorptive lesion at the cementoenamel junction. Cervical burnout, or adumbration, results from the different densities at the neck of the tooth, where the cervical region of the crown and cervical region of the root (covered by alveolar bone) meet. Cervical burnout is also accentuated on radiographs by excessive horizontal angle of the x-ray beam. The cervical region normally is slightly concave; on radiographs, this concavity is evident at the mesial and distal surfaces of the region. In addition, a portion of the cervical region of the crown is not covered by alveolar bone. Jointly, this portion of the region and the natural concave surface affect penetration by the x-ray beam. The result is a radiolucent collar at the cementoenamel junction (Figure 11-14). Radiographic defects caused by cervical burnout disappear when other radiographs taken from different angles are generated. Cervical defects that are associated with external resorption persist in each radiographic view but can be detected on clinical examination using an explorer.[4,5] Table 11-4 compares some of the radiolucent characteristics of radiolucent defects of teeth, including those associated with lesions, caries, external resorption, and internal resorption, and the characteristics of cervical burnout.

RESTORATIONS

Amalgam, composite, and glass ionomer restorative materials are popular in veterinary dentistry for the restoration of crown defects. Any defect that is restored should be monitored on radiographs for signs of failure, such as continuing resorption, progressive caries, or pulpitis. The radiopacity of the materials depends on their composition. Radiographs that are taken at some angles may present a radiolucent halo around a restoration if a radiolucent base, such as calcium hydroxide, is used or if the material is not very radiopaque and has a thin margin.[13,14]

SUMMARY

The treatment of acquired defects depends on thorough clinical examination and good-quality radiographs. The radiographs that follow can serve as guidelines in identifying the important features of the various defects. To assist readers in properly interpreting the radiographs, examples of some of the defects listed in the G. V. Black classification system (Figures 11-15 to 11-23) and the stages of resorption (Figures 11-24 to 11-38) have been provided. Figures 11-39 to 11-63 offer radiographic examples of various mixed lesions, and Figures 11-64 to 11-67 show follow-up radiographs of stable and failed restorations.

TABLE 11–4
DIFFERENTIATING RADIOLUCENT DEFECTS OF THE TEETH[4,5,10,11]

Lesion	Tactile Assessment	Radiographic Margins	Expansion of the Pulp Chamber	Present on Other Radiographic Views
Caries	Soft, sticky	Diffuse	No	Yes
External Resorption	Hard	Sharp or scalloped (diffuse if invasive defect or in replacement resorption or ankylosis)	No	Yes
Internal Resorption	Not detectable	Sharp or scalloped	Yes	Yes
Cervical Burnout	Not detectable	Diffuse	No	No

G. V. BLACK CLASSIFICATION FOR DEFECTS OF THE CROWN

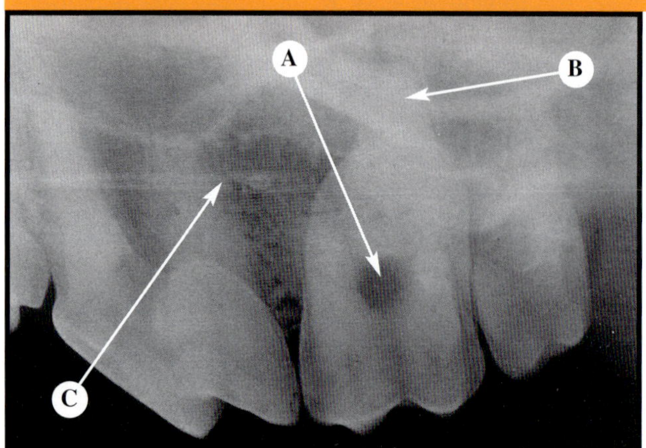

FIGURE 11-15 Class I

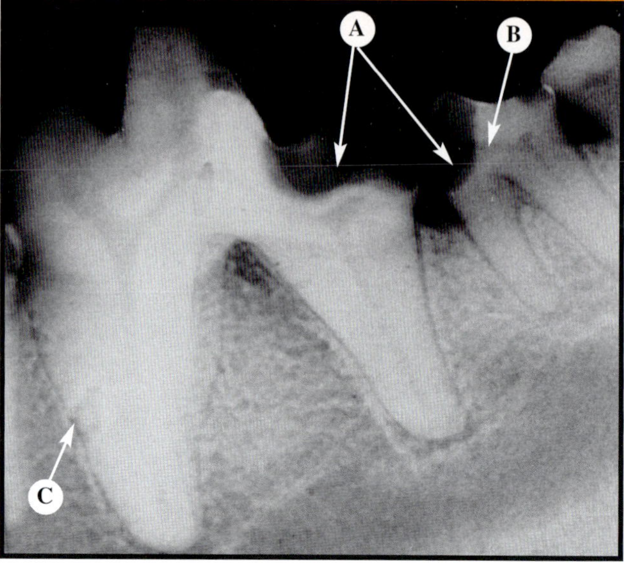

FIGURE 11-16 Class I

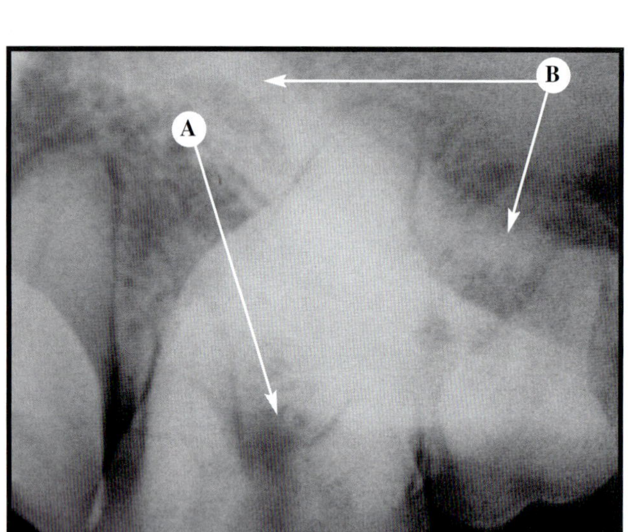

FIGURE 11-17 Class I

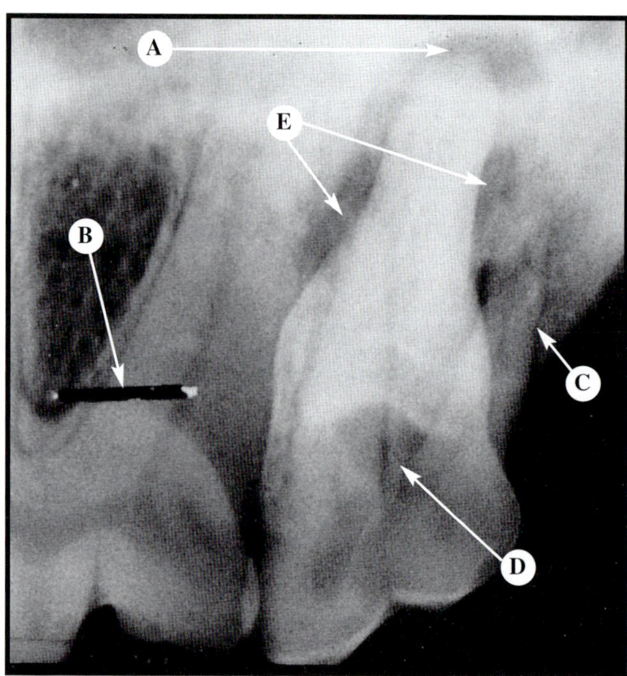

FIGURE 11-18 Class I

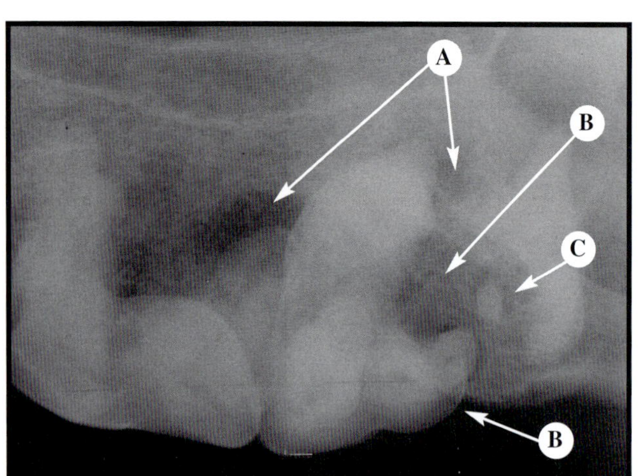

FIGURE 11-19 Class I and II

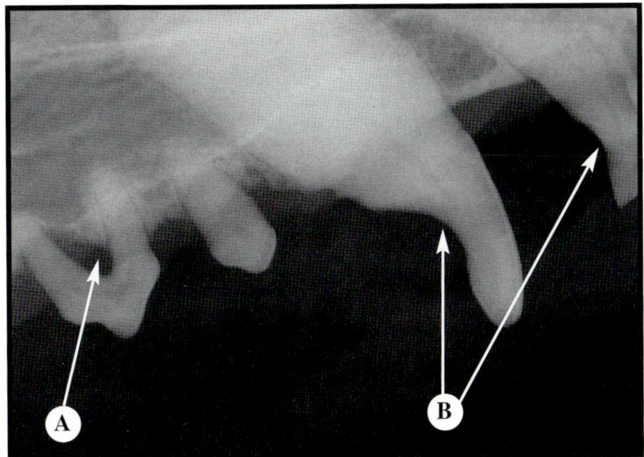

FIGURE 11-20 Class II and III

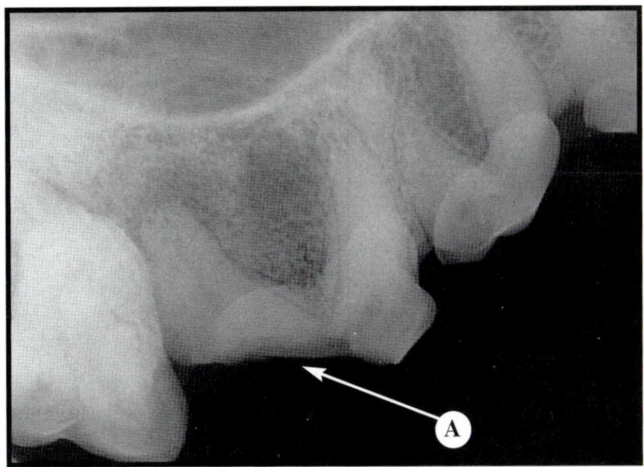

FIGURE 11-21 Class VI

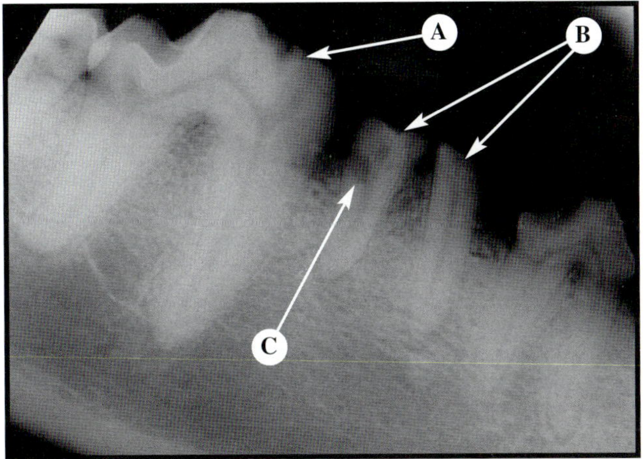

FIGURE 11-22 Class VI

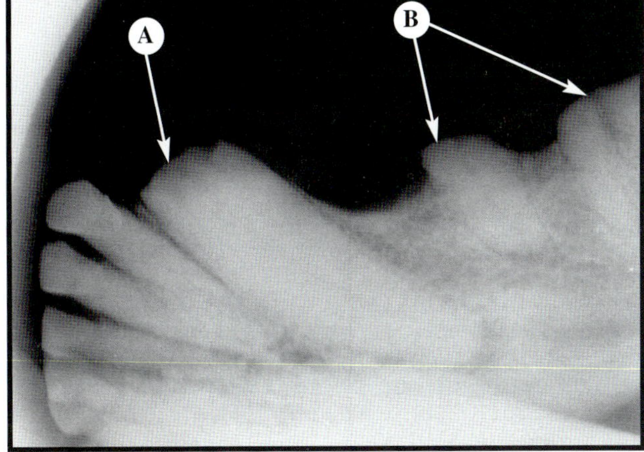

FIGURE 11-23 Class VI

Figure 11-15 *Example of Class I defect of the crown. Chemical artifacts can either obscure images or mimic lesions. (A) Caries, (B) zygomatic arch, and (C) roller mark or fixer line.* **Figure 11-16** *Example of Class I defect of the crown. Oral examination should verify that the pulp is exposed. The root defects that lie within the bone are more likely the result of resorption than carious processes. Note the film artifact above the furcation on the crown of the molar. (A) Caries, (B) pulp exposure, and (C) root resorption.* **Figure 11-17** *Class I defects of the crown. Radiography can be used to supplement the surface inspection of carious lesions. In this radiograph, detail is obscured by superimposition of structures in the interproximal regions of the molars. (A) Occlusal caries and (B) zygomatic arch.* **Figure 11-18** *Class 1 defects of the crown. In addition to caries, note the combined endodontic–periodontal lesion of the molar. (A) Periapical radiolucency, (B) scratched film emulsion, (C) periodontal bone loss, (D) occlusal caries, and (E) wide periodontal space at the palatal root.* **Figure 11-19** *Class I and II defects of the crown of both molars. Advanced loss of tissue may cause the tooth to fracture if extraction is performed. Endodontic disease of the molars and fourth premolar is evident. (A) periapical radiolucency, (B) caries, and (C) apical root fragment.* **Figure 11-20** *Class II and III defects of the crown. A wear facet can weaken the tooth structure, thereby predisposing it to a crown fracture. The dental abrasion was caused when the animal chewed on a fence or the bars of a cage. (A) Bone loss at the furcation and (B) wear facets on the canine tooth and incisor.* **Figure 11-21** *Class VI defect of the crown. The traumatic lesion may be the result of chewing on hard objects. The exposed pulp would require treatment.[14] (A) Fracture line and missing crown cusp.* **Figure 11-22** *Class VI defects of the crown. The distal root of the premolar may fracture if the tooth is extracted. Chronic endodontic exposure results in apical periodontitis and resorptive lesions of the root.[15-17] (A) Worn and chipped crown, (B) total loss of the crown, and (C) external and internal resorption.* **Figure 11-23** *Class VI defect of the crown. The abrasion is probably caused by the dog's chewing vice. Reparative dentin usually prevents direct pulp exposure if the dentin forms during gradual wearing.[14] (A) Worn and chipped canine tooth and (B) worn premolars.*

SUPRAOSSEOUS RESORPTION

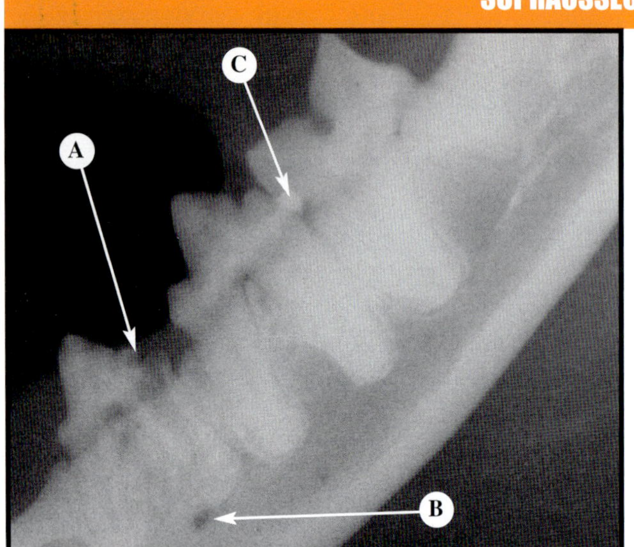

FIGURE 11-24

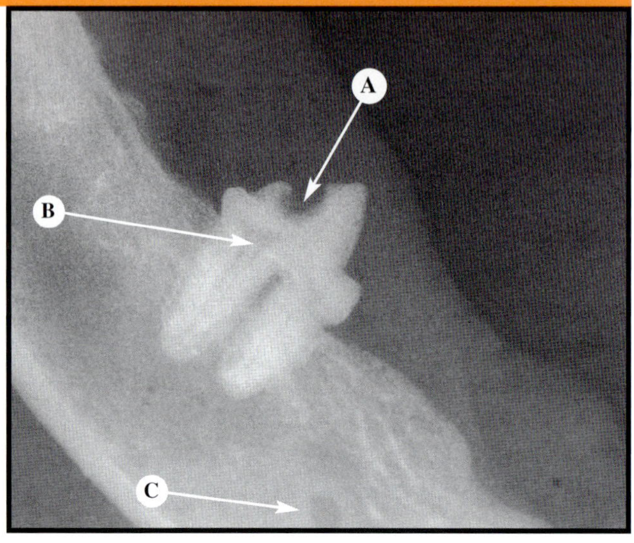

FIGURE 11-25

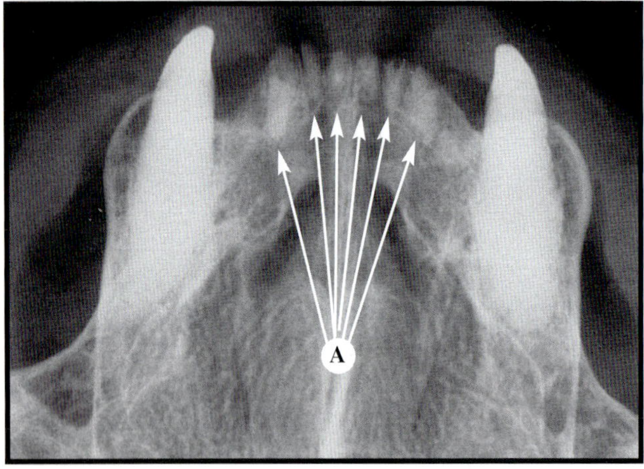

FIGURE 11-26

Figure 11-24 *Stage 2 supraosseous resorption. Internal resorption of the distal root of the third premolar is evident. This tooth would be a poor candidate for restoration. (A) Resorption of the crown with a wide root canal, (B) mental foramen, and (C) superimposition of the crowns.* **Figure 11-25** *Stage 3 supraosseous resorption. This animal also would be a poor candidate for restoration. (A) Stage 3 supraosseous resorption, (B) internal resorption, and (C) mental foramen.* **Figure 11-26** *Stage 4 supraosseous resorption. Retained roots can act as a nidus for infection or inflammation.[10,18] (A) Retained root fragments. (Figure 11-24 reprinted with permission from William CA, Aller SA: Gingivitis/stomatitis in cats, in Harvey CE (ed): The Veterinary Clinics of North America, Small Animal Practice: Feline Dentistry, vol 22, no. 6. Philadelphia, WB Saunders Co, 1992, pp 1361–1383.)*

INTRAOSSEOUS RESORPTION

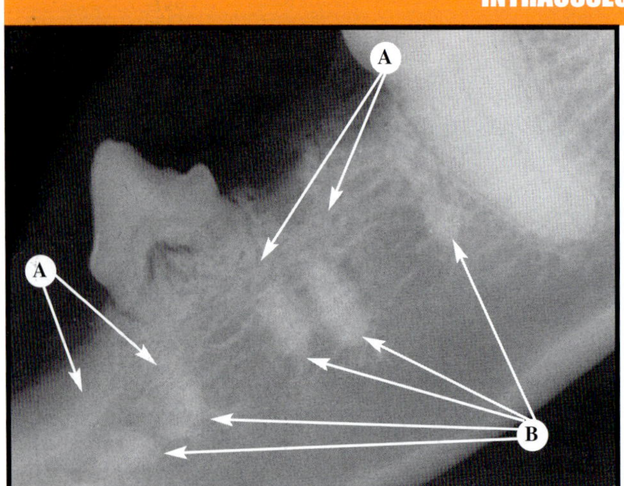

FIGURE 11-27

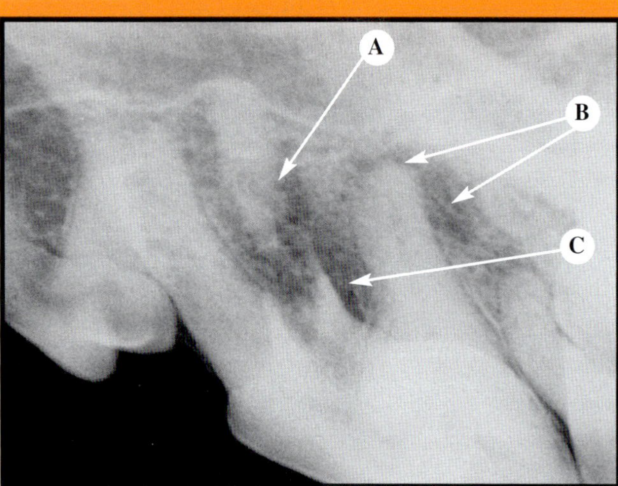

FIGURE 11-28

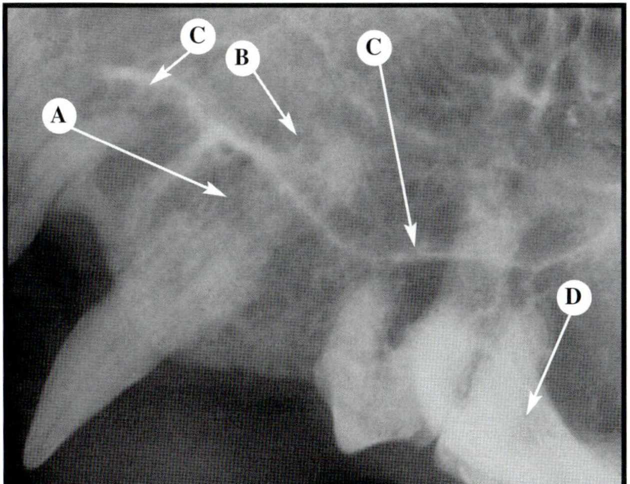

FIGURE 11-29

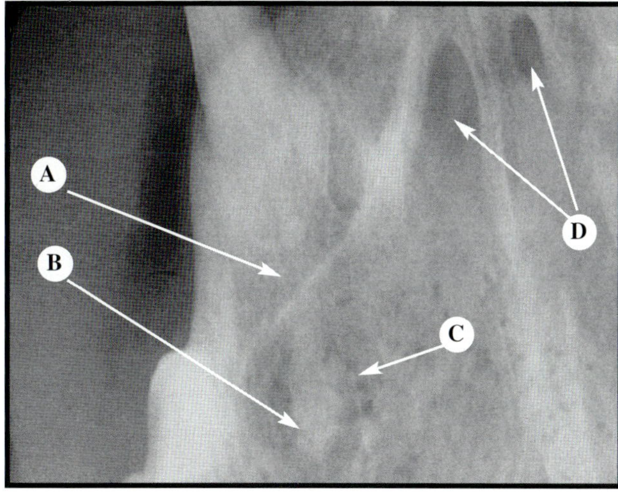

FIGURE 11-30

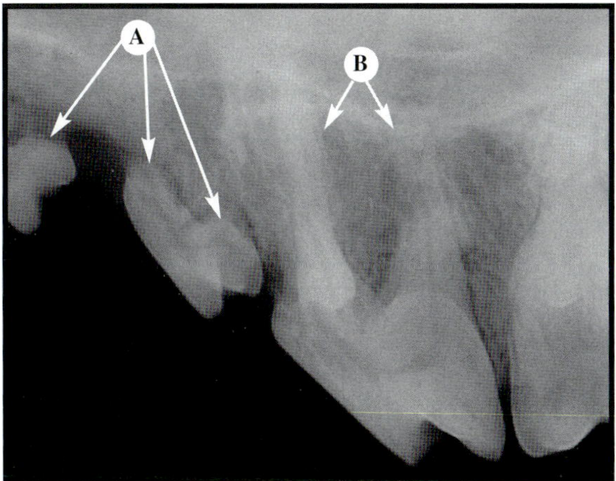

FIGURE 11-31

Figure 11-27 *The coronal half of the roots have been resorbed, thereby causing loss of the crowns because of the lack of support. Exfoliation of the premolar is imminent.* (A) *Replacement resorption and* (B) *root fragments.* **Figure 11-28** *In addition to intraosseous resorption, endodontic disease is evident. Note the sequestration of the palatal root.* (A) *intraosseous resorption,* (B) *periapical radiolucency, and* (C) *periradicular radiolucency.* **Figure 11-29** *Replacement resorption and ankylosis of the canine root are apparent.* (A) *Intraosseous resorption of the root of the canine tooth,* (B) *apex of the canine tooth,* (C) *confluence of the palatal and maxillary bones, and* (D) *zygomatic arch.* **Figure 11-30** *Replacement resorption and ankylosis are evident. The sequestration at the root apex should be removed.*[10] (A) *Intraosseous resorption,* (B) *apex of the tooth,* (C) *bone rarefaction, and* (D) *palatine fissures.* **Figure 11-31** *The root structure has been replaced by bone. Spontaneous exfoliation of the third premolar is imminent. Note the resorption around the roots of the fourth premolar.* (A) *Complete resorption of the roots and* (B) *palatal and distal roots of the fourth premolar.* (Figure 11-30 reprinted with permission from William CA, Aller SA: Gingivitis/stomatitis in cats, in Harvey CE (ed): The Veterinary Clinics of North America, Small Animal Practice: Feline Dentistry, *vol 22, no. 6. Philadelphia, WB Saunders Co, 1992, pp 1361–1383.)*

CRESTAL RESORPTION

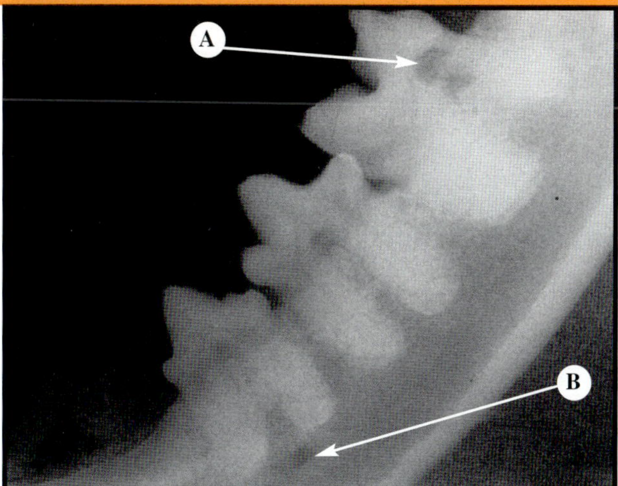

FIGURE 11-32

FIGURE 11-33

Figure 11-32 *Stage 2 crestal resorption and periodontal disease. The extent of crestal lesions are often larger than suspected by clinical evaluation alone.* (A) *Large, shallow resorptive defect,* (B) *jagged alveolar crestal bone, and* (C) *horizontal bone loss.* **Figure 11-33** *Resorptive loss of the distal root of the molar is more significant because of its small diameter.* (A) *Resorptive lesion at the furcation and* (B) *mental foramen.*

CRESTAL RESORPTION *(continued)*

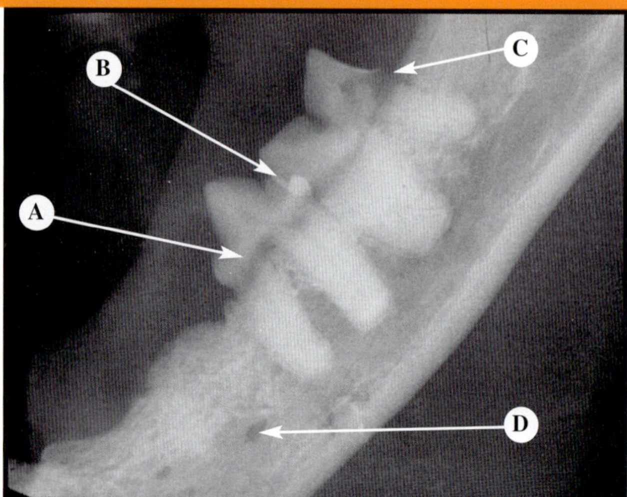

FIGURE 11-34

FIGURE 11-35

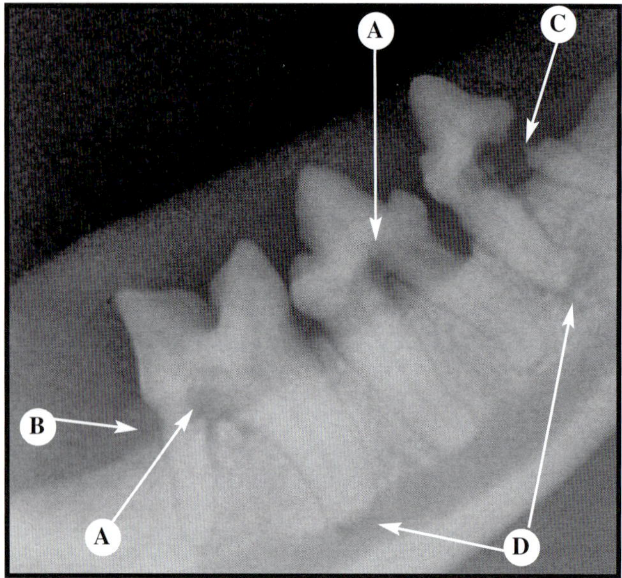

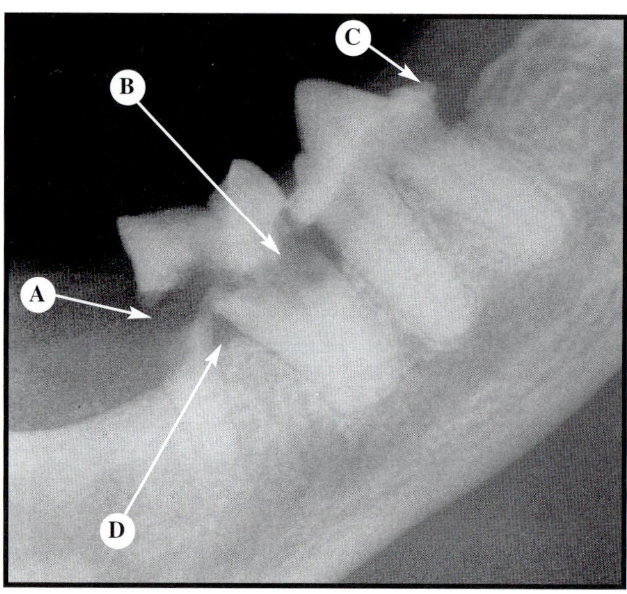

FIGURE 11-36

FIGURE 11-37

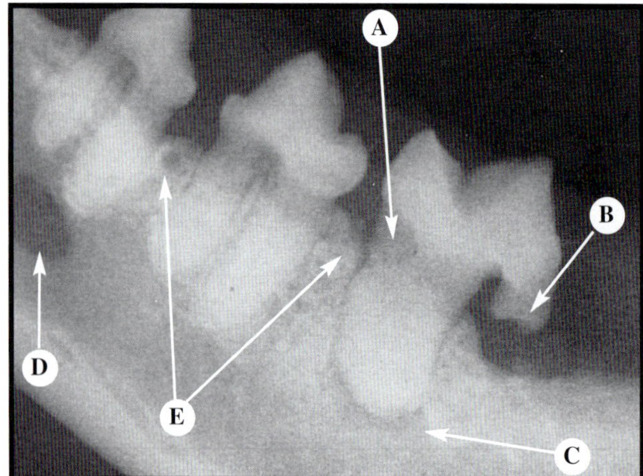

FIGURE 11-38

Figure 11-34 *Both teeth in this radiograph have signs of horizontal bone loss.* (A) *Resorptive lesions and* (B) *mandibular canal.* **Figure 11-35** *The cervical burnout is an artifact that mimics resorption. Additional radiographic views should be taken.* (A) *Cervical burnout,* (B) *superimposition,* (C) *stage 3 crestal resorption, and* (D) *mental foramen.* **Figure 11-36** *Periodontal recession of all the mandibular teeth and endodontic disease of the molar and third premolar are evident.* (A) *Stage 2 crestal resorption,* (B) *horizontal bone loss,* (C) *stage 4 crestal resorption, and* (D) *periapical radiolucency.* **Figure 11-37** *Horizontal bone loss at the periodontium is evident. The calculus overhangs accentuate the cervical burnout and make the cervical area appear to have resorptive lesions.* (A) *Stage 4 crestal resorption,* (B) *stage 3 crestal resorption,* (C) *calculus, and* (D) *bone loss at the furcation.* **Figure 11-38** *The distal root of the molar has undergone resorption. Endodontic disease of the molar and periodontal disease of the premolars and molar are evident.* (A) *Stage 2 crestal resorption,* (B) *stage 4 crestal resorption,* (C) *periapical radiolucency,* (D) *mental foramen, and* (E) *horizontal bone loss with irregular contours on crestal bone.*

MIXED RESORPTIONS—CRESTAL AND INTRAOSSEOUS RESORPTION

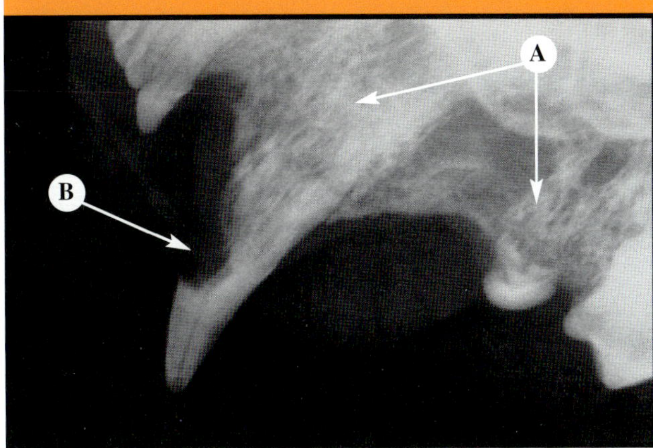

FIGURE 11-39

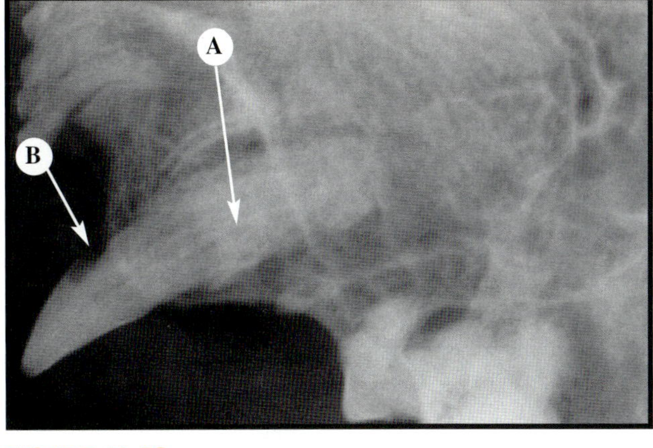

FIGURE 11-40

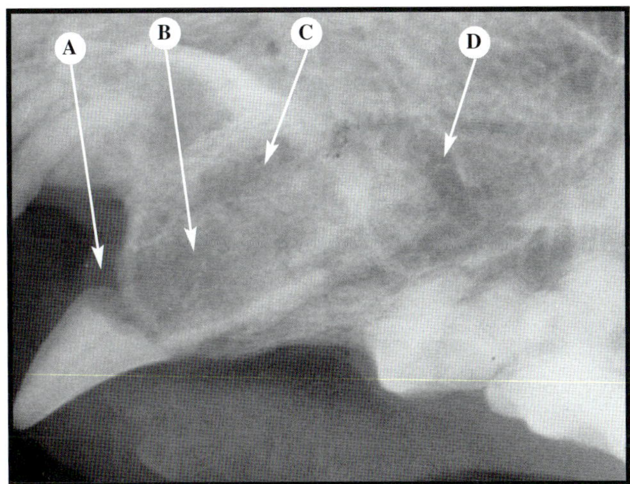

FIGURE 11-41

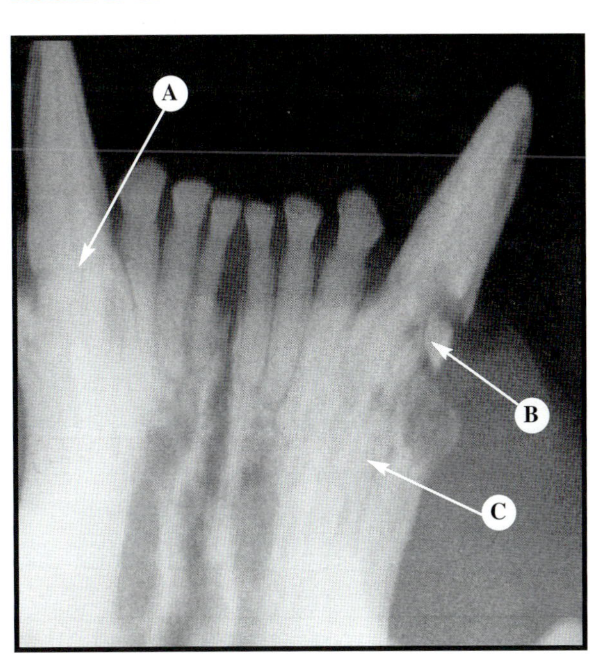

FIGURE 11-43

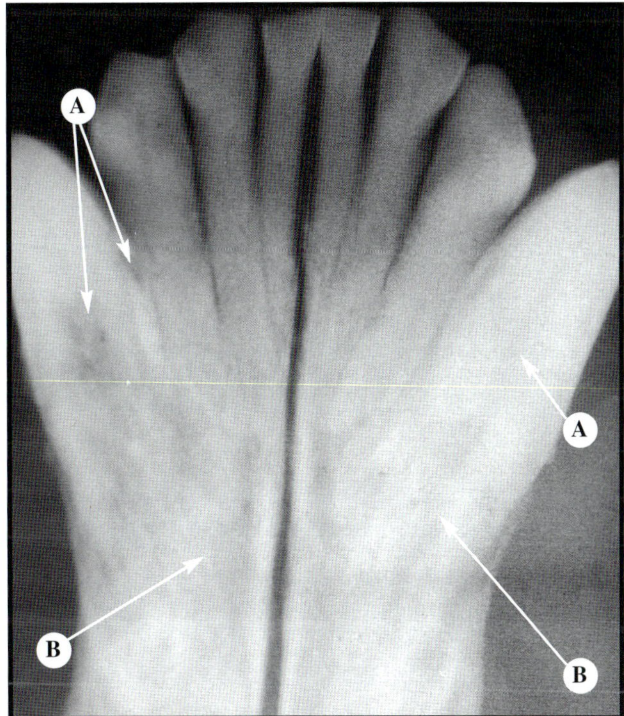

FIGURE 11-42

Figure 11-39 *Exfoliation of the crowns of the canine tooth and premolar is imminent. Extraction would be indicated. (A) Intraosseous resorption and (B) stage 3 crestal resorption with pulp exposure.* **Figure 11-40** *Long-term restoration of the crestal lesion would likely be unsuccessful because of the root lesion. (A) Intraosseous resorption and (B) stage 2 crestal resorption.* **Figure 11-41** *Although the crown appears intact, the tooth is in end-stage resorption. Sequestration is evident on the root undergoing resorption. Extraction would be indicated. (A) Stage 3 crestal resorption, (B) intraosseous resorption, (C) wide periodontal space, and (D) periapical bone rarefaction.* **Figure 11-42** *Exfoliation of the crowns of the canine teeth is imminent. (A) Stage 2 and stage 3 crestal resorptions and (B) intraosseous resorption.* **Figure 11-43** *Periodontal bone loss at the incisors and canine teeth and ankylosis of the canine teeth are evident. (A) Stage 2 crestal resorption and infrabony pocket, (B) stage 3 crestal resorption, and (C) intraosseous resorption.*

MIXED RESORPTIONS—CRESTAL AND INTRAOSSEOUS RESORPTION *(continued)*

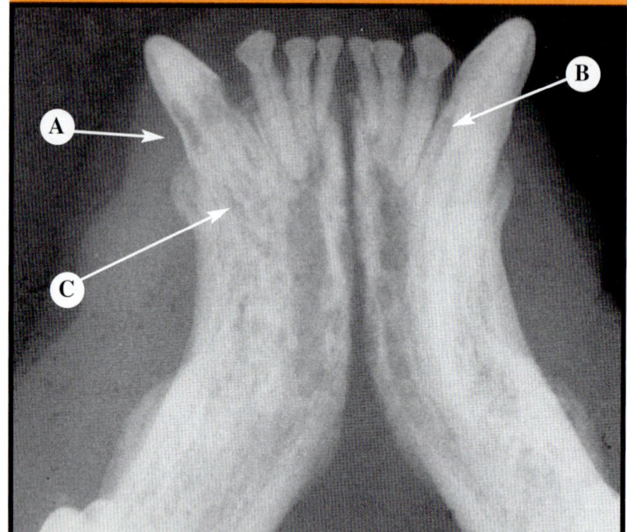

FIGURE 11-44

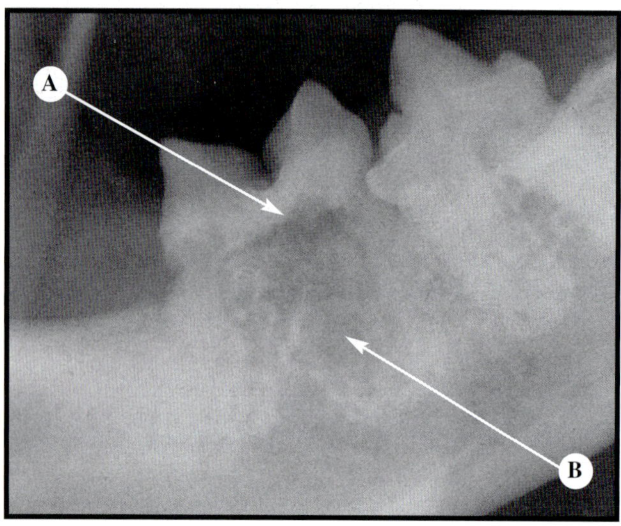

FIGURE 11-45

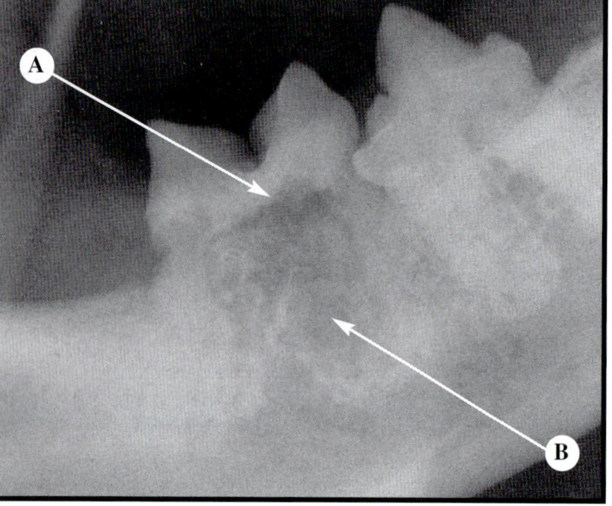

FIGURE 11-46

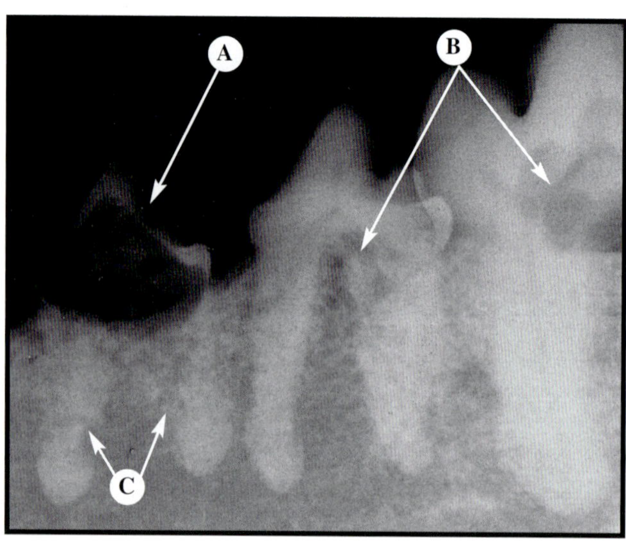

FIGURE 11-47

Figure 11-44 *Horizontal bone recession at the incisors is evident. Exfoliation of the canine tooth on the left is imminent. (A) Stage 3 crestal resorption, (B) stage 2 crestal resorption, and (C) advanced intraosseous resorption.* **Figure 11-45** *This animal would be a poor candidate for restoration. (A) Stage 3 crestal resorption, (B) intraosseous resorption, and (C) mandibular canal.* **Figure 11-46** *Replacement resorption and ankylosis of the root system of the molar are apparent. The molar should be removed to at least the level of the mandibular bone. (A) Stage 2 crestal resorption and (B) intraosseous resorption.* **Figure 11-47** *Replacement resorption and ankylosis of the third premolar are evident. Extraction of the premolar would be indicated. (A) Stage 3 crestal resorption and bone loss at the furcation, (B) intraosseous resorption, and (C) stage 2 crestal resorption.* **Figure 11-48** *All the teeth viewed in this radiographs are severely affected by resorption. (A) The premolar is undergoing resorption at all three regions, (B) stage 2 crestal resorption, and (C) intraosseous resorption.*

FIGURE 11-48

MIXED RESORPTIONS—SUPRAOSSEOUS AND CRESTAL

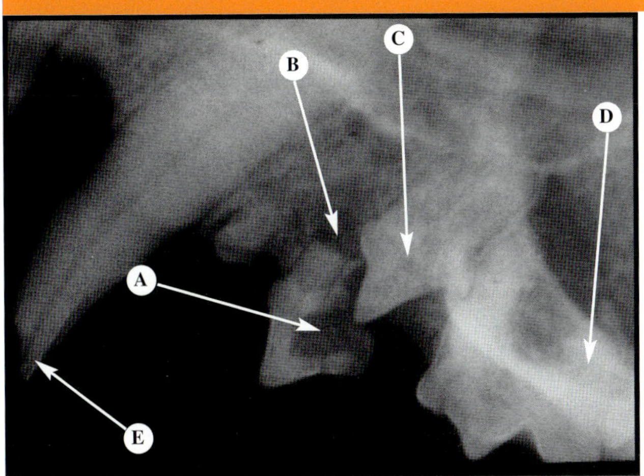

FIGURE 11-49

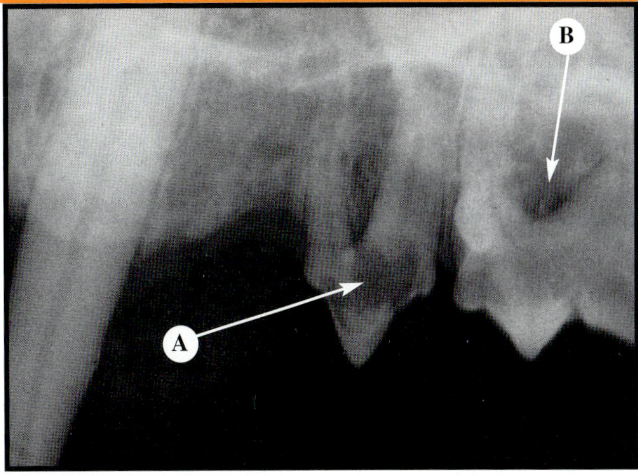

FIGURE 11-50

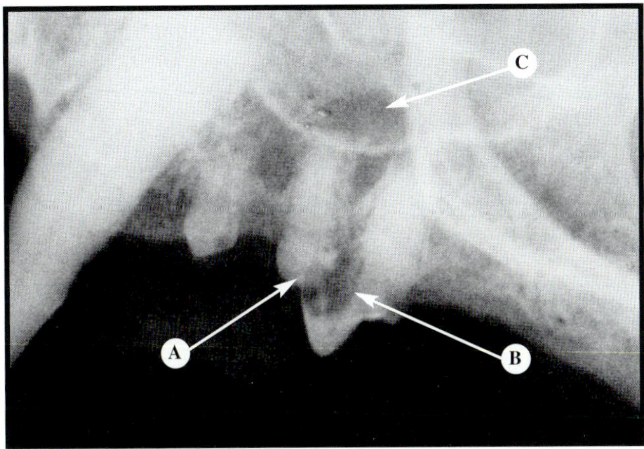

FIGURE 11-51

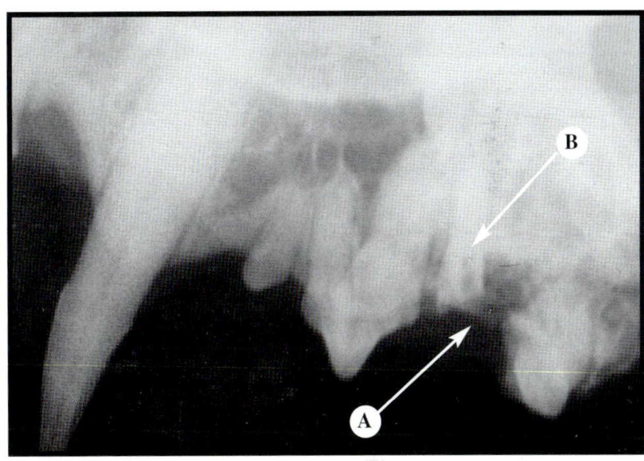

FIGURE 11-52

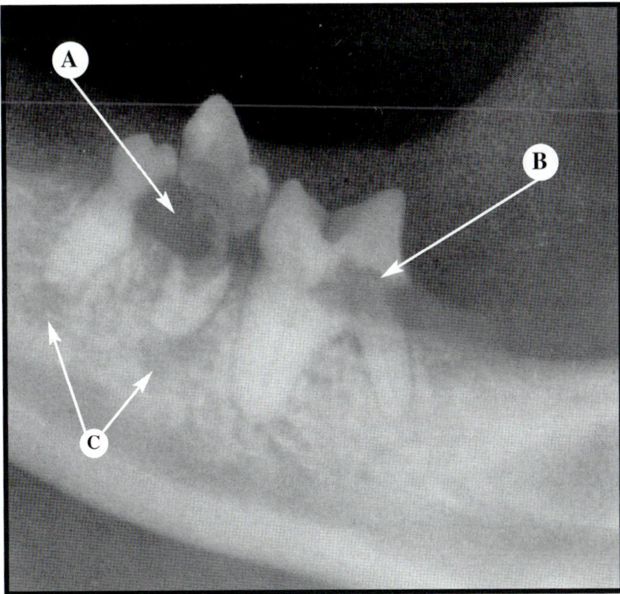

FIGURE 11-53

Figure 11-49 *Deep resorption has created the endodontic exposure in the premolar and secondary apical resorption. (A) Stage 3 supraosseous and crestal resorption, (B) apical resorption and rarefaction, (C) supernumerary third premolar, (D) zygomatic arch, and (E) G. V. Black Class VI defect.* **Figure 11-50** *Advanced loss of tissue around the crown and endodontic exposure are visible. (A) Stage 3 supraosseous and crestal resorption and (B) bone loss at the furcation.* **Figure 11-51** *Endodontic disease of the third premolar is apparent. (A) Stage 2 supraosseous resorption, (B) stage 3 crestal and supraosseous resorption, and (C) bone rarefaction.* **Figure 11-52** *Extraction would be indicated. (A) Stage 4 crestal and supraosseous resorption in the mesial portion and (B) retained mesial roots of the fourth premolar.* **Figure 11-53** *Both teeth have endodontic involvement beyond salvation. Extraction would be indicated. (A) Stage 4 crestal and supraosseous resorption, (B) stage 3 crestal resorption, and (C) periapical radiolucency.*

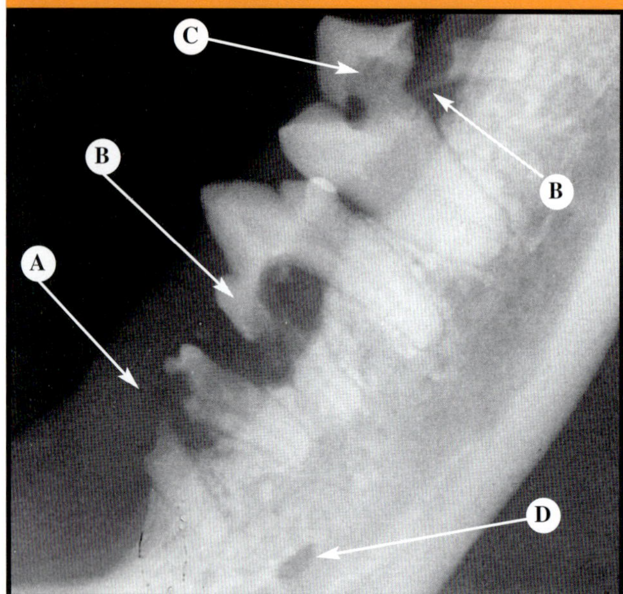

Figure 11-54 *Advanced resorption and horizontal bone loss are evident. Multiple extractions would be indicated. (A) Stage 4 supraosseous and crestal resorption, (B) stage 4 crestal resorption, (C) stage 3 and stage 4 supraosseous resorption, and (D) mental foramen.*

FIGURE 11-54

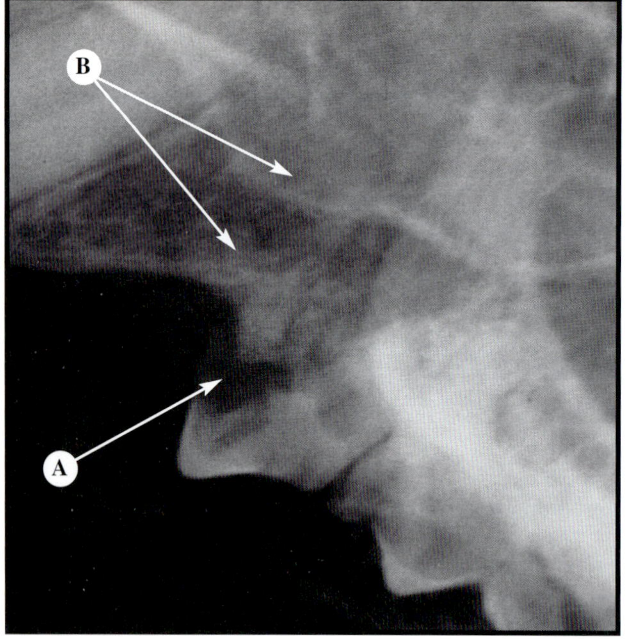

FIGURE 11-55

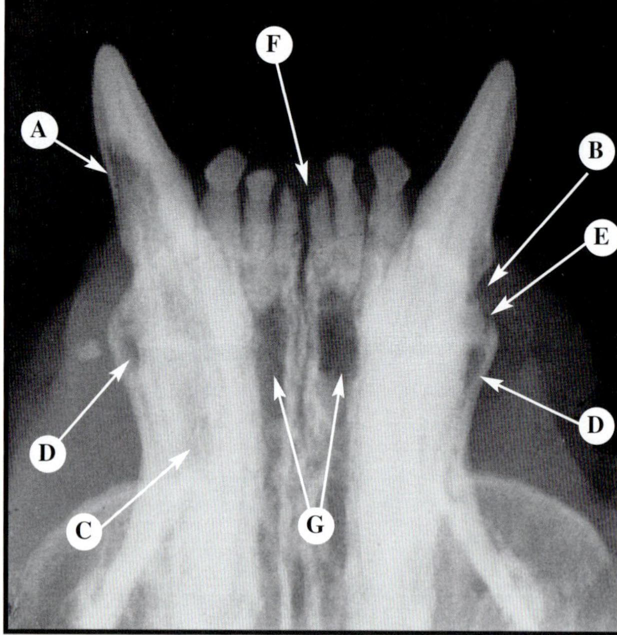

FIGURE 11-56

Figure 11-55 *End-stage intraosseous resorption is apparent. Extractions would be indicated. (A) Stage 4 crestal and supraosseous resorption and (B) mesial root fragments.* **Figure 11-56** *Advanced periodontal and endodontic disease as well as resorption are evident in all the teeth of this patient. (A) Stage 2 crestal and supraosseous resorption, (B) crestal resorption, (C) intraosseous resorption, (D) infrabony pockets, (E) calculus at the infrabony pocket, (F) stage 4 supraosseous resorption, and (G) periapical radiolucent defects overlapping the nutrient canals.*

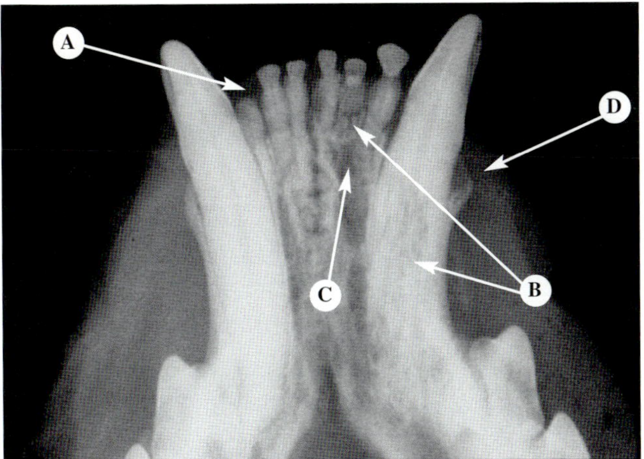

FIGURE 11-57

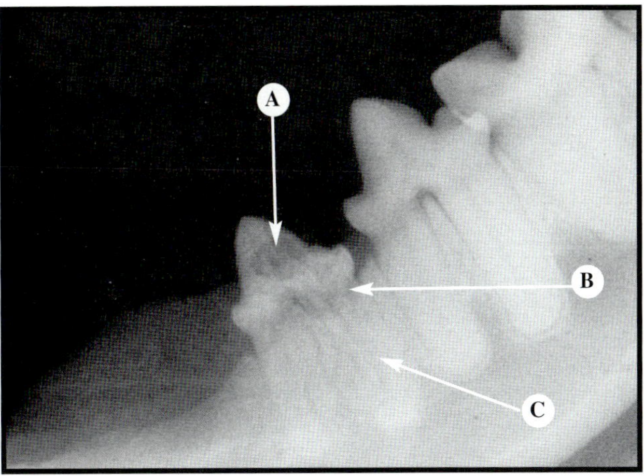

FIGURE 11-58

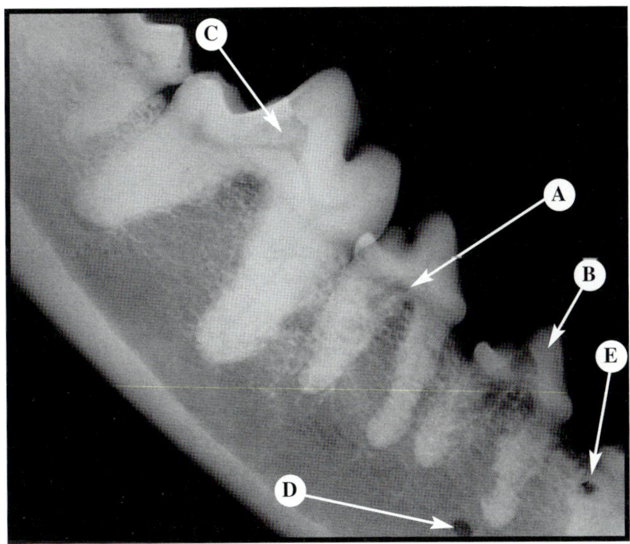

FIGURE 11-59

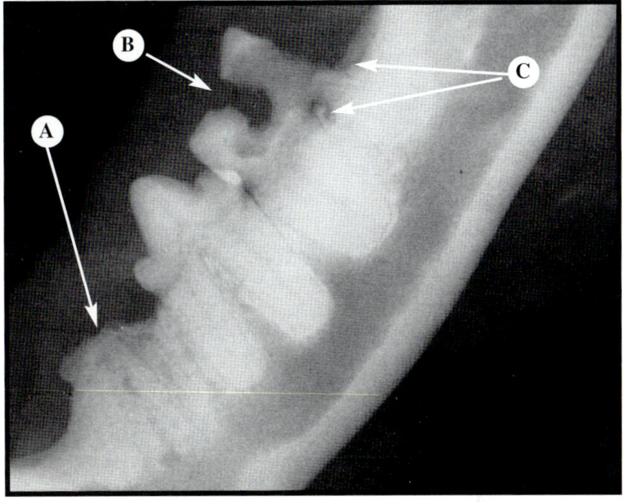

FIGURE 11-60

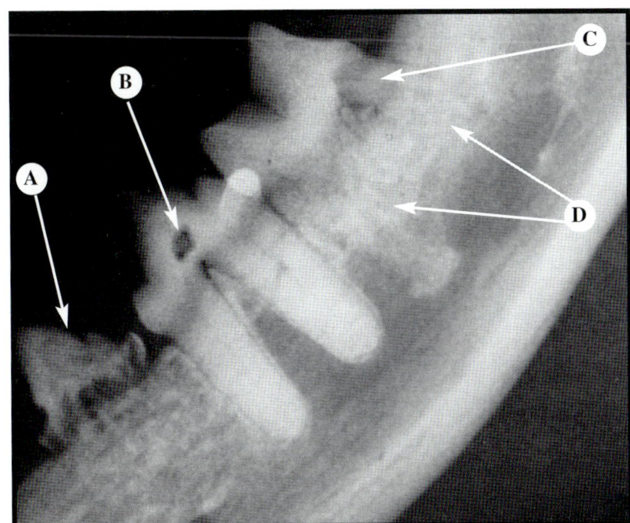

FIGURE 11-61

Figure 11-57 *Endodontic disease of the right intermediate and left corner incisors and periodontal disease of the incisors and right canine tooth are visible. Note the bulbous appearance of the incisors, which is possibly caused by a buildup of calculus or hypercementosis. In addition, there is ankylosis of the right canine tooth. (A) Stage 4 supraosseous and crestal resorption, (B) intraosseous resorption, (C) periapical radiolucency, and (D) infrabony pocket.* **Figure 11-58** *This tooth would be a poor candidate for restoration even though the lesion on the crown appears to be shallow. (A) Stage 2 supraosseous resorption, (B) stage 2 crestal resorption, and (C) intraosseous resorption.* **Figure 11-59** *Note how the film artifacts can mimic radiographic defects. Numerous resorptions are evident. (A) Stage 2 crestal resorption, (B) resorption of all three regions of the tooth, (C) a spot artifact, (D) mental foramen, and (E) an artifact caused by the dimple on the film packet.* **Figure 11-60** *Replacement resorption of the third premolar is present. Extraction of the molar would be indicated. (A) Intraosseous and stage 4 supraosseous resorption. (B) stage 4 supraosseous resorption, and (C) stage 3 crestal resorption and bone defects.* **Figure 11-61** *Note the replacement resorption evident in the third premolar and molar. (A) Resorption of all regions, (B) stage 3 supraosseous resorption, (C) stage 2 crestal resorption, and (D) intraosseous resorption.*

MIXED RESORPTIONS—SUPRAOSSEOUS, CRESTAL, AND INTRAOSSEOUS *(continued)*

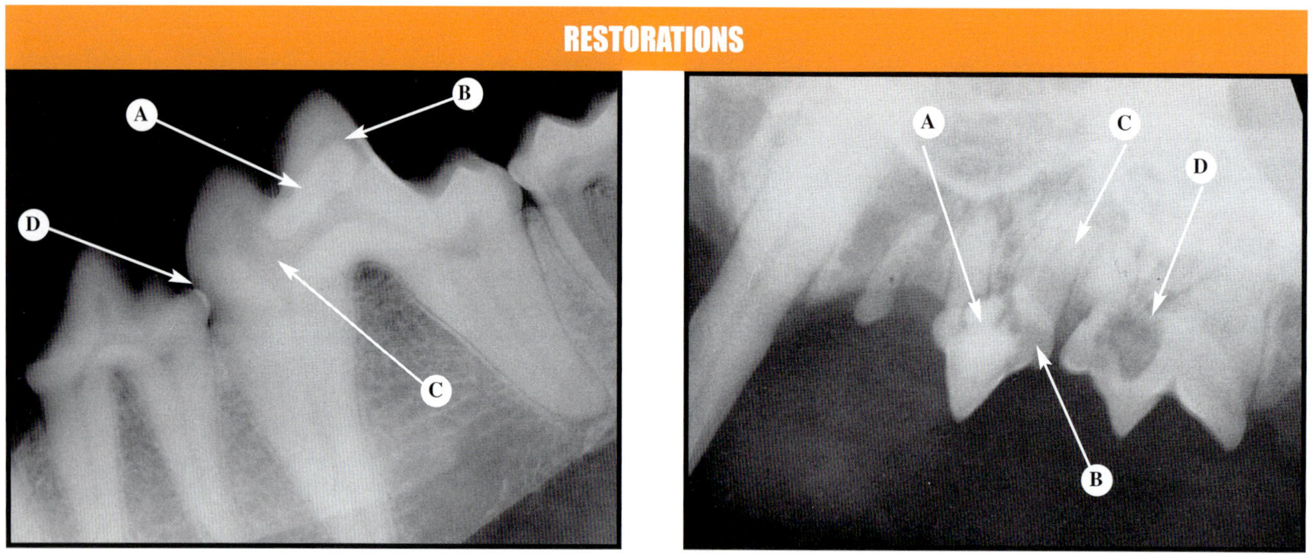

FIGURE 11-62

FIGURE 11-63

Figure 11-62 *This radiograph shows evidence of endodontic–periodontal disease of the molar and resorption.* (A) *Supraosseous resorption,* (B) *bone loss at the furcation,* (C) *horizontal bone loss,* (D) *stage 3 crestal resorption,* (E) *stage 2 crestal resorption and intraosseous resorption,* (F) *apical resorption and periapical radiolucent defects (molar).* **Figure 11-63** *This supraosseous lesion is external to the pulp chamber and the contours of the pulp chamber are discernible, thereby indicating that internal resorption is not part of the lesion (see Chapter 10).* (A) *Stage 2 supraosseous resorption,* (B) *stage 2 crestal resorption, and* (C) *intraosseous resorption.*

RESTORATIONS

FIGURE 11-64 Four Years After Restoration

FIGURE 11-65 Two Years After Restoration

Figure 11-64 *This radiograph was taken four years after tooth restoration. Note that the restoration is stable; however, monitoring should continue.* (A) *Composite restoration,* (B) *radiolucent halo,* (C) *pulp chamber, and* (D) *superimposition.* **Figure 11-65** *The prognosis two years after restoration would be poor on the basis of this radiograph. Resorption of the premolar is progressing underneath the restoration. Extraction of the fourth premolar would also be indicated.* (A) *Glass ionomer restoration of the third premolar,* (B) *radiolucent halo is larger than acceptable,* (C) *intraosseous resorption, and* (D) *stage 3 supraosseous and crestal resorption.*

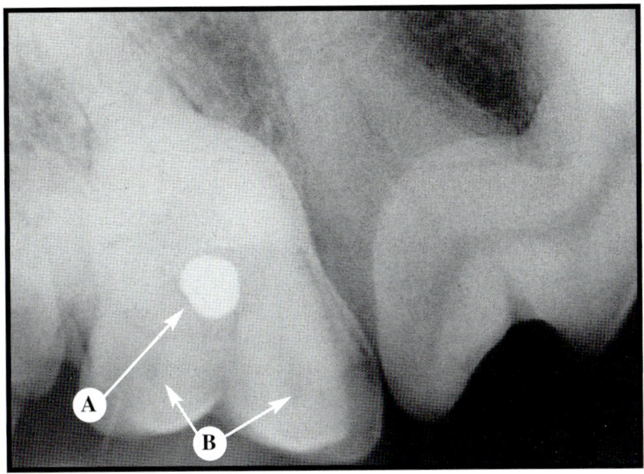

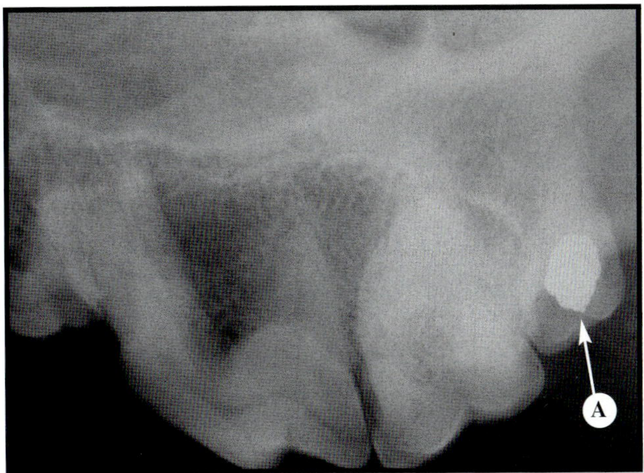

FIGURE 11-66 Three Years After Restoration

FIGURE 11-67 Four Years After Restoration

Figure 11-66 *Three years after this tooth was restored, the restoration appears to be stable. Monitoring should continue, however.* (A) *Amalgam filling and* (B) *pulp chamber.* **Figure 11-67** *The restoration shown in this radiograph also is stable after four years.* (A) *Amalgam filling.*

REFERENCES

1. Roberson TM, Sturdevant CM, Barton RE, Wall JT: Fundamentals in cavity preparation, in Sturdevant CM, Roberson TM, Heymann HO, Sturdevant JR (eds): *The Art and Science of Operative Dentistry*, ed 3. Baltimore, CV Mosby Co, 1995, pp 300 303.

2. Bojrab MJ, Tholen M: *Small Animal Oral Medicine and Surgery*. Philadelphia, Lea & Febiger, 1990, pp 194-205.

3. Cawson RA, Eveson JW: *Oral Pathology and Diagnosis*. Philadelphia, WB Saunders Co, 1987, pp 3.2-3.5.

4. Razmus TF: Caries, periodontal disease, and periapical changes, in Miles DA, Van Dis ML (eds): *The Dental Clinics of North America. The Clinical Approach to Radiological Diagnosis*, vol 38, no. 1. Philadelphia, WB Saunders Co, 1994, pp 13-31.

5. Goaz PW, White SC: *Oral Radiology: Principles and Interpretation*, ed 3. Baltimore, CV Mosby Co, 1994, pp 306-325.

6. Lyon KF: Subgingival odontoclastic resorptive lesions, in Harvey CE (ed): *The Veterinary Clinics of North America. Small Animal Practice: Feline Dentistry*. Philadelphia, WB Saunders Co, 1992, pp 1417-1432.

7. Harvey CE, Emily PP: *Small Animal Dentistry*. Philadelphia, CV Mosby Co, 1993, pp 213-214.

8. Frank AL: Extracanal invasive resorption: An update. *Compend Continu Educ Dent* 16(3):250-262, 1995.

9. Reichart PA, Durr UM, Triadan H, et al: Periodontal disease in the domestic cat. *J Periodontal Res* 19:67-75, 1984.

10. Williams CA, Aller SA: Gingivitis/stomatitis in cats, in Harvey CE (ed): *The Veterinary Clinics of North America. Small Animal Practice: Feline Dentistry*, vol 22, no. 6. Philadelphia, WB Saunders Co, 1992, pp 1361-1383.

11. Okuda A, Harvey CE: Etiopathogenesis of dental resorptive lesions, in Harvey CE (ed): *The Veterinary Clinics of North America, Small Animal Practice: Feline Dentistry*, vol 22, no. 6. Philadelphia, WB Saunders Co, 1992, pp 1385-1404.

12. Ohba S, Kiba H, Kuwabara M, Yoshida H, Koide F, Takeishi M: Contact microradiographic analysis of feline tooth resorptive lesions, *J Vet Med Sci* 55(2):329-332, 1993.

13. Goaz PW, White SC: *Oral Radiology: Principles and Interpretation*, ed 3. Baltimore, CV Mosby Co, 1994, pp 149-150.

14. Van Swol R, Eslami A, Sadeghi EM, Ellinger RF: A new treatment for furcation defects involving strategic molars. *Intl J Period Restor Dent* 9(3):185-195, 1989.

15. Cohen S, Burns RC: *Pathways of the Pulp*, ed 6. Philadelphia, CV Mosby Co, Philadelphia, 1994, pp 332-333, 364-365, 442-443.

16. Emily P: Endodontic disease, in Manfra Marretta S (ed): *Problems in Veterinary Medicine: Dentistry*, vol 2, no. 1. Philadelphia, WB Saunders Co, 1990, p 170.

17. Walton RE, Torabinejad M: *Principles and Practice of Endodontics*. Philadelphia, WB Saunders Co 1989, p 24.

18. Wood NK, Goaz PW, Lehnert JF: Periapical radiopacities, in Wood NK, Goaz PW (eds): *Differential Diagnosis of Oral Lesions*, ed 4. Philadelphia, Mosby Year Book, 1991, pp 549-550.

19. Rossman LE, Garber DA, Harvey CE: Disorders of the teeth, in Harvey CE (ed): *Veterinary Dentistry*. Philadelphia, WB Saunders Co, 1985.

Chapter Twelve

TRAUMA

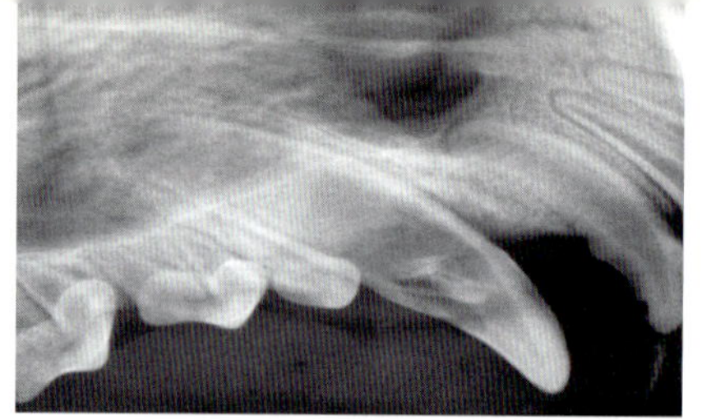

The oral anatomy of dogs and cats can be subjected to various pathologic conditions that are the result of trauma. Oral trauma has the potential of damaging the structures of the tooth and periodontium, the bone, or soft tissue. Injuries associated with trauma can occur when the animal is involved in normal activities or from accidents, complications of veterinary procedures, or dental disease.

Radiographic evaluation of traumatic injuries should consider the maturity of the teeth; degree of tooth displacement; presence of fractures of the crown, root, alveolus, or bone; pulp exposure; and preexisting disease. The interrelations between the teeth and bone and the stage of development are paramount in determining treatment measures and prognosis for an injury. We recommend that several radiographic views be taken because the appearance of fracture lines and tooth displacement depends on the angle of the primary x-ray beam.

MISSING AND FRACTURED TEETH

An apparent missing tooth does not necessarily mean that the tooth has been completely lost to the environment. Possibilities for missing teeth include intrusion, avulsion, or root fracture with avulsion of the crown. Missing teeth or their fragments can be inhaled, swallowed, or embedded in soft tissue or may have been caught in the patient's fur. These differentials must be ruled out before making a definitive diagnosis of a lost tooth or empty alveolus.[1,2]

Teeth that have a normal intact crown and normal gingival attachment but increased mobility may have a root fracture or subluxation. Root fractures are not always detectable on each radiographic view or may not be apparent on any view, especially if minimal displacement has occurred. If root fractures are evident on radiographs, their appearance can vary according to the angle of the x-ray beam. A single fracture line is accurately depicted if the x-ray beam is parallel to the entire plane of the fracture; however, on other radiographic views, a single fracture line can be mistaken for multiple fracture lines[1–3] (Figures 12-1 and 12-2). Some additional guidelines for interpreting radiographs of root fractures are discussed in Chapter 10 on endodontic disease.

Crown fractures are classified according to their location and depth in a manner similar to the classification of defects described in Chapter 11. Fractures may involve the enamel only; however, if both enamel and dentin are involved, the injury can be compounded by direct or indirect pulp exposure. Enamel infractions occur when there is an incomplete fracture without loss of tooth structure; the tooth is cracked, but no enamel is missing. Fractures of both the crown and root involve enamel, dentin, and cementum. The fracture line extends from the crown obliquely to some point below the gingival crest, which may or may not involve the pulp. Injuries that have direct pulp exposure are referred to as complicated fractures, whereas those without direct pulp exposure are called uncomplicated fractures. The apical extent of an oblique fracture line can be difficult to diagnose accurately. If the plane of the fracture line happens to be perpendicular to the x-ray beam, the fracture line on the radiograph may appear indistinct or may not be present at all. For that reason, multiple radiographic views must be generated before the absence or presence of a fracture line or its apical extension can be confirmed.[4,5]

Follow-up radiographs that monitor the healing of root fractures should be taken three weeks, six weeks, and three months after the injury was sustained and then as indicated by clinical judgment and response to treatment. Increased radiolucency at the fracture line on initial follow-up radiographs may be indicative of a distraction of fragments from hemorrhage, granulation tissue, masticatory forces, resorption within the root canal, or a change in the angle of the x-ray beam. Resorption within the root canal at the fracture line is known to be a transient stage of healing for a few months after the pulp has sustained trauma. Subsequently, some mineralization should occur and the fracture should heal by either the deposition of hard tissue or repair of connective tissue. Resorption within the bone at the fracture site, however, can indicate pulp necrosis, which may involve only the coronal region. In some cases, calcification of the root canal and obliteration eventually occur.[2,3,6]

INJURIES TO THE PERIODONTIUM

Teeth can become displaced when the periodontal ligament is torn or the alveolar bone is fractured. The vasculature to the pulp may be disrupted by the injury and thereby com-

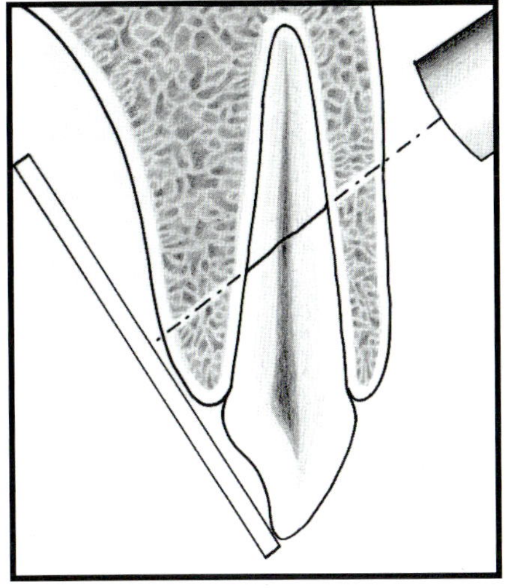

FIGURE 12-1A

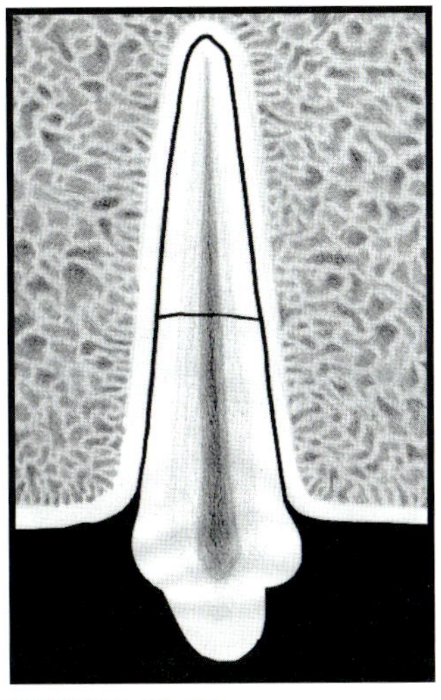

FIGURE 12-1B

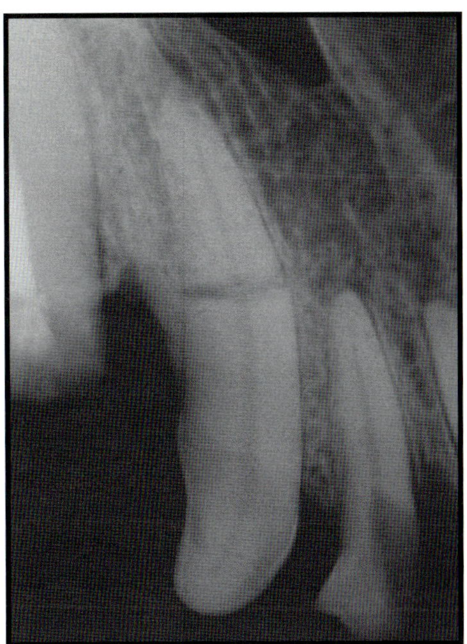

FIGURE 12-1C

Figure 12-1A *This illustration depicts how the angle of the x-ray beam should be parallel to the fracture line.* **Figure 12-1B** *The illustration here shows how the fracture line will appear as a single line on the radiograph.* **Figure 12-1C** *The resultant radiograph accurately images a single fracture line.*

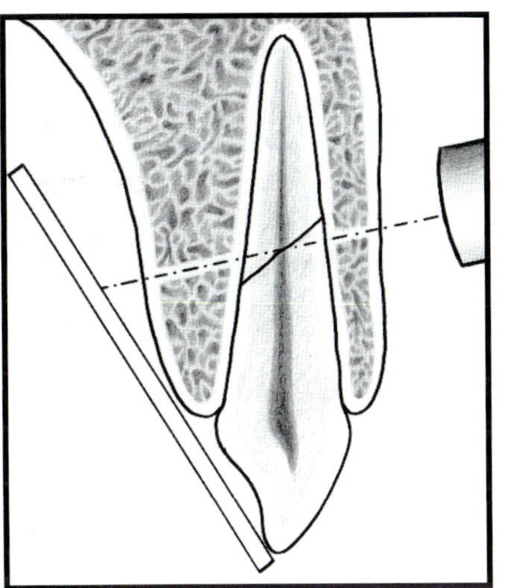

FIGURE 12-2A

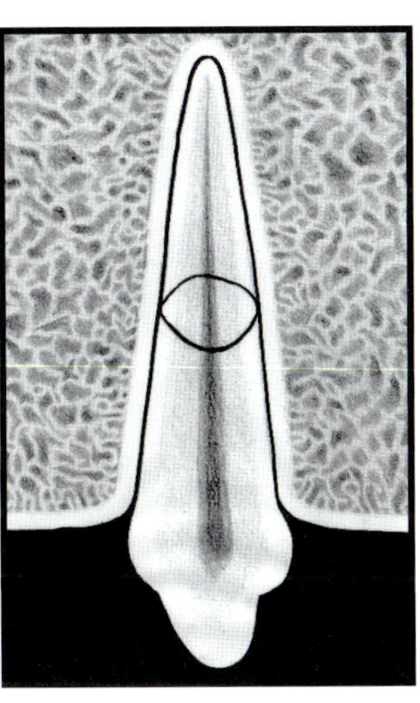

FIGURE 12-2B

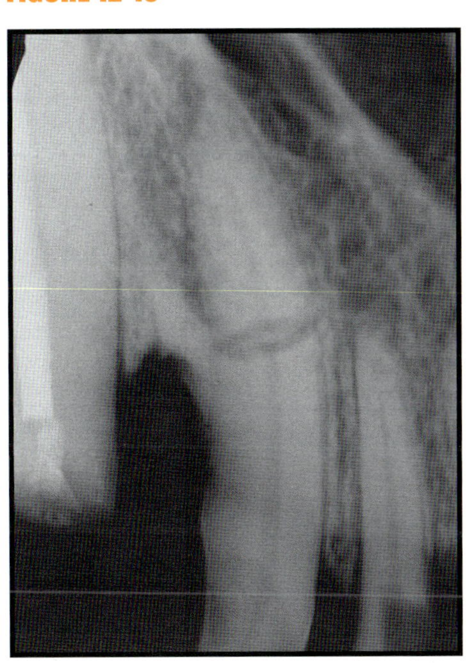

FIGURE 12-2C

Figure 12-2A *This illustration demonstrates the angle of the x-ray beam when it is not parallel to the fracture line.* **Figure 12-2B** *The image that will be captured will show both the near and far borders of the fracture line.* **Figure 12-2C** *The resultant radiograph inaccurately depicts more than one radiolucent line for a single fracture line.*

promise pulp vitality, although in young teeth, pulp revascularization and reinnervation can occur. Displacement is classified according to the degree and direction of tooth displacement.[2]

A relatively minor injury to the periodontal ligament can be caused by concussion with resultant hemorrhage and edema within the periodontal ligament. Although the affected tooth may be sensitive to percussion and mastication, it can appear normal without any evidence of bleeding from the gingival sulcus. If some of the periodontal ligament fibers are actually torn or ruptured, subluxation has occurred. The tooth then may have increased mobility but no displacement, and some bleeding may be observed at the gingival sulcus. On radiographs, concussion and subluxation injuries may not present as radiographic abnormalities[2] (Figure 12-3).

Luxation injuries can displace a tooth and damage the periodontal ligament, alveolar bone, and pulp. An extrusive luxation displaces the tooth on its long axis away from the alveolus; the tooth may be very loose because the gingival fibers are usually the only remaining intact attachment. A lateral luxation

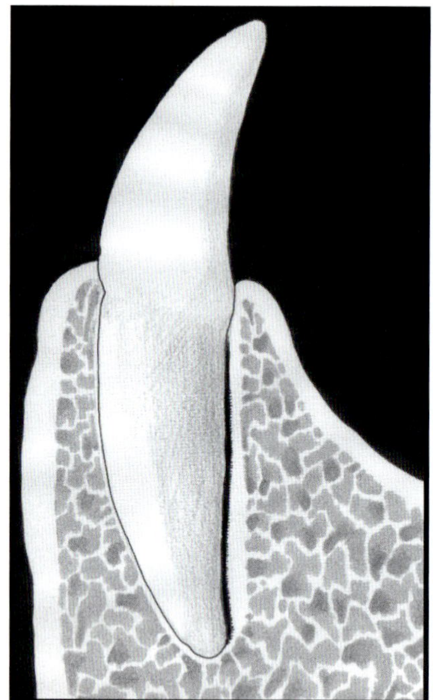

FIGURE 12-3 Subluxation

Subluxation injuries can cause the periodontal ligament to tear or loosen even though the tooth remains in place.

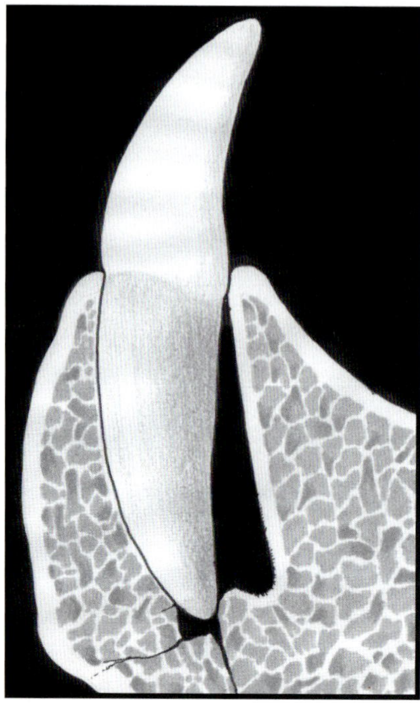

FIGURE 12-4A Lateral Luxation

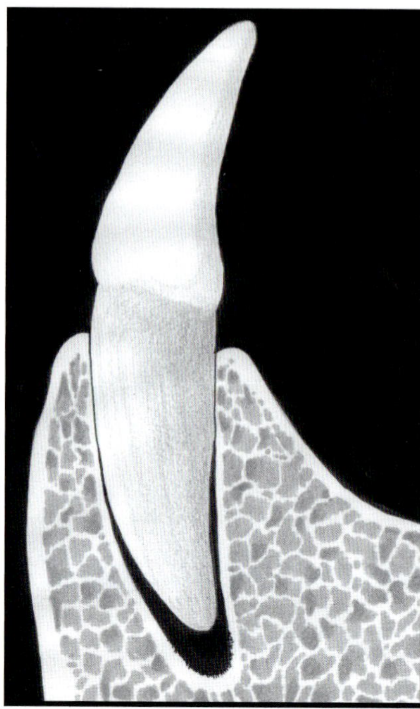

FIGURE 12-4B Extrusive Luxation

Figure 12-4A *The tooth is partially dislodged horizontally from the alveolus.* **Figure 12-4B** *The tooth is partially dislodged vertically from the alveolus.*

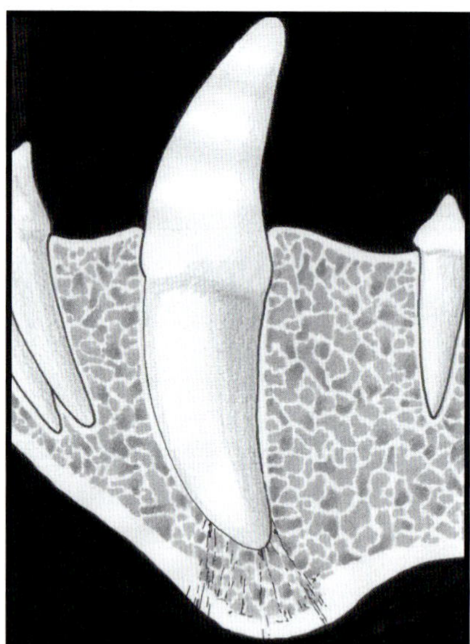

FIGURE 12-5 Intrusion

The tooth is forced apically down into the tissue.

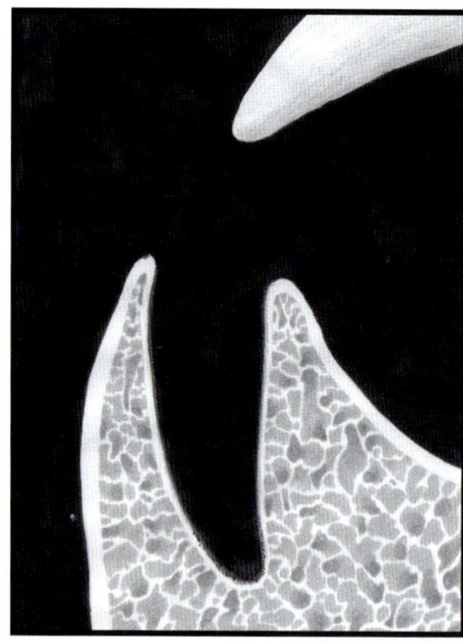

FIGURE 12-6 Avulsion

All attachments to the tooth are severed, and the tooth is completely dislocated.

can obscure the abnormally increased space in the periodontal ligament, which can be created by tooth displacement, several radiographic views are recommended[2] (Figure 12-4).

Intrusion is a crushing injury that occurs when a tooth is driven apically into alveolar bone. Severe damage to the surrounding bone, periodontal ligament, and vascular supply to the pulp can be sustained. Displacement is evident on radiographs[2] (Figure 12-5).

Avulsion injuries occur when a tooth is completely disarticulated from the alveolus. Such injuries are most commonly found in the teeth of young animals with incomplete root development or teeth with periodontal disease. The periodontal ligament, pulp, and any gingival attachments

displaces the tooth horizontally, along with a plate of fractured alveolar bone. In both instances, the pulp and periodontal ligament are severed. On radiographs, the displacement can usually be detected; however, the degree of pathologic involvement apparent on radiographs primarily depends on the angle of the x-ray beam. Because superimposition of the root over the bone can be completely severed; and an alveolar wall may be fractured[2] (Figure 12-6).

Teeth that have been replaced to their correct anatomic position must be monitored on radiographs for a long period because pulp necrosis and root resorption are potential sequelae. Suggested follow-up intervals for luxation injuries are

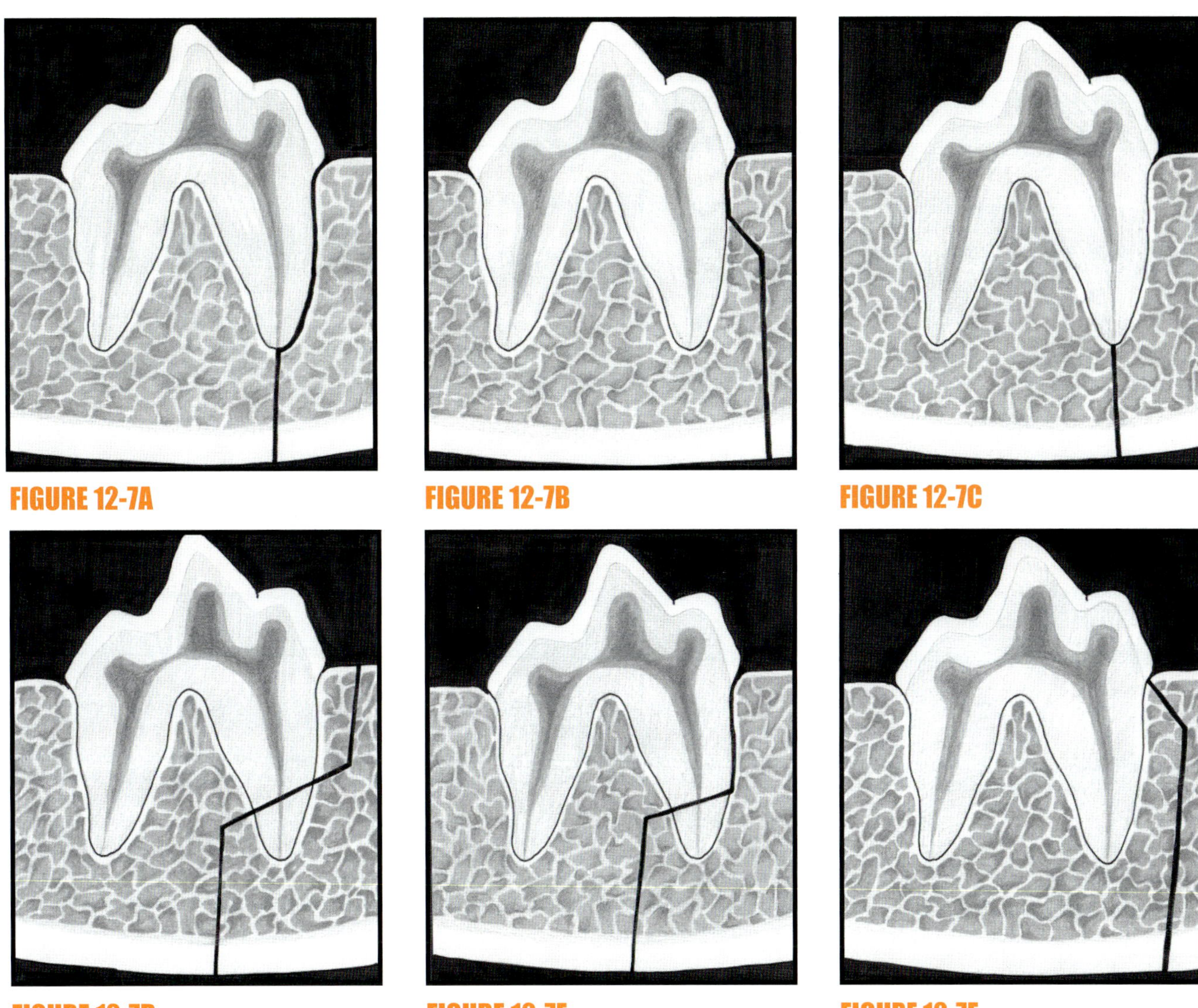

FIGURE 12-7A **FIGURE 12-7B** **FIGURE 12-7C**

FIGURE 12-7D **FIGURE 12-7E** **FIGURE 12-7F**

The relation between the fracture line and the periodontal ligament and alveolus is depicted in these illustrations. **Figure 12-7A** *The fracture line compromises the periodontal ligament and pulp vasculature. The patient would be at risk for developing a periodontal–endodontic lesion.* **Figure 12-7B** *The fracture line involves the periodontal ligament. The patient would be at risk for developing a periodontal lesion.* **Figure 12-7C** *The fracture line is not open to the oral cavity but could potentially compromise the pulp vasculature. The patient would be at risk for developing an endodontic lesion.* **Figure 12-7D** *The fracture line is superimposed over the root. Another radiographic view should be taken to rule out root fracture or involvement of the periodontal ligament.* **Figure 12-7E** *The fracture line is superimposed over the root and involves the periodontal ligament. Another radiographic view should be taken to rule out root fracture. The patient might be at risk for developing a periodontal lesion.* **Figure 12-7F** *The fracture line does not directly involve either the periodontal ligament or the pulp. (Figures 12-7A through 12-7F adapted with permission from Schloss AJ, Manfra Marretta S: Prognostic factors affecting teeth in the line of mandibular fractures.* J Vet Dent *7(4):7–9, 1990.)*

three weeks, six to eight weeks, and then on an annual basis if healing is routine. Because the results of replantation after avulsion can vary considerably, long-term follow-up radiographs after replantation also are recommended. Follow-up intervals for replantations should be one week, three weeks, eight weeks, three months, six months, and then on an annual basis or as indicated by clinical judgment. Follow-up radiographs of intrusion injuries should be taken for several years because root resorption and ankylosis are possible, even if the initial treatment was successful.[2]

INJURIES TO THE BONE

A unique feature of maxillary or mandibular fractures is that teeth are often in the fracture line. The healing of these fractures can be complicated or delayed by the presence of teeth or tooth fragments. Fractures are often open to the oral cavity and can therefore become contaminated with oral fluids, bacteria, and debris. Additional risk for complication occurs if the blood supply to the pulp or periodontium is damaged. If a fracture occurs along the root surface from the gingival margin to the apical region, healing of the subsequent

ANATOMY OF THE MANDIBLE

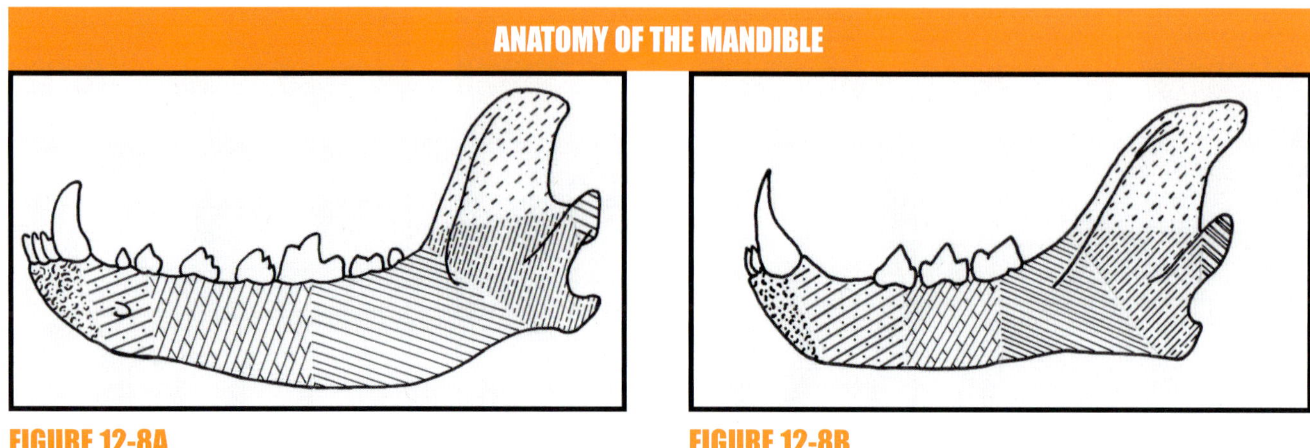

FIGURE 12-8A

FIGURE 12-8B

Figures 12-8A and 8B *The anatomic regions of the mandible in dogs (12-8A) and in cats (12-8B). (Figures 12-8A and 12-8B adapted with permission from Harvey CE (ed):* Veterinary Dentistry. *Philadelphia, WB Saunders Co, 1985, p 153.)*

DIRECTION OF THE FRACTURE LINE

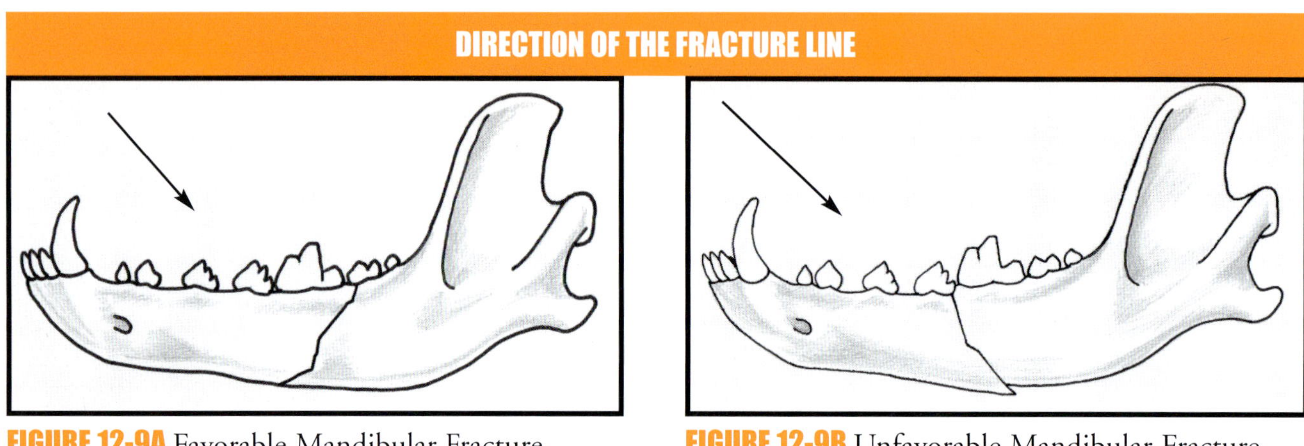

FIGURE 12-9A Favorable Mandibular Fracture

FIGURE 12-9B Unfavorable Mandibular Fracture

Figure 12-9A *When the muscle traction reduces the fracture naturally for a return to proper alignment, the fracture is called favorable.* **Figure 12-9B** *When the muscle traction distracts the ends of the fracture and the fracture line follows a caudal course, the fracture is called unfavorable.*

periodontal–endodontic lesion is likely to be affected by complications. When the fracture line involves the apex, blood supply to the pulp can be disrupted. Periodontal lesions can develop if the periodontal ligament is in the fracture line, especially if the site is open to the oral cavity. Radiographs (along with clinical evaluation) are therefore useful for determining the orientation of a fracture line and whether the periodontium, pulp, or both have been damaged. For example, if a fracture line occurs over a root, concomitant root fracture can be ruled out by altering the angle of the x-ray beam. If the fracture line moves in relation to the affected tooth, the bone only is probably fractured. If the relation between the fracture line and root remain the same despite the angle of the x-ray beam, the root is probably also fractured[2,7,8] (Figure 12-7).

For mandibular fractures, the location and nature of the fracture are valuable information. The mandible is divided into different regions according to the teeth present, bone contours, joint structures, and existing biomechanical forces during mastication and occlusion. In dogs, the first region consists of the incisors to the canine tooth and contains the symphysis. The next region goes from the canine tooth to the second premolar, where the root of the canine tooth occupies the bony space. The third region consists of the second premolar to the first molar, is the straightest region, and contains the roots of these teeth and the mandibular canal. The next region travels from the first molar to the last molar, contains the mandibular canal, and is more curved. The final three regions in dogs extend from the last molar to the angle of the mandible and include the angular process, the condyloid process, and the coronoid process (Figure 12-8). In cats, the structures within all these regions are comparable to those in dogs, except cats have fewer teeth and a shorter mandible.

Direction of the fracture line and its interaction with the muscles of mastication also influence the degree of displacement in a mandibular fracture. If the fracture line in the body of the mandible courses from the alveolar ridge in a rostral aspect to the ventral border of the mandible, the fracture is referred to as being favorable because the muscular traction tends to hold the ends of the fracture in apposition. If the fracture line courses in a caudal direction, muscular traction tends to distract the ends of the fracture (Figure 12-9). Together, these factors (location, nature, and direction) determine the fixation options for stabilizing the fracture. The structures that primarily limit appropriate placement of internal-fixation devices are the roots and mandibular canal because of the need to preserve their structural integrity.[7,9,10]

In many instances, effective fixation of mandibular or maxillary fractures depends on the teeth in the area of the fracture. The extraction of teeth in a fracture line, even when indicated, often further destabilizes the fracture. For that reason, extraction often is not immediately elected. Teeth can aid fracture repair because they can serve as a guide for fracture reduction through occlusion. Teeth also provide anchor points for fixation devices, which is especially valuable when repairing mandibular fractures because the tension side of the bone is on the dorsal border at the crestal bone. The piercing of a portion of

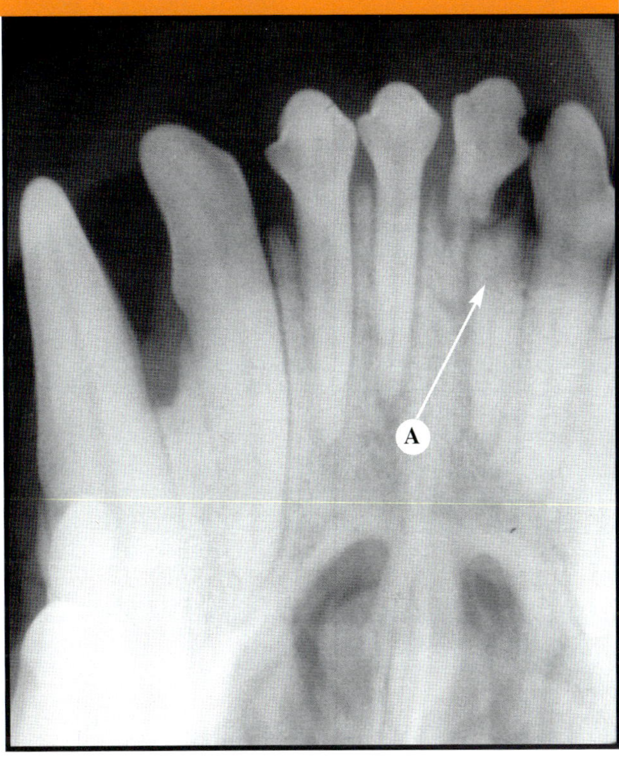

FIGURE 12-10 Loose and Missing Incisors

FIGURE 12-11 Transverse Fracture of the Root

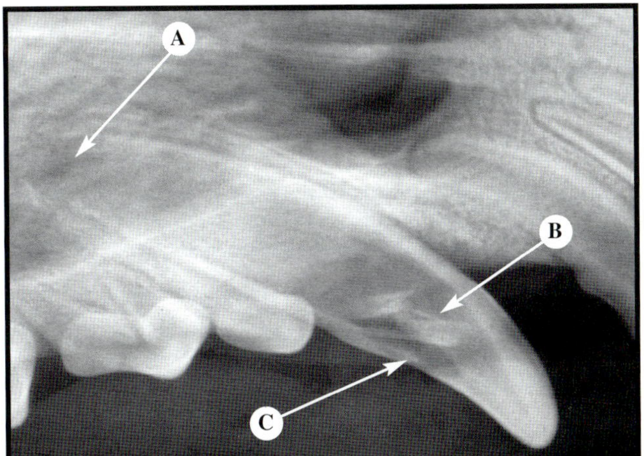

FIGURE 12-12 Impaction Fracture of the Canine Tooth

Figure 12-10 *Possible diagnostic differentials would be trauma with secondary inflammatory bone resorption, preexisting periodontal disease with loss of bone, and neoplastic destruction of the bone with subsequent trauma. (A) Missing crown on the central incisor, (B) a fracture of the root, (C) retained root segments, and (D) loss of mandibular bone.* **Figure 12-11** *Nonvital pulp in a coronal segment is likely. (A) Root fracture with displacement of the crown.* **Figure 12-12** *Endodontic treatment would be complicated by the tooth fragments in the pulp. (A) Incomplete apex, (B) fragments from the tooth wall present in the pulp chamber, and (C) a defect in the enamel wall.*

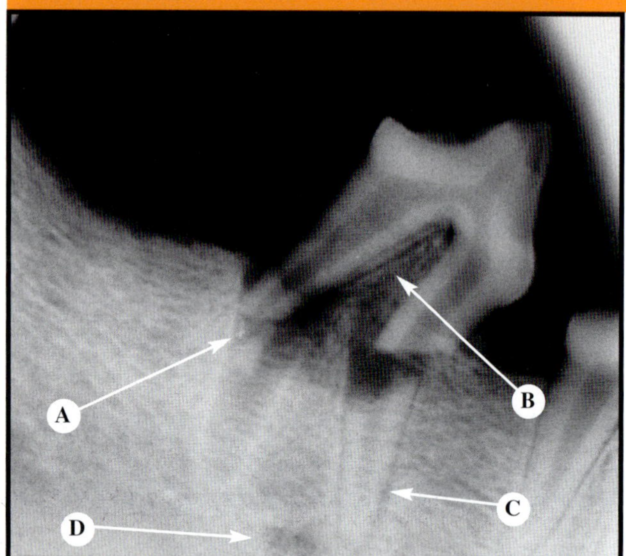

FIGURE 12-13 Fractured Premolar Roots

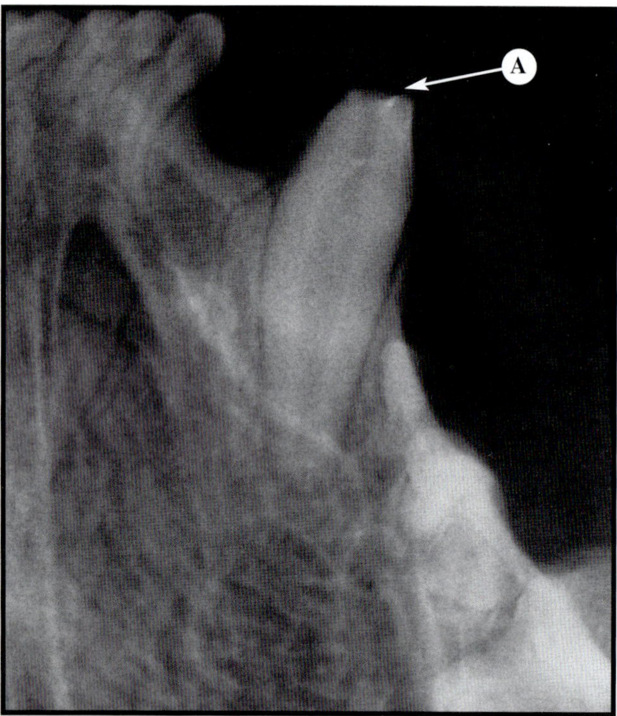

FIGURE 12-15 Fractured Canine Tooth in a Cat

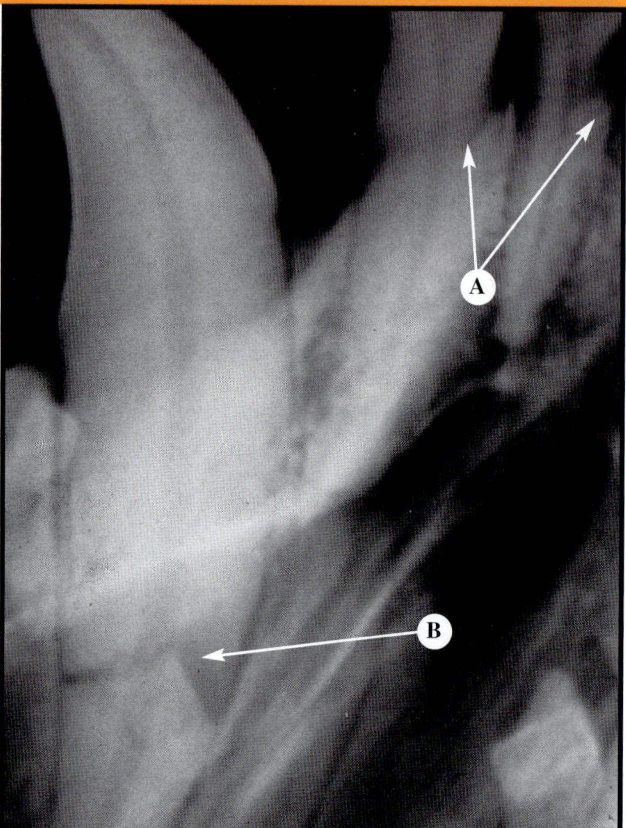

FIGURE 12-14 Fractured Root in the Canine Tooth

Figure 12-13 *Radiographs often must distinguish between luxations and fractured roots. (A) Fractured distal root apex, (B) bone at the furcation was displaced along with the tooth, (C) retained mesial root segment, and (D) mental foramen.* **Figure 12-14** *More than one fracture line may be present in the canine tooth, as demonstrated by the variable width. An alternative would be that the fracture line deviates in the same angle and direction. Additional radiographic views would be necessary to make a definitive diagnosis. (A) Fractures of the incisors and (B) a fracture in the apical one third of the tooth.* **Figure 12-15** *Direct pulp exposure is apparent. The apex is formed; however, the wide root canal suggests a young age. (A) Fracture of the crown. (Figure 12-14 reprinted with permission from Manfra Marretta S (ed):* Problems in Veterinary Medicine: Dentistry, *vol 2, no. 1. Philadelphia, JB Lippincott Co, 1990, p 164.)*

BONE FRACTURES OF PATHOLOGIC ORIGIN

Fractures of pathologic origin are a special category of bone fractures because they are the result of preexisting disease. A common cause of pathologic fractures in adult dogs is chronic periodontal disease, especially in toy breeds. The large size of the teeth in relation to the skeletal size allows extensive bone destruction when periodontal recession occurs. The weakened bone structure can be fractured spontaneously during normal activities, such as eating or playing. A predilection site for mandibular fractures with pathologic involvement is the mandibular first molar and canine teeth for dogs and the

the actual tooth structure or the mandibular canal with fixation devices is contraindicated because it can be painful, disrupt the blood and nerve supply, delay healing, and predispose bone and teeth to inflammatory resorption. Follow-up radiographs of bone fractures that involve the teeth are recommended for at least one year after repair because root resorption, pulpitis, or pulp necrosis are potential sequelae.[7—9,11]

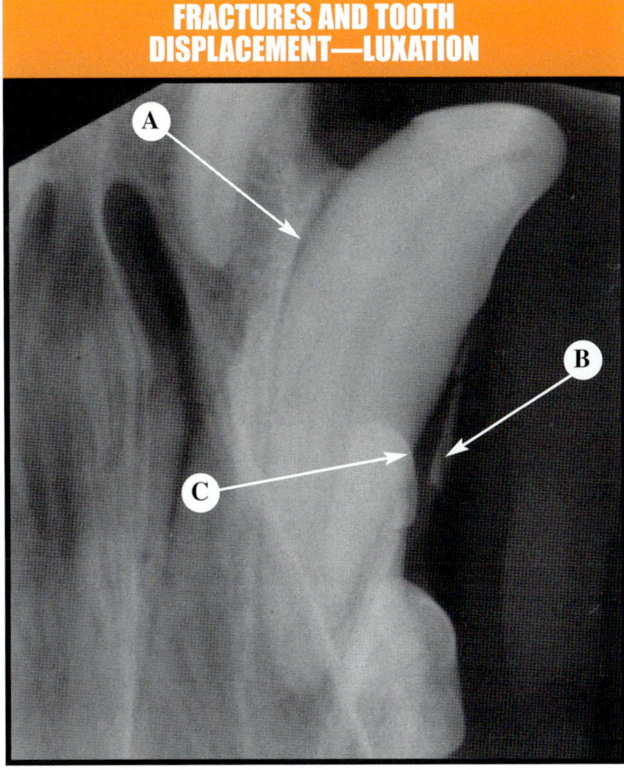

FIGURE 12-16

Displacement of the tooth may be more obvious at other radiographic angles. The tooth may have been displaced, may have fractured the bone, and subsequently was replaced, leaving fragments of fractured bone. (A) Periodontal ligament space, (B) fractured buccal alveolar plate, and (C) widened periodontal ligament space.

mandibular canine teeth in cats. When these teeth are candidates for extraction, mandibular fractures can be a complication of the extraction. If extraction of a mandibular molar or canine tooth is contemplated in a small-breed dog, radiographs should be taken before and after surgery to evaluate the risk for fracture and verify successful extraction.[7,9,12]

RADIOGRAPHIC INTERPRETATION OF TRAUMATIC INJURY

In most instances, veterinary dentistry involves a shared partnership between clinical examination and radiography. When a tooth sustains traumatic injury, however, radiographs are the only diagnostic tool that can implement effective surgical treatment. Fracture lines and tooth displacement that are indiscernible on clinical examination must be captured as radiographic images. We have emphasized throughout this book how accurate imaging relies on proper positioning of the patient, film, and angle of the primary x-ray beam (see Chapters 1, 2, 3, and 4). In addition to their diagnostic value, radiographs are essential for proper follow-up monitoring of the

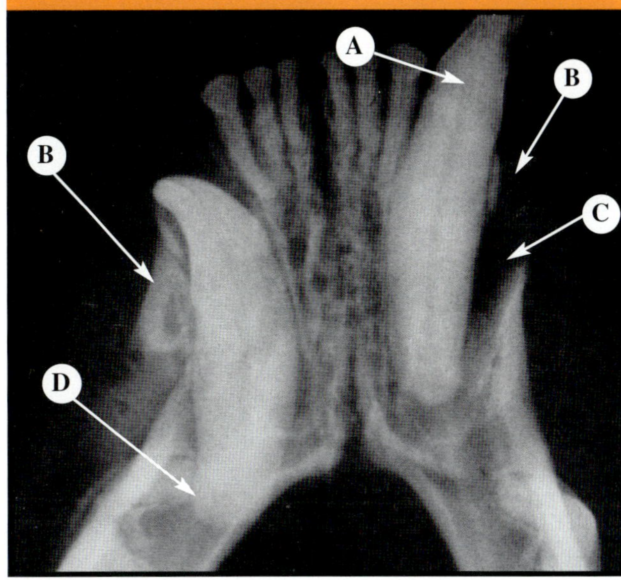

FIGURE 12-17A

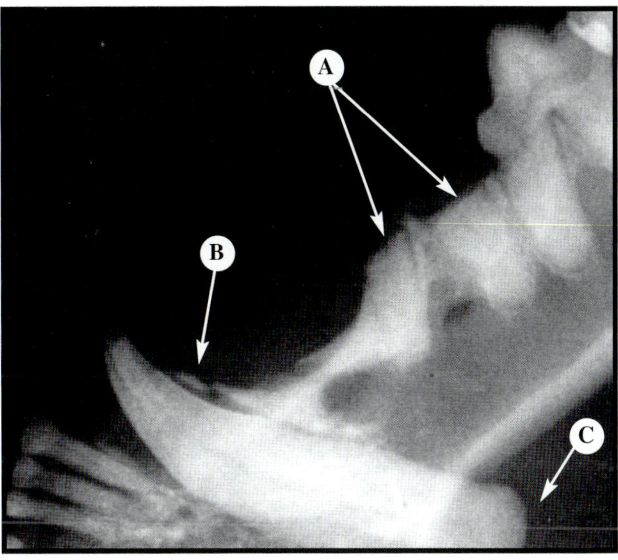

FIGURE 12-17B

Figure 12-17A *Note the intrusion of one canine tooth and luxation of its counterpart. (A) Luxation of the canine tooth, (B) fractured alveolar bone, (C) widened periodontal ligament space, and (D) apical intrusion of the canine tooth through the alveolus.* **Figure 12-17B** *In this lateral view, the apex of the tooth has exited apically beyond the ventral border of the mandibular cortex. (A) Retained roots of the third premolar, (B) fractured alveolar bone, and (C) tooth apex.*

surgical correction. The radiographs provided in Figures 12-10 to 12-15 can assist readers in evaluating the integrity of the teeth, and those in Figures 12-16 to 12-29 offer examples of the various injuries that can be sustained by dogs and cats along with examplcs that evaluate case presentations.

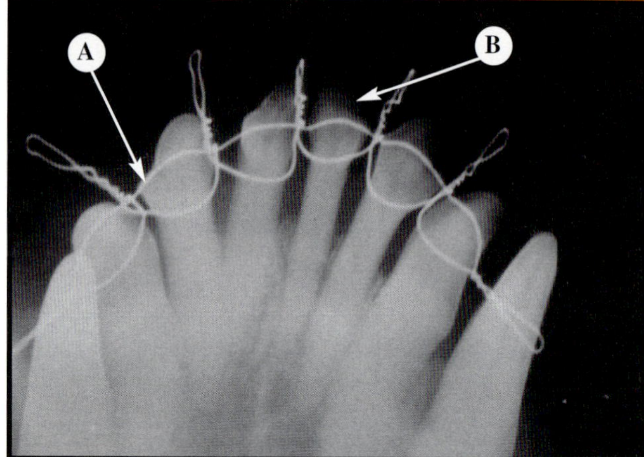

FIGURE 12-18A Radiograph After Replantation

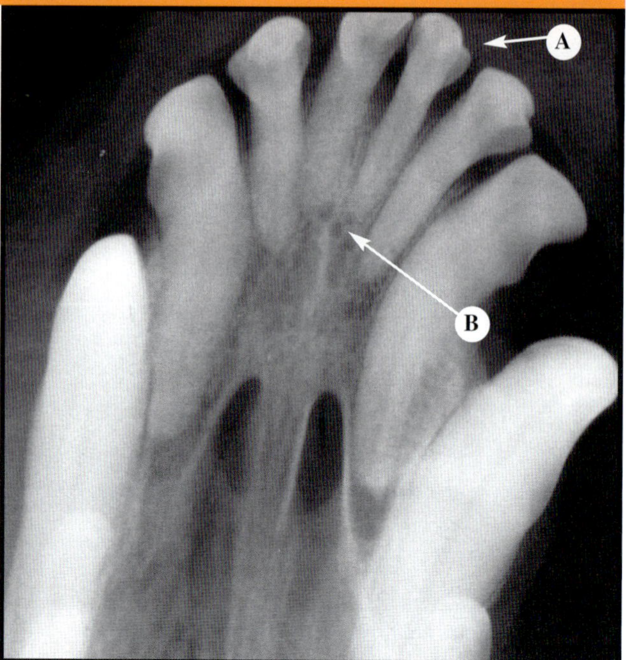

FIGURE 12-18B Radiograph One Month After Replantation

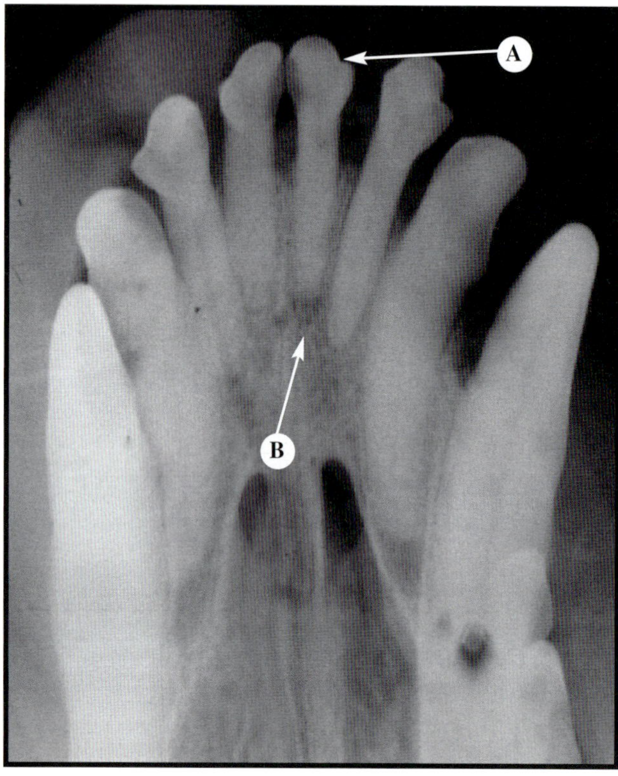

FIGURE 12-18C Radiograph Nine Months After Replantation

Figure 12-18A *Radiolucent acrylic was formed over the wire undersupport. The tooth was disarticulated for four hours before being replanted.* (A) *Interdental wiring and* (B) *avulsed central incisor (replaced.)* **Figure 12-18B** *The periapical radiolucent defects suggest endodontic involvement.* (A) *Replanted incisor and* (B) *multiple small radiolucent defects.* **Figure 12-18C** *Although the crown is clinically normal, the tooth shows endodontic involvement and root resorption.* (A) *Replanted incisor and* (B) *periapical radiolucency and apical resorption.*

REFERENCES

1. Manfra Marretta S, Schrader SC, Matthiesen DT: Problems associated with the management dend treatment of jaw fractures, in Manfra Marretta S (ed): *Problems in Veterinary Medicine: Dentistry*, vol 2, no. 1. Philadelphia, JB Lippincott Co, 1990, pp 220–247.

2. Andreasen JO, Andreasen FM: *Essentials of Traumatic Injuries to the Teeth*. Copenhagen, Munksgaard, 1992, pp 21–154.

3. Fountain SB, Camp JH: Traumatic injuries, in Cohen S, Burns, RC (eds): *Pathways of the Pulp*, ed 6. Philadelphia, CV Mosby Co, 1994, pp 436–485.

4 Andreasen JO, Andreasen FM: *Textbook and Color Atlas of Traumatic Injuries to the Teeth*, ed 3. Copenhagen, Munksgaard, 1994, pp 279–284.

5. Sowray JH: Localized injuries of the teeth and alveolar process, in Williams JH (ed): *Rowe and Williams' Maxillofacial Injuries*. New York, Churchill Livingstone, 1994, pp 257–262 .

6. Hovland EJ: Horizontal root fractures: Treatment and repair, in *The Dental Clinics of North America*, vol 36, no. 2. Philadelphia, WB Saunders Co, 1992, pp 509–516.

7. Schrader SC: Dental orthopedics, in Bojrab MJ, Tholen M: *Small Animal Oral Medicine and Surgery*. Philadelphia, Lea & Febiger, 1990, pp 241–264.

8. Schloss AJ, Manfra Marretta S: Prognostic factors affecting teeth in the line of mandibular fractures. *J Vet Dent* 7(4):7–9, 1990.

9. Weigel JP: Trauma to oral structures, in Harvey CE (ed): *Veterinary Dentistry*. Philadelphia, WB Saunders Co, 1985, pp 140–155.

10. Harvey CE, Emily PP: *Small Animal Dentistry*. Philadelphia, CV Mosby Co, 1993, pp 323–335.

11. Kern DA, Smith MM, Stevenson S, Moon ML, Saunders GK, Irby MH, Dyer KR: Evaluation of three fixation techniques for

FRACTURES OF THE AVEOLUS—IDENTIFICATION

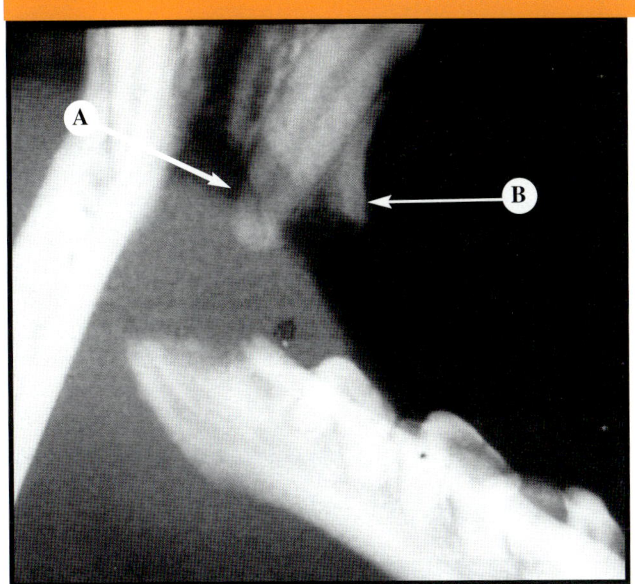

FIGURE 12-19 Fracture of the Mandible with Tooth Involvement

FIGURE 12-20 Fracture of the Aveolus with Tooth Involvement

Figure 12-19 *The vasculature to the pulp is severed, and the pulp is probably not vital. (A) Exposed tooth apex and (B) fracture through the alveolus.*
Figure 12-20 *This radiograph shows how a fracture through the alveolus can involve the tooth. (A) Fracture of the alveolus, (B) fracture of the exposed distal root, (C) curved root apex, and (D) apical root fragment.*

FRACTURES OF THE ALVEOLUS—CASE EVALUATION

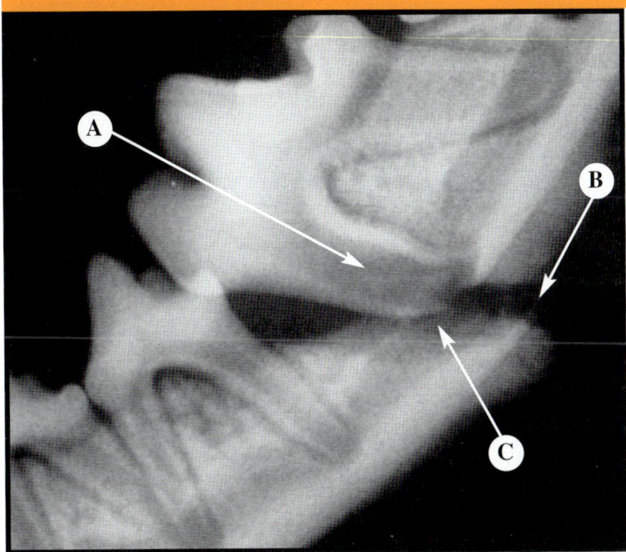

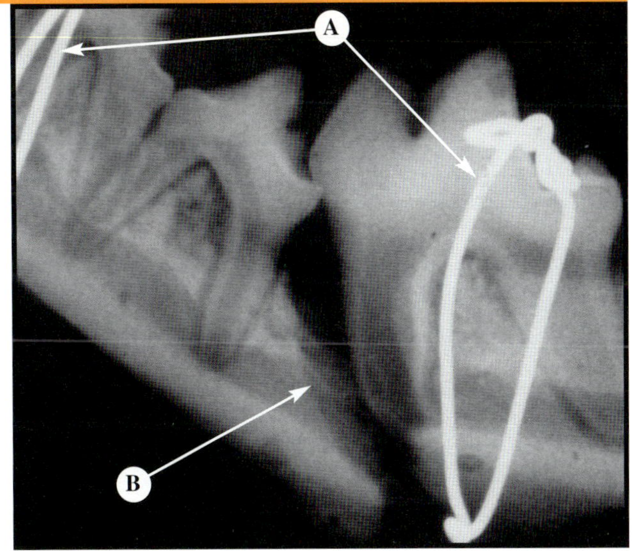

FIGURE 12-21A Fracture of the Mandible

FIGURE 12-21B Radiograph After Surgery

Figure 12-21A *The fracture involves the periodontal space and apex of the mesial root of the molar. This patient is at risk for developing a periodontal–endodontic lesion. (A) Wide canal diameter, (B) fracture line and displaced mandible, and (C) immature apex.* **Figure 12-21B** *Immediately after surgery, the prognosis is guarded for maintaining a vital molar; however, the young age of the patient may be helpful. (A) The cerclage wires are used for stabilization of an intraoral dental acrylic splint, which is not visible because it is radiolucent and (B) the reduction shown is not optimum.*

(continues on next page)

repair of mandibular fractures in dogs. *JAVMA* 206(12):1883–90, 1995.

12. Kapatkin AS, Manfra Marretta S, Schloss AL: Problems associated with basic oral surgical techniques, in Manfra Marretta S (ed): *Problems in Veterinary Medicine: Dentistry*, vol 2, no. 1. Philadelphia, JB Lippincott Co, 1990, pp 85–109.

FRACTURES OF THE ALVEOLUS—CASE EVALUATION *(continued)*

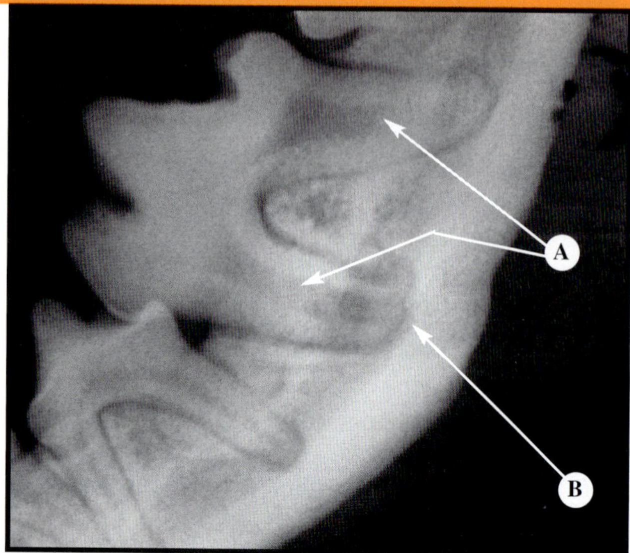

FIGURE 12-21C Follow-up Radiograph at Two Months

FIGURE 12-21D Follow-up Radiograph at Four Months

Figure 12-21C *The bone fracture appears to have healed. (A) Cerclage wires are used to stabilize an intraoral dental acrylic splint, which is not visible because it is radiolucent and (B) developer artifacts.* **Figure 12-21D** *The fracture line cannot be seen on the radiograph; the repair is successful. The root canals are equal in width, and the apex is maturing; however, continued monitoring would be recommended. (A) Root canals and (B) mesial root apex.*

FRACTURES OF THE BONE—IDENTIFICATION

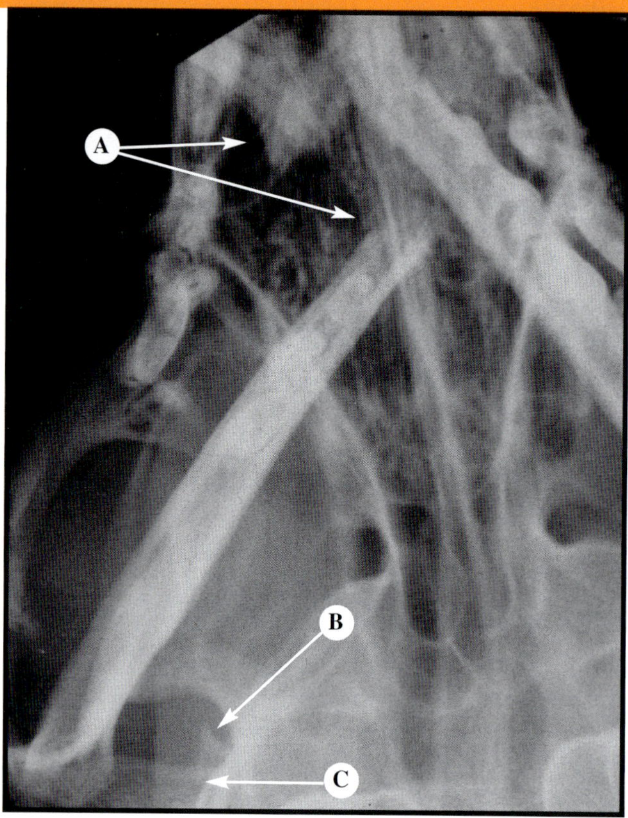

FIGURE 12-22 Fracture of the Maxilla in a Cat

FIGURE 12-23 Fracture of the Mandible with Luxation of the Temporomandibular Joint

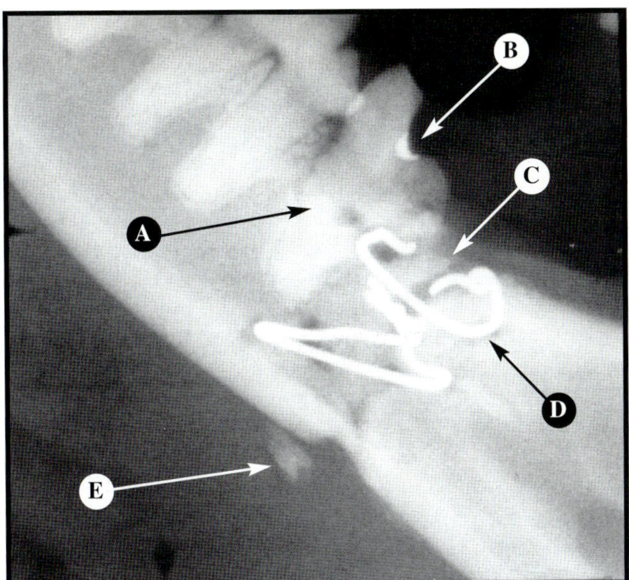

FIGURE 12-24 Mandibular Fracture Repair

Figure 12-22 *Extraction of the fourth premolar could further promote fracture instability; however, if root fragments are left in the fracture line, healing could be affected. The palatal root was extracted, with the remaining fragments being extracted after the fracture healed. (A) Lateral luxation of the canine tooth, (B) bone fracture of the alveolus, (C) palatal midline separation and displacement, (D) fracture of the zygomatic arch, (E) fracture of the fourth premolar with separation of the palatal root, and (F) fracture line.* **Figure 12-23** *Fracture stabilization could be hindered because of lack of support from the temporomandibular joint (TMJ). (A) Fracture of the mandible, (B) bone fragment, and (C) TMJ luxation.* **Figure 12-24** *An example of inappropriate repair. The root fragment in the fracture line could delay healing. Drilling through the teeth to place fixation devices is contraindicated. (A) Drill hole in mesial root, (B) pulp cap restoration, (C) fractured or severed distal root segment of the molar in the fracture line, (D) broken cerclage wire, and (E) cortical bone fragment.*

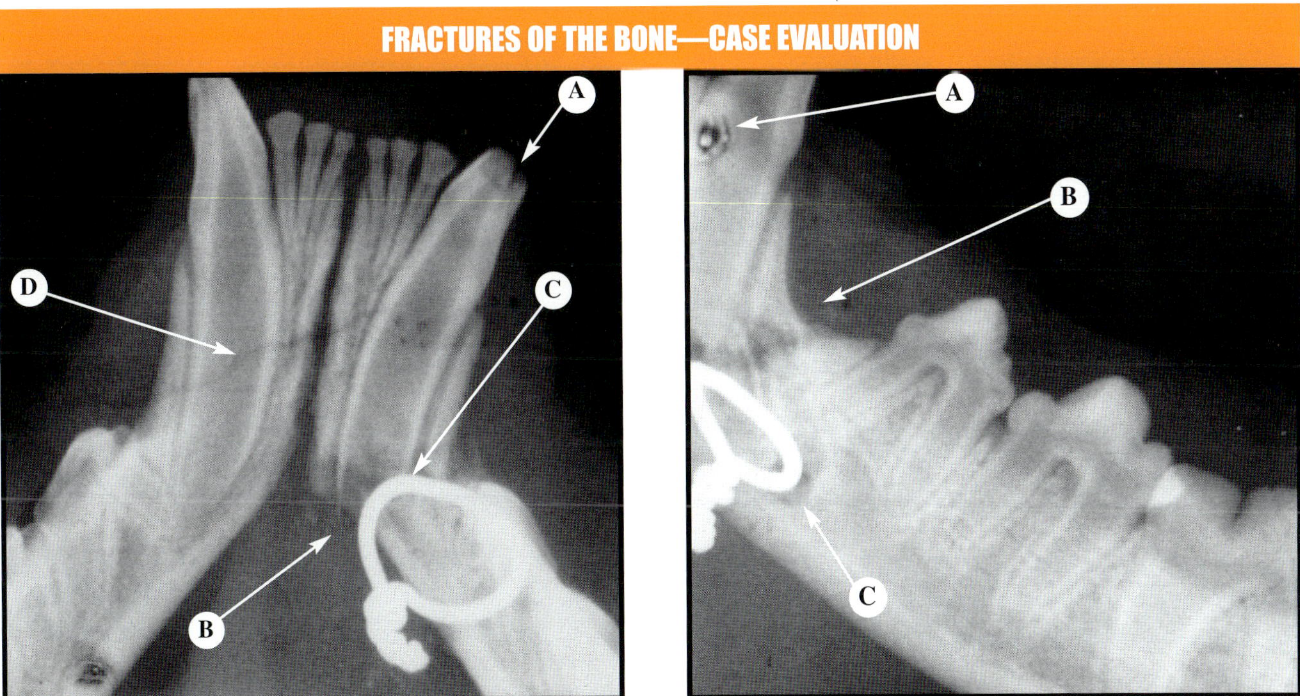

FIGURE 12-25A Fracture of the Mandible

FIGURE 12-25B Lateral View After Initial Surgery

Figure 12-25A *The apical region of the tooth could be further damaged by the technique used for fixation. (A) Fracture of the crown with pulp exposure, (B) fracture, (C) wire impinging on the tooth apex, and (D) artifact.* **Figure 12-25B** *An example of inappropriate repair. The wire is lending no support to segments and is piercing structures in the mandibular canal and tooth apex. (A) Artifact from film hanger, (B) fracture line of alveolar bone, and (C) wire invading the mandibular canal.*

FRACTURES OF THE BONE—CASE EVALUATION *(continued)*

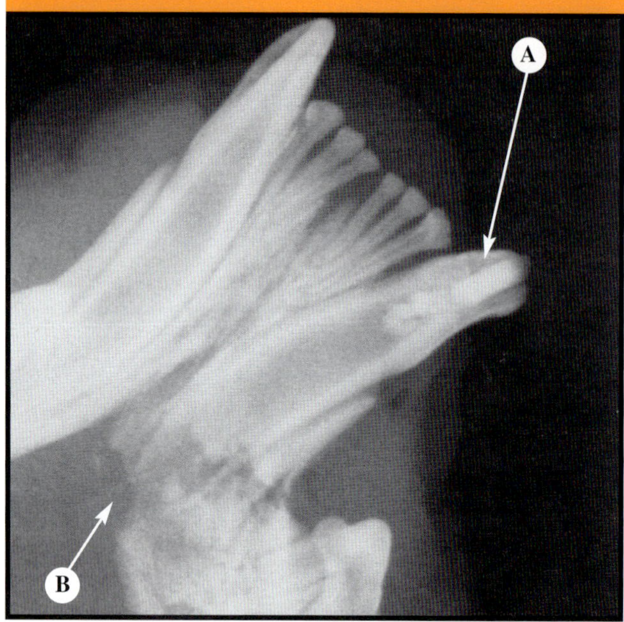

Figure 12-25C *A dental acrylic splint will be placed for fixation. The endodontic prognosis is poor because of the apical damage sustained to the fracture and the iatrogenic damage to periapical tissue. The cat, however, is very young; and the apex is immature. Reinervation and revascularization are possible. Continued monitoring would be recommended. (A) Vital pulpotomy and (B) site of removed wire.*

FIGURE 12-25C Evaluation of Second Surgery

FRACTURES OF PATHOLOGIC ORIGIN—CHARACTERISTICS

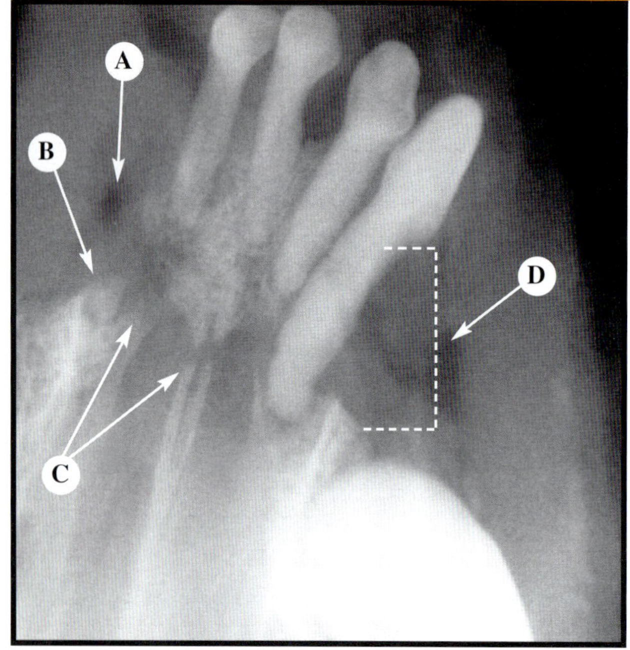

FIGURE 12-26 Fracture of the Premaxilla

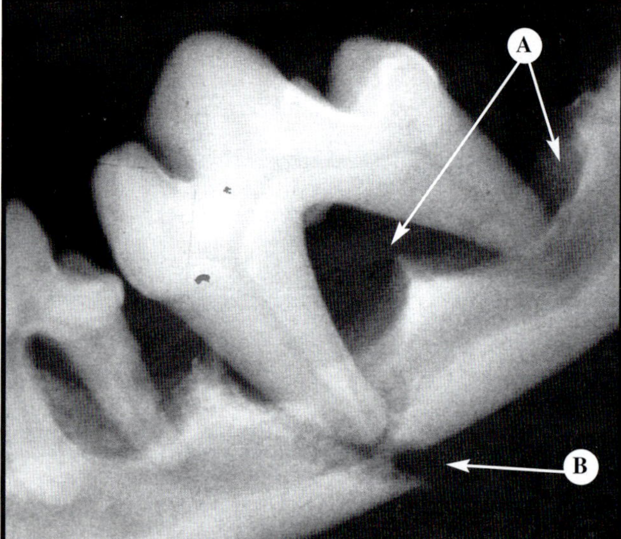

FIGURE 12-27 Fracture of the Mandible

Figure 12-26 *The fracture shown here is secondary to periodontal bone recession around the incisors. (A) Artifacts, (B) root tip of the central incisor, (C) fracture lines, and (D) periodontal bone loss.* **Figure 12-27** *Periodontal disease can cause even normal chewing activities to fracture supportive bone spontaneously. (A) Periodontal bone recession and (B) fracture of the mandible.*

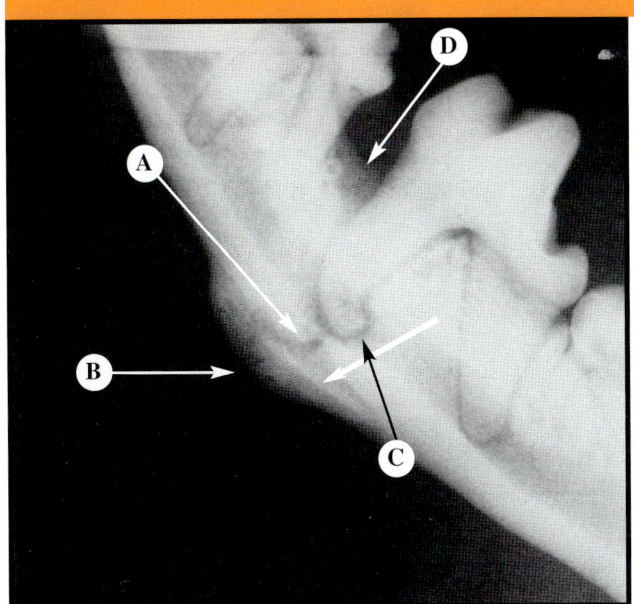

FIGURE 12-28A Fracture of the Mandible

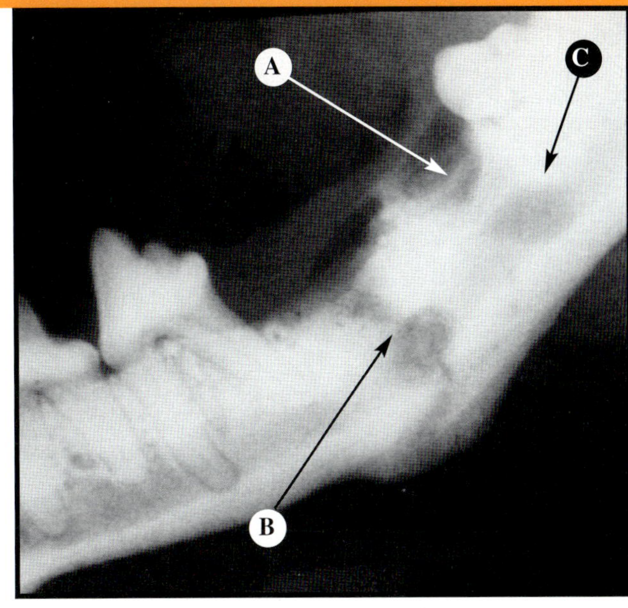

FIGURE 12-28B Radiograph After Extraction

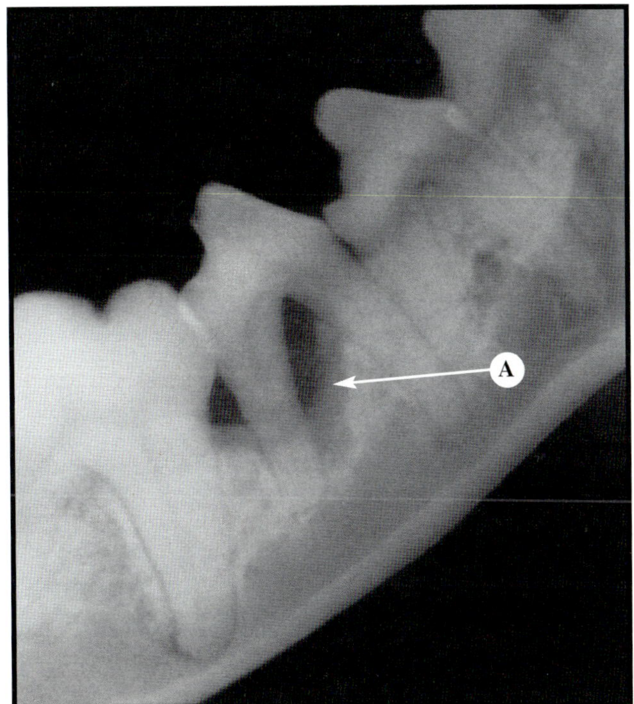

FIGURE 12-29A Periodontal Bone Loss

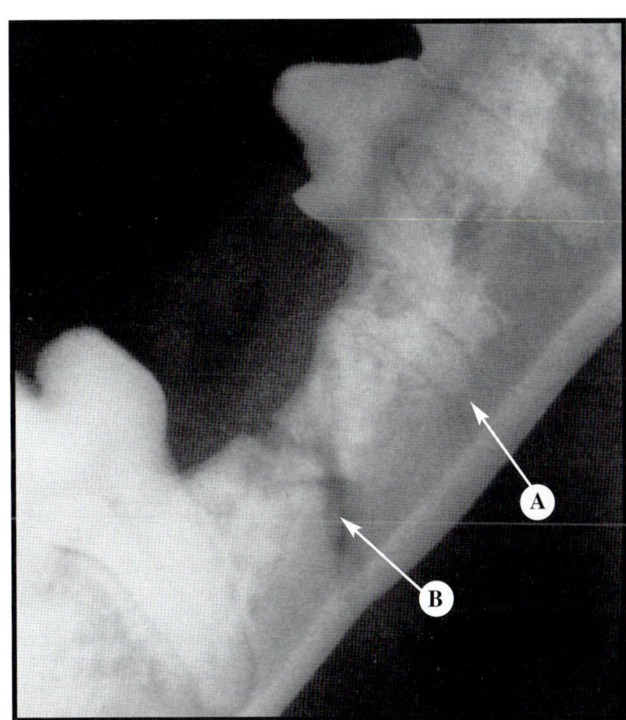

FIGURE 12-29B Resultant Iatrogenic Trauma

Figure 12-28A *The hooked apices present an extraction challenge. The callus indicates attempted healing of the mandible. (A) Fracture line, (B) bony callus, (C) apex, and (D) periodontal bone loss.* **Figure 12-28B** *Radiograph taken after extraction. The fracture line does not appear to have worsened because of extraction. The bony callus appears intact. (A) Periodontal bone loss, (B) fracture line and mesial alveolus, and (C) distal alveolus.* **Figure 12-29A** *Radiograph taken before extraction. The strength of the mandible is compromised secondary to bone loss. (A) Severe angular loss of periodontal bone.* **Figure 12-29B** *This is an example of poor technique. Extraction of the fourth premolar without previous hemisection increased the risk for both a fracture of mandibular bone at the site of the distal root and a fracture of the mesial root, which resulted in a retained root tip. (A) Retained root tip and (B) fracture of the mandible.*

Chapter Thirteen

OSSEOUS LESIONS

Radiographic evaluation of the bones of the skull and jaw plays an important role in defining and treating pathologic conditions of the jaw. Definitive diagnosis, however, is rarely possible using radiographs only because many disparate lesions share the same radiographic characteristics. Therefore, patient history, clinical signs, and histopathologic involvement are essential for an accurate assessment.

CHARACTERICS OF JAW LESIONS

The pathologic characteristics of jaw lesions should initially be classified according to radiographic deviation from normal appearance rather than by categorizing them by mass, such as cyst or tumor, or by type of infection (Table 13-1). A lesion should be characterized on radiographs according to its site, size, shape, symmetry, borders, content, association with other structures, and effect on those structures. When lesions affect the teeth, involvement of the crown or root should be noted. The contents of a lesion can be differentiated by its relative radiolucency or radiopacity on radiographs. Many lesions, however, are mixed and contain both radiolucent and radiopaque elements. The appearance of a lesion can also vary with the stage of pathogenesis because many radiopaque lesions are initially radiolucent.[1-3]

TABLE 13 – 1[1]
CHARACTERISTICS OF LESIONS AFFECTING THE JAW

Characteristic	*Description*
Type	Solitary, multiple, or generalized
Shape	Unilocular, multilocular, or nonlocular
Borders	Well defined or poorly defined
Radiographic Density	Radiolucent, radiopaque, or mixed pattern
Tooth Involvement	No tooth involvement or active tooth involvement
Adjacent Structures	Displaced, eroded, or no effect

COMPARISON OF AGGRESSIVE AND NONAGGRESSIVE LESIONS

Discrete, slow-growing lesions are usually cystic or benign tumors; whereas rapid-growing lesions are usually infectious, inflammatory, or malignant (Table 13-2). Generalized bone involvement may indicate the presence of a metabolic, endocrine, or nutritional disorder.[1]

On radiographs, nonaggressive lesions have a homogeneous appearance with distinct margins. Benign lesions tend to grow by expansion and eventually displace or erode adjacent structures. Teeth are often displaced to a new position. If root resorption is present, it is usually smooth and directed along the advancing edge of the lesion. If expansion against cortical bone occurs, progressive thinning or bowing of the cortex can occur. Periosteal reaction is usually characterized by a uniform and homogeneous opacity on radiographs. Bone formation along the outer cortical surface coincides with resorption of the inner cortical surface by the expanding lesion, and the cortex maintains its continuity. If the rate of expansion increases, a lamellar periosteal reaction occurs, thereby creating a layered onion-skin appearance[1] (Figure 13-1).

An aggressive lesion usually grows rapidly and invades adjacent tissue. On radiographs, an aggressive lesion appears nonuniform with indistinct or ragged margins. Bone destruction is characterized by a moth-eaten appearance. Multiple lytic areas of moderate and nonuniform size that tend to coalesce can be present. Formation of new bone that was induced by the tumor usually creates focal opacities in and around the lesion. Aggressive lesions usually grow around teeth, destroy bone, and on radiographs create the appearance of teeth floating in the air with no bony support. Clinically, the affected teeth maintain their position but develop increased mobility. If root resorption occurs, it is very irregular and involves multiple areas. Eventually, the roots acquire a pointed, spiked appearance. A rapidly growing lesion that destroys the cortex can induce an irregular periosteal reaction. If the periosteal reaction occurs parallel to the cortex, the effect on radiographs will be a lamellar or onion-skin appearance on radiographs. If rapid elevation of the periosteum causes a spicular periosteal reaction, the effect on radiographs will be a sunburst appearance[1] (Figure 13-2).

TABLE 13–2
DIFFERENTIATING AGGRESSIVE AND NONAGGRESSIVE LESIONS[3,4]

Characteristic	Nonaggressive Lesions	Aggressive Lesions
Rate of Change	Little or none	Rapid
Number of Lesions	Single or few	Multiple
Bone Destruction	Well-defined area of lysis	Lytic areas of variable size or uniformly pinpointed
Margins	Distinct, regular, smooth, or sclerotic	Indistinct and ragged
Transition from Abnormal to Normal Regions	Narrow zone from lesion edge to normal region	Wide zone from lesion edge to normal region
Cortical Involvement	Thinning or expanding of cortex	Lysis of cortex
Periosteal Formation of New Bone	Uniform opacity or lamellar onion-skin pattern	Layers of varied opacity or sunburst effect
Internal Formation of New Bone	Organized, homogenous opacity	Focal opacities
Position of Tooth	Displaced	In position
Tooth Mobility	Sometimes affected	Increased mobility
Tooth Resorption	Smooth and unidirectional along the edge of the lesion	Irregular and surrounds the root; root may appear spiked

Not every lesion can be neatly categorized as either aggressive or nonaggressive. Some lesions may exhibit both aggressive and nonaggressive qualities on radiographs. Readers are cautioned that because a continuum of attributes can be presented on radiographs, they should serve only as a guideline in the final evaluation of a lesion.

GINGIVAL GROWTHS

Tissue proliferation that involves the gingiva is popularly called an epulis, which is a general term that offers no indication as to the origin or severity of the lesion. The classification and nomenclature of veterinary epulides have been addressed many times in the literature and continue to evolve. Epulides are similar in gross appearance and frequently inflamed but vary considerably in histopathologic presentation and on radiographs. The gingival aspect of the lesion often underrepresents true size, depth, or character of the complete lesion. The biologic behavior of an epulis ranges from benign to locally invasive or even malignant with the potential for metastasis. Thorough evaluation of an epulis is therefore based on clinical signs, radiographs, and histopathologic presentation. If a lesion appears at the site of a previously diagnosed benign lesion that had resolved, the assumption should not be that it is the same recurring lesion. The new lesion should be completely evaluated, including on radiographs and by histopathologic examination, to rule out an epulis that is undergoing malignant transformation.[5-10]

EDENTULOUS SPACES

When gingivitis or gingivostomatitis, nonhealed extraction sites, fistulas, or swellings persist in edentulous spaces, radiographs would be indicated. Incidental findings can include osteosclerosis, sclerosis of the socket, and retained inactive roots. Routine healing

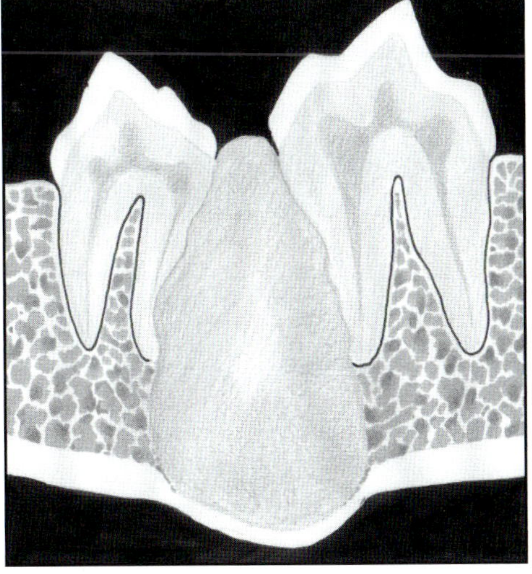

FIGURE 13-1

Nonaggressive bone lesion.

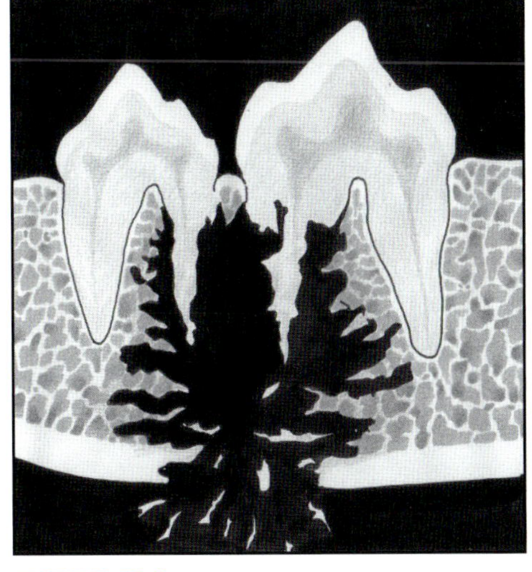

FIGURE 13-2

Aggressive bone lesion.

GENERALIZED RADIOLUCENCY

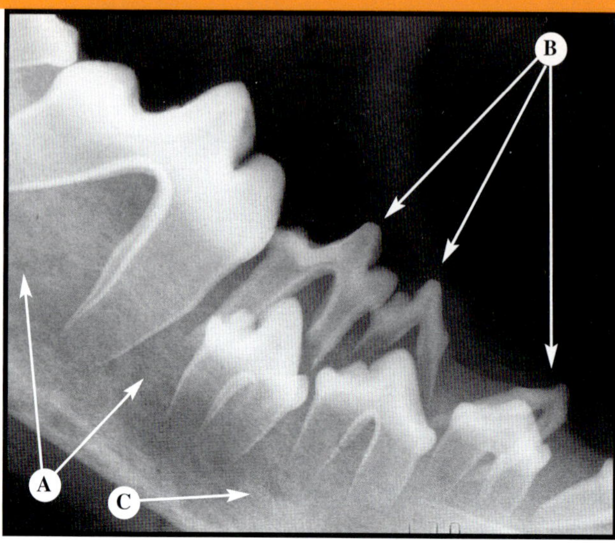

FIGURE 13-3 Renal Secondary Hyperparathyroidism of the Maxilla

FIGURE 13-4 Renal Secondary Hyperparathyroidism of the Mandible

Figure 13-3 *The absence of lamina dura and generalized demineralization of bone is typical of hyperparathyroidism. (A) A lack of pattern to bone trabeculae and (B) absence of lamina dura and periodontal space.* **Figure 13-4** *The generalized radiolucency in this radiograph gives the impression that the teeth are suspended or floating in soft tissue. Other presentations for floating teeth include severe periodontitis, neoplasia, granuloma, and metabolic bone disease. (A) Homogeneous appearance of bone, (B) deciduous teeth, and (C) indistinct mandibular canal. (Figure 13-4 from Carmicheal D, Williams CA, Aller MS: Renal dysplasia with secondary hyperparathyroidism and loose teeth in a young dog. J Vet Dent 12(4):144, 1995. Reproduced with permission.)*

MULTILOCULAR RADIOLUCENCY—AMELOBLASTOMA

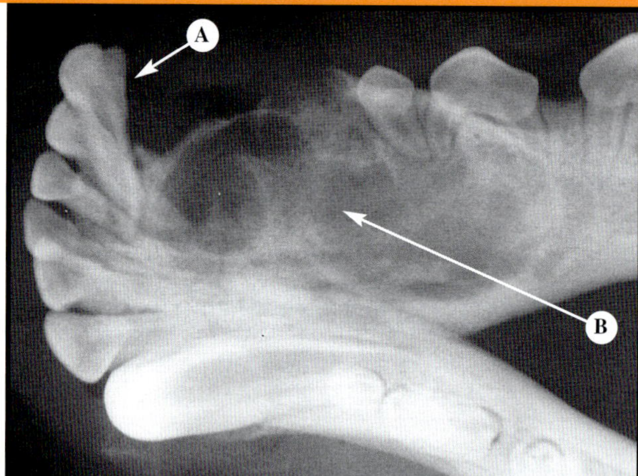

FIGURE 13-5

FIGURE 13-6

of an extraction site normally involves resorption of the lamina dura and deposition of new bone that blends with surrounding bone. Sclerosis of the socket can occur when the lamina dura persists after healing and the socket repairs itself with sclerotic bone. In humans, the condition has been associated with gastrointestinal or renal disease. If nonpathologic roots are retained, they often remain inactive or eventually undergo replacement resorption and are of no consequence. In radi-

ographs of cats, however, the most commonly encountered lesions in an edentulous mandible with mucosal inflammation are retained root fragments or small bony sequestra that often require removal to resolve the clinical signs.[11-13]

RADIOGRAPHIC INTERPRETATION OF OSSEOUS LESIONS

Radiolucent lesions can have generalized, multilocular, or

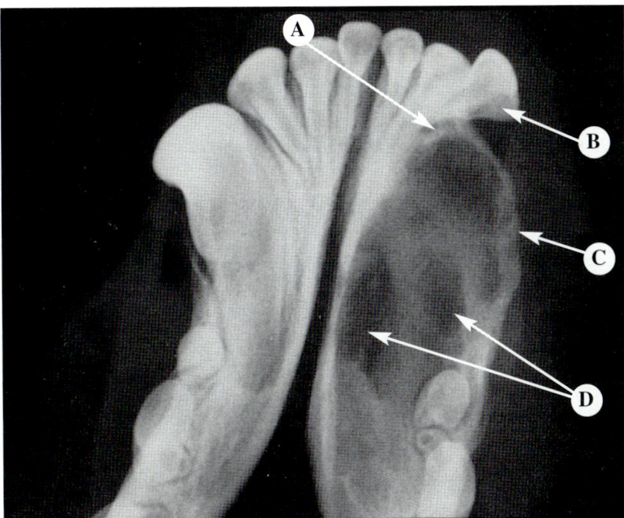

FIGURE 13-7

Figure 13-5 *The radiopaque septa gives the appearance of a soap bubble or honeycomb. The presence of septa suggests a well-recognized, locally expanding lesion. The soft tissue of the nose is being deviated by the expansion of the lesion.* (A) *Multilocular radiolucency and* (B) *the nose.*
Figure 13-6 *This lateral oblique view shows a radiolucent lesion that is compartmentalized by radiopaque patterns. Expansion of a tumor can often cause displacement of adjacent teeth. With the characteristics being shown here, the lesion is large but probably nonaggressive.* (A) *Displaced corner incisor and* (B) *multilocular radiolucency and missing canine tooth.* **Figure 13-7** *Despite the large size and expansion of this lesion, it has relatively nonaggressive characteristics: incisor displacement, compartmentalization, and cortical bowing.* (A) *Sclerotic border,* (B) *displaced corner incisor,* (C) *intact mandibular cortex, and* (D) *multilocular radiolucent areas.*

RADIOLUCENT LESIONS WITH INDISTINCT BORDERS

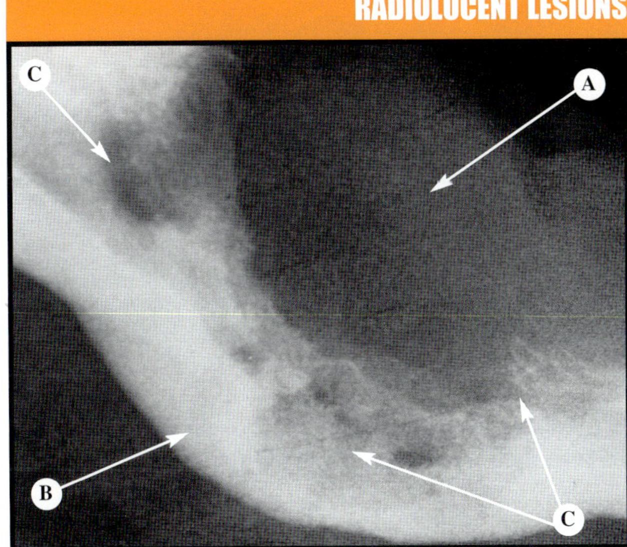

FIGURE 13-8 Mandibular Tumor

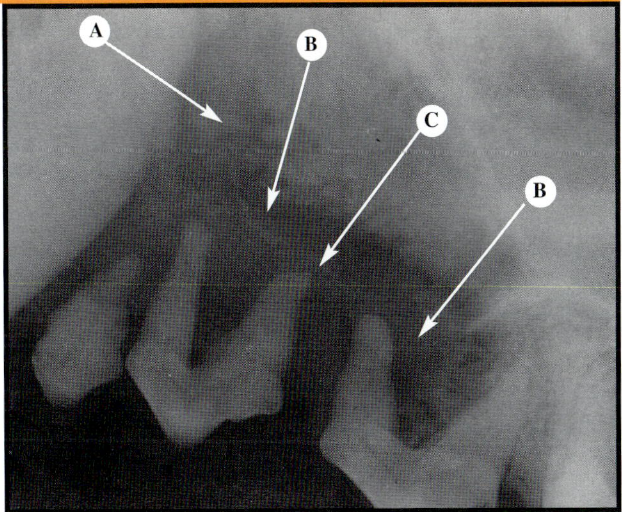

FIGURE 13-9 Fibrosarcoma

Figure 13-8 *The lesion shown in this radiograph is aggressive, and all the teeth are lost.* (A) *Large soft-tissue mass,* (B) *bowed mandible, and* (C) *irregular borders.* **Figure 13-9** *The teeth appear to be floating in the soft tissue but are not displaced. The features shown in this radiograph would indicate a diagnosis of an aggressive lesion. The diagnosis would be on the basis of histopathologic characteristics.* (A) *Irregular border,* (B) *geographic bone lysis, and* (C) *spiked, pointed root apex.*

indistinct borders (Figures 13-3 to 13-10). In addition to involving dental structures, lesions of the jaw represent a range of radiographic characteristics, such as radiolucent, radiopaque, or mixed patterns. It is interesting to note that even lesions with the same histopathologic diagnosis can vary considerably in their radiographic presentation. For example, the histopathologic classification of epulides is still evolving. As classification of the various lesions becomes more sophisticated, dissimilar lesions will not be grouped together and radiographic presentations may become more predictable. Readers

may note that the terminology used in the following figure legends may be dated; however, the terms represent the histopathologic diagnosis at the time it was made.

Lesions that demonstrate various radiolucent patterns are presented in Figures 13-3 to 13-9, whereas Figures 13-10 through 13-13 offer examples of lesions that are associated with unerupted teeth. Lesions that can affect the bone structure in diverse ways include those associated with odontoma, osteomyelitis, and numerous types of neoplasia (Figures 13-14 to 13-27). Radiopaque lesions include increased bone density

(continues on page 191)

LESIONS ASSOCIATED WITH UNERUPTED TEETH

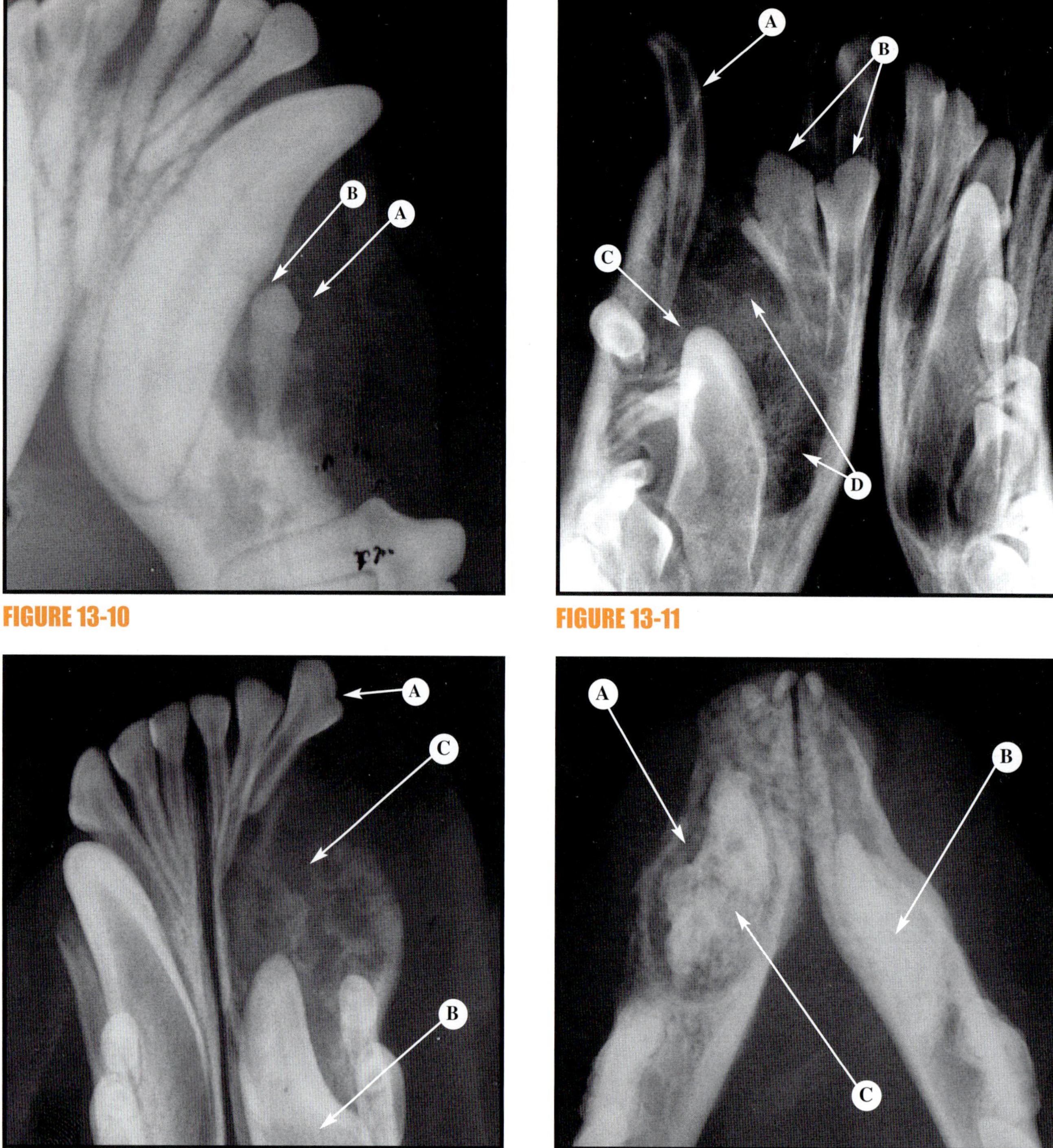

FIGURE 13-10

FIGURE 13-11

FIGURE 13-12

FIGURE 13-13

Figure 13-10 *The crowns of developing, unerupted teeth are normally surrounded by a radiolucent halo. If the tooth does not erupt, however, the pericoronal space can enlarge as it accumulates fluid. (A) Displaced unerupted first premolar and (B) radiolucent region surrounding the tooth.* **Figure 13-11** *The adult canine teeth and incisors shown in this radiograph are displaced, and the mandible has expanded. Tumors in young animals grow rapidly and can affect the growth and development of both teeth and bones. (A) Deciduous canine tooth, (B) unerupted incisors, (C) unerupted canine tooth, and (D) radiolucent soft-tissue mass (squamous cell carcinoma).* **Figure 13-12** *Oral tumors in young dogs and cats can affect the growth and development of the teeth and bones. (A) Displaced corner incisor, (B) unerupted, displaced canine tooth, and (C) multilocular mass.* **Figure 13-13** *Malformed teeth often fail to erupt. Unerupted teeth may undergo encystation, sclerosis, or resorption. Characteristics of odontodysplasia are present. (A) Pericoronal radiolucency, (B) unerupted sclerotic and malformed canine tooth, and (C) canine tooth that is unerupted and poorly mineralized or undergoing resorption.*

MIXED LESIONS—ODONTOMA

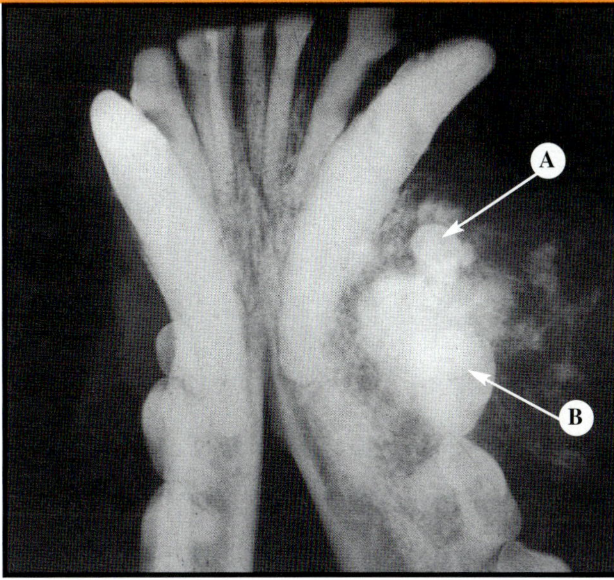

FIGURE 13-14

FIGURE 13-15

Figure 13-14 *Highly differentiated odontomas contain one or more tooth-like structures. Odontomas expand slowly. Some develop cystic structures. A diagnostic differential here would be dentigerous cyst associated with unerupted premolars. (A) Displaced canine tooth, (B) denticles (also called toothlets), (C) bowed but intact mandibular cortex, and (D) radiolucent mass.* **Figure 13-15** *The appearance of odontomas can vary. Many become more radiopaque with time. They also can prevent eruption of adjacent teeth. (A) Unerupted premolar and (B) radiopaque mass. (Courtesy of William J. Zontine, DVM, MS, Diplomate, American College of Veterinary Radiology, Vista, California.)*

LESIONS OF DENTAL ORIGIN—OSTEOMYELITIS AND ENDODONTIC DISEASE

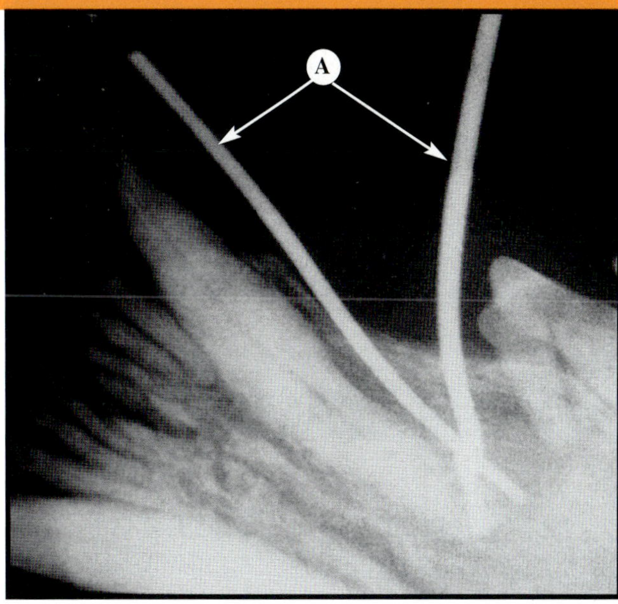

FIGURE 13-16A Ventrodorsal View

FIGURE 13-16B Lateral View

Figure 13-16A *The lesion on this radiograph is aggressive. Note that the radiolucent area just mesial to the third premolar is corresponding to a fistulous tract. (A) Endodontic exposure and fractured crown, (B) moth-eaten bony lysis, and (C) apical resorption.* **Figure 13-16B** *The gutta percha points have been inserted into the fistula to demonstrate that the origin of the tract is at the apex of the canine tooth and not at the premolar. (A) Gutta percha points.*

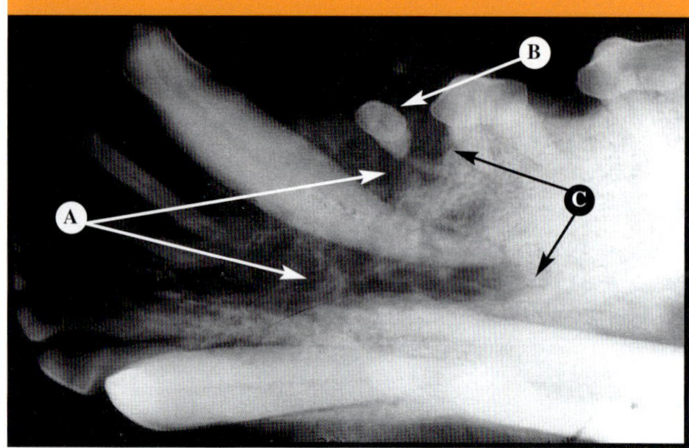

FIGURE 13-17 Oral Tumor

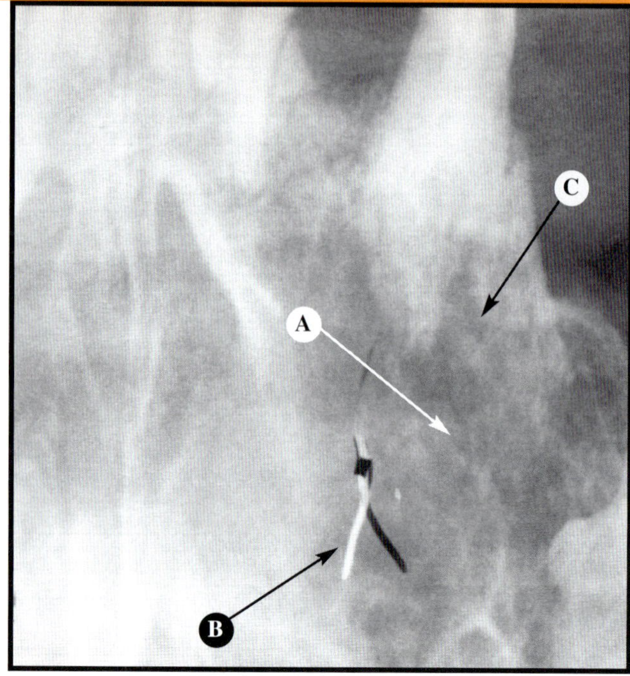

FIGURE 13-18 Fibrosarcoma

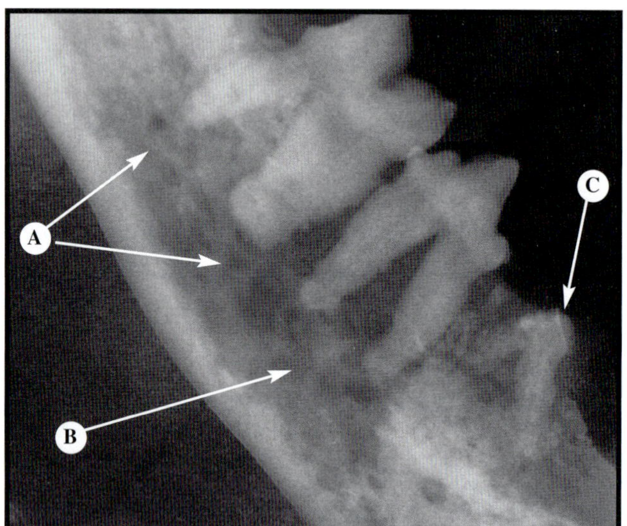

FIGURE 13-19 Squamous Cell Carcinoma

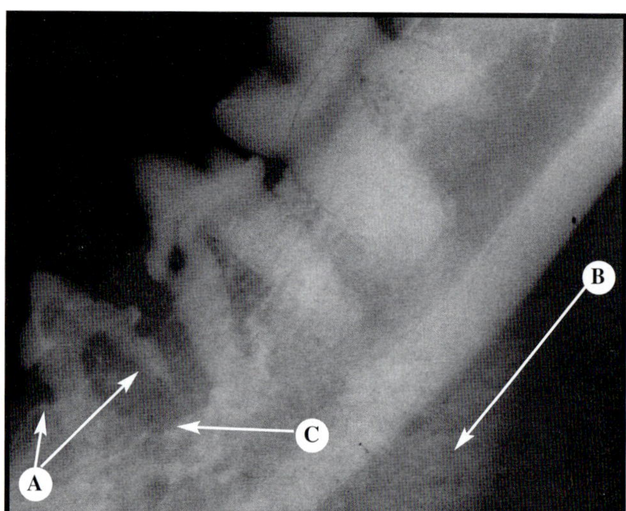

FIGURE 13-20 Squamous Cell Carcinoma

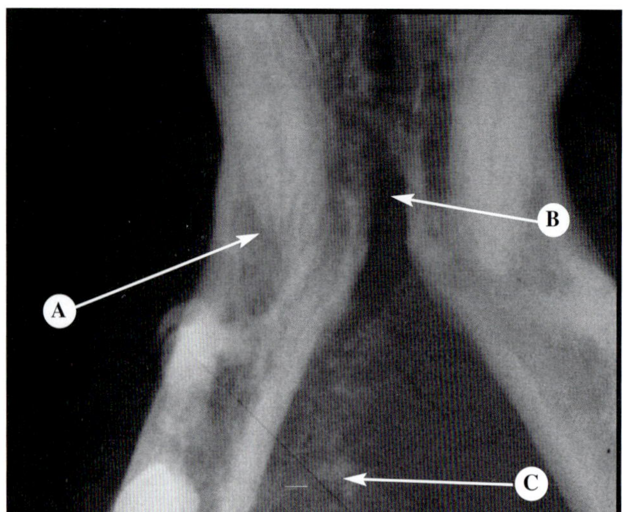

FIGURE 13-21 Squamous Cell Carcinoma

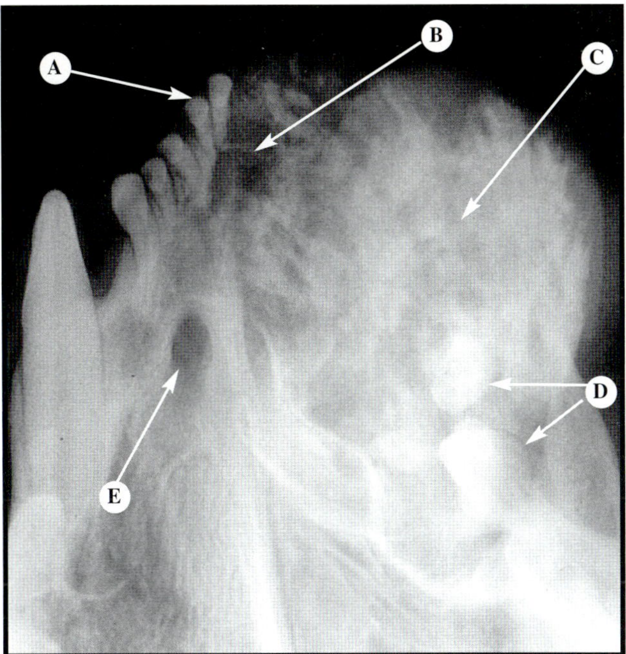

FIGURE 13-22A Squamous Cell Carcinoma

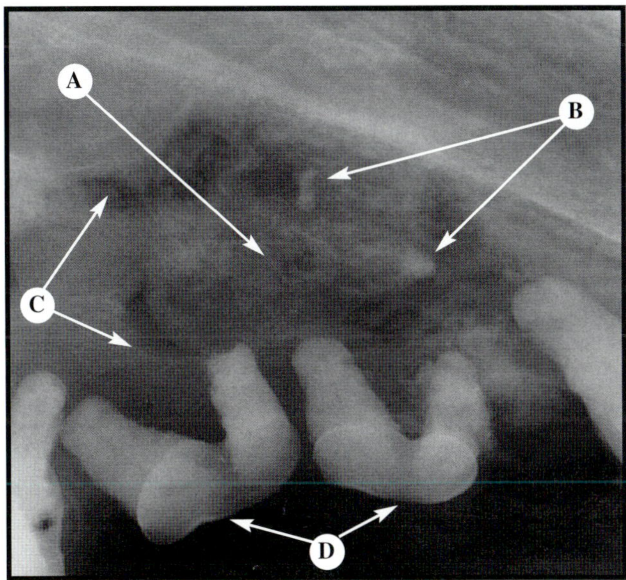

FIGURE 13-23 Osteosarcoma

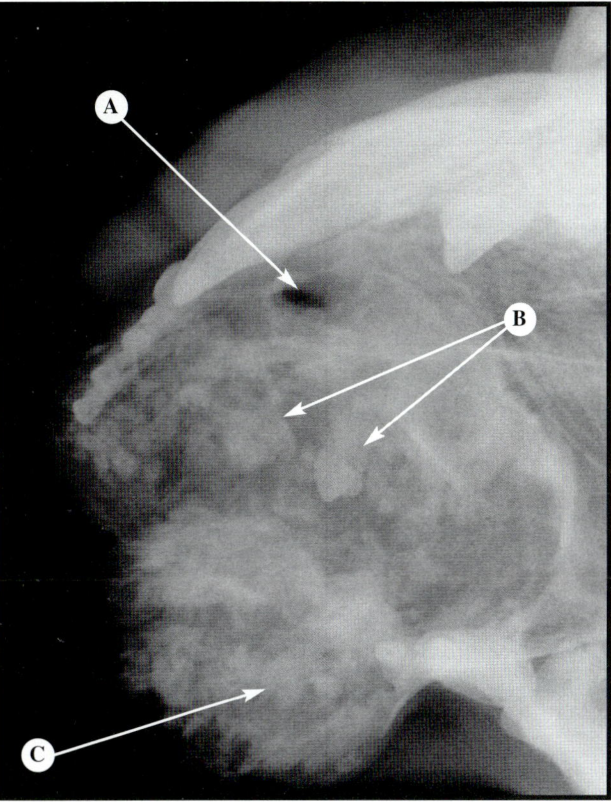

FIGURE 13-22B Squamous Cell Carcinoma

Figure 13-17 *The lesion shown on this radiograph is aggressive. (A) Moth-eaten bony lysis, (B) floating tooth, and (C) irregular border.* **Figure 13-18** *This fibrosarcoma is an aggressive lesion. The artifact is scratched film emulsion. (A) Moth-eaten bony lysis, (B) artifact, and (C) apical resorption.* **Figure 13-19** *Squamous cell carcinomas also are aggressive lesions. Note that the roots of all the teeth are undergoing varying degrees of resorption. (A) Indistinct, irregular borders, (B) moth-eaten bony lysis, and (C) third premolar undergoing resorption.* **Figure 13-20** *The borders of this aggressive squamous cell carcinoma are indistinct. The sunburst periosteal reaction superimposes the moth-eaten appearance of the bone. (A) Root resorption, (B) amorphous periosteal reaction, and (C) bony lysis with irregular borders.* **Figure 13-21** *The symphysis is separating because this squamous cell carcinoma is expanding; growth is likely to occur beyond the midline. (A) Apical resorption, (B) symphyseal separation and perforation of cortical bone, and (C) periosteal reaction.* **Figure 13-22A** *Ventrodorsal view of squamous cell carcinoma of the maxilla. (A) Displaced incisors, (B) bony lysis, (C) new bone formation with a sunburst appearance, (D) buried teeth, and (E) palatine fissure.* **Figure 13-22B** *Lateral view of same lesion shown in Figure 13-22A. (A) Palatine fissure, (B) buried teeth, and (C) new bone formation with sunburst appearance.* **Figure 13-23** *Osteosarcomas also are aggressive lesions. (A) Bony lysis, (B) radiopaque foci, (C) irregular, ragged borders, and (D) floating teeth.*

(continued from page 187)

(Figures 13-24 through 13-27), foreign bodies (Figures 13-28 through 13-33), and retained root fragments (Figures 13-34 through 13-39). Examples of bone response to inflammation, gingival growths (epulides), and lesions associated with extraction can also help readers to distinguish between aggressive and nonaggressive lesions (Figures 13-40 to 13-64). Finally, radiographs of conditions that can develop in edentulous spaces of the mandible are presented (Figures 13-65 to 13-69).

REFERENCES

1. Goaz PW, White SC: *Oral Radiology: Principles and Interpretation,* ed 3. Philadelphia, CV Mosby Co, 1994, pp 291-305, 474-476.
2. Wood NK, Goaz PW: *Differential Diagnosis of Oral Lesions,* ed 4. Philadelphia, Mosby Year Book, 1991, pp 283-613.

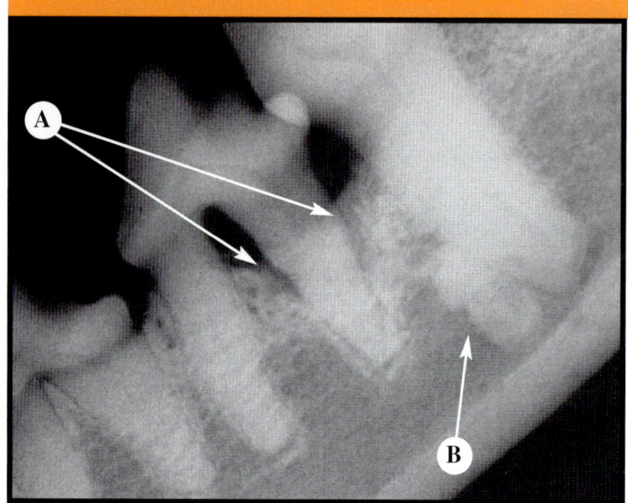

FIGURE 13-24 Condensing Osteitis

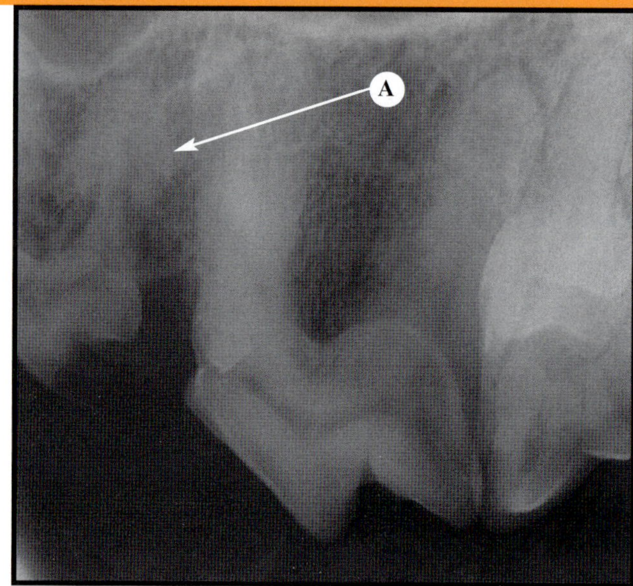

FIGURE 13-25 Osteosclerosis

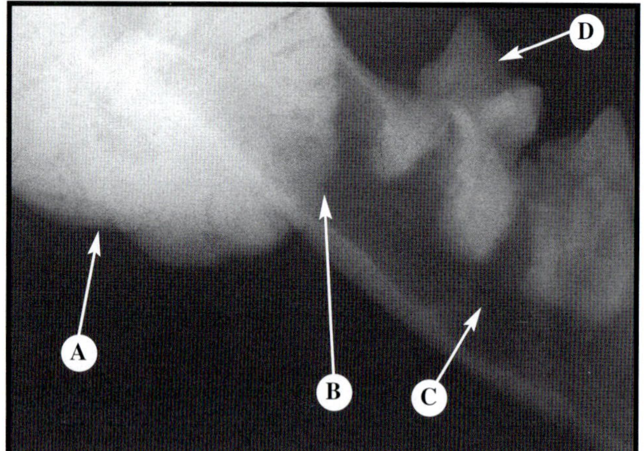

FIGURE 13-26 Osteoma

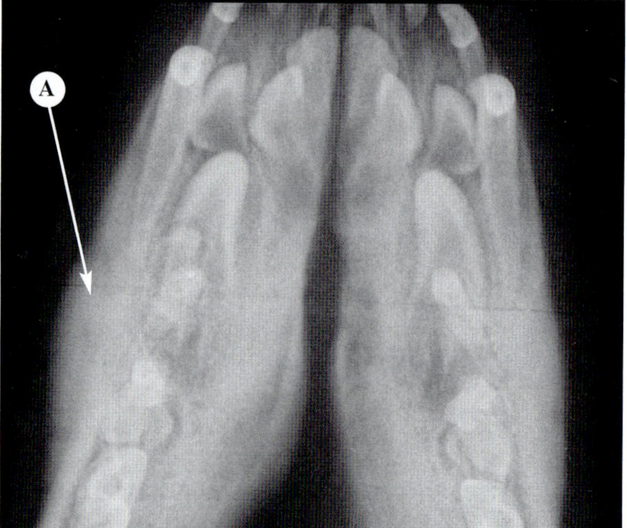

FIGURE 13-27 Osteopathy

Figure 13-24 *A solitary, dense, focal radiopacity is near the apex of a tooth, but the contour of the root is visible. Condensing osteitis is often an incidental finding and is associated with low-grade inflammation of bone. The periodontal lesion of the fourth premolar is a possible contributory factor. (A) Angular periodontal bone recession and (B) condensing osteitis.*
Figure 13-25 *The contour of the distal root is obscured by a radiopacity. In humans, osteosclerosis has been associated with trauma, bruxism, and retained deciduous teeth. (A) Osteosclerosis.* **Figure 13-26** *Although these lesions are large, osteomas are nonaggressive. (A) Single homogeneous radiopaque mass, (B) distinct border, (C) mandibular canal, and (D) molar.* **Figure 13-27** *Craniomandibular osteopathy is a predilection in West Highland white terriers. The condition is generally confined to the bones of the skull and mandible. (A) Bilaterally symmetric bony proliferation and increased bone opacity.*

FOREIGN BODIES

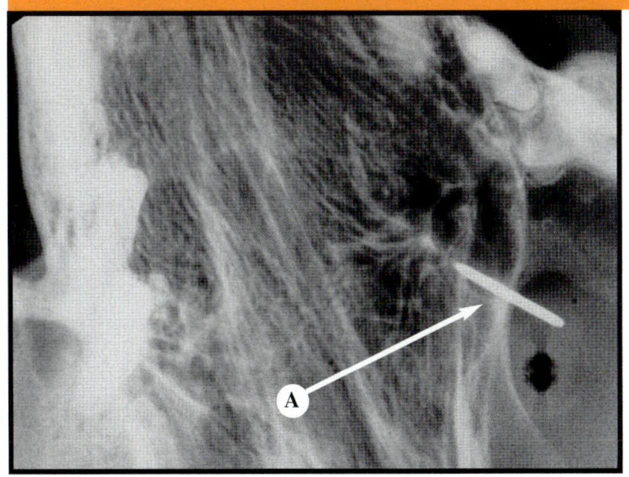

FIGURE 13-28A Ventrodorsal View of a Foreign Body in the Palate

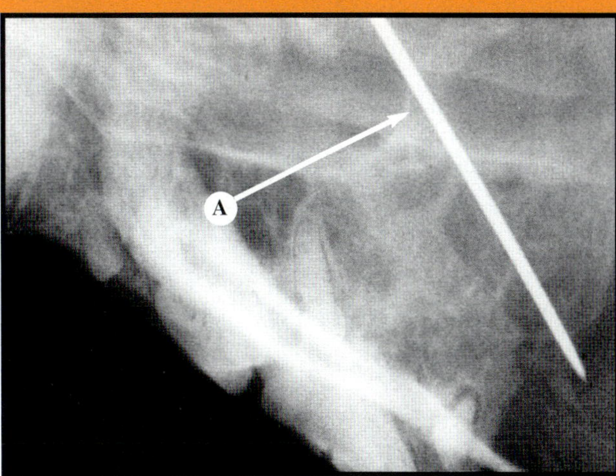

FIGURE 13-28B Lateral View of a Foreign Body in the Palate

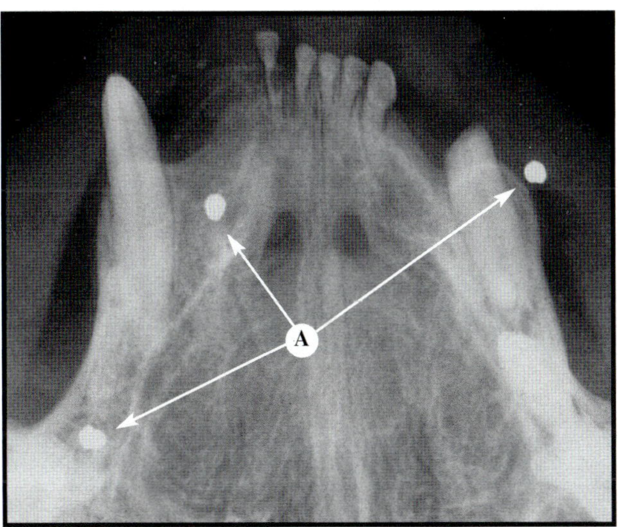

FIGURE 13-29 B-B Pellets in the Maxilla

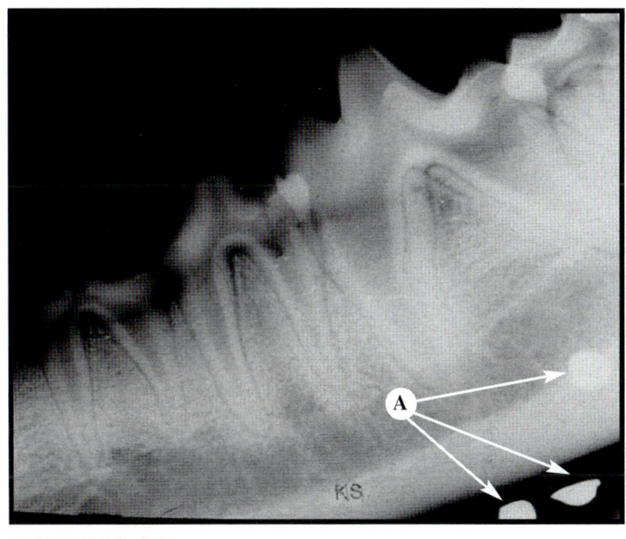

FIGURE 13-30 Buckshot in the Mandible

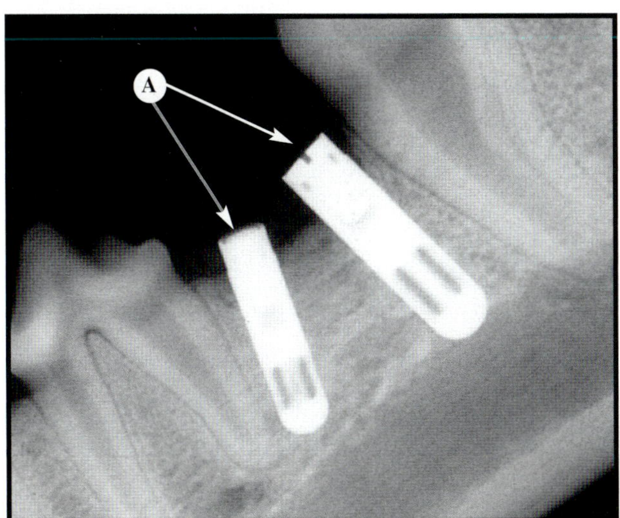

FIGURE 13-31 Implant at the Fourth Premolar

Figure 13-28A *Two views are essential in estimating the size, shape, and position of a foreign body. This ventrodorsal view of the palate identifies the foreign object. (A) Sewing needle.* **Figure 13-28B** *The lateral view of the palate gives an indication of the size. (A) Sewing needle.* **Figure 13-29** *Foreign bodies can be distracting and can cause the viewer to miss other lesions. The canine tooth on the right is fractured and has an apical lesion, whereas the second incisor on the left is undergoing resorption. The left third incisor is missing. (A) B-B pellets.* **Figure 13-30** *Often foreign bodies of this type are an incidental finding. (A) Buckshot pellets.* **Figure 13-31** *Implants are used to anchor a pontic (false tooth) in place, which may have ethical implications if the animal is a showdog. (A) Implants.*

FOREIGN BODIES *(continued)*

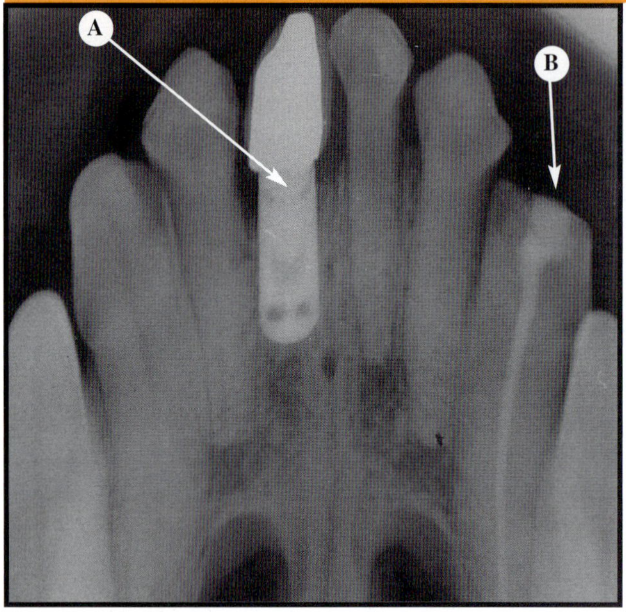

FIGURE 13-32 Implant at the Incisor Site

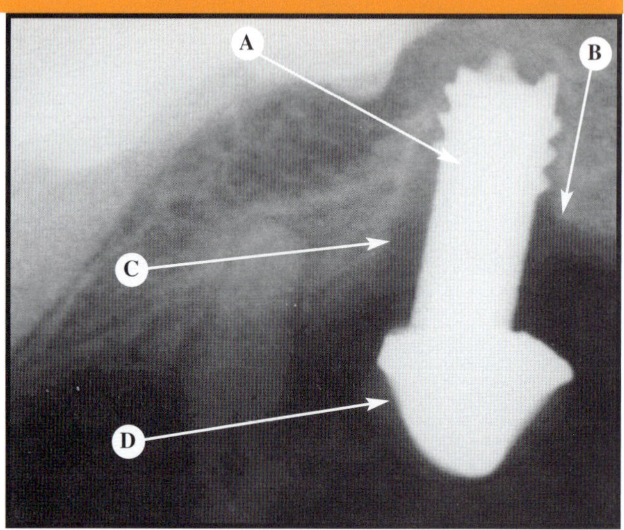

FIGURE 13-33 Failed Implantation

Figure 13-32 *The implant appears to have good osseous integration, and the pontic is in place. Again, the ethical consequences must be considered. (A) Implant and* (B) *endodontic filling.* **Figure 13-33** *Osseous integration has failed to occur. The pontic was prematurely placed or perioimplant disease has caused long-term failure. (A) Implant, (B) horizontal bone recession, (C) widened bone-to-implant space, and (D) pontic.*

RETAINED ROOTS AND FRAGMENTS

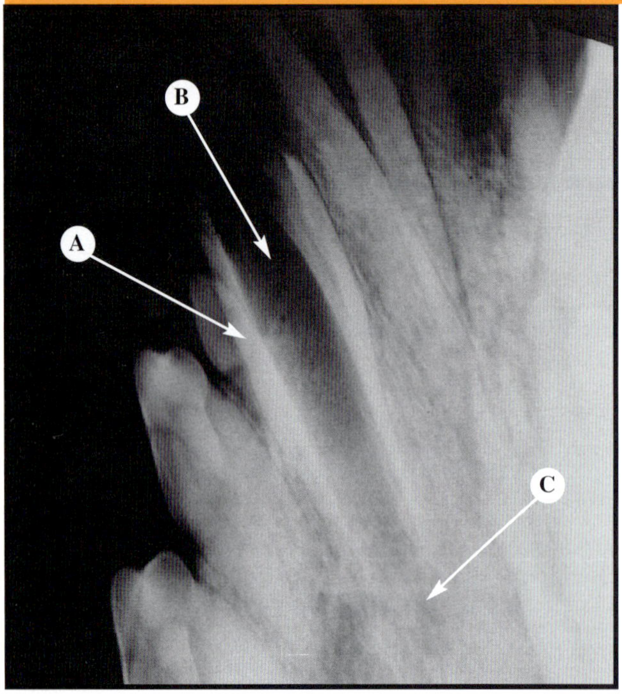

FIGURE 13-34

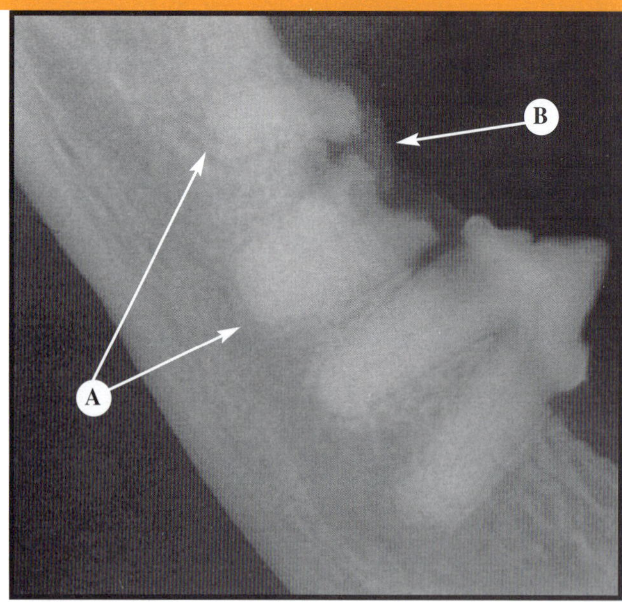

FIGURE 13-35

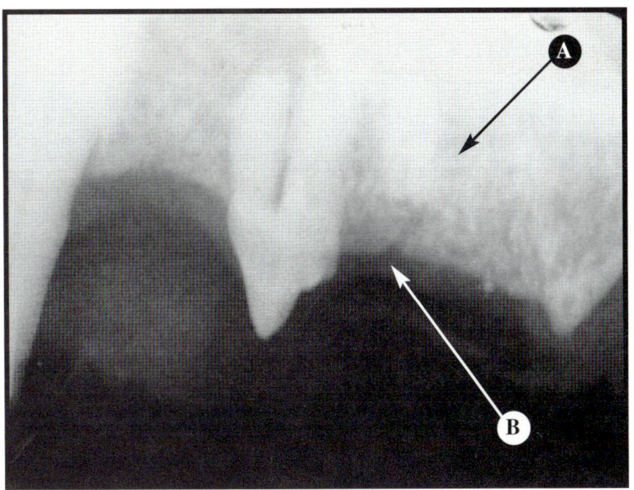

FIGURE 13-36A

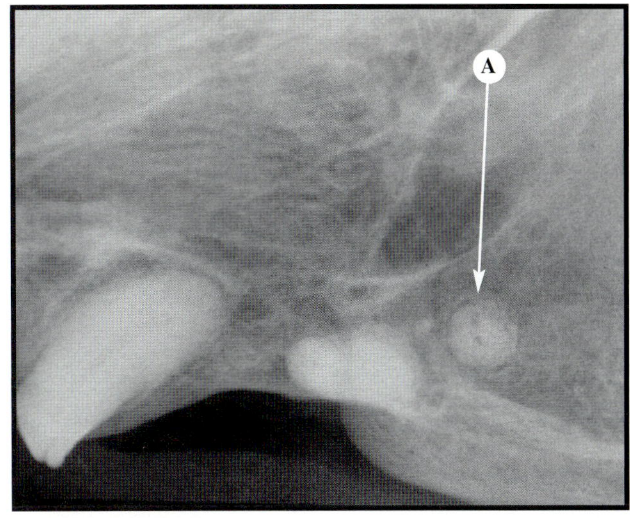

FIGURE 13-36B

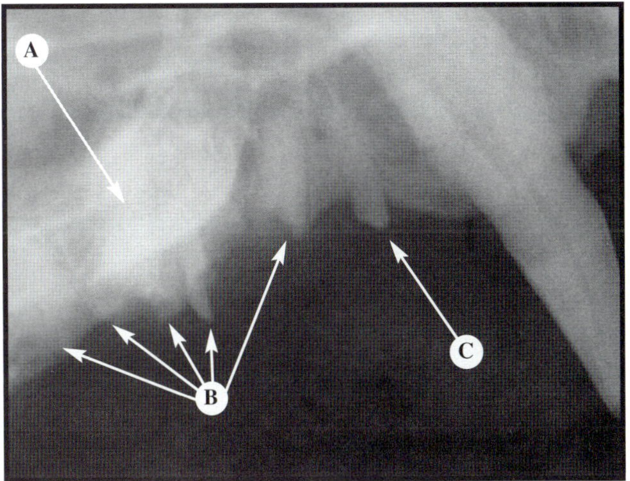

FIGURE 13-37

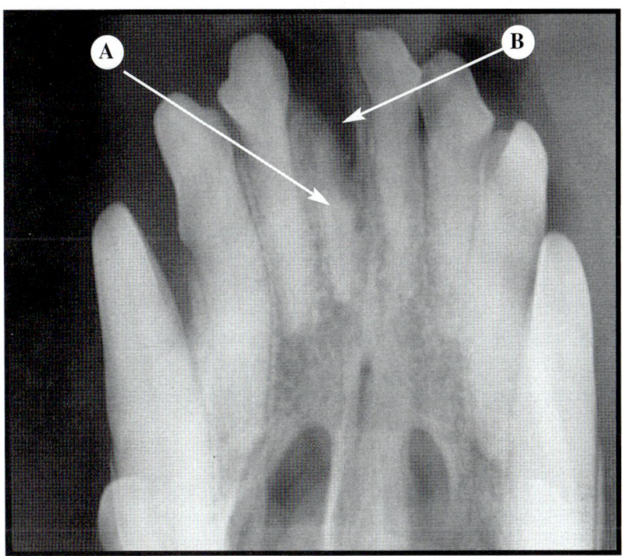

FIGURE 13-38

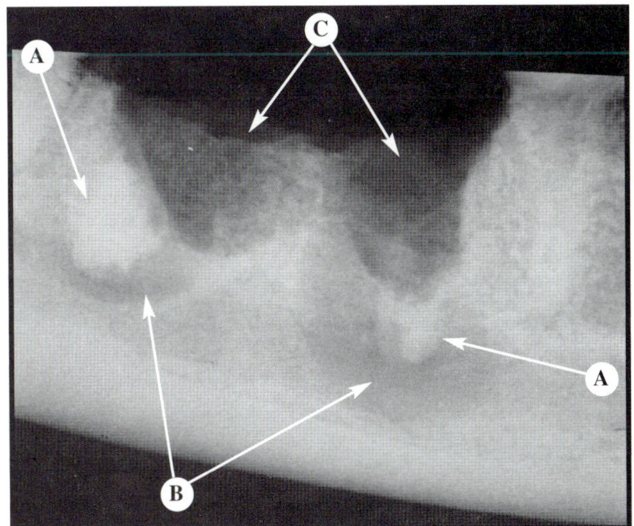

FIGURE 13-39

Figure 13-34 *The wide root canal and open apex indicate that endodontic exposure and pulp death occurred when the dog was very young. (A) Dentinal wall, (B) large pulpal space, and (C) incomplete apex.* **Figure 13-35** *When root fragments protrude above the level of bone, healing is impaired. (A) Retained molar roots and (B) bony proliferation.* **Figure 13-36A** *Lateral view of retained root fragment of the fourth premolar. The bony proliferation over the root makes localization of the root challenging. Therefore, two radiographic views would be beneficial. The bony fistula suggests that the retained fragment is infected. (A) Retained root tip and (B) radiolucent bony tract leading to the root.* **Figure 13-36B** *Ventrodorsal view of same retained root fragment shown in Figure 13-36A. This second radiographic view reveals the location of the root at the palatal site. (A) Retained root.* **Figure 13-37** *The second premolar may be mistaken for a retained root. Root fragments that protrude above the bone level or are infected can impair healing. (A) Zygomatic arch, (B) retained roots and fragments, and (C) second premolar.* **Figure 13-38** *The bone has not healed over the site. (A) Retained root of the central incisor and (B) bone cleft.* **Figure 13-39** *The extraction site has not healed. Root fragments that sequester can impair healing. (A) Retained root fragments, (B) radiolucent defects, and (C) persistent bony defects of the alveoli.*

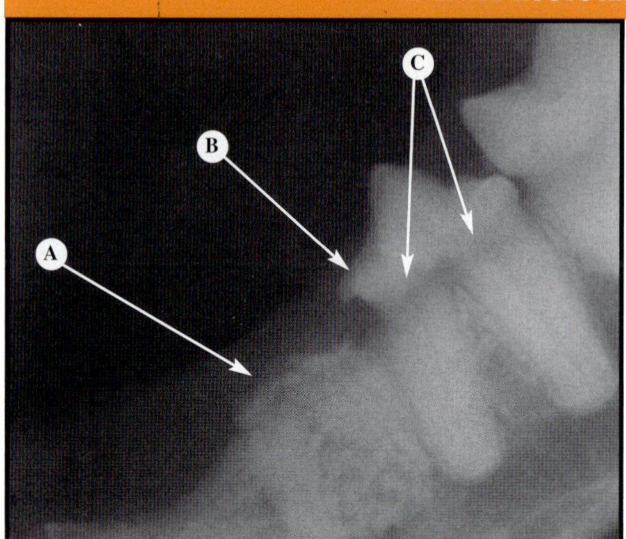

FIGURE 13-40

Figure 13-40 *Stage 4 tooth resorption. The third mandibular premolar in a cat is undergoing resorption. Calculus overhangs accentuate the concaved neck of the tooth, thereby mimicking a resorptive lesion. Cervical burnout mimics a G. V. Black Class V lesion (see Chapter 11). (A) Remnants of the roots of the third premolar, (B) calculus, and (C) cervical burnout.*

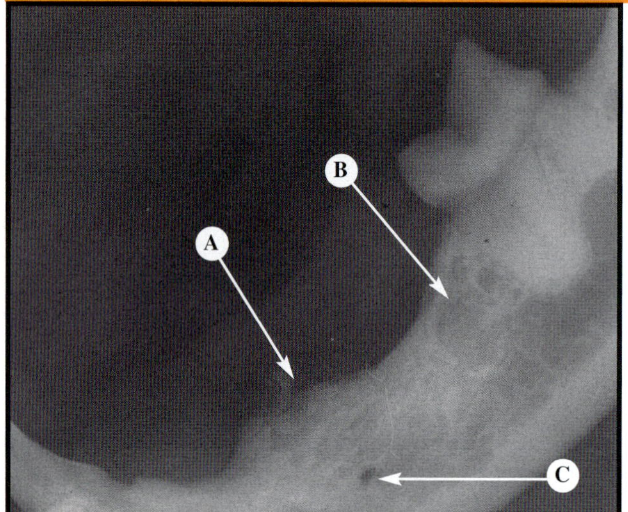

FIGURE 13-41 Crestal Bone Proliferation

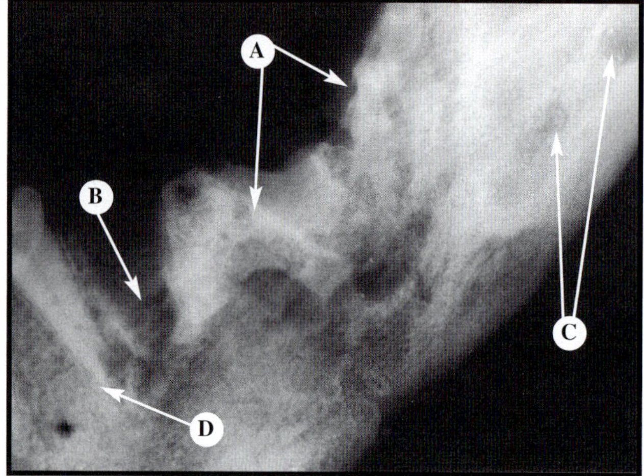

FIGURE 13-42 Sequestrum After Bone Plating

Figure 13-41 *Proliferation of the crestal bone at the third premolar site has an irregular contour, thereby suggesting that the premolar has been lost through replacement resorption (see Figure 13-40) or healed poorly after extraction. The lesion is nonaggressive. (A) Site of third premolar, (B) site of fourth premolar, and (C) mental foramen.* **Figure 13-42** *Osteomyelitis and bone sequestrum after plating. The inflammation is an aggressive lesion. Fixation devices in the mandibular canal are contraindicated. (A) Bone sequestrum, (B) moth-eaten bony lysis, (C) sites of screw fixation, and (D) spiked root resorption.*

GINGIVAL GROWTHS (EPULIDES)—OSSIFYING FIBROMA

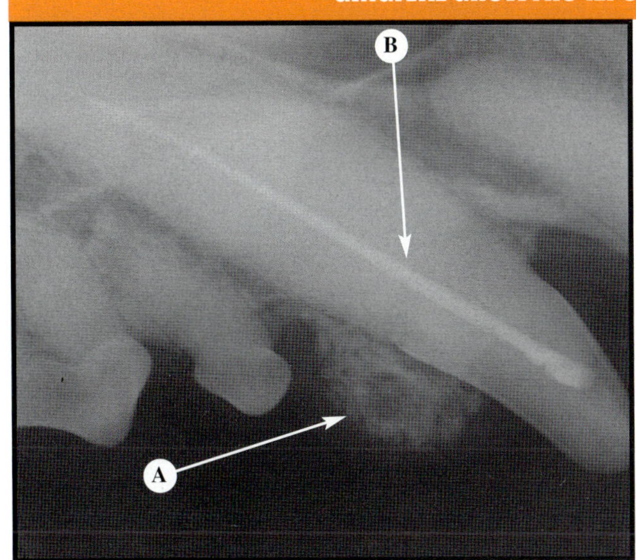

FIGURE 13-43

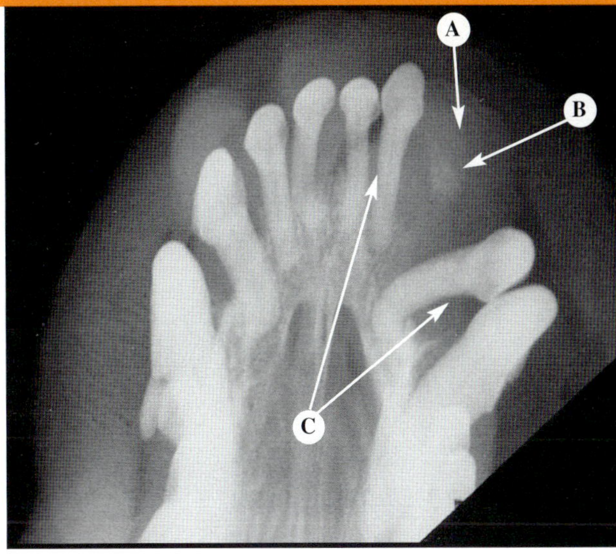

FIGURE 13-44

Figure 13-43 *The radiopacity indicates ossification. Although the mass is superficial to the gingival margin, its border extends into the periodontal tissue. The endodontic treatment is an incidental finding.* (A) *Radiopaque gingival mass and* (B) *endodontic filling.* **Figure 13-44** *Although it extends deep into the soft tissue, this gingival mass has a typical space-occupying nonaggressive appearance.* (A) *Gingival mass,* (B) *radiopaque foci, and* (C) *displaced incisors.*

GINGIVAL GROWTHS (EPULIDES)—FIBROMATOUS EPULIS

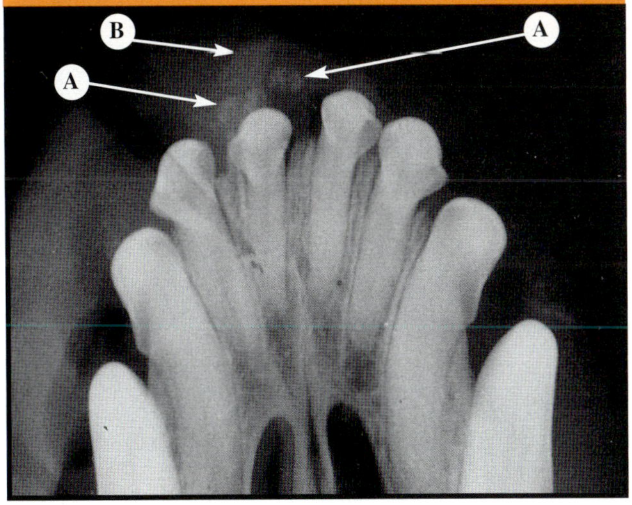

FIGURE 13-45

This lesion is nonaggressive. The central incisor that is involved appears to be slightly displaced, although this may be a positioning artifact that was created by the angle of the beam. (A) *Radiopaque foci and* (B) *gingival mass.*

GINGIVAL GROWTHS (EPULIDES)—ACANTHOMATOUS EPULIS

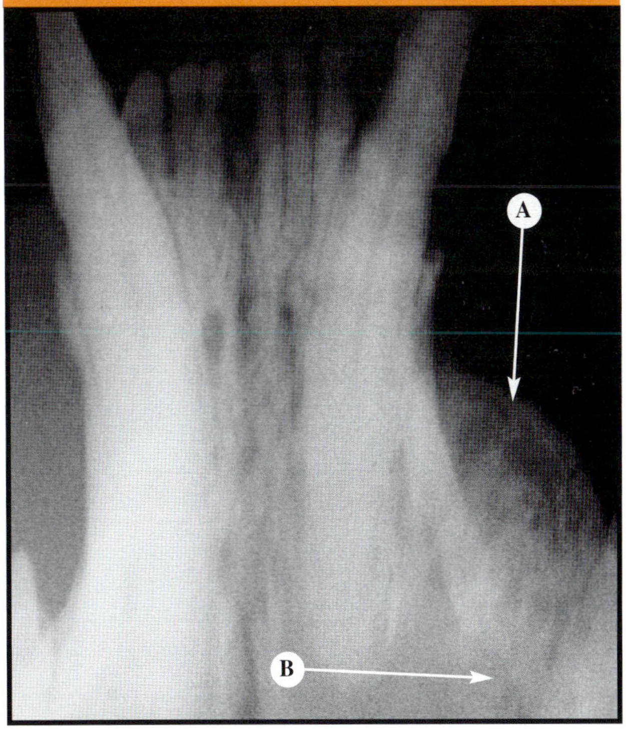

FIGURE 13-46

This acanthomatous epulis is a moderately aggressive lesion. (A) *Radiopaque gingival mass and* (B) *indistinct border with loss of mandibular cortex.*

GINGIVAL GROWTHS (EPULIDES)—ACANTHOMATOUS EPULIS *(continued)*

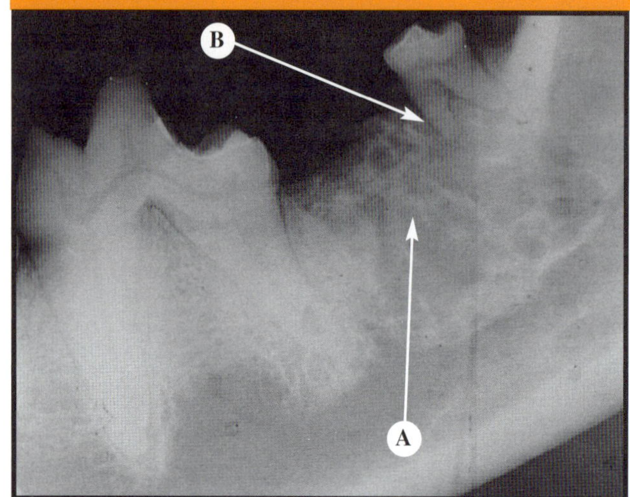

FIGURE 13-47

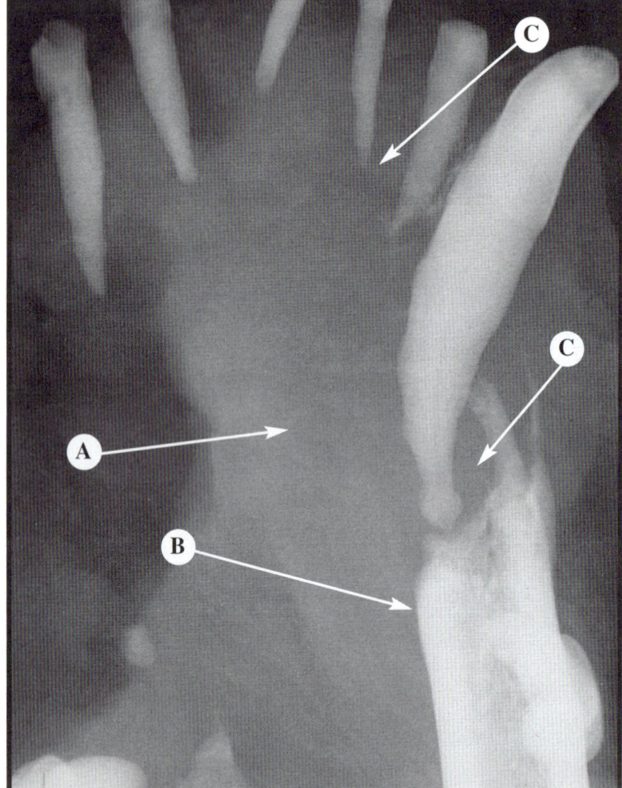

FIGURE 13-49

FIGURE 13-48

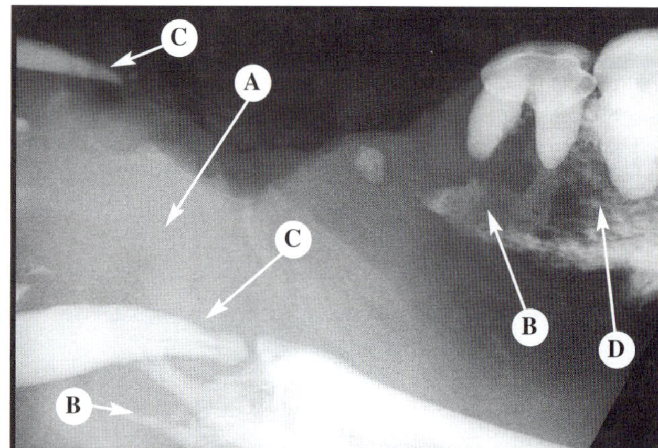

FIGURE 13-50

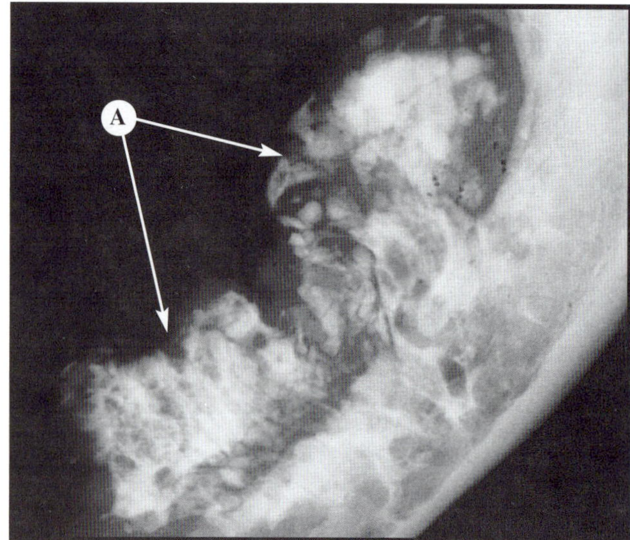

FIGURE 13-51

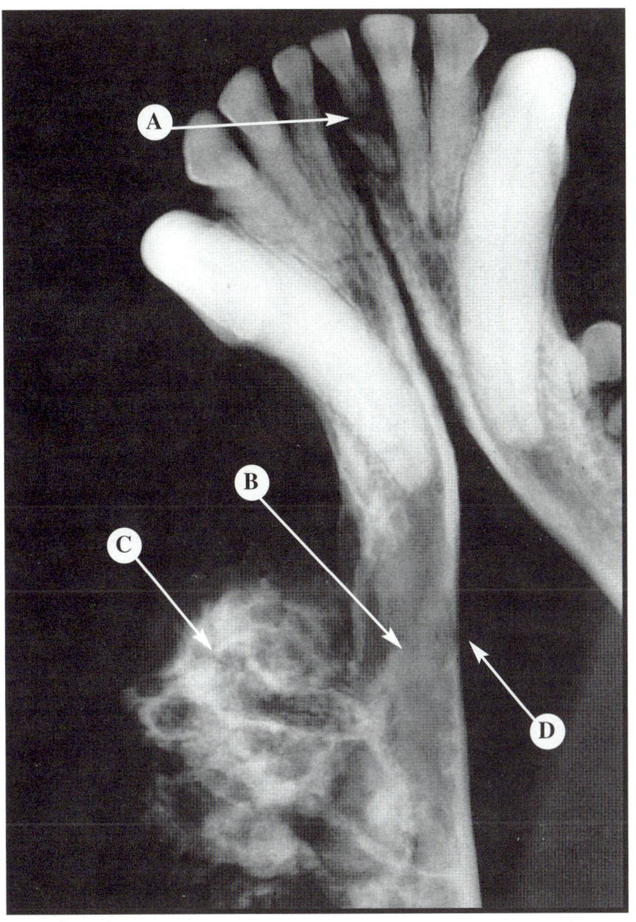

FIGURE 13-52

Figure 13-47 *This lesion is aggressive. (A) Moth-eaten bony lysis and (B) root resorption.* **Figure 13-48** *This mass is replacing the periodontium and displacing the teeth from the normal dental line. The lesion is aggressive. (A) Soft-tissue mass and (B) floating incisors.* **Figure 13-49** *This ventrodorsal view shows that an aggressive tumor has replaced the cranial osseous aspect of the mandible. (A) Large, homogeneous soft-tissue mass, (B) mandible, and (C) root resorption (or spiking).* **Figure 13-50** *The lateral view of the same tumor in Figure 13-49 confirms that the lesion is aggressive. (A) Tissue mass, (B) irregular borders, (C) root resorption, and (D) moth-eaten bony lysis.* **Figure 13-51** *This tumor has an undifferentiated bone formation. The borders are indistinct, and the tumor has numerous focal opacities. The lesion is aggressively invading the mandibular bone. (A) Mandibular tumor.* **Figure 13-52** *Another view of Figure 13-51. (A) Fracture of the incisor root, (B) bony lysis, (C) large radiopaque mass, and (D) thin cortex.*

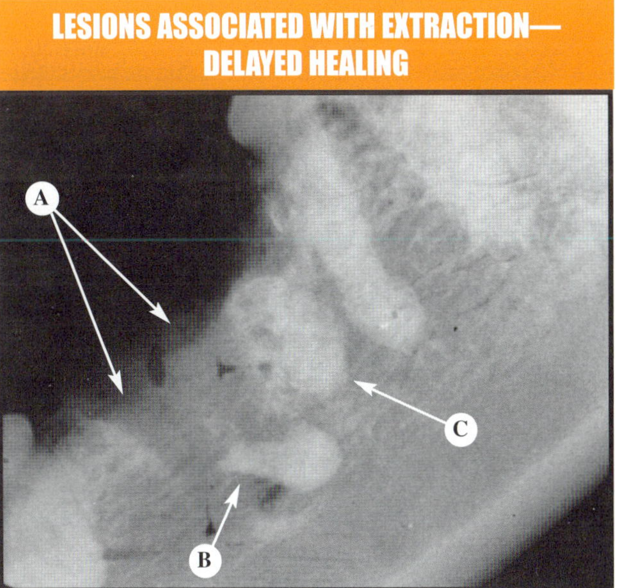

FIGURE 13-53

The lamina dura is still visible even though the alveolus is filled with bone. Alveolar sclerosis has been associated with gastrointestinal disease or early renal disease. (A) Lamina dura and (B) sclerotic plaques.

FIGURE 13-54

Retained root fragments can impair healing. The dark spots are film artifacts. (A) Unhealed alveoli, (B) displaced retained mesial root tip, and (C) sclerotic distal root tip.

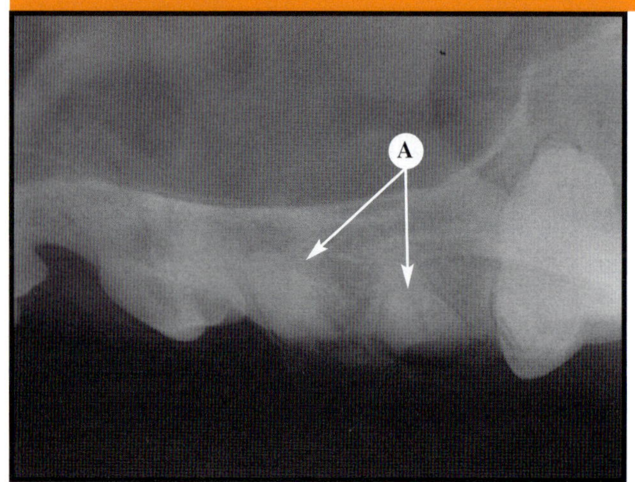

FIGURE 13-55

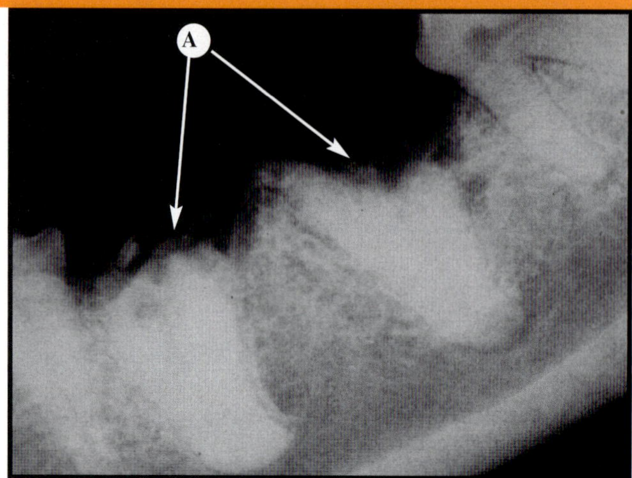

FIGURE 13-56

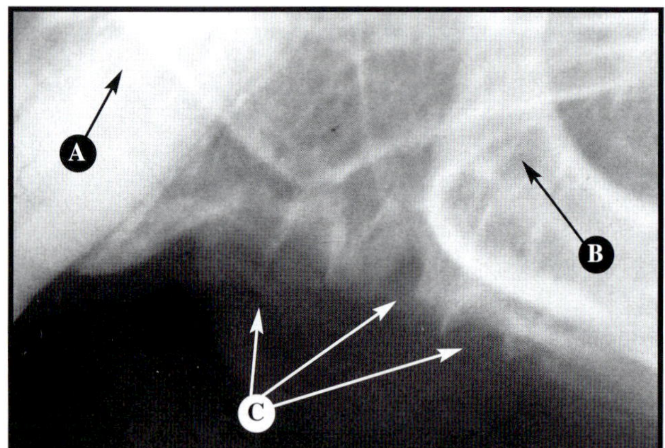

FIGURE 13-57

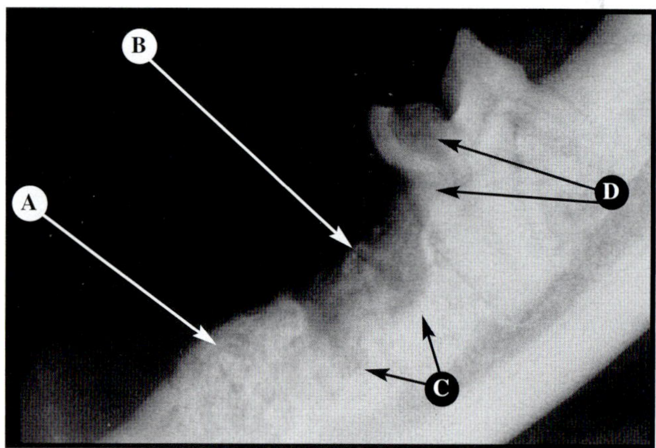

FIGURE 13-58

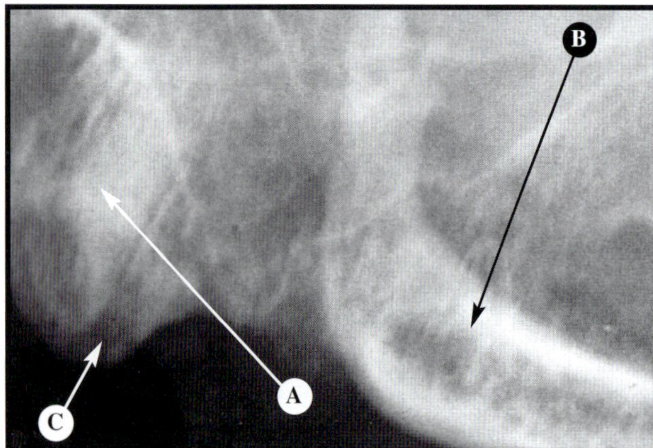

FIGURE 13-59

Figure 13-55 *Bony proliferation partially covering the root fragments can make removal more challenging. Multiple views would assist in localizing the site. (A) Retained roots of the fourth premolar.* **Figure 13-56** *Exposed retained root tips tend not to heal, especially if sharp fragments extend above the alveolar bone. (A) Retained roots.* **Figure 13-57** *No remodeling of the alveoli has occurred, and the lamina dura are still visible. (A) Canine tooth, (B) zygomatic arch, and (C) multiple extraction sites.* **Figure 13-58** *Sclerosis and bony proliferation are present at the site of the third premolar. Retained root fragments can impair healing. The molar has severe supraosseous and crestal resorption mesially. (A) Site of the third premolar, (B) unhealed site of the fourth premolar, (C) retained root fragments, and (D) resorptive lesions.* **Figure 13-59** *Retained root segments can act as a nidus for infection or can impair healing. (A) Retained root fragment from the canine tooth, (B) zygomatic arch, and (C) fistulous canal. (Figure 13-58 reprinted with permission from Williams CA, Aller MS: Gingivitis/stomatitis in cats, in Harvey CE (ed): The Veterinary Clinics of North America. Small Animal Practice, vol 22, no. 6. Philadelphia, WB Saunders Co, 1992, pp 1361–1383.)*

LESIONS ASSOCIATED WITH EXTRACTION—HEALED SITE

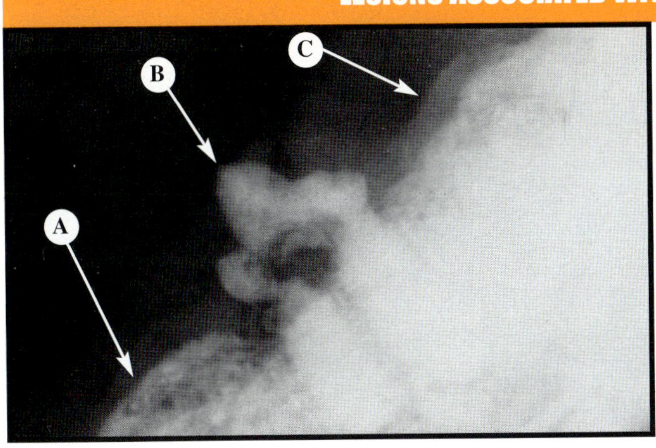

FIGURE 13-60

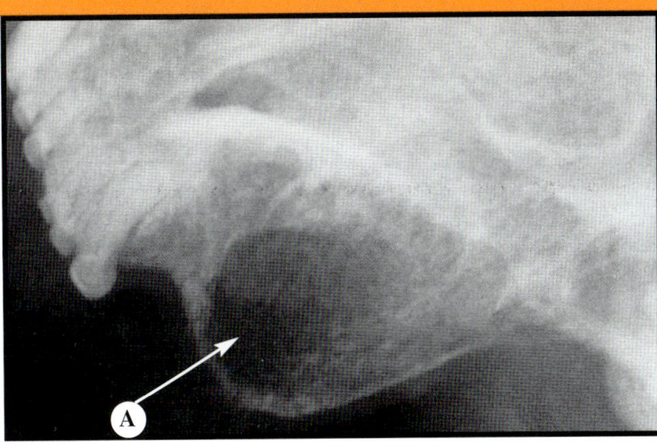

FIGURE 13-61

Figure 13-60 *The fourth premolar is undergoing advanced external resorption. The other two sites are healed. Bony proliferation is common in cats and is acceptable to some degree if healing is considered successful; however, the sites should be monitored accordingly.* (A) *Site of the third premolar,* (B) *fourth premolar, and* (C) *site of the first molar.* **Figure 13-61** *The diagnostic differentials would include fibrous healing, rarefaction caused by an inflammatory process, cystic lesion, or a neoplastic lesion.* (A) *Radiolucent defect in the alveolus.*

OTHER LESIONS ASSOCIATED WITH EXTRACTION

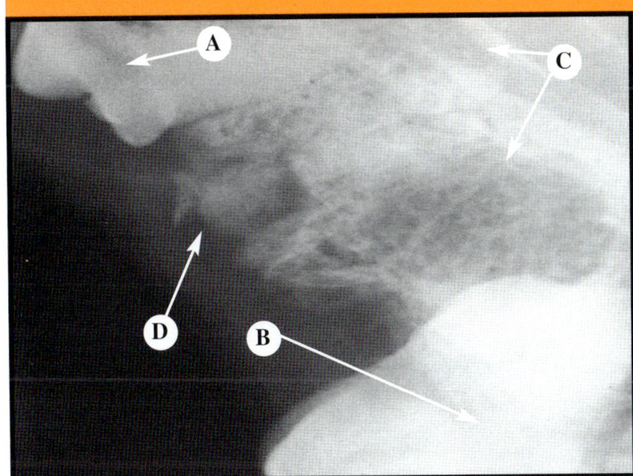

FIGURE 13-62 Fistulating Site

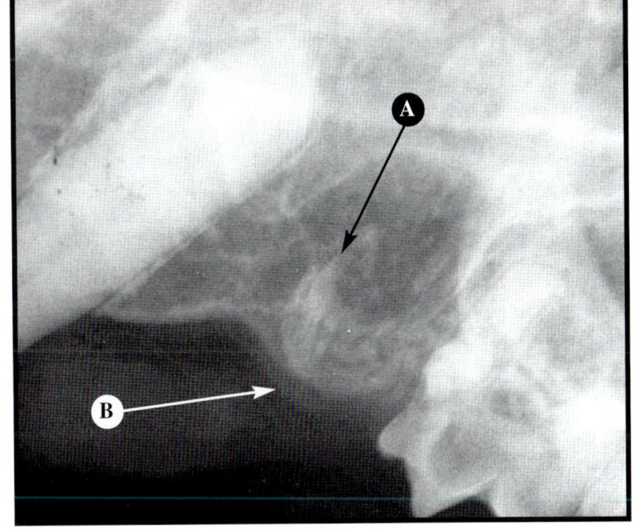

FIGURE 13-63 Retained Root Fragments

Figure 13-62 *This radiograph shows retained roots and sequestrum at the extraction site of the fourth premolar.* (A) *Third premolar,* (B) *first molar,* (C) *retained roots, and* (D) *sequestrum.* **Figure 13-63** *Unexposed, noninflamed, and inactive root fragments do not usually interfere with the healing process.* (A) *Retained root fragment of the third premolar and* (B) *healed crestal bone.* **Figure 13-64** *The extraction sites are healed. Further diagnostic procedures are needed to determine the cause of asymmetry in the nasal sinus.* (A) *Radiolucent alveoli with radiopaque foci attributable to fibrosis with osseous metaplasia,* (B) *obscured sinus, and* (C) *second premolar.*

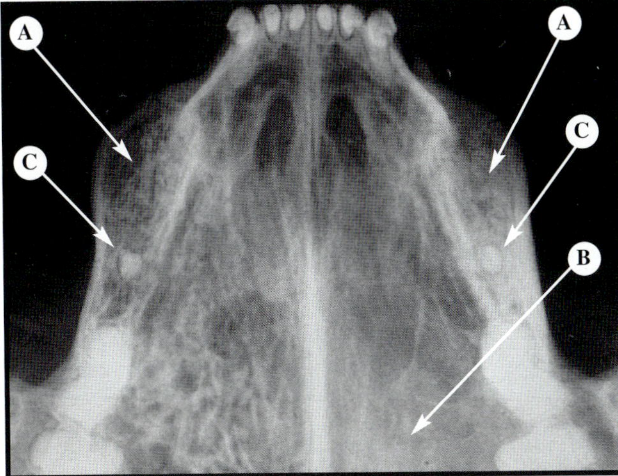

FIGURE 13-64 Fibrosis with Osseous Metaplasia

EDENTULOUS MANDIBLES

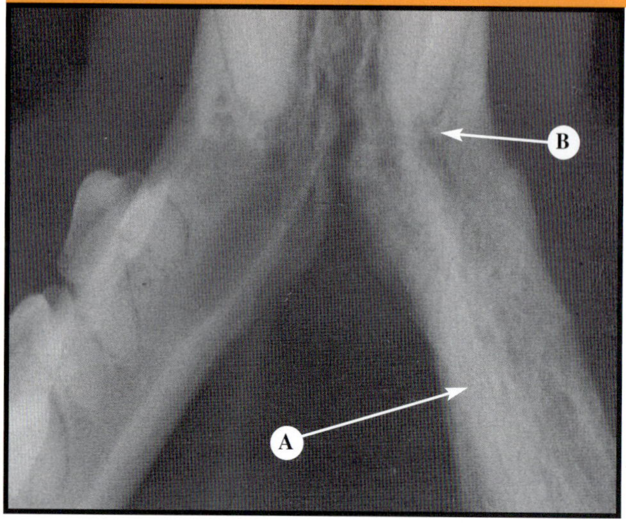

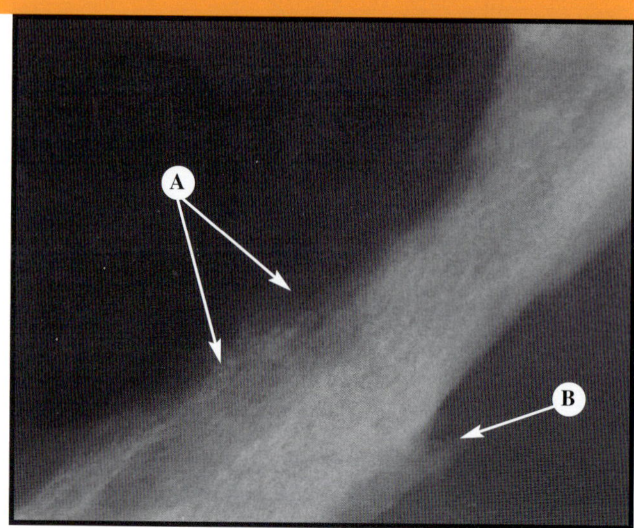

FIGURE 13-65A Chronic Osteomyletitis:
Ventrodorsal View

FIGURE 13-65B Chronic Osteomyletitis: Lateral View

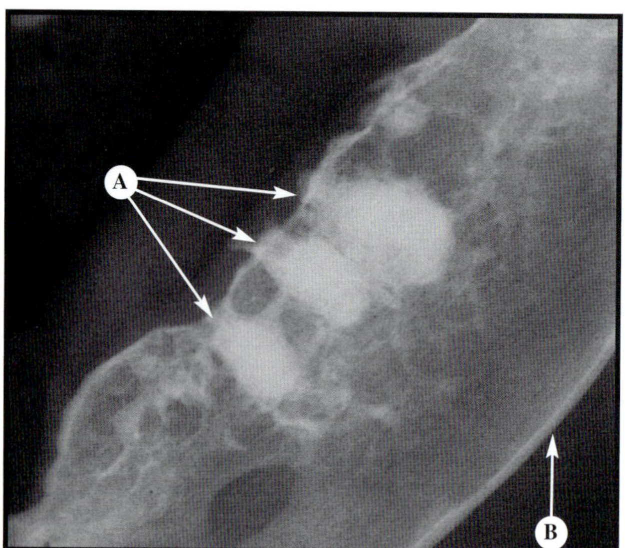

FIGURE 13-66 Fibrous Dysplasia

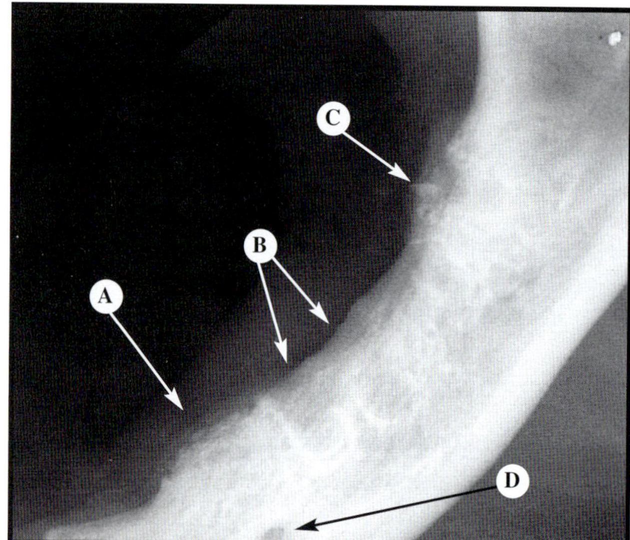

FIGURE 13-67 Bony Sclerosis

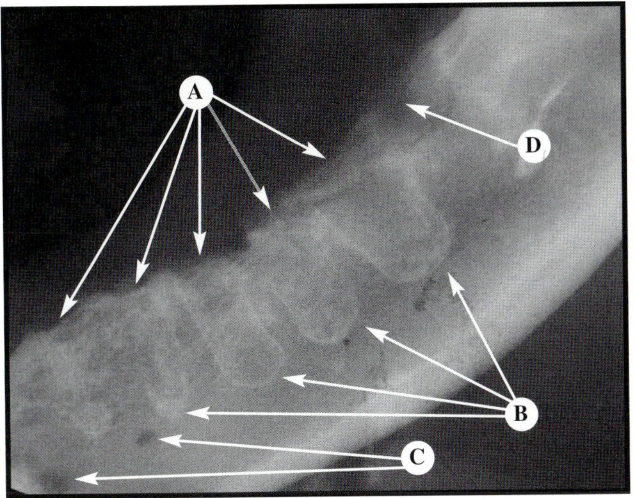

FIGURE 13-68 Alveolar Sclerosis

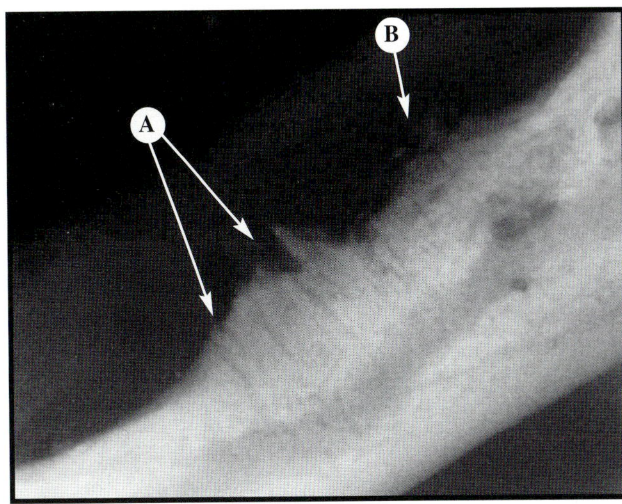

FIGURE 13-69 Retained Root Fragments

Histopathologic evaluation confirmed chronic osteomyelitis. **Figure 13-65A** *Compare the radiographic appearance of the normal mandible on the left to the diseased mandible on the right.* (A) *Sclerosis and* (B) *small periapical radiolucency.* **Figure 13-65B** *Note the generalized sclerotic appearance with some moth-eaten regions. This lesion is aggressive.* (A) *Irregular moth-eaten crestal bone and unhealed alveoli and* (B) *periosteal reaction.* **Figure 13-66** *The mandible shows generalized radiolucency. The lesion suggests possible metabolic, endocrine, or genetic condition.* (A) *Retained root fragments and* (B) *very thin cortices and a ground-glass appearance to the trabecular bone.* **Figure 13-67** *In cats, retained root fragments or small bony sequestra with sclerotic plaques are the most commonly encountered radiographic lesions of an edentulous mandible with mucosal inflammation.* (A) *Periosteal proliferation,* (B) *sclerosis of the socket,* (C) *bone and root fragments, and* (D) *mental foramen.* **Figure 13-68** *The contour of the crestal bone is very irregular and non-homogeneous. Alveolar sclerosis is associated with gastrointestinal disease or early renal disease in humans.* (A) *Irregular crestal bone,* (B) *persistent lamina dura,* (C) *mental foramen, and* (D) *unhealed extraction site.* **Figure 13-69** *The proliferative periosteum is typical of sites of missing teeth in cats. The retained roots should be removed or debrided to achieve healing.* (A) *Retained root fragments and* (B) *periosteal proliferation.*

REFERENCES *(continued from page 191)*

3. Miles DA, Van Dis M, Kaugers GE, Lovas JGL: *Oral and Maxillofacial Radiology: Radiologic/Pathologic Correlations.* Philadelphia, WB Saunders Co, 1991, pp 1–5.

4. Konde LJ: Aggressive versus nonaggressive bone lesions, in Thrall DE (ed): *Textbook of Veterinary Diagnostic Radiology*, ed 2. Philadelphia, WB Saunders Co, 1994, pp 15–22.

5. Verstraete FJM, Ligthelm AJ, Weber A: The histological nature of epulides in dogs. *J Comp Pathol* 106:169–182, 1992.

6. Reichart PA, Philipsen HP, Durr UM: Epulides in dogs. *J Oral Pathol Med* 18:92–96. 1989.

7. Walsh KM, Denholm LJ, Cooper BJ: Epithelial odontogenic tumours in domestic animals. *J Comp Pathol* 97:503–521, 1987.

8. Gardner DG, Baker DC: Fibromatous epulis in dogs and peripheral odontogenic fibroma in human beings: Two equivalent lesions. *Oral Surg Oral Med Oral Pathol* 71:317–321, 1991.

9. Gardner DG: Canine acanthomatous epulis: The only common spontaneous ameloblastoma in animals. *Oral Surg Oral Med Oral Pathol Oral Rad Endo* 79:612–615.

10. White RAS, Jefferies AR, Gorman NT: Sarcoma development following irradiation of acanthomatous epulis in two dogs. *Vet Rec* 118:668, 1986.

11. Williams CA, Aller MS: Gingivitis/stomatitis in cats, in Harvey CE (ed): *The Veterinary Clinics of North America. Small Animal Practice*, vol 22, no. 1. Philadelphia, WB Saunders Co, 1992, pp 1361–1383.

12. Dupont G: Crown amputation with intentional root retention for advanced feline resorptive lesions: A clinical study. *J Vet Dent* 12(1):9–13, 1995.

13. Monahan R: Periapical and localized radiopacities, in *The Dental Clinics of North America*, vol 38, no. 1. Philadelphia, WB Saunders Co, 1994, pp 113–136.

Chapter Fourteen

DENTAL RADIOGRAPHY: A TEST OF YOUR KNOWLEDGE

This chapter gives practitioners an opportunity to test and sharpen the interpretative skills they have learned. The first section of the chapter contains the Dental Radiography Test, which features radiographs with premises that challenge each reader's ability to be analytic by determining a correct clinical diagnosis. After completing the test, readers are given the opportunity to compare their findings with our diagnoses, which are found in the second section on Diagnoses and Significant Points.

When examining each radiograph, readers should remember the basic principles of radiographic interpretation. Some of the parameters to be considered are the species, age of the patient, anatomic location depicted on the radiograph, technique or procedure used, primary diagnosis, secondary findings, and artifacts. Readers should always assume that they are looking into the dimple when viewing intraoral radiographs.

Before they begin the test, readers should recall that throughout the book, we identified the primary characteristics or features of the radiographs but did not necessarily highlight every detail on each radiograph. Our goal was to provide the tools that practitioners need to generate and evaluate dental radiographs methodically and with consistency. By developing keen interpretative skills, we believe that our readers will be able to recognize details on radiographs. For example, although we did not mention the presence of a supernumerary microdont molar superimposed over the molar in Figure 4-26C (Chapter 4 on Intraoral Imaging Techniques), we hope that our readers recognized this detail.

Finally, Figures 14-15A, 14-15B, 14-18, and 14-33 are included in this chapter courtesy of William J. Zontine, DVM, MS, Vista, California.

We trust that this chapter will prove to be entertaining and insightful.

Thomas W. Mulligan, DVM   Mary Suzanne Aller, DVM

Charles A. Williams, DVM

DENTAL RADIOGRAPHY TEST

Instructions: Study each radiograph carefully, base your evaluations on the principles of radiographic interpretation, and determine the diagnosis or significance of findings accordingly.

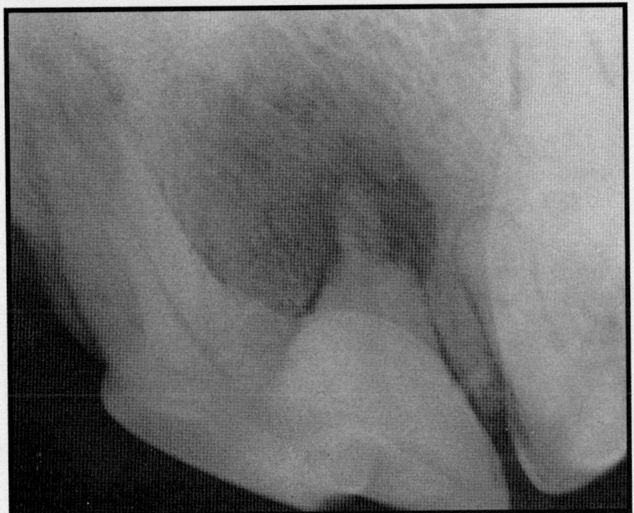

FIGURE 14-1

Premise: *Facial swelling below the eye*
Interpretation: ___________________________

____

____

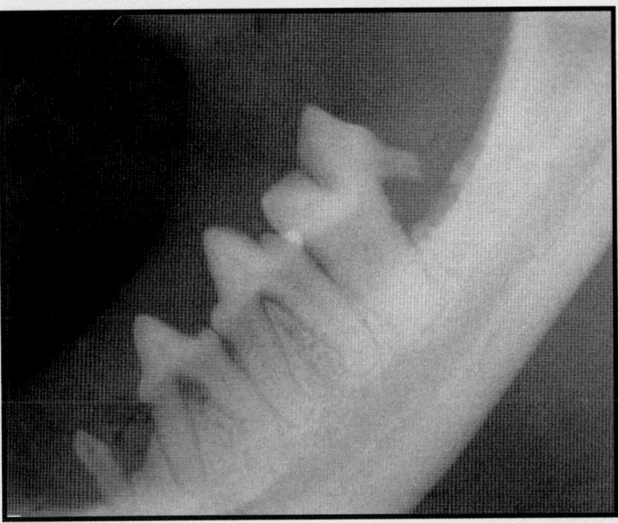

FIGURE 14-2

Premise: *Neck lesion detected during prophylaxis*
Interpretation: ___________________________

____

____

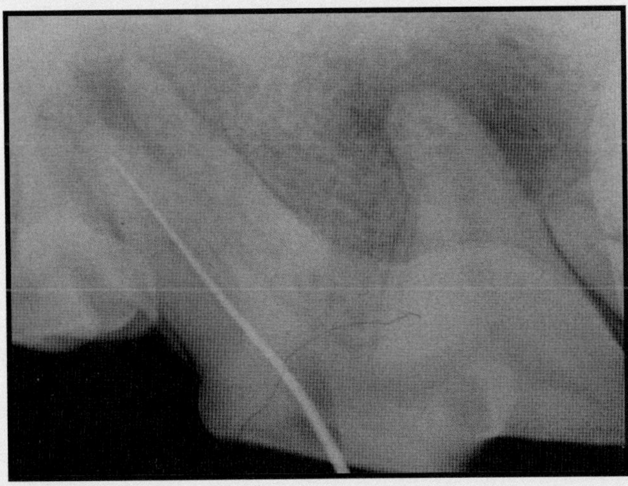

FIGURE 14-3

Premise: *Attempting to insert the file in the palatal root*
Interpretation: ___________________________

____

____

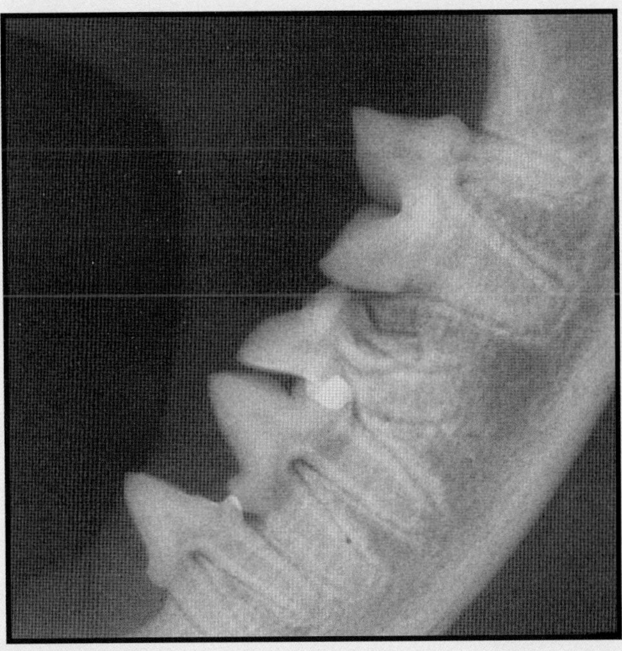

FIGURE 14-4

Premise: *Crowded dentition because of extra tooth or bifid crown?*
Interpretation: ___________________________

____

____

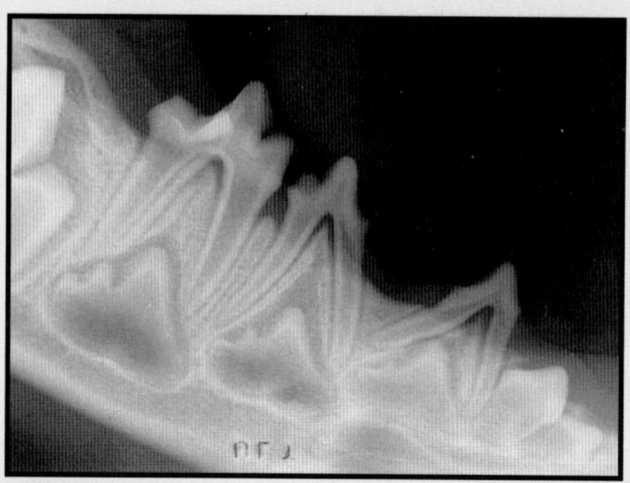

FIGURE 14-5

Premise: *Preeruption evaluation of a puppy*
Interpretation: _______________________________

____
____
____

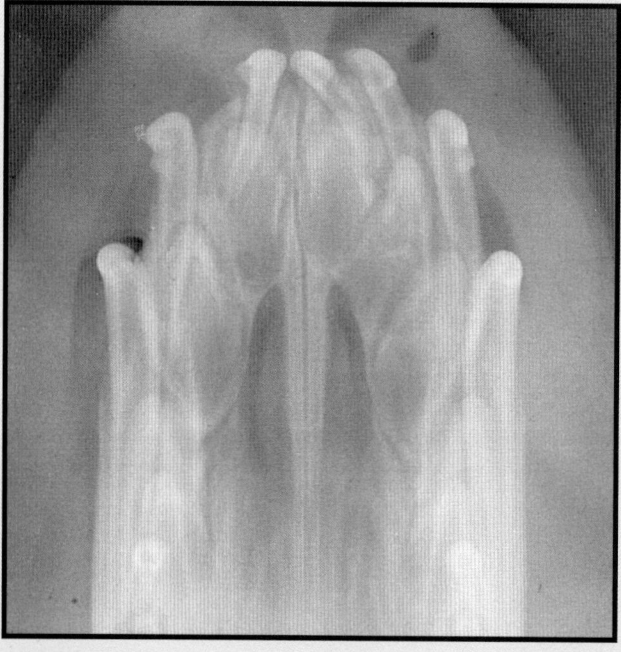

FIGURE 14-7

Premise: *Preeruption evaluation of a puppy*
Interpretation: _______________________________

____
____
____

FIGURE 14-6

Premise: *Evaluation to detect missing teeth*
Interpretation: _______________________________

____
____
____

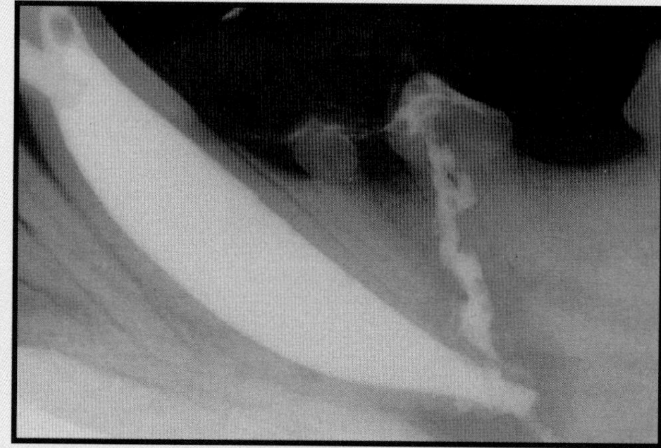

FIGURE 14-8

Premise: *Evaluation of endodontic treatment after surgery*
Interpretation: _______________________________

____
____
____

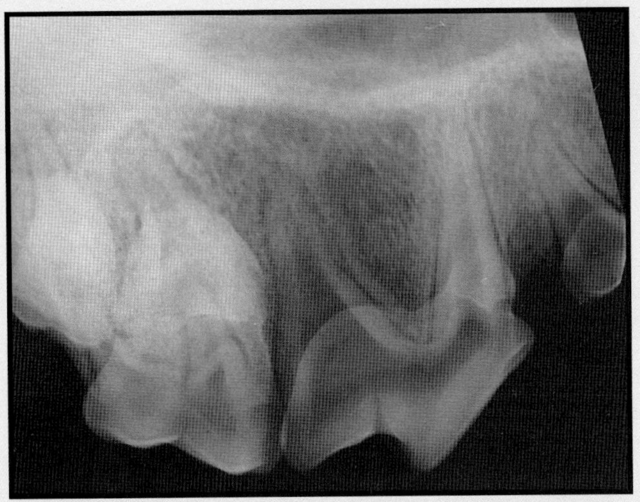

FIGURE 14-9

Premise: *Evaluation of a peg tooth at the site of the third premolar*

Interpretation: _______________________________

____

____

____

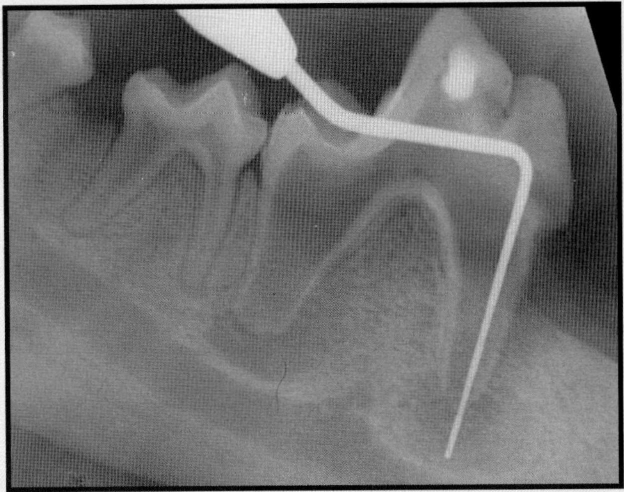

FIGURE 14-10

Premise: *Evaluation of gingivitis at the molar*

Interpretation: _______________________________

____

____

____

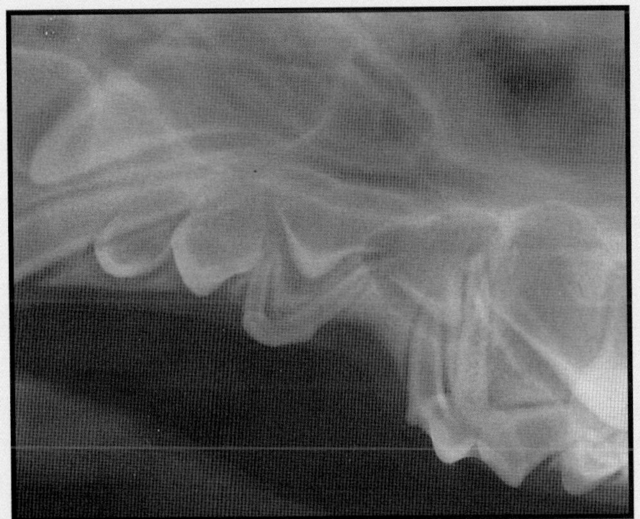

FIGURE 14-11

Premise: *Preeruption evaluation of a puppy*

Interpretation: _______________________________

____

____

____

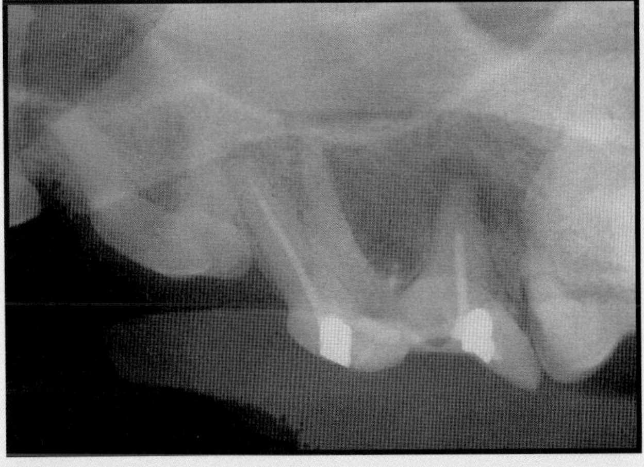

FIGURE 14-12

Premise: *Evaluation of endodontic treatment after surgery*

Interpretation: _______________________________

____

____

____

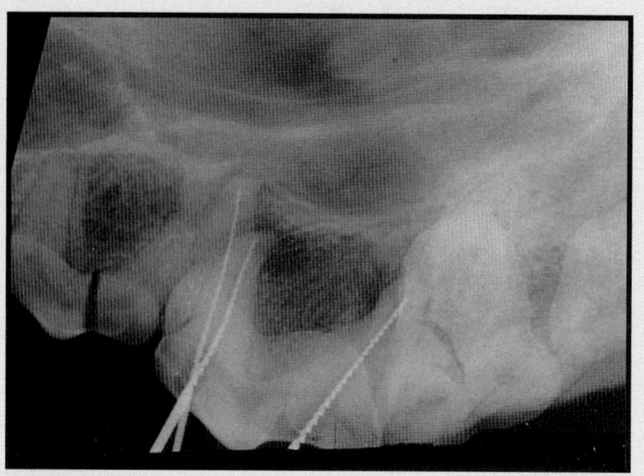

FIGURE 14-13

Premise: *Evaluation of procedure (file placement) during surgery*
Interpretation: ___________________________
____
____
____

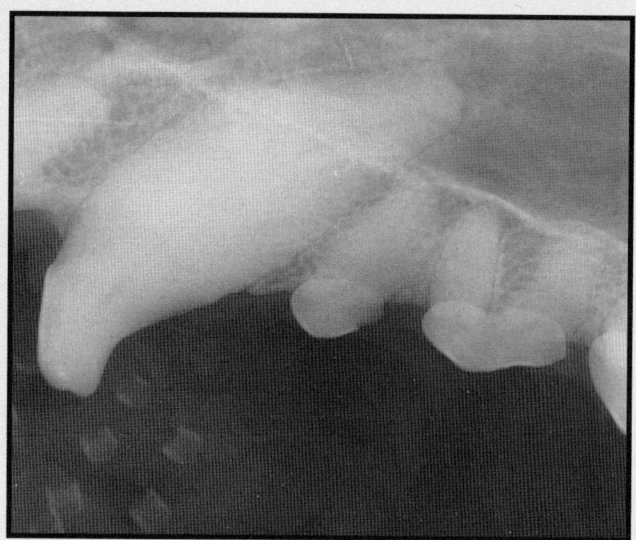

FIGURE 14-14

Premise: *Evaluation of fractured canine tooth in a geriatric dog*
Interpretation: ___________________________
____
____
____

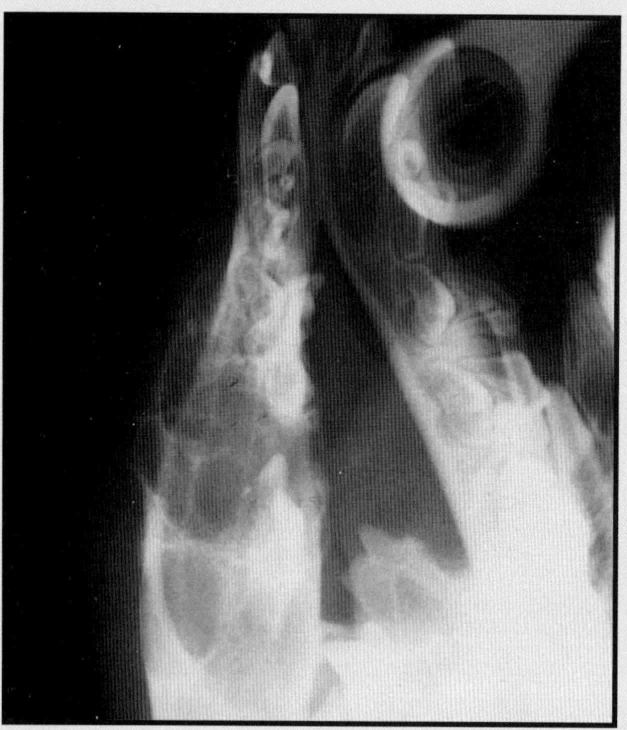

FIGURE 14-15A

Premise: *Mandibular swelling in a young dog*
Interpretation: ___________________________
____
____
____

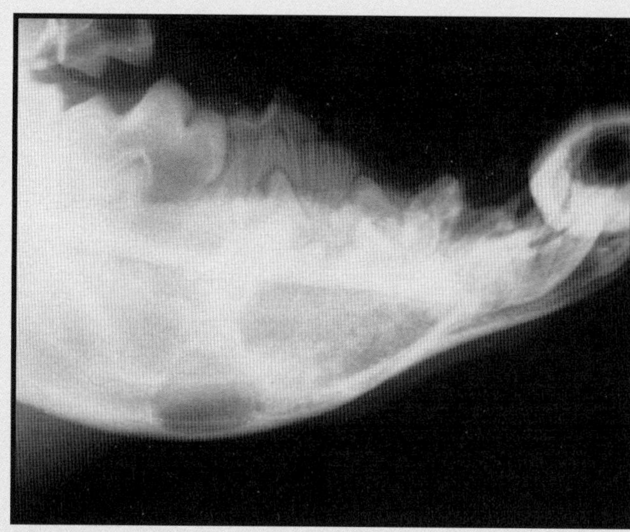

FIGURE 14-15B

Premise: *Mandibular swelling in a young dog*
Interpretation: ___________________________
____
____
____

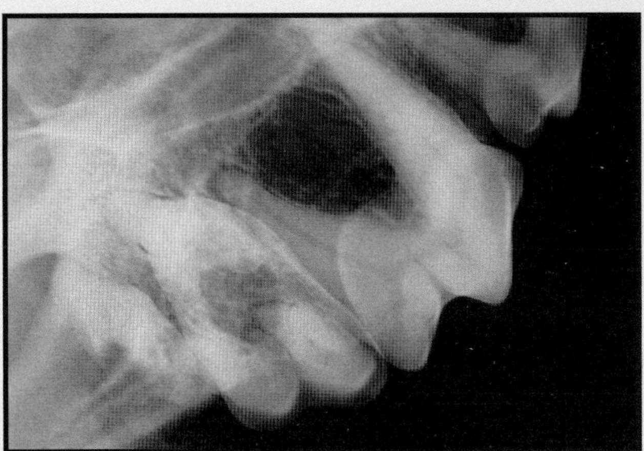

FIGURE 14-16

Premise: *Soft, sticky defects detected on the molars*

Interpretation: ______________________________

____

____

____

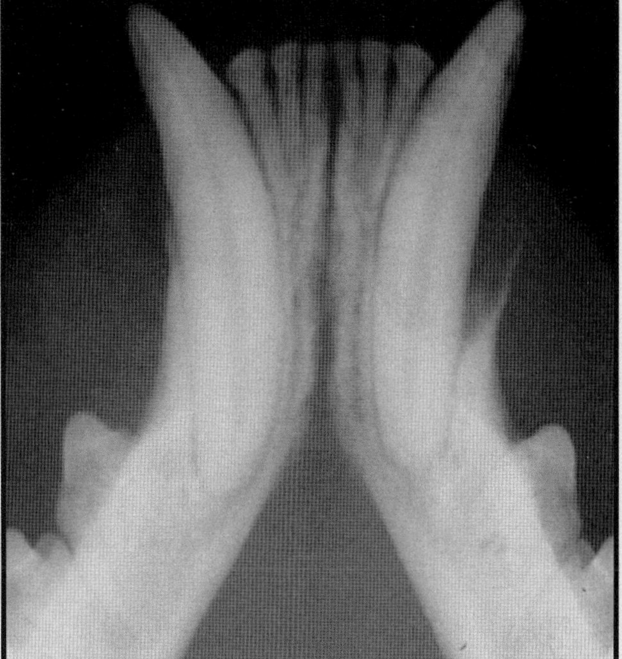

FIGURE 14-17

Premise: *A periodontal pocket was found at the canine tooth during prophylaxis*

Interpretation: ______________________________

____

____

____

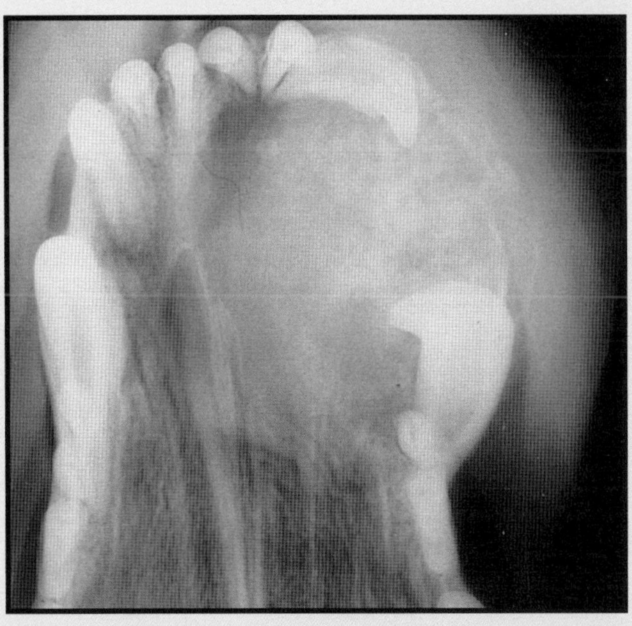

FIGURE 14-18

Premise: *Swelling of the premaxilla and maxilla*

Interpretation: ______________________________

____

____

____

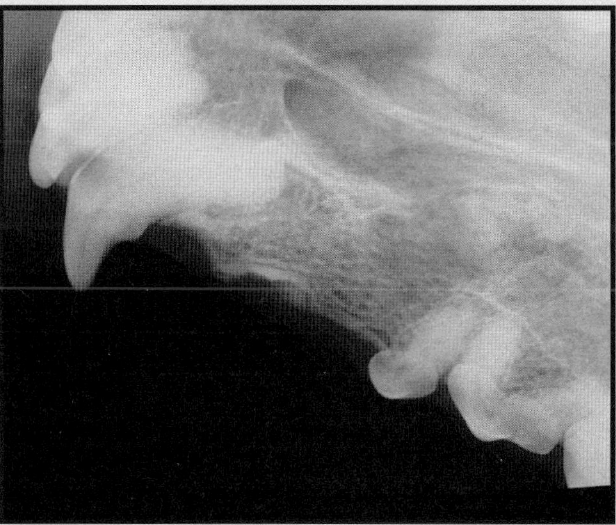

FIGURE 14-19

Premise: *Chronic draining fistula at the site of a missing maxillary canine tooth*

Interpretation: ______________________________

____

____

____

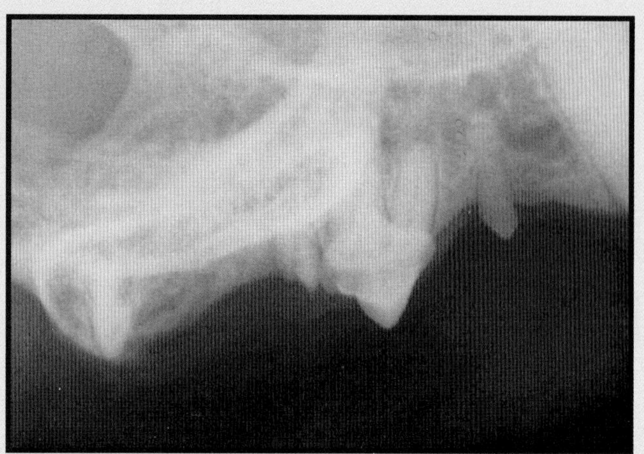

FIGURE 14-20

Premise: *Chronic gingivitis*

Interpretation: _______________________________

____

____

____

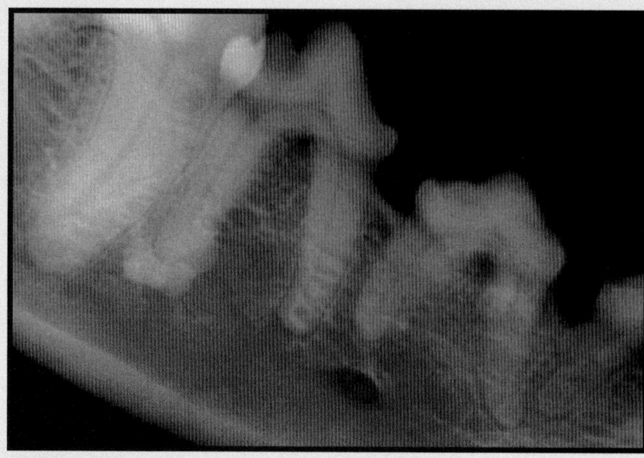

FIGURE 14-21

Premise: *Evaluation of exposures at the furcation*

Interpretation: _______________________________

____

____

____

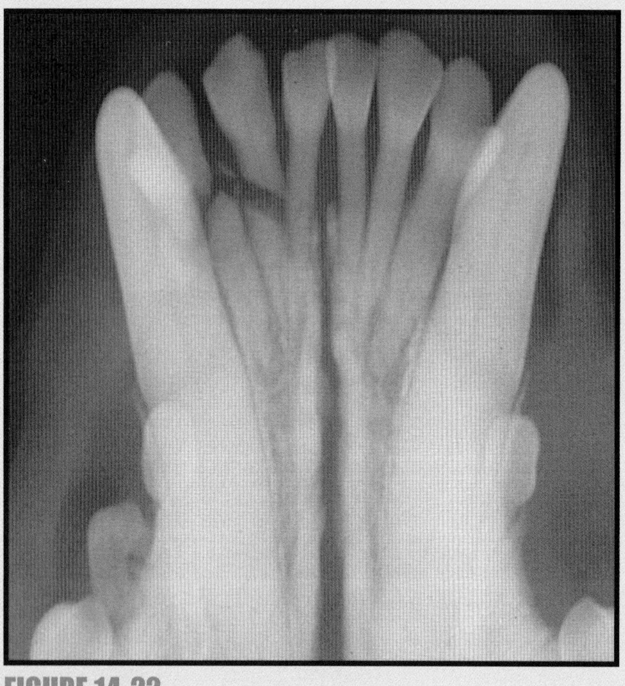

FIGURE 14-22

Premise: *Evaluation of loose incisors*

Interpretation: _______________________________

____

____

____

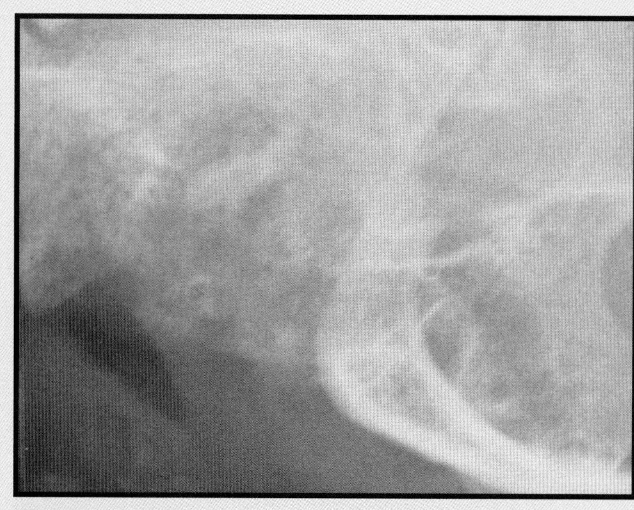

FIGURE 14-23

Premise: *Chronic unilateral nasal discharge*

Interpretation: _______________________________

____

____

____

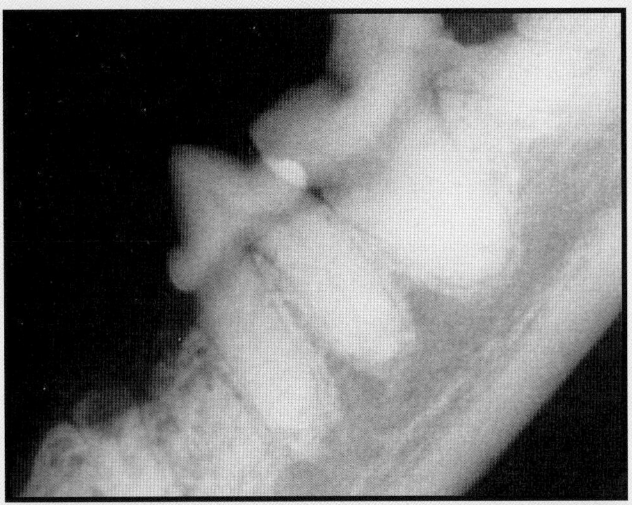

FIGURE 14-24

Premise: *Gingival swelling at the third premolar and a neck lesion on the molar*

Interpretation: _______________________________

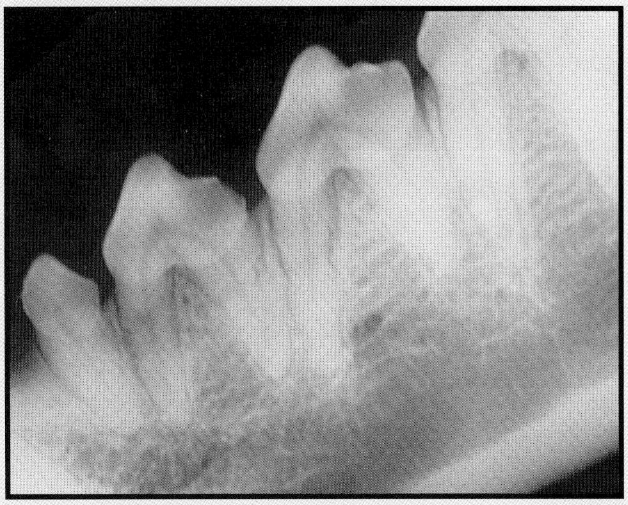

____

____

____

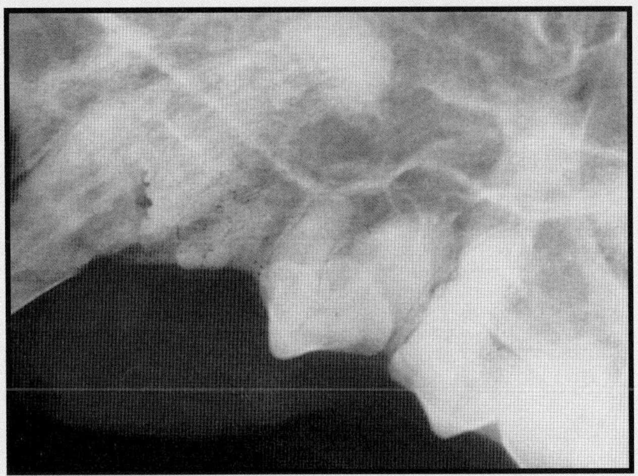

FIGURE 14-26

Premise: *Defect detected at the neck of the canine tooth*

Interpretation: _______________________________

____

____

FIGURE 14-25

Premise: *Evaluation of chipped premolars*

Interpretation: _______________________________

____

____

____

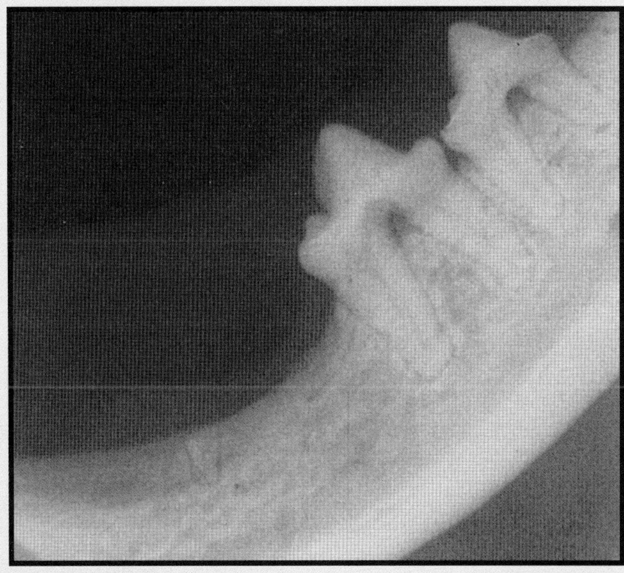

FIGURE 14-27

Premise: *Defect detected at the necks of the premolars*

Interpretation: _______________________________

____

____

____

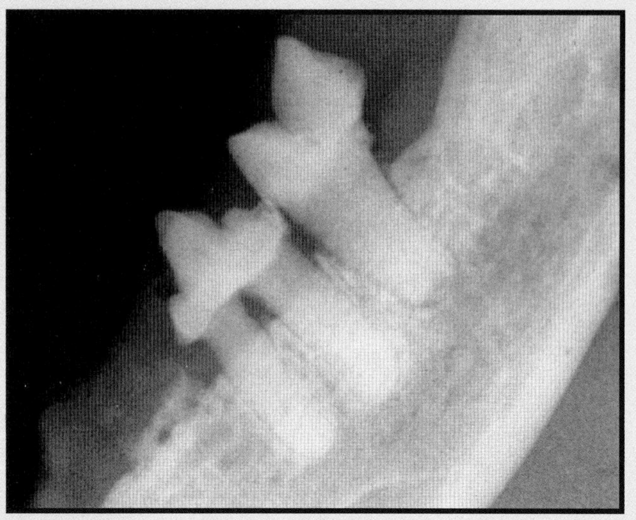

FIGURE 14-28

Premise: *Evaluation of exposures at the furcation and of the defect detected on the molar*

Interpretation: ___________________________

____

____

____

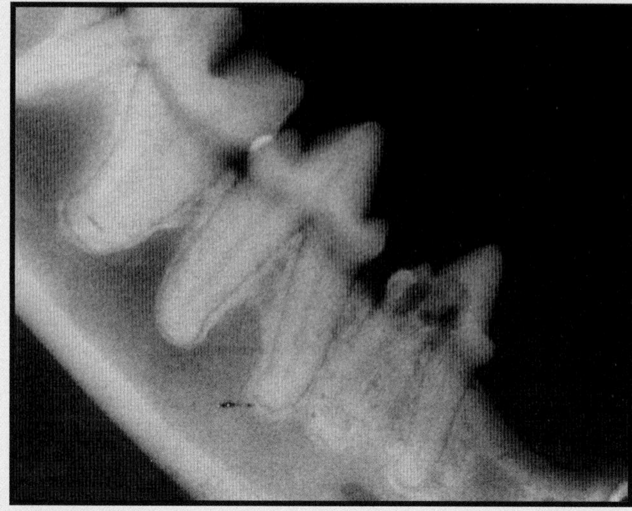

FIGURE 14-29

Premise: *Defect detected on the neck of the premolar*

Interpretation: ___________________________

____

____

____

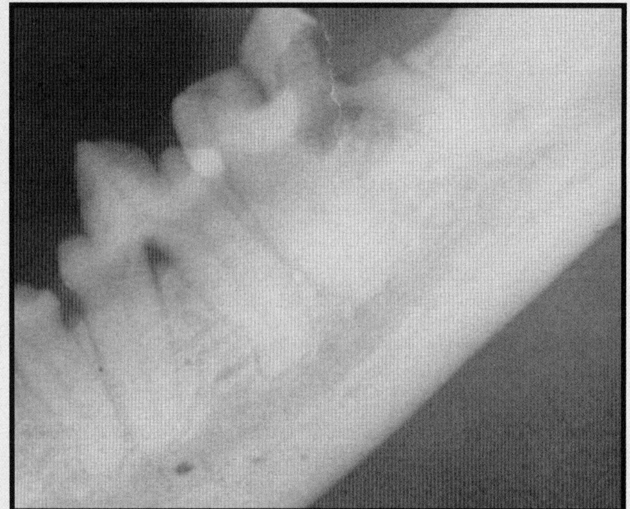

FIGURE 14-30

Premise: *Defects detected on the necks of the molar and premolar*

Interpretation: ___________________________

____

____

____

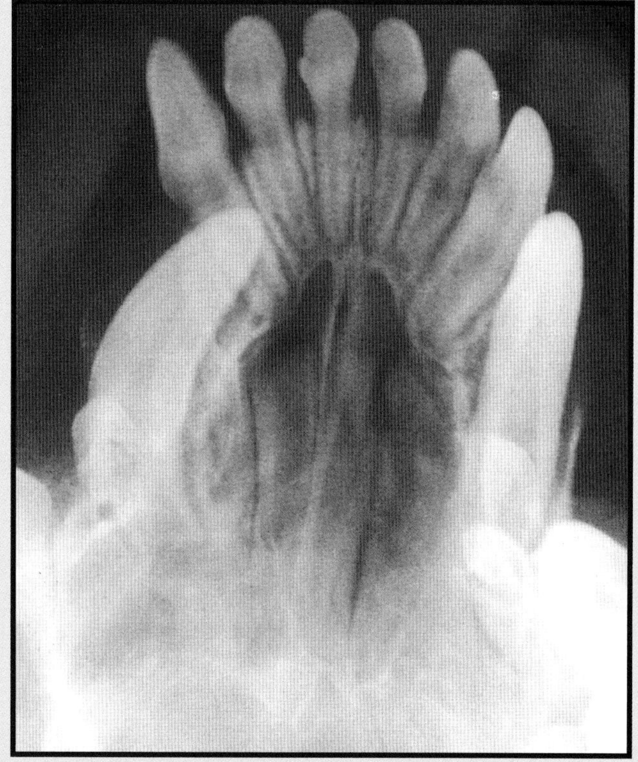

FIGURE 14-31

Premise: *Missing permanent canine tooth*

Interpretation: ___________________________

____

____

____

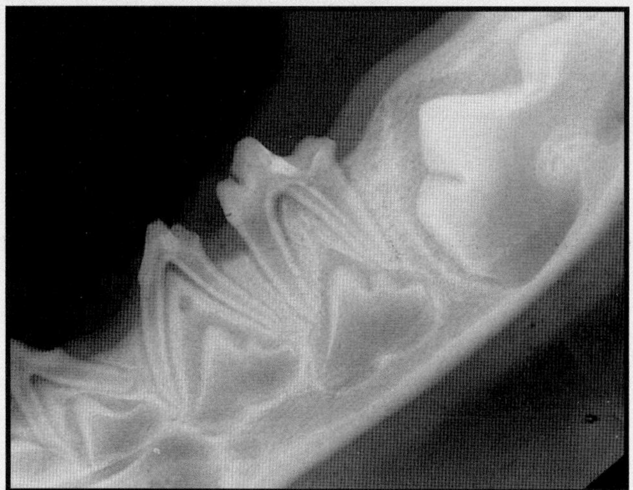

FIGURE 14-32
Premise: *Preeruption evaluation of a puppy*
Interpretation: _______________________

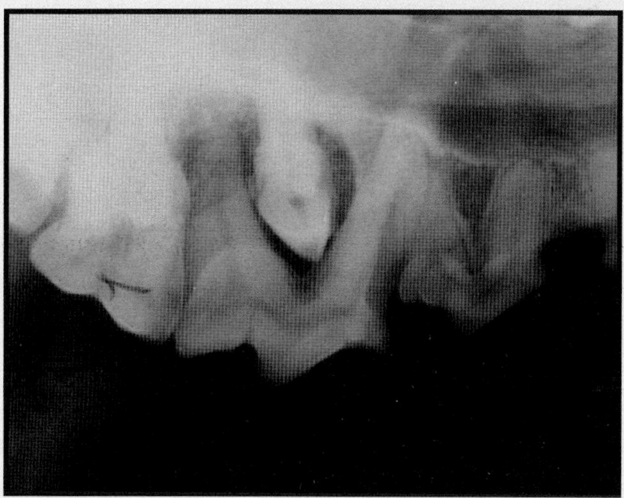

FIGURE 14-33
Premise: *Swelling at the fourth premolar*
Interpretation: _______________________

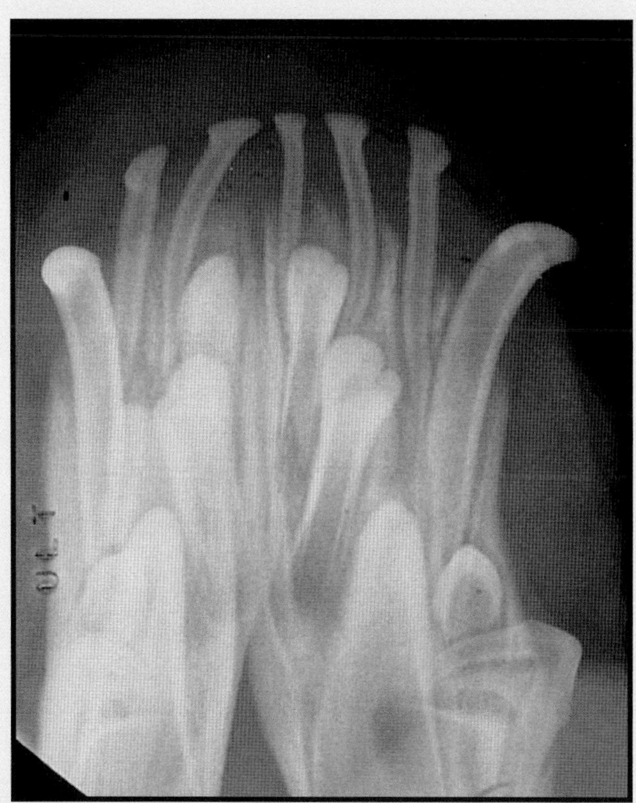

FIGURE 14-34
Premise: *Preeruption evaluation of a puppy*
Interpretation: _______________________

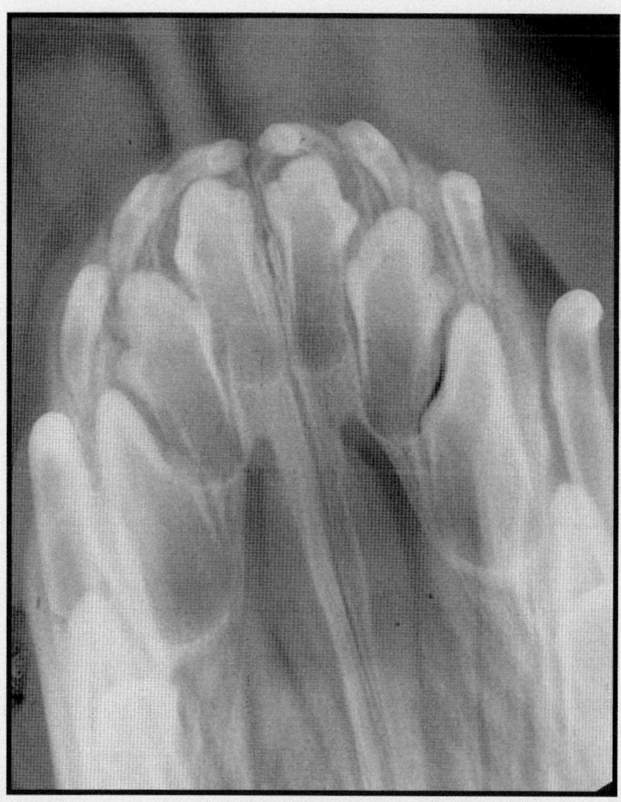

FIGURE 14-35
Premise: *Preeruption evaluation of a puppy*
Interpretation: _______________________

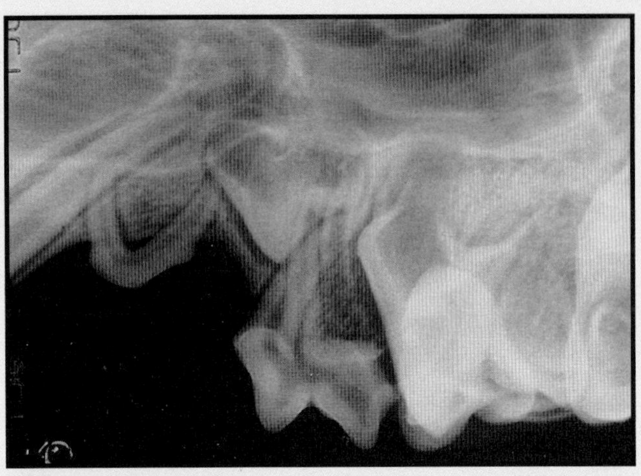

FIGURE 14-36

Premise: *Preeruption evaluation of a puppy*
Interpretation: ___________________________

____

____

____

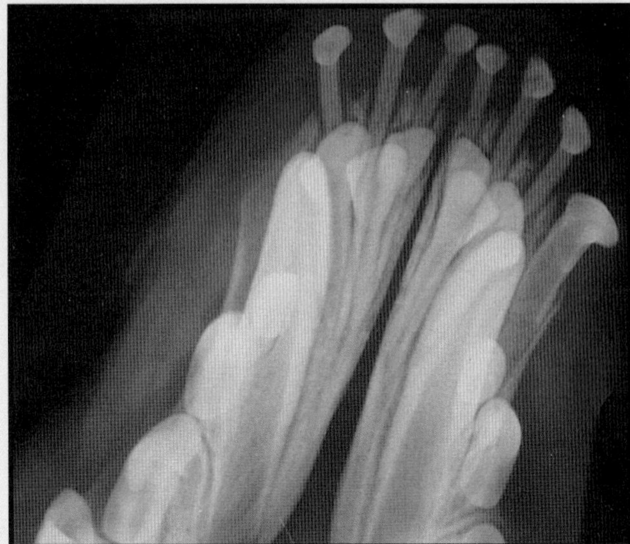

FIGURE 14-37

Premise: *Evaluation of eruption in an 8-month-old dog*
Interpretation: ___________________________

____

____

____

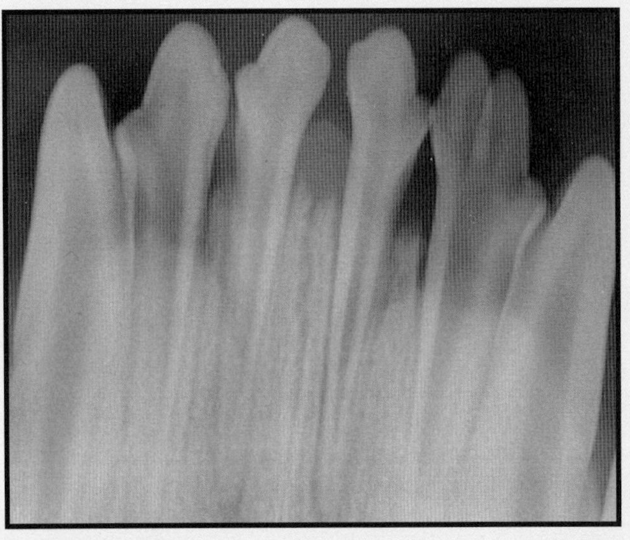

FIGURE 14-38

Premise: *Bifid crown of an incisor*
Interpretation: ___________________________

____

____

____

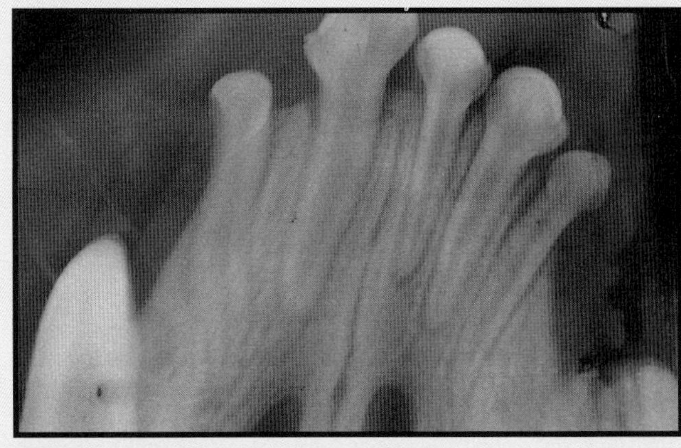

FIGURE 14-39

Premise: *Evaluation of a missing incisor and bifid crown*
Interpretation: ___________________________

____

____

____

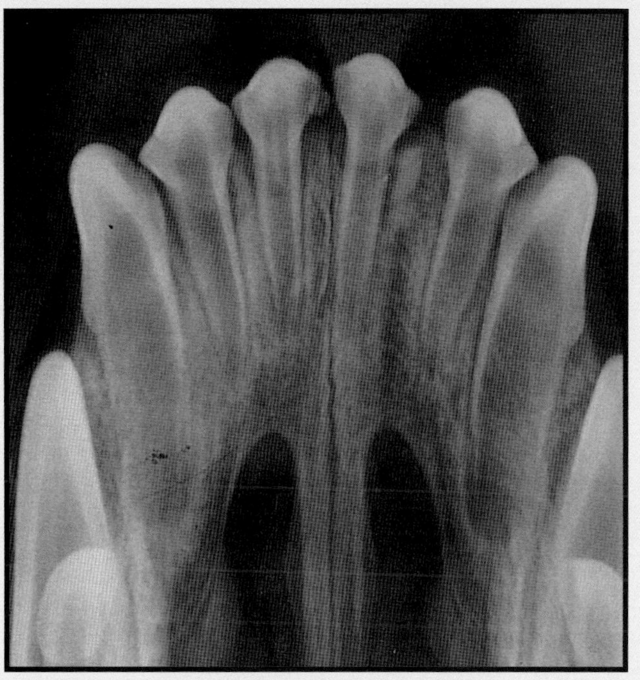

FIGURE 14-40

Premise: *Gingival swelling between the incisors*

Interpretation: _______________________________

____

____

____

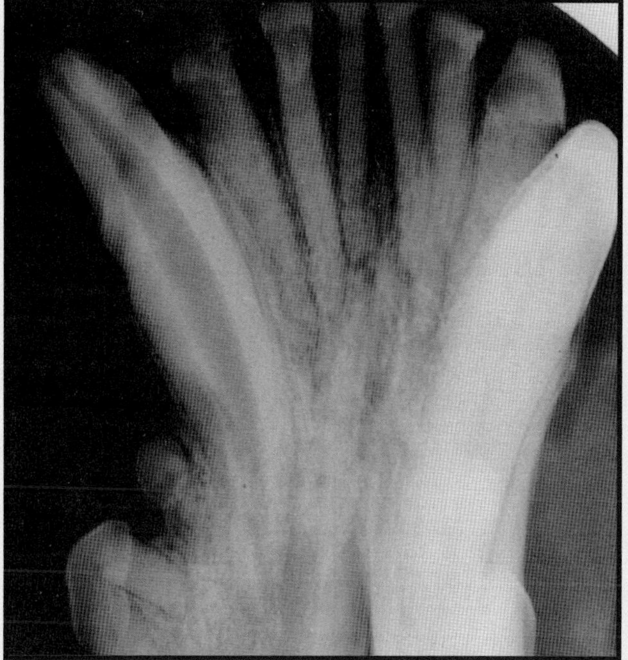

FIGURE 14-41

Premise: *Evaluation of a fractured canine tooth*

Interpretation: _______________________________

____

____

____

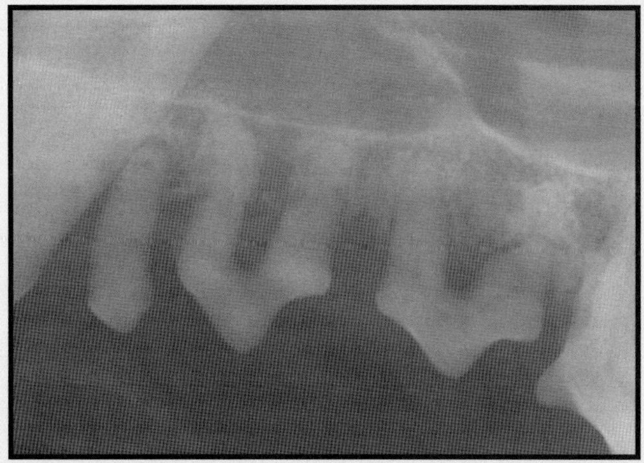

FIGURE 14-42

Premise: *Evaluation of loose teeth and periodontal recession*

Interpretation: _______________________________

____

____

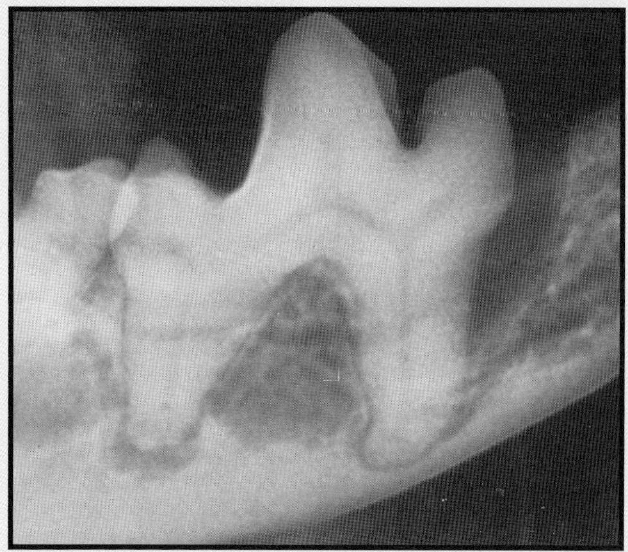

FIGURE 14-43

Premise: *Evaluation of a periodontal pocket found during prophylaxis*

Interpretation: _______________________________

____

____

____

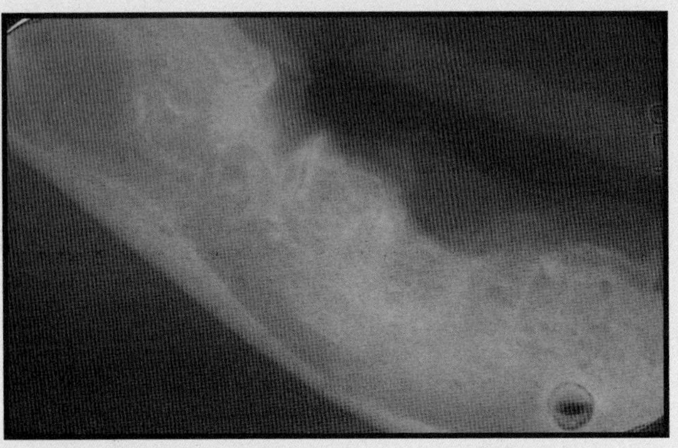

FIGURE 14-44

Premise: *Evaluation of gingivitis in the edentulous space of a cat*
Interpretation: _______________________________________

____

____

____

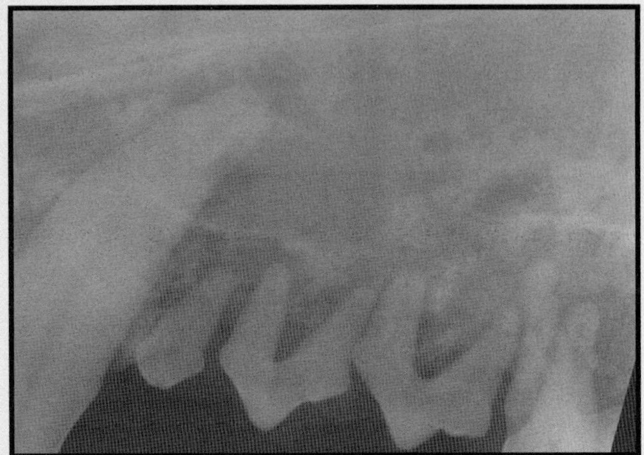

FIGURE 14-45

Premise: *Evaluation of gingivitis and an epulis*
Interpretation: _______________________________________

____

____

____

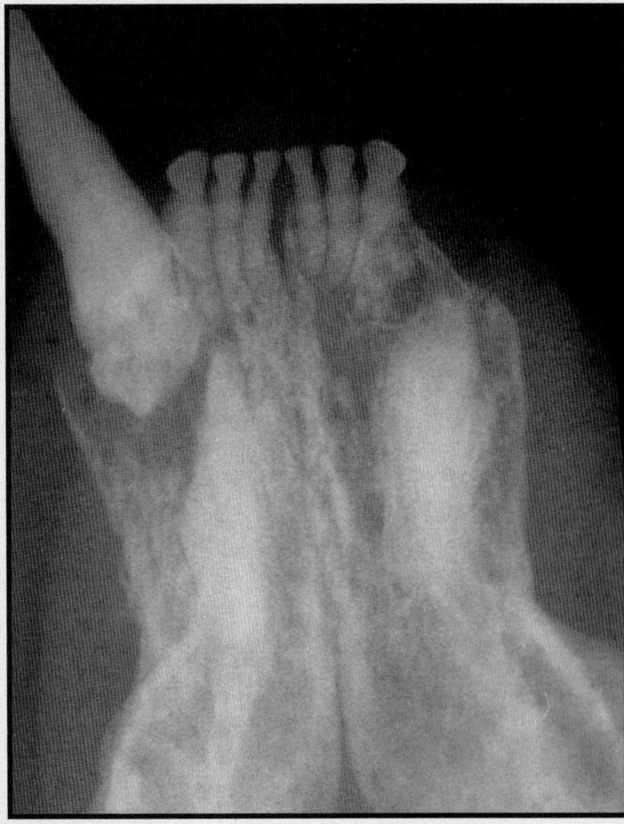

FIGURE 14-46

Premise: *Evaluation of a neck lesion (crestal defect) on a premolar*
Interpretation: _______________________________________

____

____

____

FIGURE 14-47

Premise: *Evaluation of a loose canine tooth and periodontal pocket*
Interpretation: _______________________________________

____

____

____

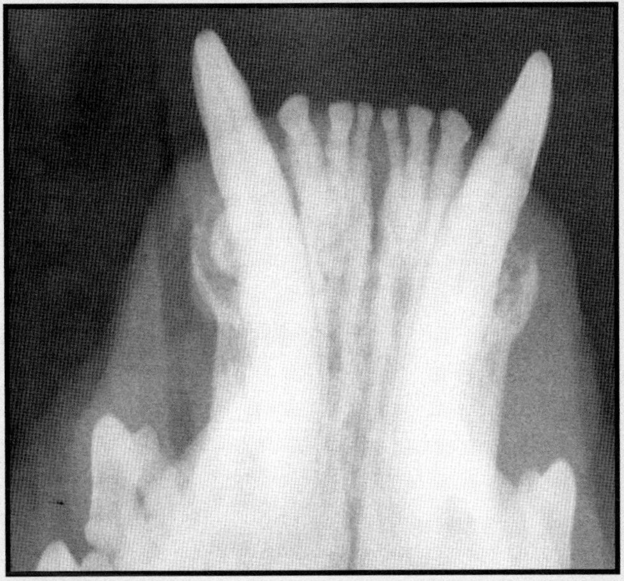

FIGURE 14-48

Premise: *Evaluation of a periodontal pocket around the canine tooth*
Interpretation: _______________________________________
____
____
____

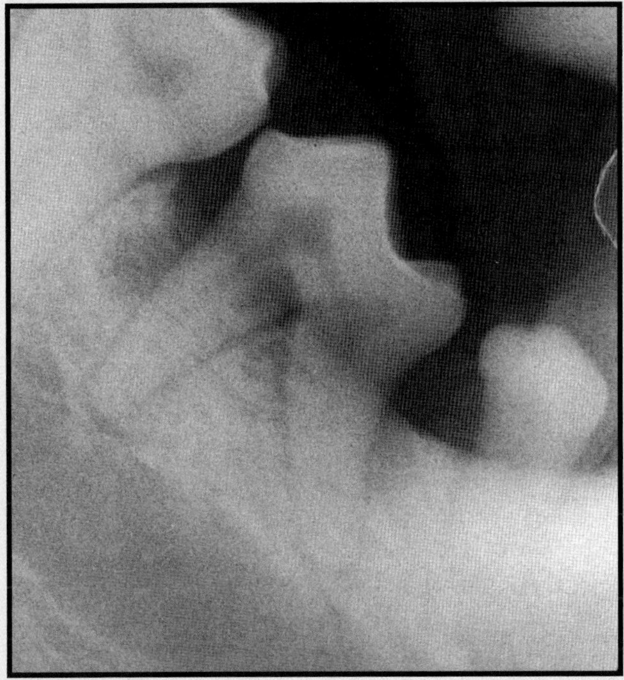

FIGURE 14-49

Premise: *Evaluation of gingivitis at a molar*
Interpretation: _______________________________________
____
____
____

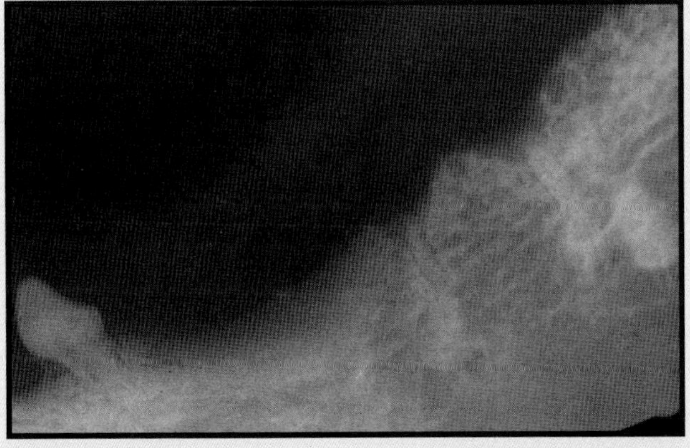

FIGURE 14-50

Premise: *Evaluation of gingivitis in the edentulous space of a dog*
Interpretation: _______________________________________
____
____
____

DENTAL RADIOGRAPHY TEST INTERPRETATIONS AND SIGNIFICANT POINTS

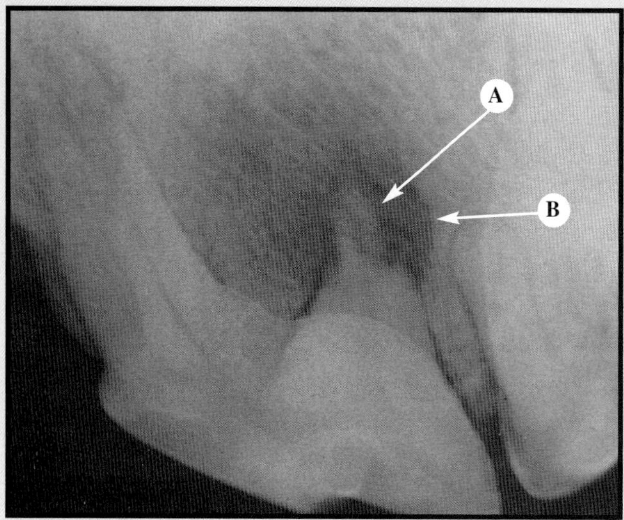

FIGURE 14-1

Interpretation: (A) *Apical resorption and* (B) *periapical radiolucency*
Significance: *Presence of an endodontic lesion. Conventional nonsurgical endodontic treatment may not be adequate to arrest or resolve the lesion.*

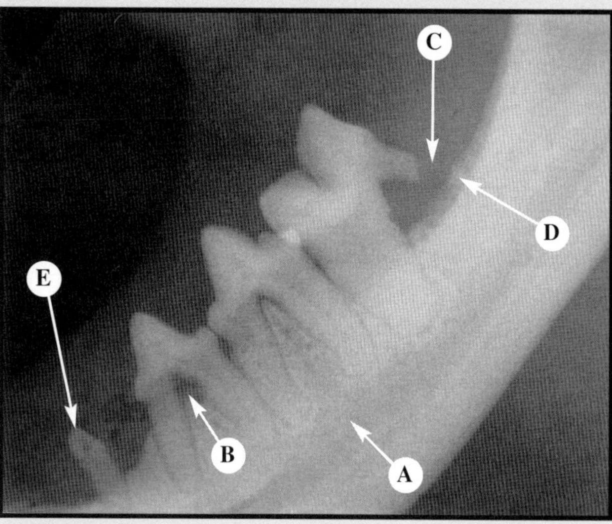

FIGURE 14-2

Interpretation: (A) *Mandibular canal,* (B) *external resorption with bone loss,* (C) *resorption completely through the distal root with pulp exposure,* (D) *apical fragment of the distal root, and* (E) *supernumerary microdont with horizontal bone loss.*
Significance: *Periodontal-endodontic lesion of the molar with root resorption. Extraction of the molar would be indicated. Remember to retrieve the apical fragment.*

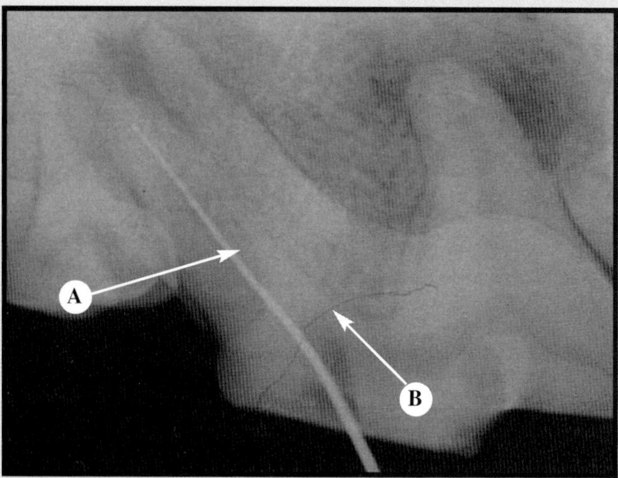

FIGURE 14-3

Interpretation: (A) *Mesiobuccal root and a file and* (B) *artifact (static electricity).*
Significance: *A file still needs to be placed in the palatal root. The film packet was opened too fast.*

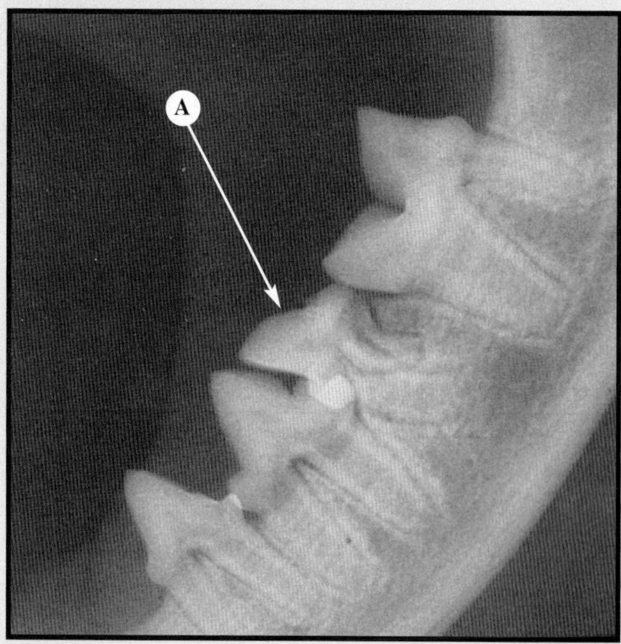

FIGURE 14-4

Interpretation: (A) *Rotated supernumerary premolar with separate root systems.*
Significance: *Separate extraction of the rotated supernumerary premolar is possible if elected.*

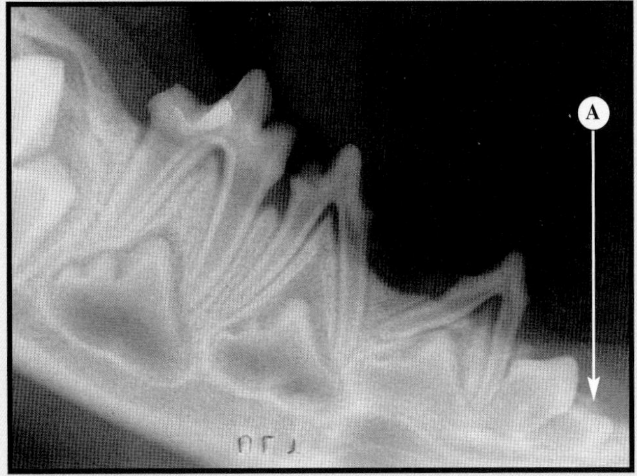

FIGURE 14-5

Interpretation: (A) *Extra mandibular first premolar*
Significance: *Presence of an extra premolar would be undesirable for showdogs or breeding stock. In addition, the extra premolar could be problematic if dentition is overcrowded. Otherwise, the extra premolar has no consequence.*

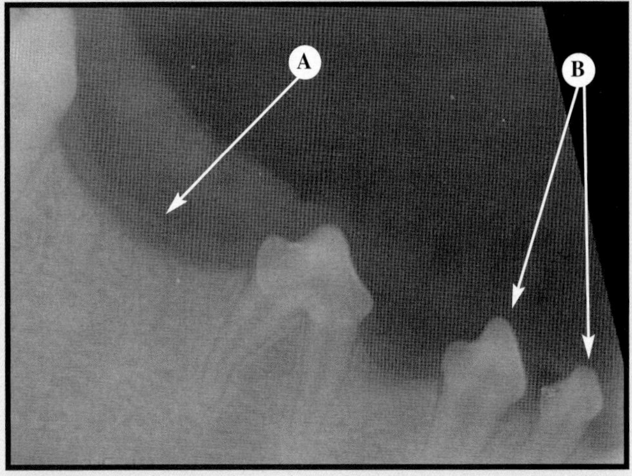

FIGURE 14-6

Interpretation: (A) *Missing fourth premolar and* (B) *two first premolars with missing second premolar*
Significance: *A differential of the interpretation would be normal first premolar with a single-rooted second premolar. Missing teeth and extra premolars would be undesirable for showdogs or breeding stock. Otherwise, they have no significance.*

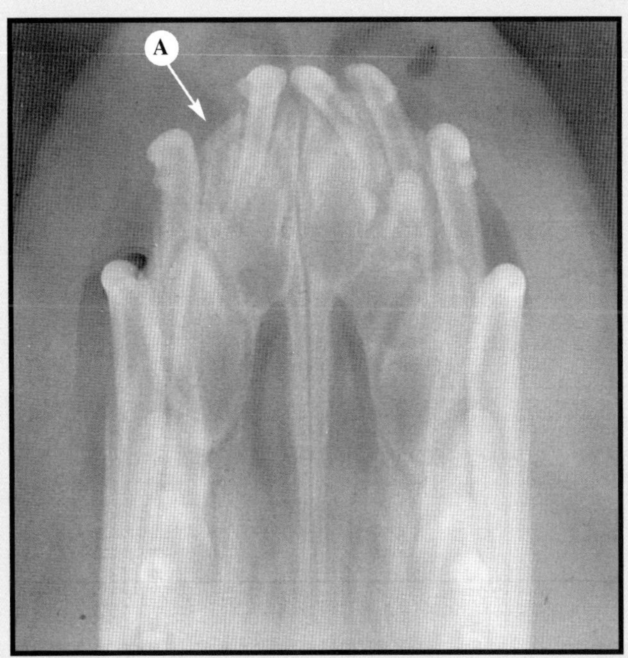

FIGURE 14-7

Interpretation: (A) *Missing permanent and deciduous upper intermediate incisors (unilateral).*
Significance: *When the deciduous precursor is congenitally absent, the corresponding permanent tooth is usually also absent.*

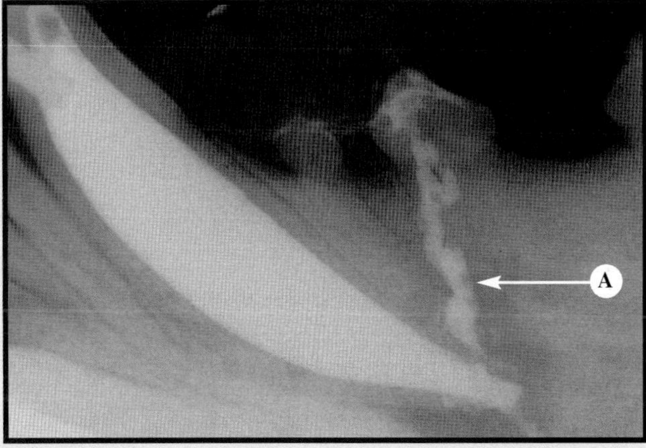

FIGURE 14-8

Interpretation: (A) *Leakage of endodontic cement from an open apex*
Significance: *The cement is following the line of the periapical fistula, which indicates extensive advanced periapical pathologic involvement. Further treatment may be necessary.*

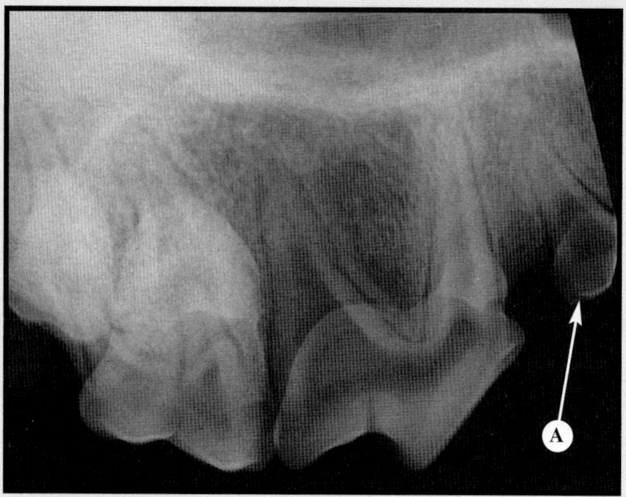

FIGURE 14-9

Interpretation: (A) *A single-rooted tooth in the position of the third upper premolar, which normally has two roots*
Significance: *The tooth is possibly a microdont. If the tooth matures normally, however, the event is inconsequential unless the patient is a showdog or used as breeding stock.*

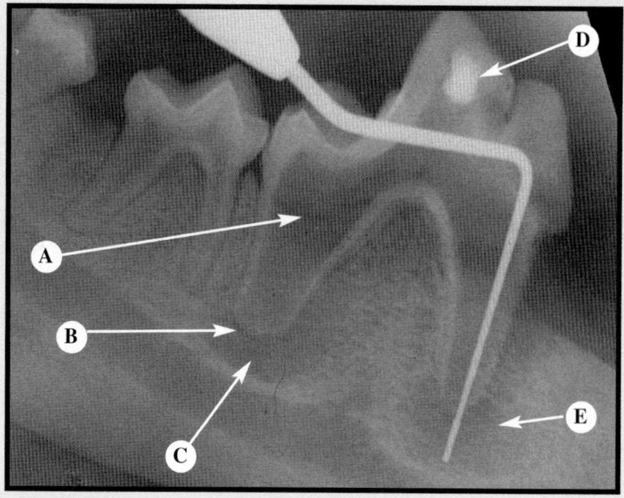

FIGURE 14-10

Interpretation: (A) *An immature tooth,* (B) *incomplete formation of the apex,* (C) *periapical radiolucency,* (D) *pulpotomy restoration, and* (E) *widened periodontal space and loss of integrity in the lamina dura*
Significance: *Failed pulpotomy and endodontic–periodontal lesion.*

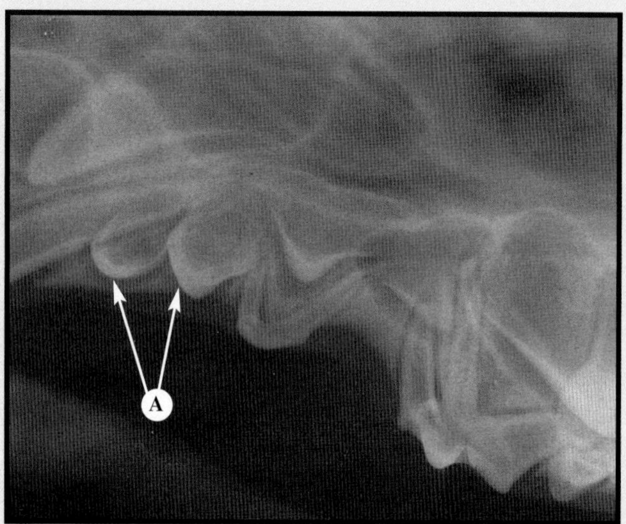

FIGURE 14-11

Interpretation: (A) *Extra upper first premolars*
Significance: *Extra premolars have no consequence unless the dentition is crowded; however, they would be undesirable for showdogs or breeding stock.*

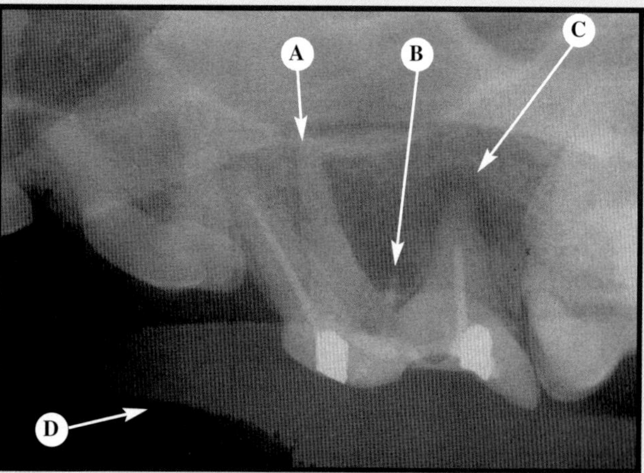

FIGURE 14-12

Interpretation: (A) *The palatal root is not obturated,* (B) *cement has leaked into the furcation,* (C) *a periapical radiolucent defect, and* (D) *film artifacts*
Significance: *Incomplete endodontic treatment. Leakage of cement into the furcation space indicates the possible presence of root perforation or lateral canal.*

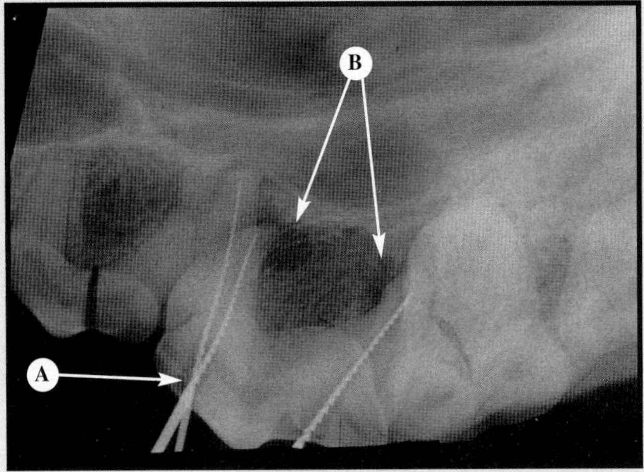

FIGURE 14-13

Interpretation: (A) *Files and* (B) *periapical radiolucent defects*
Significance: *Radiographs that were taken during surgery confirmed proper placement of files.*

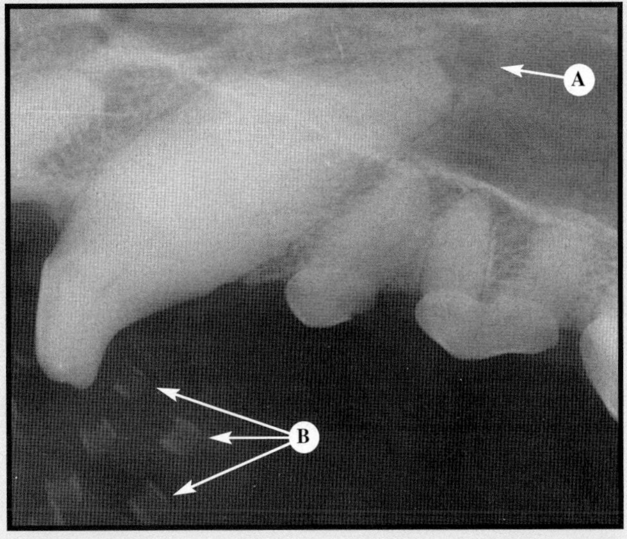

FIGURE 14-14

Interpretation: (A) *Periapical radiolucency of worn and chipped canine tooth and* (B) *vibrissae follicles*
Significance: *The periapical radiolucent areas indicate endodontic disease. The follicles are an incidental finding. Conventional endodontic treatment would be difficult because of the atretic canal.*

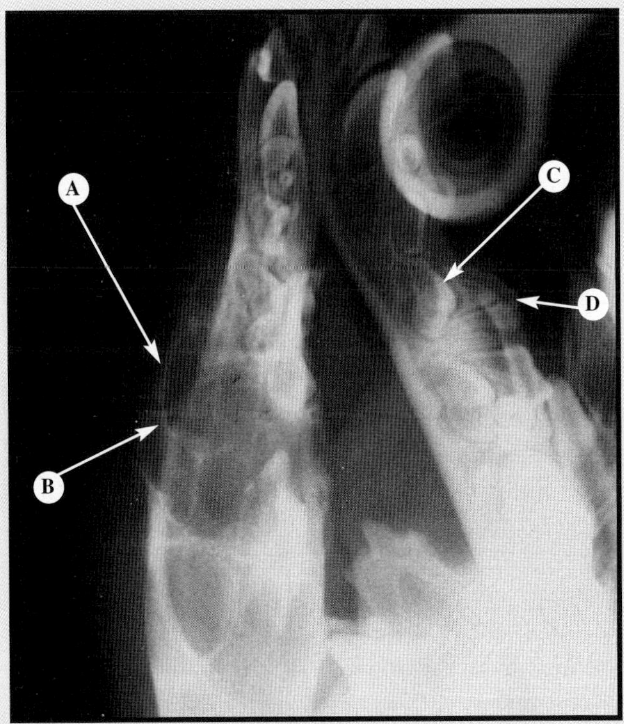

FIGURE 14-15A

Interpretation: (A) *Multilocular radiolucent bone lesion,* (B) *thinning and lysis of the cortex,* (C) *permanent tooth bud, and* (D) *deciduous teeth*
Significance: *A moderately aggressive lesion with advanced localized invasion was diagnosed as an adamantinoma, which is now known as ameloblastoma.*

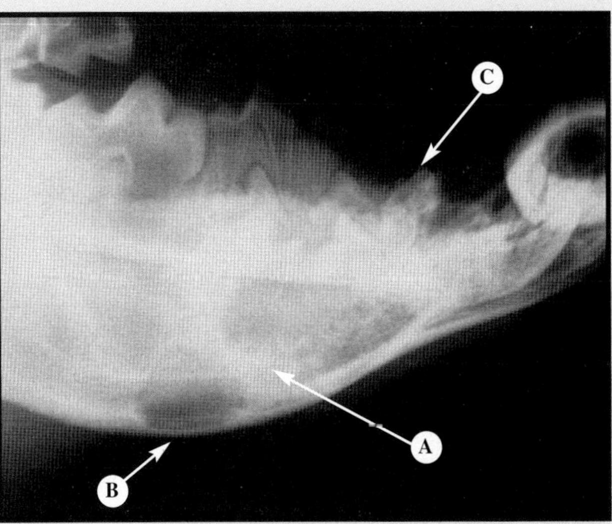

FIGURE 14-15B

Interpretation: (A) *Large, multilocular radiolucent bone lesion,* (B) *thin expanded cortex, and* (C) *superimposition of the teeth from both mandibles*
Significance: *Too much superimposition exists on this radiograph to judge the relationship of the teeth to the mass. A second intraoral view would be necessary.*

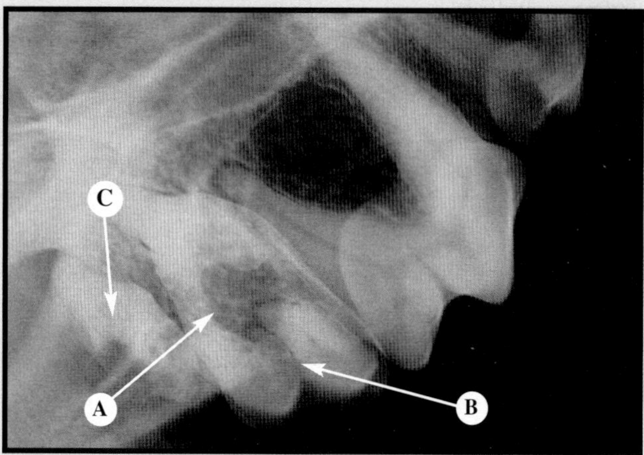

FIGURE 14-16

Interpretation: (A) *Caries defect,* (B) *developmental groove, and* (C) *defects of the crown and root*

Significance: *The caries is extensive, and there is endodontic involvement of the second molar.*

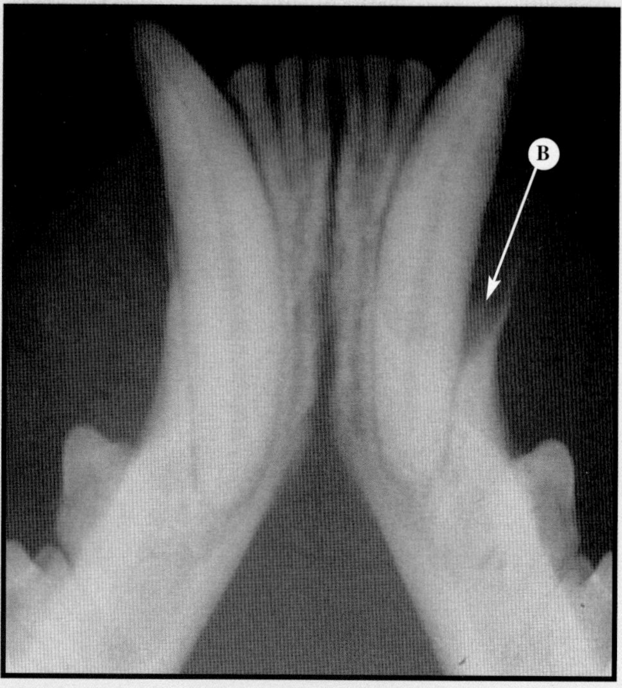

FIGURE 14-17

Interpretation: (A) *Bone loss and infrabony pocket*

Significance: *Infrabony pockets, which are an ideal environment for substantial plaque accumulation, are not grossly evident on clinical examination and therefore can elude diagnosis. In addition, once they have been discovered, successful treatment can be difficult because it primarily involves extensive follow-up home care.*

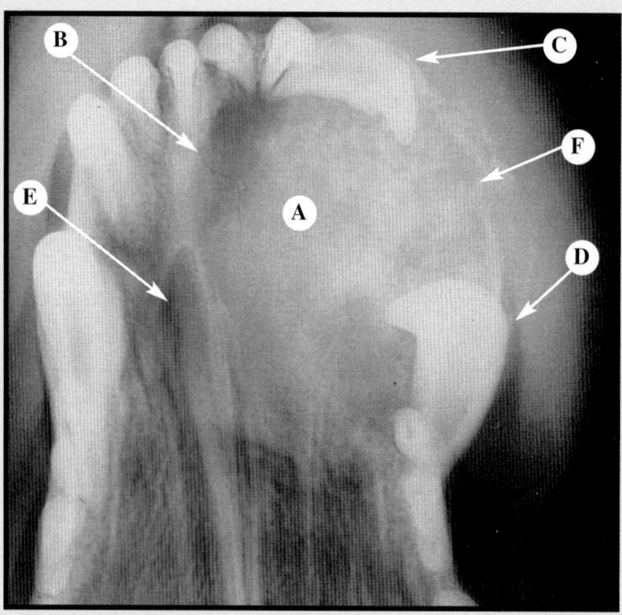

FIGURE 14-18

Interpretation: (A) *Bone lesion with focal opacities,* (B) *irregular margin,* (C) *displaced incisor,* (D) *displaced canine tooth,* (E) *displaced and eroded midline bone, and* (F) *eroded cortical bone*

Significance: *An aggressive lesion, diagnosed as an adamantinoma (now called ameloblastoma), is advancing with localized invasion.*

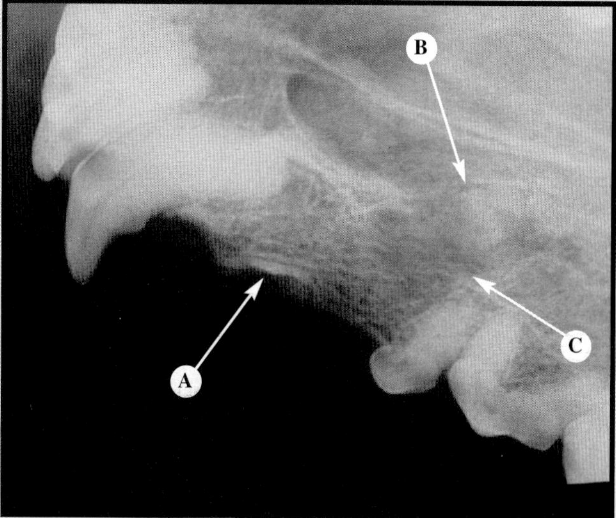

FIGURE 14-19

Interpretation: (A) *Retained root fragment at crestal bone,* (B) *retained fragments of the apical root, and* (C) *bone rarefaction*

Significance: *Removal of the root fragments would be indicated.*

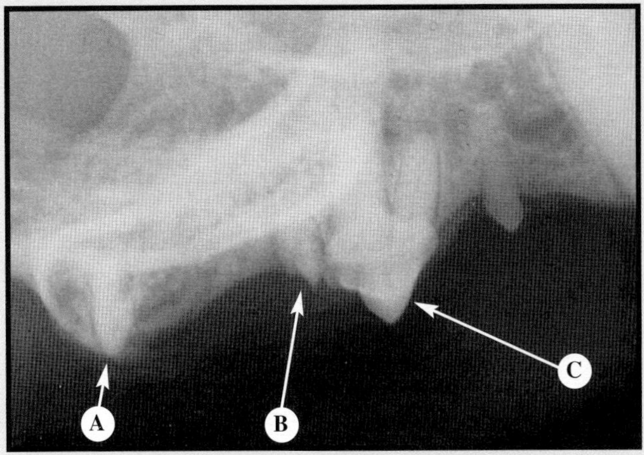

FIGURE 14-20

Interpretation: (A) *Molar,* (B) *retained mesial root fragment of the maxillary left fourth premolar, and* (C) *the third premolar*

Significance: *Removal of the retained root fragment would be indicated. The small molar should not be mistaken for a retained root. After extraction of the root fragment, another radiographic view is indicated to rule out additional retained roots.*

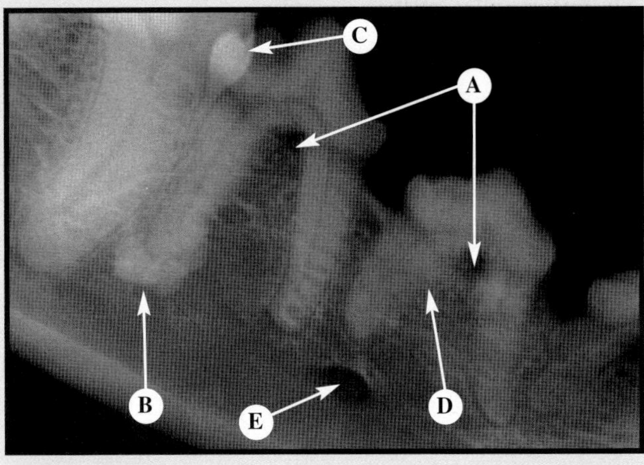

FIGURE 14-21

Interpretation: (A) *Decreased bone density at the furcations,* (B) *condensing osteitis of the distal root of the fourth premolar,* (C) *superimposition of the crowns of the premolar and molar,* (D) *loss of periodontal ligament space in the third premolar, and* (E) *mental foramen*

Significance: *The third premolar is undergoing replacement resorption. Condensing osteitis is often asymptomatic but associated with chronic low-grade inflammation. Mild periodontal disease is present.*

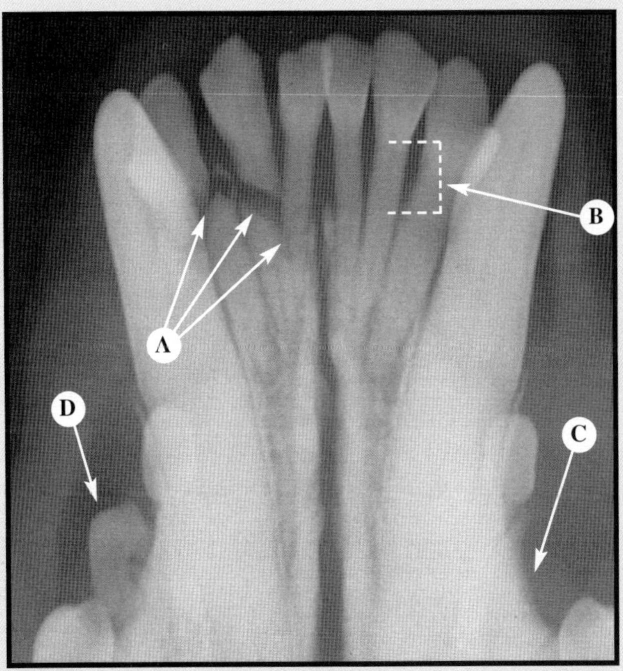

FIGURE 14-22

Interpretation: (A) *Fracture lines,* (B) *horizontal bone loss,* (C) *missing premolar, and* (D) *retained deciduous premolar*

Significance: *Diagnostic differentials for loose teeth are root fracture and periodontal bone loss. Both are possible. Teeth that have less alveolar bone support are more likely to sustain root fractures.*

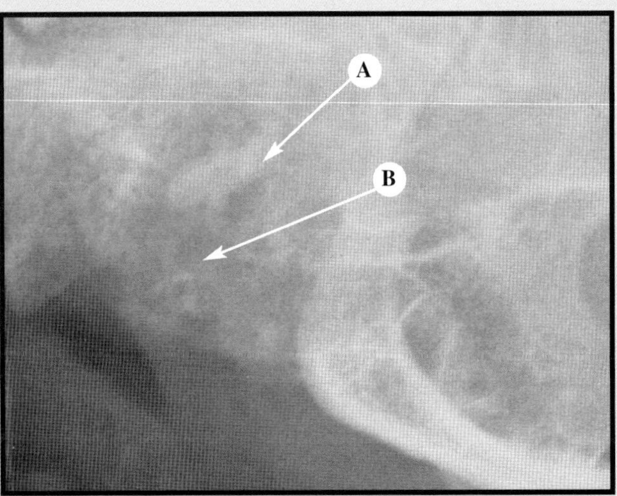

FIGURE 14-23

Interpretation: (A) *Retained root fragment of the canine tooth and* (B) *bone rarefaction with loss of the white-line landmark*

Significance: *The root fragment should be removed. A nasal fistula associated with the root fragment may complicate healing after the fragment is extracted.*

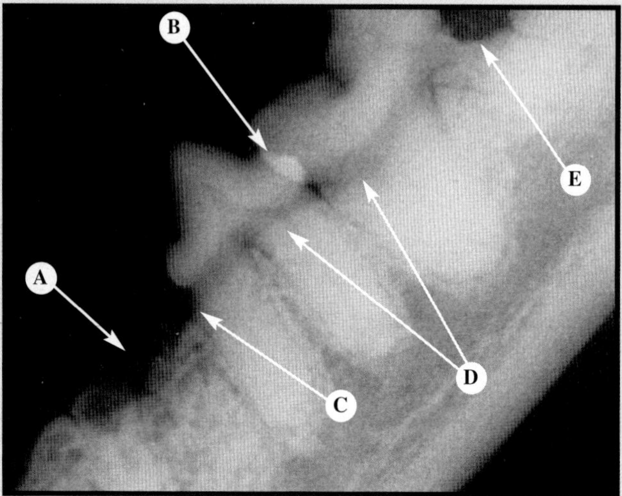

FIGURE 14-24

Interpretation: (A) *Root fragments of the third premolar,* (B) *superimposition,* (C) *horizontal bone loss,* (D) *cervical burnout, and* (E) *crestal root resorption and bone loss*

Significance: *Root resorption is associated with periodontal disease of the molar. The gingival swelling is the result of tissue reaction to the retained root fragments.*

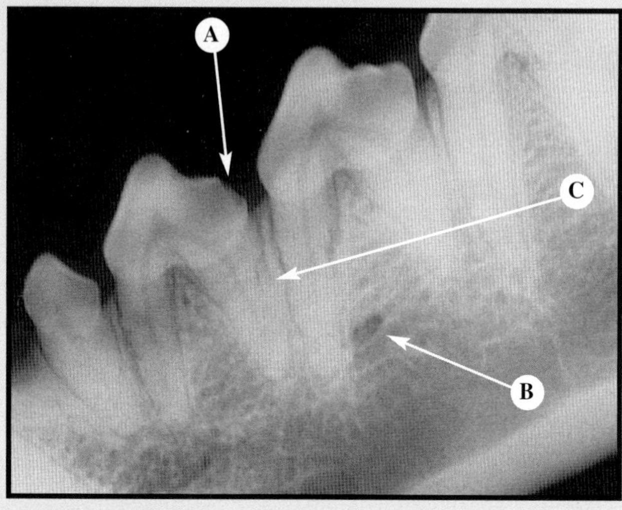

FIGURE 14-25

Interpretation: (A) *Chipped crown,* (B) *mental foramen, and* (C) *external root resorption*

Significance: *Root resorption is associated with the evidence of trauma (i.e., the chipped crown).*

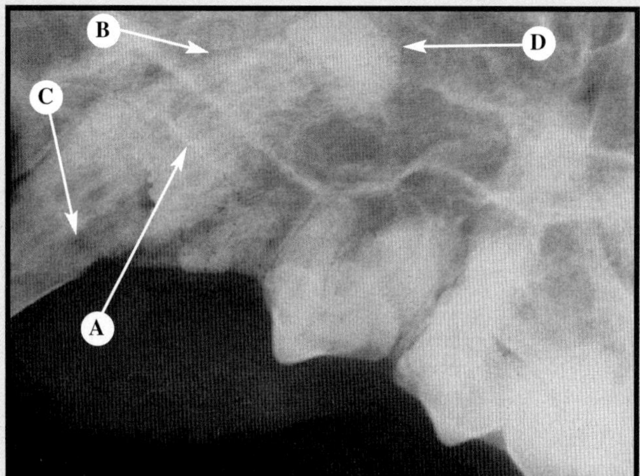

FIGURE 14-26

Interpretation: (A) *Intraosseous root resorption,* (B) *loss of periodontal ligament space,* (C) *crestal resorption, and* (D) *apex*

Significance: *The crestal resorption extends apically and involves most of the root. Extraction will be challenging because of the presence of ankylosis.*

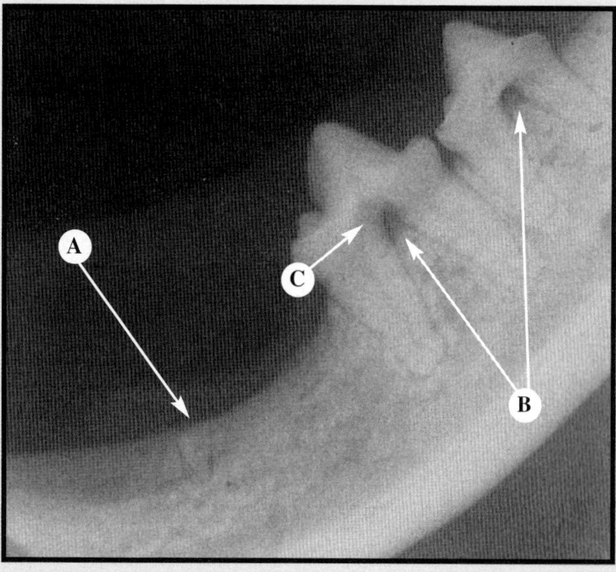

FIGURE 14-27

Interpretation: (A) *Retained distal root fragment of the molar,* (B) *bone loss at the furcation, and* (C) *radiolucency*

Significance: *Root fragments that have no clinical lesions or lesions evident on radiographs and that are at or below the level of crestal bone can be monitored, but they do not usually require treatment. The defects are caused by bone loss at the furcation, which represents mild periodontal bone loss. Reevaluate fourth premolar. Repeat clinical examination, and take a radiograph from a different angle to rule out presence of a resorptive lesion.*

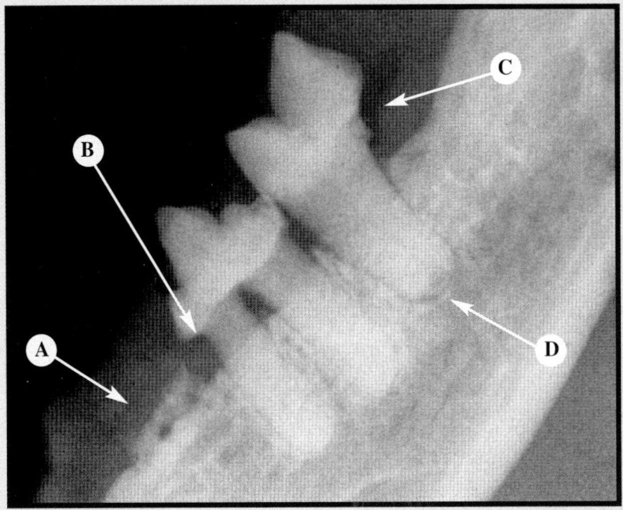

FIGURE 14-28

Interpretation: (A) *Tooth remnants or periosteal bone reaction,* (B) *horizontal bone loss and calculus overhang,* (C) *root resorption and bone loss, and* (D) *apical resorption*

Significance: *An endodontic lesion of the molar is associated with root resorption and moderate periodontal bone loss.*

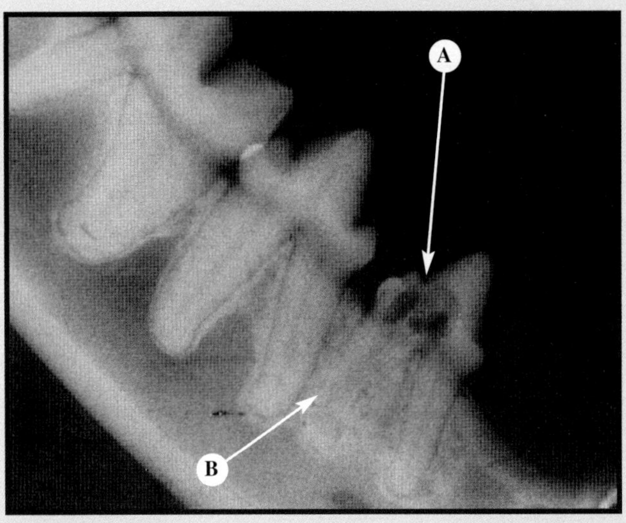

FIGURE 14-29

Interpretation: (A) *Crestal and supraosseous resorption and* (B) *intraosseous resorption of the distal root*

Significance: *The crown is undergoing internal resorption, and ankylosis of the distal root of the premolar is present.*

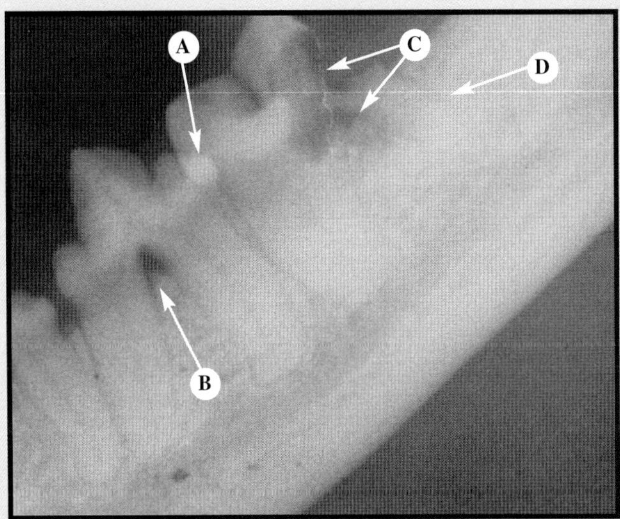

FIGURE 14-30

Interpretation: (A) *Superimposition,* (B) *bone loss at the furcation,* (C) *crestal and supraosseous resorption, and* (D) *distal root of the molar*

Significance: *Extraction of the molar is indicated. During clinical examination by probing, bone loss at the furcation can be clinically confused with feline resorptive lesions.*

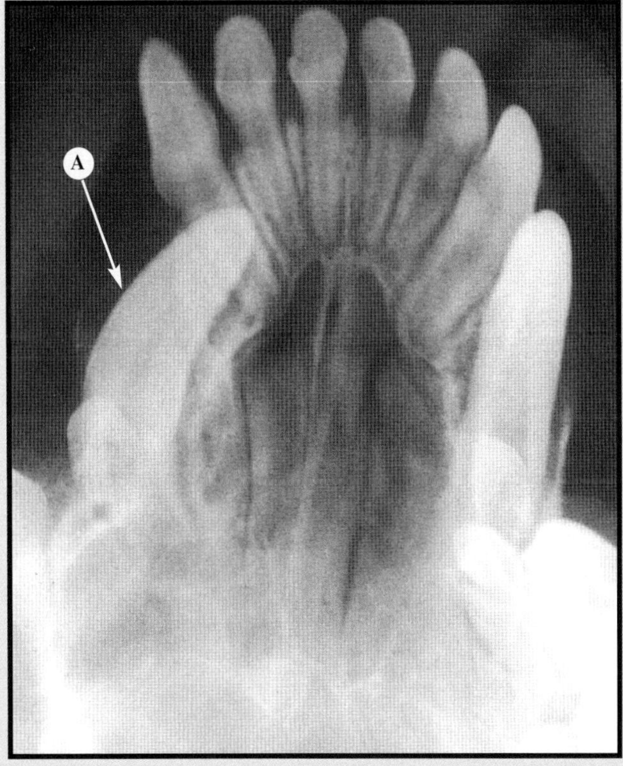

FIGURE 14-31

Interpretation: (A) *Unerupted ectopic permanent canine tooth exhibiting excessive curvature (dilaceration)*

Significance: *Dilaceration can be caused by impaction. Ectopically directed teeth may not complete eruption.*

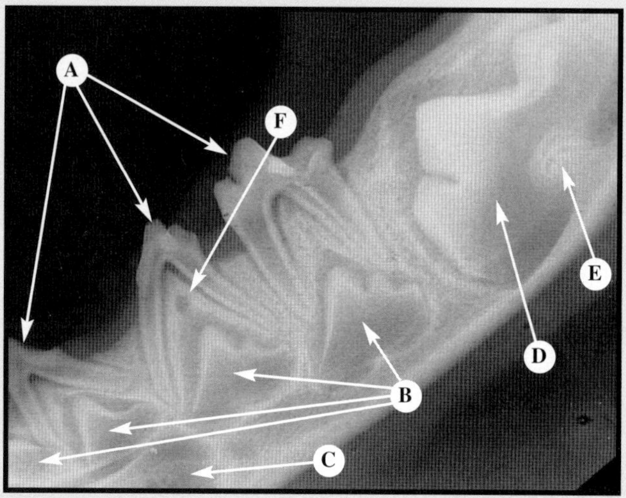

FIGURE 14-32

Interpretation: (A) *Deciduous teeth,* (B) *permanent premolars,* (C) *permanent canine tooth,* (D) *molar,* (E) *furcation of the molar, and* (F) *gubernaculum (eruption pathway)*
Significance: *All of the permanent teeth are present.*

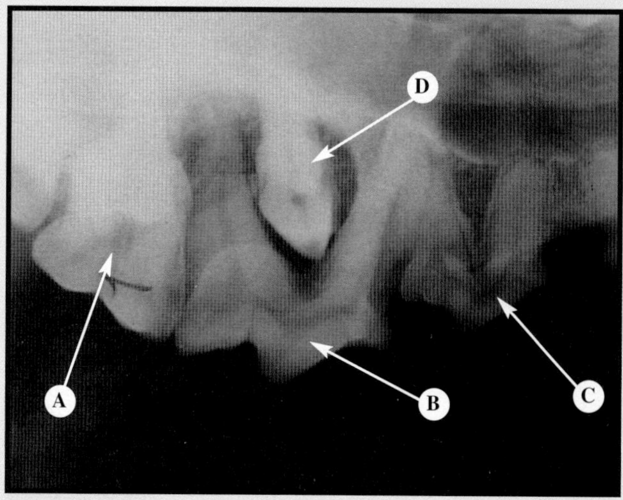

FIGURE 14-33

Interpretation: (A) *Maxillary first molar,* (B) *fourth premolar,* (C) *third premolar, and* (D) *unerupted supernumerary premolar*
Significance: *The swelling is being caused by the unerupted tooth. Swelling around the fourth premolar is not always caused by an abscess.*

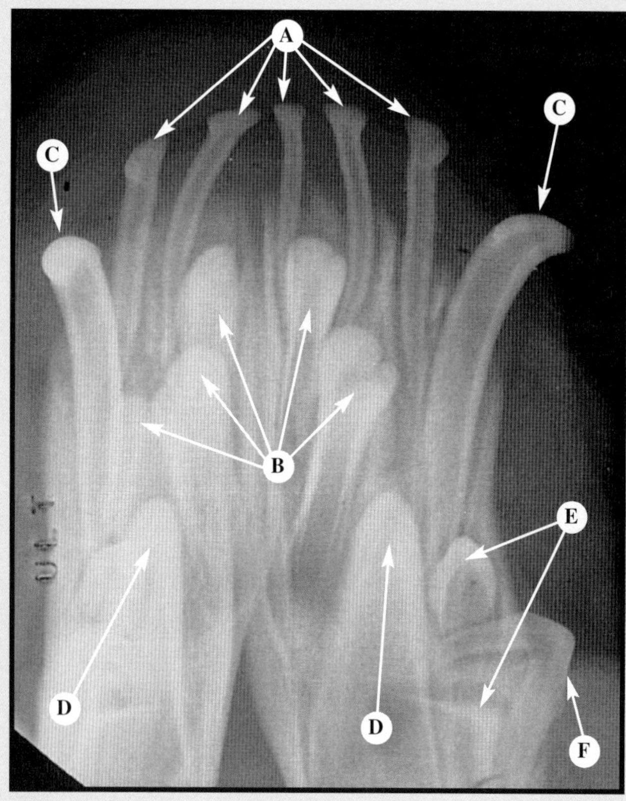

FIGURE 14-34

Interpretation: (A) *Five deciduous incisors are present,* (B) *permanent incisors,* (C) *deciduous canine teeth,* (D) *permanent canine teeth,* (E) *permanent premolars, and* (F) *deciduous premolars*
Significance: *One permanent incisor may be missing. A second radiographic view should be taken to verify and rule out superimposition.*

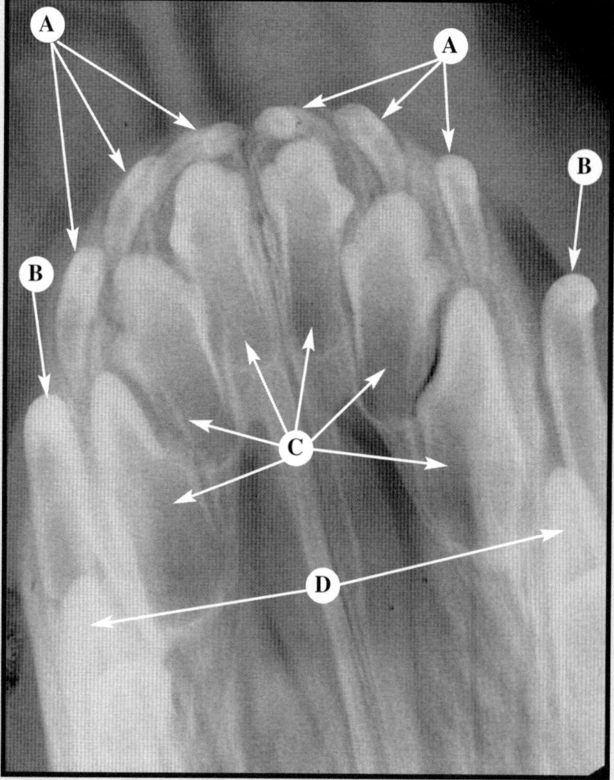

FIGURE 14-35

Interpretation: (A) *Deciduous incisors,* (B) *deciduous canine teeth,* (C) *permanent incisors, and* (D) *permanent canine teeth.*
Significance: *All of the permanent teeth are present in this region.*

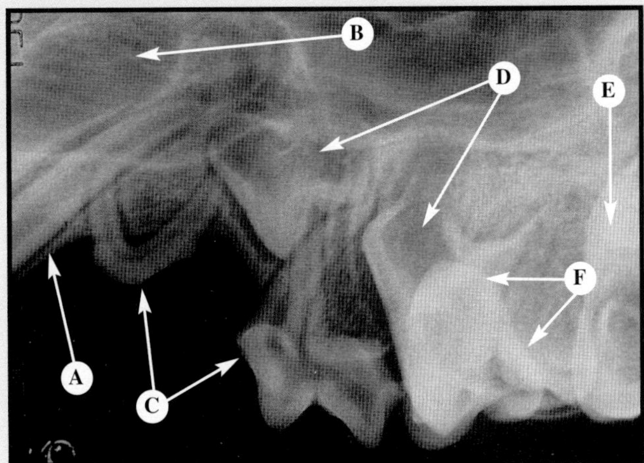

FIGURE 14-36

Interpretation: (A) *Deciduous canine tooth,* (B) *root developing in the permanent canine tooth* (C) *deciduous premolars,* (D) *permanent premolars,* (E) *molar, and* (F) *superimposed deciduous premolar*
Significance: *The first and second permanent premolars are missing.*

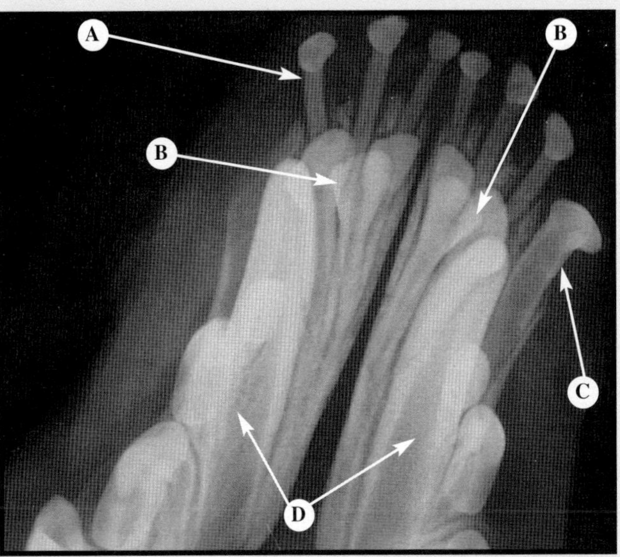

FIGURE 14-37

Interpretation: (A) *Deciduous incisors,* (B) *unerupted fully formed permanent incisors,* (C) *deciduous canine tooth, and* (D) *permanent canine teeth nearly formed*
Significance: *Deciduous retention and delayed eruption of the permanent teeth are evident. The distance between the two permanent canine teeth may be narrow, and canine malocclusion could possibly develop because the current lingual orientation may persist.*

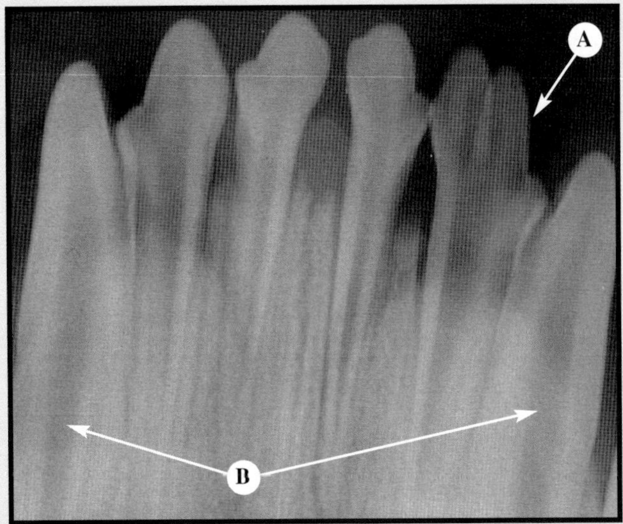

FIGURE 14-38

Interpretation: (A) *Bifid crown and shared root system and* (B) *third incisors*
Significance: *The elongated view makes the third incisors appear conical like a canine tooth. Gemination of the incisor is evident.*

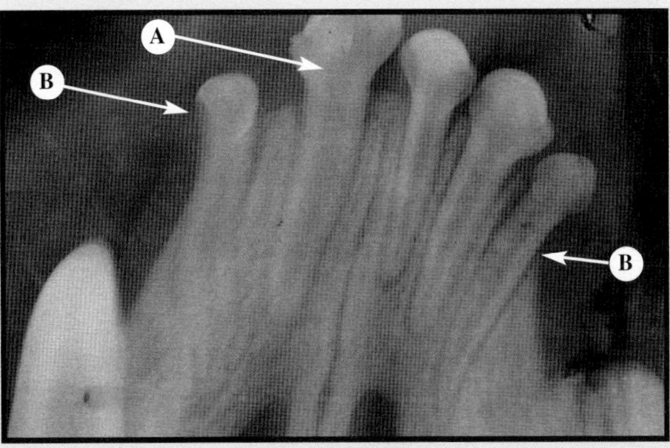

FIGURE 14-39

Interpretation: (A) *Bifid crown and shared root system and* (B) *microdonts located at the positions of the corner incisors*
Significance: *There is fusion of the incisor, thereby creating the bifid crown.*

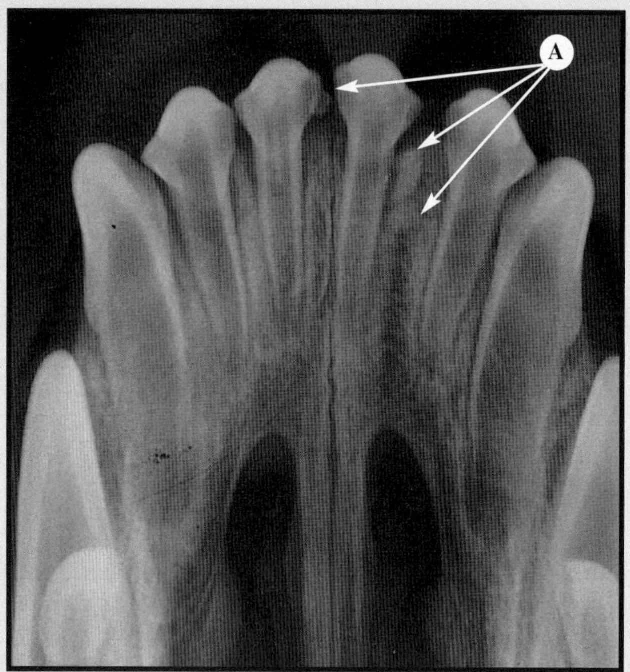

FIGURE 14-40

Interpretation: (A) *Mesiodens*

Significance: *A mesiodens occupies space and can disrupt normal tissue despite its small size. More than one mesiodens is present. Radiographs of different views should be taken before and after the mesiodens have been removed to ensure complete removal.*

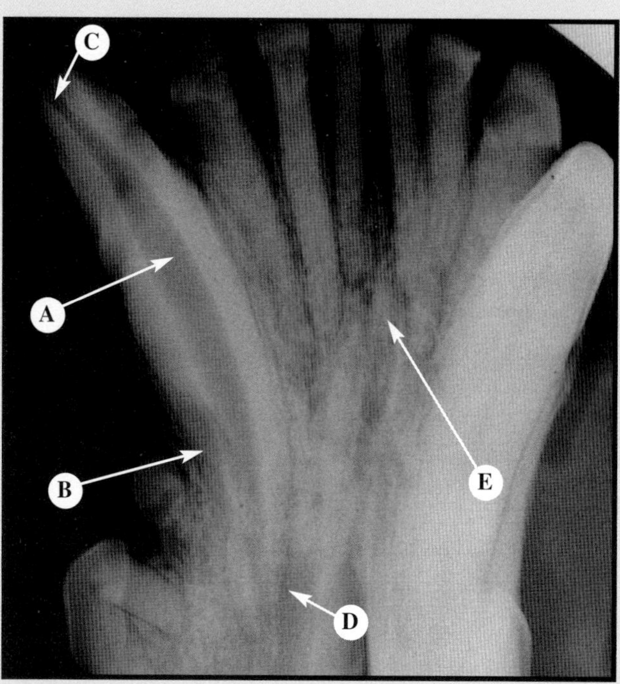

FIGURE 14-41

Interpretation: (A) *Wide root canal lumen,* (B) *root defect,* (C) *exposed pulp chamber,* (D) *open apex and periapical radiolucency, and* (E) *possible artifact or fracture line*

Significance: *Pulp necrosis occurred before maturation was complete. Resorption is occurring. The prognosis is poor for successful endodontic treatment. Another radiographic view is recommended to evaluate the integrity of the incisor roots.*

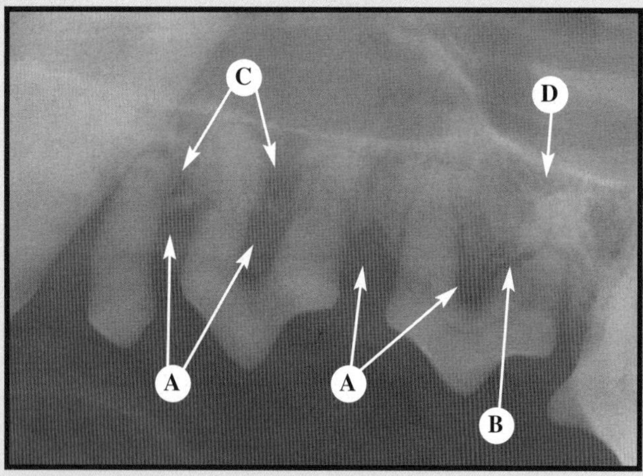

FIGURE 14-42

Interpretation: (A) *Jagged alveolar crests and bone loss,* (B) *root fracture,* (C) *root defects, and* (D) *mild sclerosis*

Significance: *Moderate to severe horizontal bone recession has occurred. Root fractures and periodontal recession are a cause of increased mobility and periodontal disease; both conditions are present. Diagnostic differentials for root defects would include root caries and resorptive lesions.*

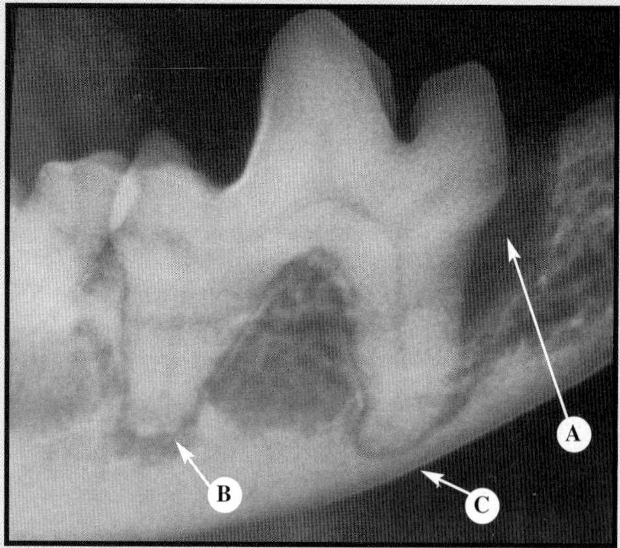

FIGURE 14-43

Interpretation: (A) *Vertical bone loss,* (B) *wide periapical periodontal space, and* (C) *root apex is close to cortical border*

Significance: *A periodontal–endodontic lesion is evident. Care should be taken if extraction is elected. The patient would be at risk for pathologic or iatrogenic fracture of the mandible.*

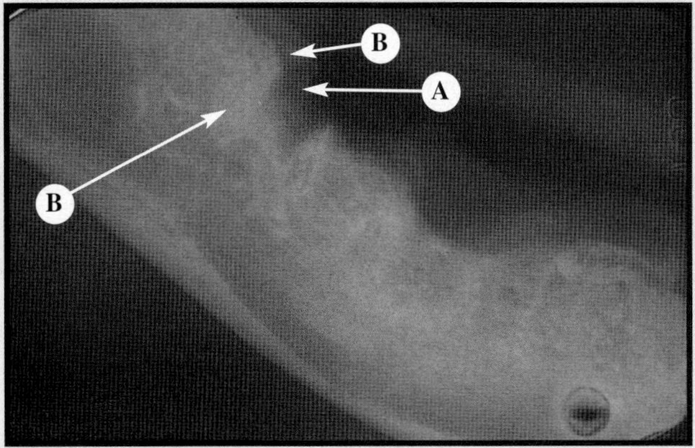

FIGURE 14-44

Interpretation: (A) *Unhealed extraction site and* (B) *retained distal root of the molar*

Significance: *The retained root fragment must be removed to promote healing of the extraction site.*

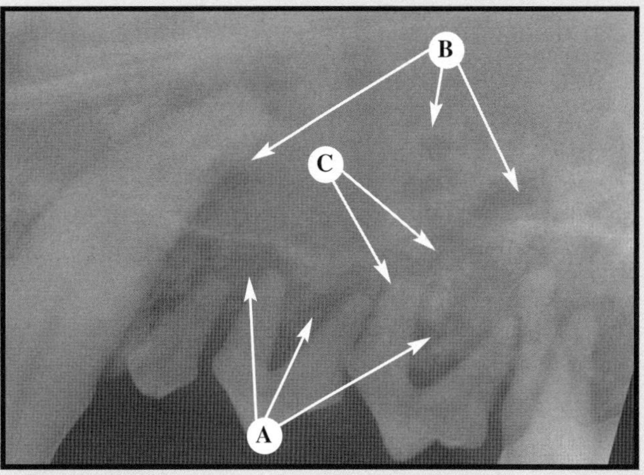

FIGURE 14-45

Interpretation: (A) *radiolucent bone defects,* (B) *moth-eaten bone lysis,* (C) *loss of integrity of the white-line landmark*

Significance: *The epulis is an aggressive lesion. A deep biopsy should be submitted for histopathologic evaluation. The diagnosis was amelanotic melanoma.*

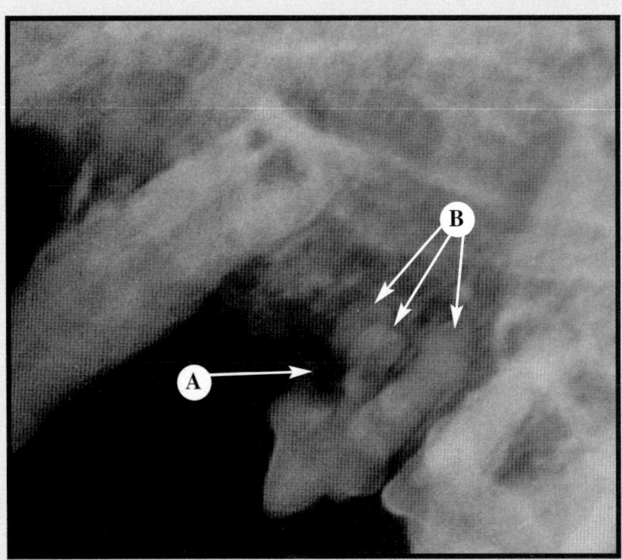

FIGURE 14-46

Interpretation: (A) *Deep crestal root resorption and* (B) *three premolar root apices*

Significance: *Extraction is indicated. This tooth normally has two roots. If the third root is not removed because it was not noticed on the radiographs, the extraction will be incomplete. Note the periodontal bone recession and apical resorption of the canine tooth.*

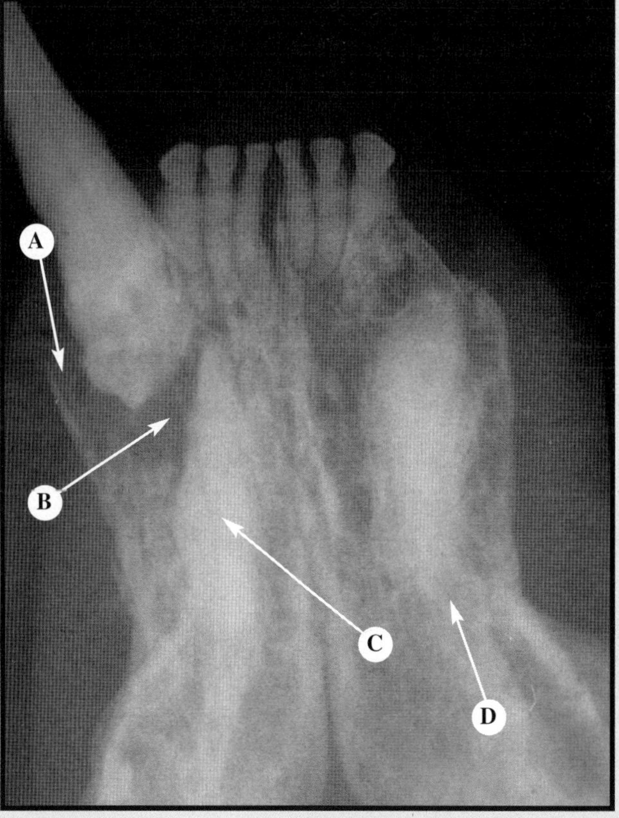

FIGURE 14-47

Interpretation: (A) *Infrabony pocket,* (B) *root fracture,* (C) *root apex, and* (D) *retained root*

Significance: *The increased mobility is being caused by the root fracture. There is endodontic involvement of the tooth.*

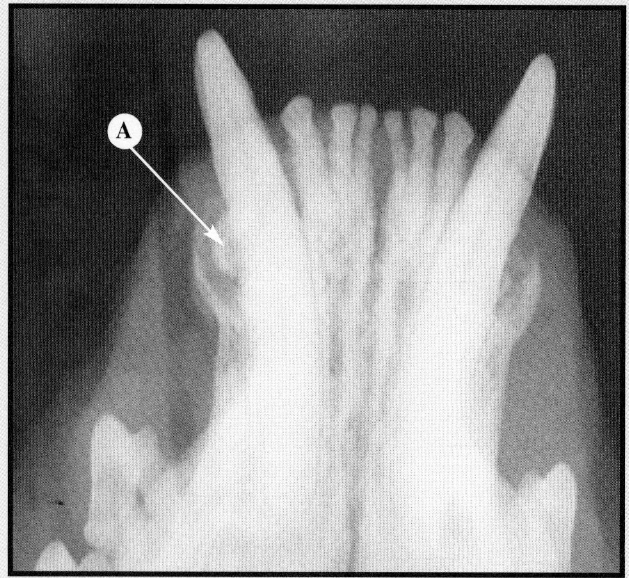

FIGURE 14-48

Interpretation: (A) *Calculus in the infrabony pocket*

Significance: *If the tooth is to be treated and maintained, root planing is indicated to remove the calculus. The probe will measure a deeper pocket after the calculus has been removed; the calculus may be obstructing insertion of the probe.*

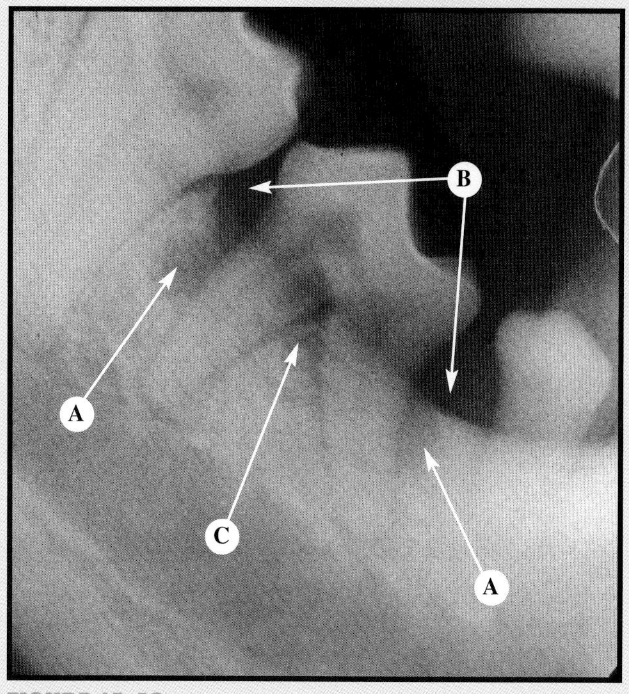

FIGURE 14-49

Interpretation: (A) *Infrabony pocket,* (B) *height of the crestal bone, and* (C) *radiolucency at the furcation*

Significance: *The furcation should be considered when planning periodontal treatment of the molar.*

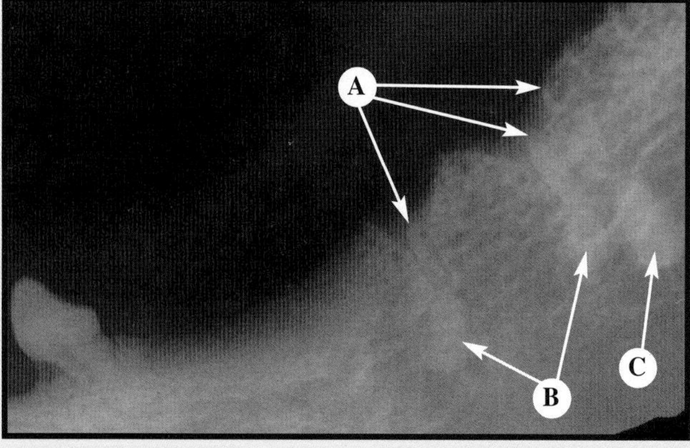

FIGURE 14-50

Interpretation: (A) *Root fragments associated with defects in the contour of the crestal bone,* (B) *root apices, and* (C) *condensing osteitis or another sclerotic root fragment from an adjacent tooth*

Significance: *Removal of the exposed root fragments may help to resolve the gingivitis.*

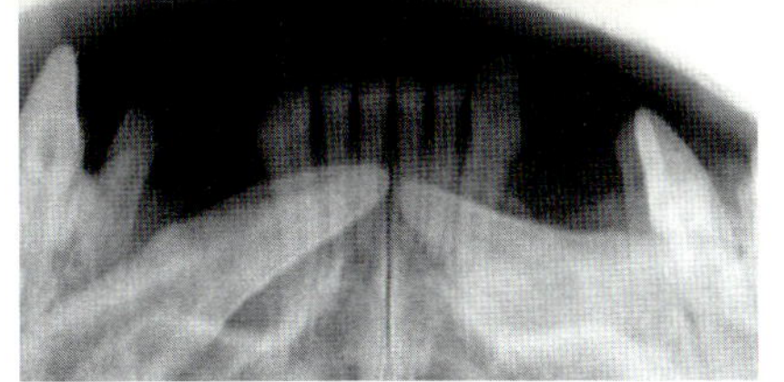

GLOSSARY

ABSCESS—A localized collection of pus in a cavity formed by the disintegration of tissues.

ABRASION—Wearing away of tooth substance by mechanical forces, such as chewing and mastication.

ACANTHOMATOUS EPULIS— An epithelial odontogenic tumor.

ACCESS—A surgical preparation allowing entrance to a treatment site, such as instrumentation of a root canal.

ACCESSORY ROOT CANAL—A lateral branching of the main root canal providing a pathway other than the apex through which the blood vessels can enter and exit the canal. Also known as the lateral canal.

ADAMANTINOMA—See ameloblastoma.

ADUMBRATION—A lack of sharpness of the x-ray shadow.

ALIGNMENT—Arrangement of teeth in relation to their supporting bone, adjacent teeth, and opposing dentition.

ALVEOLAR— Pertaining to an alveolus.

ALVEOLAR BONE—The compact bone that composes the alveolus (tooth socket). It is supported by trabecular bone. Fibers of the periodontal ligament insert into it.

ALVEOLAR CREST—The most coronal portion of alveolar bone, seen on radiographs as the ridge of bone between the teeth.

ALVEOLAR PROCESS—The part of the bone that surrounds and supports the teeth. It consists of trabecular bone within vestibular and oral cortical plates.

ALVEOLUS—The bony socket of a tooth.

AMELOBLASTOMA—An epithelial neoplasm that resembles the enamel organ.

AMELOGENESIS IMPERFECTA—The malformation or absence of enamel.

AMPERE—The unit of quantity of electric current, produced by one volt acting through a resistance of one ohm.

ANGULATION—The direction of the primary beam of radiation in relation to the subject and film.

ANODONTIA—Missing teeth from lack of formation. Partial anodontia is the absence of some teeth; total anodontia is the absence of all teeth.

ANOMALY—Any marked deviation from that which is ordinary or normal. In a tooth, abnormality in size or shape.

APEX—The end of the tooth root.

APEXIFICATION—An endodontic treatment designed to induce root development or apical closure of the root by hard tissue deposition.

APEXIGENESIS—An endodontic treatment that allows the completion of root development by Hertwig's sheath.

APICAL —Toward the apex (tip) of the root.

APICAL FORAMEN—Refers to the aperture in the root of a tooth, allowing for passage of blood vessels and nerves.

APICOECTOMY—The surgical removal of the apex or apical portion of a root.

ARCH—The dentition together with the alveolar ridge, regardless of the number of teeth present.

ARTIFACT—A blemish or image in the radiograph that is not part of the true radiographic image of the subject.

ATTACHED GINGIVA—That part of the gingival tissue that is firm, resilient, and tightly bound to the underlying periosteum of the alveolar bone. Located along the gingival border.

ATTENUATION—The process by which a beam of radiation is reduced in energy when passing through some material.

ATTRITION—The loss of tooth substance resulting from friction caused by occlusal forces.

AVULSION—The complete displacement of a tooth from its alveolar socket caused by trauma.

AXIS—The central lengthwise line through the crown and the root.

BIFURCATION—Division into two parts or branches, as any two roots of a tooth.

BITE—The act of incision of a morsel of food. The amount of pressure developed in closing the jaws. Lay term for occlusion.

BONE, ALVEOLAR—See Alveolar Bone.

BONE DENSITY—The degree of compactness of bone, influenced by the concentration of inorganic salts.

BONE, INTERPROXIMAL—The bone between the teeth.

BONE LAMELLA—Bone that has the appearance of layers, which is produced by inactive periods alternating with new bone formation.

BONE RAREFACTION—Decreased density in bone.

BORDER—The margin or edge of a structure.

BUCCAL—Referring to or adjacent to the cheek.

CALCIFIC METAMORPHOSIS—Partial or complete obliteration of the pulp chamber and canal, commonly secondary to trauma.

CALCULUS—Mineralized deposits on teeth.

CALLUS—The tissue adjacent to the broken fragments of a bone that is involved in its repair.

CANAL—The part of the root that contains the pulp tissue and is surrounded by dentin.

CANINE—In dentistry, one of the four, long, pointed teeth at the fourth position of each quadrant.

CARIES—Progressive destruction of tooth substance by demineralization of enamel or cementum.

CAVITY—The loss of tooth structure caused by dental caries.

CELLULITIS—A diffuse inflammatory process that spreads along fascial planes and through tissue spaces.

CEMENTOENAMEL JUNCTION—The cervical line marking the junction of enamel and cementum of a tooth.

CEMENTUM—The calcified connective tissue that covers the tooth root and forms an attachment with the periodontal ligament.

CERVICAL—Referring to the neck region of the tooth, where the crown meets the root.

CINGULUM—The convex protruberance at the cervical third of incisor crowns on the lingual or palatal aspects.

CLEFT LIP—A congenital anomaly of the face caused by the failure of fusion between the embryonic maxillary and medial nasal processes.

CLEFT PALATE—A congenital anomaly of the oral cavity caused by the failure of fusion between the embryonic palatal shelves.

CLINICAL CROWN—That part of the tooth coronal to the crestal bone. If there has been bone recession, some root structure may be included.

CLINICAL ROOT—That part of the tooth apical to the crestal bone. If there has been bone recession, the entire root length may not be included.

COLLIMATION—The peripheral containment of the useful x-ray beam.

COLLIMATOR—A device or system made of absorbing material designed to limit the dimensions and direction of an x-ray beam.

CONCRESCENCE—The fusion after eruption of two teeth at their cemental surfaces.

CONDENSATION—The insertion and compression of dental materials into a cavity or root canal.

CONDENSATION, LATERAL—Compression of dental material against the side of the root canal wall.

CONDENSATION, VERTICAL—Compression of dental material into a root canal toward the apex.

CONE—An accessory device on a dental x-ray machine designed to indicate the direction of the central axis of its beam and the source-to-film distance.

CONTRAST, RADIOGRAPHIC—The differences in radiographic image density produced by structural composition of the subject or by varying amounts of radiation.

CONTRAST, LONG SCALE—An increased number of gray tones that range from black to white on a radiograph.

CONTRAST, SHORT SCALE—A minimum number of grays that range from black to white on a radiograph.

CORONAL—Referring to the tooth crown.

CROWN—That part of the tooth covered by enamel.

CURETTAGE—The scraping or cleaning of the walls of a cavity or surface by means of a curet.

CUSP—A pointed or rounded area on or near the masticating surface of a tooth.

CUSPID—A synonym for the canine tooth.

CYST—A space in bone or soft tissue often lined by epithelium.

CYST, DENTIGEROUS—A fluid-filled sac that surrounds the crown of an unerupted tooth.

CYST, ERUPTION—A dentigerous cyst that causes a bulging of the alveolar ridge.

DECIDUOUS— Referring to the first set of teeth, which will be shed and replaced by the permanent teeth.

DEFINITION—Referring to the sharpness, distinctness, or outline of a radiographic image. Also, the property of projected images relating to their sharpness, distinctness, or clarity of outline. Penumbra width is a measure of definition.

DENS IN DENTE—A tooth anomaly characterized by invagination of the enamel, giving the radiographic appearance of a tooth inside another tooth.

DENSITY, RADIOGRAPHIC—The degree of darkening of x-ray film.

DENTICLE—A calcified body that is composed of irregular dentin or an ectopic calcification of pulp tissue.

DENTIN—The portion of the tooth that lies subjacent to the enamel and cementum, consisting of an organic matrix on which mineral salts are deposited.

DENTIN, SECONDARY—Dentin formed or deposited on the walls of pulp chambers and root canals after tooth formation.

DENTINOCEMENTAL JUNCTION—The line where the dentin and cementum meet.

DENTINOGENESIS IMPERFECTA—A defect in dentin formation.

DENTITION, MIXED—A combination of permanent and deciduous teeth.

DENTITION, PRIMARY—Deciduous dentition.

DIASTEMA—A space between two adjacent teeth usually in the same dental arch. *Anterior diastema* refers to the space between the maxillary central incisors.

DILACERATION—Severe angular distortion of the tooth form.

DISTAL—That portion of the tooth farthest from the midpoint of the dental arch.

DOSIMETRY—The determination of the radiation exposure during a given time.

DYSPLASIA—Developmental abnormalities.

DYSPLASIA, DENTINAL—A genetic disturbance of the dentin characterized by early calcification of the pulp chambers and root canals and by root resorption.

ECTOPIC—Arising from an abnormal site or tissue.

EDENTULOUS—Without teeth.

ELONGATION—A radiographic image of a subject that is longer than the actual subject, because of beam angulation.

EMBEDDED—A tooth, root tip, or foreign body that is covered in bone.

ENAMEL—The hardest substance in the body. It covers the crown of the tooth down to approximately the floor of the gingival sulcus.

ENDODONTIC—Referring to the dental pulp.

ENDODONTICS—The branch of dentistry concerned with the diagnosis and treatment of diseases of the dental pulp.

ENDODONTIC SYSTEM—The pulpal component of the tooth. It is comprised of both the pulp chamber and the root canal.

ENDODONTIC TECHNIQUES—Procedures used to treat the pulp tissue.

ENDODONTOLOGY—That part of dentistry that studies treatment and diseases of the dental pulp and their sequelae.

EPULIS, ACANTHOMATOUS—A gingival growth.

EROSION—A wearing of tooth substance by mechanical or chemical means.

ERUPTION—The movement of a tooth from within its follicle into the oral cavity.

ERUPTION CYST—A dentigerous cyst that causes a bulging of the overlying alveolar ridge.

ERUPTION, ECTOPIC—Abnormal direction of tooth eruption.

EXFOLIATION—The physiologic loss of a tooth.

EXODONTICS—The practice of tooth extraction.

EXOTOSIS—A bony growth projecting from a bony surface.

EXTRACORONAL—Pertaining to that which is outside or external to the coronal portion of a tooth.

EXTRAORAL—Outside of the mouth, referring to radiographic film placement.

EXTRUSION—Movement of teeth beyond their natural occlusal placement.

FACET—A flattened wear pattern on a tooth.

FAUCES—The soft tissue archway between the pharyngeal and oral cavities.

FIBROMA—A benign tumor composed mostly of fibrous connective tissue.

FIBROMA, PERIPHERAL ODONTOGENIC—A fibrous, connective-tissue tumor associated with gingival margin and believed to originate from the periodontium. Often contains areas of calcification; fibromatous epulis.

FIBROSARCOMA—A malignant mesenchymal tumor; the main cell type is a fibroblast.

FILE, ROOT CANAL—A small metal instrument designed to clean and shape the root canal.

FILLING—A material used to fill a space. A lay term for restoration.

FILLING, RETROGRADE—A filling material placed in the apical portion of a tooth root to seal the apical portion of the root canal.

FILLING, ROOT CANAL—Material placed in the root canal system to seal the space previously occupied by the dental pulp.

FILM FAULT, DYSCHROIC FOG—A fogging of the radiograph characterized by the appearance of a green surface when the film is seen by reflected light, usually caused by an exhaustion of the fixing solution or incomplete fixation.

FILM FAULT, RETICULATION—A network of corrugations produced by an excessive difference in temperature between any two of the three darkroom solutions.

FILM SPEED—The amount of exposure to light or x-rays required to produce a given image density. Films are classified in groups designated by letters, starting with A. Between each group there is a twofold increase in film speed. The faster film requires less x-ray exposure to produce an image.

FISSURE—A deep groove or cleft.

FISTULA—An abnormal tract connecting two body surfaces or organs or leading from a pathologic or natural internal cavity to the surface. The tract may be lined with epithelium.

FIXATION, RADIOGRAPHIC—The chemical removal of all the undeveloped salts of the film emulsion so that only the developed or reduced silver will remain as the permanent image.

FOG, RADIATION—Film darkening caused by radiation from sources other than the primary beam.

FORAMEN—A natural opening in the bone or other structure.

FURCATION—The area where roots branch from the neck in multirooted teeth.

FUSION—Two teeth united during development by the union of their tooth germs.

GEMINATION—Teeth with bifid crowns and shared root canals resulting from incomplete division of the enamel organ during tooth development; also called tooth twins.

GINGIVA—That part of the gum tissue and mucous membrane that immediately surrounds the tooth and is continuous with its periodontal ligament and with the mucosal tissues of the mouth.

GINGIVAL SULCUS—The extremely important small moat that surrounds each tooth and is formed by the gingiva (gum). This small area is the most biologically active in the mouth.

GINGIVITIS—The inflammation of the gingiva.

GUBERNACULUM—The structure that guides the erupting permanent tooth into position.

GUTTA PERCHA—A radiopaque rubberlike material that is used for filling root canals.

HERTWIG'S FIBERS—The fibers located in the apical portion of the periodontal membrane, responsible for root lengthening and apical closure.

HORN, PULP—A small projection of vital pulp tissue directly under a cusp.

HYPERCEMENTOSIS—Excessive formation of cementum on the roots of one or more teeth.

HYPOCALCIFICATION—Reduced calcification of the enamel.

HYPOPLASIA—Defective or incomplete development.

HYPOPLASIA, ENAMEL, CHRONOLOGIC—A prenatal or postnatal systemic hypoplasia that affects amelogenesis occurring at the time of a systemic disorder. The affected areas form circumferential bands on the teeth.

IMPACTION—A tooth unable to erupt normally.

IMPLANT—A device that is surgically inserted into or onto the jawbone.

IMPULSE—A surge of electric current of short time span.

INCISAL EDGE—The biting surface of an anterior tooth.

INCISORS—The anterior teeth that are used for cutting.

INTERRADICULAR—Relating to the area between the roots of a multirooted tooth.

INTERPROXIMAL—Between the proximal surfaces of neighboring teeth.

INTRAORAL—Inside the mouth.

INVERSE-SQUARE LAW—A principle that defines the strength of radiation and states that radiation from a source varies inversely with the square of the distance.

INVERSION—Upside down.

JUNCTION, CEMENTOENAMEL—The junction of the enamel of the crown and the cementum of the root of a tooth; also commonly referred to as the neck.

JUNCTION, DENTINOCEMENTAL—The area where the cementum and dentin meet in a tooth.

JUNCTION, DENTINOENAMEL—The area where the enamel and dentin meet in a tooth.

JUVENILE PERIODONTITIS—A distinct form of aggressive periodontal disease, either localized or generalized, in prepubertal patients.

LABIAL—The lip side (of an anterior tooth).

LAMINA DURA—A radiographic term that refers to the thin plate of compact bone adjacent to the periodontal membrane.

LATITUDE—The range between minimum and maximum film exposures to radiation that yield useful images discernible under normal viewing conditions.

LENTULA—Small metal spiral used to introduce material into the root canal.

LIGAMENT, PERIODONTAL—The attachment of the tooth to the alveolus, consisting of fibrous connective tissue, vessels, and nerves.

LINGUAL—The side of the mandibular teeth next to the tongue.

LINGUOVERSION—A malocclusion in which the crown is displaced or tipped toward the tongue or lingual aspect.

LOWER—Referring to the mandible or mandibular teeth.

LUCENCY—See Radiolucency.

LUMEN—The space within a tubular structure, used here when referring to the interior of the root canal.

MACRODONTIA—Large teeth.

MALOCCLUSION—A problematic deviation in the position of the teeth.

MANDIBLE—Lower jaw.

MANDIBULAR—Pertaining to the lower jaw.

MAXILLA—A skull bone forming part of the upper jaw.

MAXILLARY—Pertaining to the upper jaw.

MESIAL—Toward the center point of the dental arch.

MESIODENS—A supernumerary tooth at the midline of the dental arch, either erupted or unerupted.

MICRODONTIA—Abnormally small teeth.

MICRODONT—An abnormally small tooth, often conical or peg-shaped; may or may not be supernumerary.

MOLAR—A tooth adapted for grinding by having a broad and ridged occlusal surface.

OBTURATION—The filling and sealing of a root canal.

OCCLUSAL—Pertaining to the surface of the tooth that faces or contacts a tooth in the opposing arch.

OCCLUSION—The state of contact between the upper and lower teeth when the mouth is closed.

ODONTOBLASTS—The cells that form the dentin of the tooth.

ODONTODYSPLASIA—A developmental anomaly characterized by deficient enamel and dentin development. Also known as "ghost teeth."

ODONTOGENESIS—The process of tooth formation.

ODONTOMA—A tumor of hard tissue containing enamel, dentin, cementum, and pulp tissue arranged in the form of teeth; when small, also called toothettes or toothlets.

OLIGODONTIA—Having only a few teeth.

ORONASAL—Pertaining to the mouth and nose or nasal sinus.

OSSEOUS INDUCTION—A technique whereby bone-grafting material is placed in a periodontal or infrabony pocket. Bone formation in the pocket is induced by the presence of the material.

OSTEITIS—Inflammation of the bone.

OSTEITIS, CONDENSING—A chronic inflammation of the bone resulting in abnormally dense bone.

OSTEOBLAST—The cell involved with growth and development of bone.

OSTEOCEMENTUM—A hard, bonelike cementum deposited after root formation is completed.

OSTEOCLAST—A large, multinucleated giant cell involved in the resorption of bone.

OSTEOMA—A benign neoplasm of bone or bone tissue.

OSTEOMYELITIS—An inflammation of the bone, marrow, and endosteum.

PALATAL—The side of the maxillary teeth next to the palate.

PERIDENS—A supernumerary tooth located buccal, lingual, or distal to a normal tooth.

PARALLAX—The apparent change in position of an object when viewed from two different angles. When two or more radiographs are made from slightly different positions, the direction and amount of shift of the object from one radiograph to the next helps to determine its true position.

PERIAPICAL—The area around the apex of the root.

PERIODONTAL—Anything that is situated or occurs around a tooth.

PERIODONTITIS—Inflammation of the periodontium.

PERIODONTITIS, JUVENILE—Periodontitis present in prepubertal or pubertal patients.

PERIODONTIUM—The tissues that surround and support the teeth.

PERIODONTOLOGY—The scientific study of the periodontium and periodontal diseases.

PERIOSTEUM—The connective tissue that covers the bone.

PLANE, AXIAL—A hypothetical surface parallel to the long axis of an object.

PLANE, HORIZONTAL—A plane that is parallel to the horizon.

PLANE, VERTICAL— A plane that is perpendicular to the horizon.

POCKET—A diseased periodontal attachment in which there is a space next to the tooth bordered by either gingiva or bone.

POCKET, INFRABONY—A periodontal pocket in bone where the base is apical to the alveolar crest.

PONTIC—An artificial tooth on a fixed partial denture; usually occupies the space previously occupied by the natural crown.

PREMOLAR—The teeth between the canine teeth and molars.

PROXIMAL—The mesial or distal surface of a tooth.

PULP—The soft tissue that occupies the central portion of teeth, which is composed of blood vessels, nerves, and other cells, including odontoblasts that form dentin.

PULP CHAMBER—The portion of the endodontic system located coronal to the cementoenamel junction.

PULPAL—Relating to the pulp.

PULPECTOMY—The complete removal of pulp from the pulp chamber and root canal.

PULPITIS—Inflammation of the pulp.

PULPOTOMY—Surgical exposure and amputation of the dental pulp.

RADIATION, SCATTERED, OR BACKSCATTER—Radiation of altered direction that may include secondary radiation.

RADIATION, SECONDARY—New radiation created by primary radiation acting on or passing through matter.

RADICULAR—Pertaining to the root.

RADIOGRAPH—An image produced on a radiation-sensitive film emulsion by exposure to ionizing radiation directed through an area.

RADIOLUCENCY—A radiographic representation of decreased tissue density, relative to surrounding tissues.

RADIOLUCENT—Allowing the passage of radiation with relatively less change by absorption. The radiographic image will be in shades of gray to black.

RADIOPACITY—Attenuation of an x-ray beam as it passes through a subject.

RADIOPAQUE—Allowing passage of radiation after attenuation of the beam by absorption. The radiographic image will range from gray to white.

RAREFACTION—A decreased density of bone.

RECAPITULATION—The sequential reintroduction and reuse of each instrument in cleaning and shaping the root canal during endodontic treatment.

RECESSION, BONE—Loss of the alveolar bone level in an apical direction.

RECTIFICATION—Conversion of electric current from alternating to direct current.

RECTIFIER—A device used for converting an alternating current to direct current, or limiting the flow of current to one direction.

RECTIFIER, FULL-WAVE—A device that rectifies the entire wave of an alternating current in an x-ray machine.

RECTIFIER, HALF-WAVE—A device used in rectifying half of the sine wave of an alternating current in x-ray machines.

RESIN, COMPOSITE—A resin used for restorations to which inorganic fillers, such as glass or quartz beads, have been added.

RESORPTION—Physiologic or pathologic loss of bone or root substance.

RESORPTION, HORIZONTAL—A pattern of bone recession in which the alveolar crestal bone margins between the teeth remain level.

RESORPTION, INTERNAL—Root resorption occurring inside the pulp cavity.

RESORPTION, VERTICAL—A pattern of bone loss in which the alveolar bone next to a tooth is lost without simultaneous crestal bone loss.

RETROFILL—Obturation of the apex of a tooth root by surgical access.

RIDGE, ALVEOLAR—The bony ridge of the upper and lower jaws that contains the alveoli.

RIDGE, CRESTAL—The highest continuous surface of the ridge but not neccessarily the center of the ridge.

ROOT—The part of the tooth covered by cementum.

ROOT CANAL—The central cavity of the root through which the vital structures (nerves, arterioles, venules) pass into the pulp chamber.

ROOT PLANING—The smoothing of roughened root surfaces by the use of scalers and curettes.

ROSTROVERSION—A malocclusion in which the crown is displaced or tipped mesially, toward the rostral aspect.

RUGA—The irregular ridges in the mucous membrane covering the anterior part of the hard palate.

SCLEROSIS—Increased calcification and radiopacity.

SEQUESTRUM—A piece of devitalized bone that has become separated from vital bone.

SILHOUETTE SIGN OR EFFECT—When two structures of the same radiopacity are in contact, their contacting margins cannot be distinguished radiographically so they appear to be a single silhouette or image.

SPICULE—A small, thin, needle-shaped object.

STOMATITIS—Inflammation of the soft tissues of the mouth.

SUBLUXATION—Incomplete dislocation of a tooth from its normal position.

SUMMATION—Radiographic image that appears either relatively more radiolucent or more radiopaque, resulting from the overlapping or superimposition of structures on a film.

SUPERERUPTION—The condition in which the teeth have erupted beyond the normal occlusal plane.

SUPERNUMERARY—Extra.

TUBEHEAD—The structure of the x-ray machine that is handled when aligning the x-ray beam with the subject and film. It contains the x-ray tube and transformers in an oil bath. A device for collimating the emergent beam is attached.

UPPER—Referring to the maxilla or maxillary teeth.

VIBRISSAE—Long, coarse hairs, such as those occurring about the muzzle of an animal.

X-RAY—A type of electromagnetic radiation, with wavelengths between 1000 A and 0.0001 A.

ZINC OXIDE-EUGENOL—Two substances that are mixed together as a paste that later hardens; used for root canal fillings and restorative work.

BIBLIOGRAPHY

Carranza FA: *Glickman's Clinical Periodontology*, ed 7. Philadelphia, WB Saunders Co, 1990.

Cohen S, Burns RC: *Pathways of the Pulp*, ed 6. Philadelphia, CV Mosby Co, 1994.

Evans HE, Christensen GC: *Miller's Anatomy of the Dog*, ed 2. Philadelphia, WB Saunders Co, 1979.

Raxmus TF, Williamson GF: *Current Oral and Maxillofacial Imaging*. Philadelphia, WB Saunders Co, 1996.

Zwemer TJ: *Boucher's Clinical Dental Terminology*, ed 4. Philadelphia, CV Mosby Co, 1993.

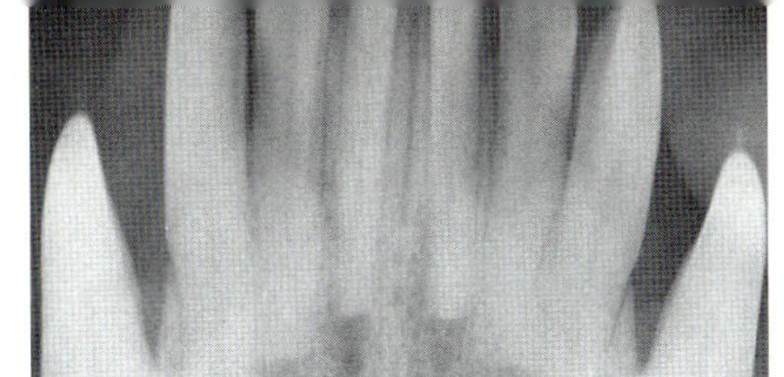

INDEX

A

Abscess, definition of, 231 (*See also* Phoenix abscess)
Acanthomatous, 197, 231
Acquired changes, 153–169
Adaptations, 28
Aging process, 72–73, 75–90, 106–107, 135–136
Alternating current (AC), 7
rectification of, 237 (*See also* Self-rectification)
Alveolar
bone, 68, 104, 232
crest, 104, 106, 228
definition of, 231
mucosa, 231
process, 231
ridge, 238
socket, 68, 222
Alveolus
definition of, 231
fractures of, 179–180, 231
sclerosis of, 199, 202
unhealed, 199
Amalgam debris, 62, 148–149 (*See also* Fillings)
Ameloblastomas, 187–188, 231
Ampere, 7, 231
Anachoresis, 130
Anatomy, dental, 68–90
Anepulis, 185
Anesthesia, 14
Angle of beam
adjusting, 7–8, 19, 156, 170–171
caudorostral, 35
horizontal, 16–17, 157
parallax effect, 237
rostrocaudal, 35
Ankylosis, 155, 157, 164, 224–225
Anomalies, 91–103, 99, 129–130, 231

deformed teeth, 134
dens in dente, 233
dens invaginatus, 233
evaluation of, 130
fossa, 235
Apex, 69, 231 (*See also* Root)
apexigenesis, definition of, 231
defects at, 114
malformed, 100–101
open, 124
periapical, definition of, 237
Apexification, 140, 231
Hertwig's fibers, 235
Apical
delta, 72, 125, 137–138
foramen (*See* Foramina)
superimposition, 55
Apicoectomy, 140, 149, 231
Arcades, viewing, 29, 55
Arch, adjusting beam along, 15, 231
Asymmetry, 66
Attached gingiva (*See* Gingiva)
Avulsion, 170, 172, 178, 232
Axis, 15–16, 232

B

Backscatter (*See* Radiation, scattered)
Beam angle (*See* Angle of beam)
Bifid (*See* Crowns)
Bifurcation, 99, 232 (*See also* Crowns)
Bilateral symmetry (*See* Symmetry)
Bisecting angle technique, 15–21, 29
problems with, 45, 50–52, 144
Bite blocks, 13–14
Black, G.V. classification system (*See* G.V. Black classification system)
Bone
cleft in, 113
defects, 114–117, 122
density, 106, 233

fractures (*See* Fractures)
interproximal, 232
plating, 196
rarefaction, 122, 125, 133, 222, 237
reparation, 126
resorption (*See* Resorptions)
Bone loss
angular, 116, 120
at the furcation, 104–106, 111, 113, 118, 135, 223–225
horizontal, 117–118, 135, 223–225
periodontal, 183
radicular, 116, 238
at roots, 116–118
vertical, 114, 116–118, 228
Buccal, 21, 232
opposite, 20–22
plates, 106
Buccal object rule (*See* Tube-shift technique)

C

Calcification, 71, 126, 133
of canal, 125
endodontic, 126
Calcium hydroxide layer, 140–141, 157
Calculus, 166, 201, 230, 232
overhangs, 117, 162, 195
Canals (*See also* Root canals)
accessory, 125
apical, 138
atretic, 126, 136, 221
drainage of, 134
fistulous, 189, 200
gubernacular, 72
incisive, 68
infraorbital, 68
lateral, 125, 144, 220

mandibular, 68–69, 112, 116
nutrient, 166
obturation of, 125, 143, 149
Canine teeth, 69–72, 76–77, 79–80,
    83–84, 87–88, 232
    cuspids, 233
    missing, 92–93
    projections for, 30, 32–33, 36–37,
        40–44
    retained, 95, 97
    rostroversion of, 96
Carcinomas, 187, 190–191
Caries, 106, 132, 153, 156, 222, 232
    cavity, definition of, 232
Carrier transport system, 11
Cats
    bone recession in, 111–113,
        115–116
    dental formula for, 69
    eruption of teeth in, 72
    normal dentition of, 69, 73–75,
        83–90
    positioning, 24–25, 30–31, 39–44
Cementoenamel junction, 68–69,
    108–109, 232
Cementum, 68, 106, 232
    osteocementum, definition of, 236
Cerclage wires, 179–181
Cervical
    burnout, 45, 53, 114, 156, 195,
        224
    line, 68–69
Chipped teeth, 211, 224
Chronologic hypoplasia (*See*
    Hypoplasia)
Cingulum, 154, 232
Clark's Rule (*See* Tube-shift technique)
Cleft lip and palate, 102–103, 232
Collimation, 7, 232
Concrescence, 99, 232
Condensation, 146–147, 232
Condyloid process, 174
Cone (*See* X-ray cone)
Contrast, radiographic, 28, 232
    long scale versus short scale, 28,
        233
    problems with, 46–47
Conversions, 28
Coronal cusp, 135
Coronoid process, 174
Cortical bone, 68
    bowing, 184

thinning, 121, 184, 221
Crater defects, 116
    formation of, 105–106
Crestal bone, 68, 104, 106, 222, 230
    defects in, 156
    jagged, 108–109, 112, 228
    loss of, 111–113
    proliferation of, 196
Crests, 68
Crowns, 68–69, 233
    bifid, 99, 101, 214, 227
    bifurcated, 99
    cleft, 99
    defects in, 158–159
    dilaceration of, 101
Crown-to-root ratio, 109
Cusps, 68, 159, 233
Cystic
    degeneration, 95
    tumors, 184
Cysts
    dentigerous, 91, 233
    eruption, 233
    periapical, 233
    periodontal, 233

D

Deciduous teeth, 69, 71–72, 221, 233
Defects (*See* Fractures; Resorptions)
Delayed eruption (*See* Eruption)
Demineralization process, 153
Deformed teeth, 134
Dens in dente, 233
Dens invaginatus, 233
Density
    bony, 106
    radiographic, 233
Dental care (*See also* Plaque)
    abrasion, 231
    crevicular brushing technique, 233
    erosion, 234
    facet wear, 158–159, 234
Dental film (*See* Film)
Dental follicles, 72
Dental x-ray machines (*See* X-ray
    machines)
Denticles, 189, 233
Dentigerous cysts (*See* Cysts)
Dentin, 68, 126–127, 153, 233
    borders of, 232

bridges in, 141–142, 151
dentinocemental junction, 233
dysplasia of, 99–100
reparative, 159
secondary, 233
walls of, 69, 124, 126, 136, 194
Dentinogenesis imperfecta, 99, 233
Dentition (*See also* Anomalies;
    Avulsion; Cats; Dogs;
    Edentulous; Eruption;
    Exfoliation; Extrusions;
    Interdental spaces; Intrusion)
    anodontia, definition of, 231
    bite, definition of, 232
    caudal, obscured, 25
    chipped, 211, 224
    crowded, 91, 95, 97–98, 113, 205
    deciduous, 69, 71–72, 221, 233
    distal, 16, 29, 233
    embedded, 95, 234
    integrity of, 175–176
    lateral, 51–53
    loose, 170–172, 210, 216, 223,
        228–229
    macrodontia, 235
    malformed, 99–101
    malpositioned, 96–97
    microdontia, 92–93, 98–99, 220,
        236
    misaligned, 97
    missing, 91–94, 170, 206, 214,
        219, 223, 226–227
    odontogenesis, definition of, 237
    osteoblasts, definition of, 237
    permanent, 69, 72
    pitted, 154
    remnants of, 225
    rostral, 51–52
    succedent, 72
    supernumerary, 93, 97–99, 238
    unstable, 105–106
Developing (*See* Processing)
Developmental groove, 69, 222
Developmental phase, 71–72, 75–90
    problems during, 91–99
Diagnosis and treatment sequence, 65
Dilacerations, 91, 96, 99–101, 233
    (*See also* Crowns)
Diphyodonts, 69
Disparities, 102
Displacements, 171, 177, 222

lingual, 97
rostral, 95
Distal, 16, 233
roots, 68, 137, 218
teeth, 29
Distilled water for rinsing, 62
Distortion, minimizing, 15
Dogs
bone recession in, 111–114
dental formula for, 69
eruption of teeth in, 72
normal dentition of, 69–71, 73, 75–83
positioning, 24–25, 30–39
Dorsal recumbency (*See* Positions)
Drainage, coronal, 134
Dysplasia
dentinal, definition of, 234
fibrous, 202

E

Ectopic pathways, 91, 96–97, 99, 226, 234 (*See also* Eruption)
Edentulous, 217, 234
anodontia, 231
oligodontia, 237
spaces, 185–187
Elongation, 15–16, 18, 45, 50–52, 234
Enamel, 156, 234
amelogenesis imperfecta, 231
cementoenamel junction, 68–69, 108–109, 233
invaginations of, 134
Endodontic
anatomy, 124–125
calcification, 126, 133
cement, 219
filling material, 72
involvement, 35, 117, 165, 229
leakage, 149
lesions, 218
Endodontic disease, 221
causes of, 127–129
clinical signs of, 124
interpretation of, 124–152
radiographic evaluation of treatment, 137–151
radiographic signs, 125–127, 131–136

techniques used, 233
Endodontic–periodontal lesions, 128–129, 133–135, 143, 151, 220
Endodontic system, hidden, 124, 233
Endodontology, definition of, 233
Endosseous implants (*See* Implants)
Endotracheal tube, 60
Epulides, 185, 187, 196–198, 216, 229, 231, 234
fibromatous, 197–198
Eruption, 71–72, 234 (*See also* Cysts)
delayed, 91, 94–96, 227
ectopic, 91, 96–97, 234
impactions, 91, 235
supereruption, 105, 108, 118, 238
tooth buds, 91, 221
unerupted teeth, 186, 188–189
Exfoliation, 69, 72, 234
imminent, 161, 163
spontaneous, 101, 161
Exposure times, 24–25, 27–28
dosimetry, 233
errors in, 58–59
latitude in, 235
problems with, 46, 59
Extensions, 68
overextensions, 146–147
Extractions, 183, 185–186
delayed healing of, 199
exodontics, definition of, 234
healed sites of, 200–201
incomplete, 195
indicated, 151, 163, 165, 225, 229
multiple indicated, 166
requiring care, 120–121, 177, 228
unhealed sites of, 187, 195, 199–200, 229
Extraoral radiographs
exposure techniques, 24–26
imaging techniques, 23–26
limitations of, 23
problems with, 54–55
Extrusions, 105, 118, 139, 143, 225, 234
overextrusion, 145

F

Fibromas
ossifying, 196

peripheral odontogenic, 234
Fibrosarcomas, 187–188, 190, 234
Fibrosis, 201
Fibrous joints, 68
Files
carrier, 143
root canal, 137–139, 205, 208, 212, 218, 221, 225, 234
Fillings
amalgam, 169
fractured, 144, 148–149
insufficient, 145–147
root canal, definition of, 237
Film, 7–9, 15 (*See also* Exposure times; Processing)
badges, wearing, 14
bending slightly, 31–35, 38–39, 56
cassettes, 24
density, problems with, 46
distance (*See* Focal film distance)
errors in handling, 56–57
errors in positioning, 57–58
extraoral, definition of, 234
flexibility of, 27
intraoral, definition of, 235
labeling and storing, 12
mounts, 13, 47
opening too quickly, 48, 218
orientation of, 8–9, 28–29, 65–66
packets, 9
positioning, 14–15, 23–44
selecting, 8
speeds of, 8–10, 24, 27–28, 234
storage, 47, 64
ultra-speed, 8–10
Fissures, 154, 234
palatine, 68
Fistulation, 142, 146–147, 201, 235 (*See also* Oronasal fistulas)
draining, 209
Fixation, radiographic (*See* Processing)
Fixation devices, 175, 178–182, 196
radiolucent acrylic, 178–180, 182
Focal film distance, 7, 15, 24
adumbration, 231
attenuation, 232
Fog, radiation, 47, 234
Follicle (*See* Dental follicles)
Foramina
anterior, 68
apical, 124–125, 231
infraorbital, 68
mental, 68, 218, 224
Foreign bodies, 187, 192–193

Foreshortening, 15–16, 45, 51–53
Fractures, 105–106, 120, 131, 153–183, 208, 215, 223 (*See also* Fixation devices; Reduction; Root fractures)
    alveolar, 179–180, 231
    coronal, 136, 156, 170
    of cusps, 154
    iatrogenic, danger of, 120–121, 182–183
    mandibular, 111, 174, 180–183
    maxillary, 180–181
    pathologic, 176–177
    premaxillary, 182
    superficial, 131
Furcations, 68–69, 125, 220, 226, 230, 234
    exposures at, 210–212
    interradicular, 235
Fusion, 99, 102, 227, 234

G

Gemination, 99, 102, 234
    incomplete, 227
Gingiva, 104, 234
    attachments to, 107, 110, 170, 231
    growths on, 185, 196–198
    hiding angular defects, 114
    pockets in, 228
Gingivitis, 104, 185, 207, 216–217, 235
    chronic, 210
    juvenile onset, 112
Gingivostomatitis, 185
Glare, eliminating, 12
Grain size, 7–8
Gray, shades of (*See* Contrast, radiographic)
Gubernaculum, 226, 235
Gutta percha, 189, 235
    cone, 145–147
    exposed, 150
    plastic, 143–144, 146–147
G.V. Black classification system, 153–159, 225

H

Hanging clips, 10, 57, 63
Hazardous materials, 14

Hemimandible, 44
Hertwig's fibers, 235
Horn, pulp, 68–69, 235
Histopathologies, 184, 195–196
Hypercementosis, 99, 106–108, 235
Hyperparathyroidism, 186
Hyperplasia, 104
Hypocalcification, 99, 235
Hypomaturation, 99
Hypoplasia, 99–101, 235
    of the enamel, chronologic, 236

I

Imaging techniques
    extraoral, 23–26
    intraoral, 27–44
    principles of interpretation, 65–68
Impactions, 91, 235
Implants, 178, 193, 236 (*See also* Replantations)
Implements, fractured, 150
    drills, 139
    files, 144, 148–149
Impulse, 7, 235
Incisors, 69–72, 76–79, 83–86, 235
    missing, 92–93
    projections for, 30–31, 36, 39, 42
    supernumerary, 97–98
Induction, osseous, 121–122
Inflammation, 122
    cellulitis, definition of, 232
    mucosal, 187
    stomatitis, definition of, 238
Infrabony pockets, 104–106, 115, 121, 133, 166, 229–230
    classifying, 105–106
    delineating, 106
Integrity (*See* Dentition; Lamina dura)
Intensifying screens, 8, 23–24
Interdental spaces, 53
    diastema, 233
    interproximal, 16, 37, 43, 53
Interlock, adverse, 102
Interpreting radiographic images, 65–67
Interproximal
    bone, 43
    defects, 154
    recession, 110
Interradicular bone, 69

loss of, 116
Intraoral radiographs
    exposure techniques, 27–28
    imaging techniques, 27–44
    problems with, 45–48, 53
Intrusion, 170, 172, 177
Invagination (*See* Enamel)
Inverse-square law, 28, 235

J

J-hook, 100–101
J-shaped lesion, 128

L

Lamellated periosteal reaction, 184
Lamina dura, 68–69, 133, 235
    integrity of, 66, 104, 125–126, 220
    loss of, 110
    obscured, 73, 106
    regional thickening of, 126
    resorption of, 186
Landmarks, anatomic, 68, 223, 229
Lateral
    canthus, 34, 41
    luxation, 172
    wall, perforation of, 150
Ledging, 139
Lesions, 23, 153, 165–166, 222
    aggressive versus nonaggressive, 184–185, 188–191, 198
    associated with extraction, 199–201
    characterizing, 66–67, 105, 128–129, 184–187
    combined, 116–119, 128, 135
    extracting, 121
    jaw, 184
    mixed, 153
    neck, 205, 211–212
    occult, 124
    osseous, 184–203
Ligaments (*See* Periodontal)
Lingual, 20–22, 235
    displacement, 97
    plates, 106
Lumen (*See* Root canals, diameter of)
Luxation, 172, 177, 180
    extrusive, 172
    subluxation, 171, 239

Lysis, 67, 142, 184, 188–189, 229
  false, 62

M

Malocclusions, 91–93, 96–97, 227, 235
  secondary, 103
Mandibles, 68, 235
  anatomy of, 174
  bowed, 188–189
  canine teeth in, 30, 36–37, 42–44, 76–77, 83–84, 87
  edentulous, 201–202
  extraoral oblique view of, 73, 75
  incisors in, 30, 36, 42, 76–77, 83–84
  molars in, 15, 30, 38–39, 43–44
  oblique lateral view of, 25
  premolars in, 15, 30, 38, 43–44, 82, 89
  rostral view of, 71, 74
  symphysis of, 68
  tumors of, 188
Material safety data sheets (MSDS), 14
Maturational phase, 72–73
Maxilla, 68, 235
  canine teeth in, 30, 32–33, 40–41, 70, 88
  extraoral oblique view of, 70, 74
  feline, 17, 73
  fourth premolar in, 17, 19, 34–35
  incisors in, 30–31, 39, 70, 78–79, 85–86
  molars in, 31, 34, 70
  oblique lateral view of, 24
  premolars in, 31, 33–34, 40–41, 70, 81, 90
Mental foramina (*See* Foramina)
Mesially tipped canine teeth, 91
Mesiodens, 91, 94–95, 99, 228, 236
Metaplasia, osseous, 201
Microdonts, 92–93, 98–99, 220, 236
Mineralization, 67, 104, 170
  disturbances to, 99–101
Mobility (*See* Dentition, loose)
Molars, 69–72, 83, 236
  projections for, 30–31, 34, 38–39, 43–44

Monitoring treatment measures, 137–141
  obturation, 140
  use of instruments, 138–139
Mouth gags, metal, 13–14

N

Nasal (*See also* Oronasal fistulas)
  discharge, chronic, 210
  septum, 68
  sinus, 68, 105, 161
Necrosis, 100–101, 127 (*See also* Pulp)
Neoplasias, ruling out, 114
Nidus, 160, 200

O

Obturation, 125, 140, 236
  inadequate, 150
  overobturation, 139
  reobturation, 145–147
  retrofill, 237
Occlusion, 105, 175, 236
  attrition, 232
  surfaces of, 53, 153
Odontoblasts, 126, 236
Odontodysplasia, 99–100, 188, 236
Odontogenic fibromas (*See* Fibromas)
Odontomas, 91, 94–95, 187, 189, 236
Oral anatomy and function (*See also* Dentition; Edentulous; Orientation)
  alignment, 231
  canals, 232
  fauces, 234
  foramina, definition of, 235
  junctions, 235
  palate, 68
  periodontal, definition of, 237
  ruga, 237
Orientation (*See also* Film; Planes; Positions; Projections; Views)
  apical, definition of, 231
  buccal, definition of, 232
  cervical, definition of, 232
  coronal, definition of, 233
  extracoronal, definition of, 234
  interproximal, 235
  inversion, 235

  labial, 235
  mesial, definition of, 235
  palatal, definition of, 236
  proximal, definition of, 237
Oronasal fistulas, 105, 119, 236
  increased risk of, 113, 115, 118, 223
Orthodontic conditions, 91
Osseous
  induction, 121–122
  lesions (*See* Lesions)
Ossification, 114, 196
Osteitis, 187, 237
  condensing, 126, 191, 223, 230, 236
Osteomas, 192, 237
Osteomyelitis, 187, 189, 236
  chronic, 201–202
  probability of, 151
Osteopathy, craniomandibular, 192
Osteosarcomas, 187, 191
Osteosclerosis, 125, 185, 187, 192
  idiopathic, 126

P

Palate, 68
Palatal
  fissures, 68, 191, 225
  suture line, 68
Palatine
  direction, 97
  fissures, 68, 191, 225
  roof, 35, 145, 205, 218, 220
  suture line, 68
Parallel positioning technique, 15–16, 23–26
  problems with, 52–53
Parotid gland, 222
Periapical
  cysts (*See* Cysts)
  lesions, false, 61, 68
  radiolucency, 100–101, 125–126, 130–133, 136, 140–143, 150, 218, 220–221
  rarefaction, 133, 135
  trauma, 139
Peridens, definition of, 236
Periodontal
  curettage, 233
  cysts (*See* Cysts)

ligaments, 68, 170–171, 223–224, 235

pocketing, 119, 216–217

root planing, 230, 237

spaces, 69, 104, 110, 117–119, 125–126, 131, 134

Periodontal disease, 104–123, 228

chronic, 176

evaluating treatment measures, 107–109

periodontology, 237

radiographic signs of, 109–110

Periodontal–endodontic lesions, 129, 135, 173, 228

Periodontitis

definition of, 236

juvenile-onset, definition of, 236

Periodontium, 197, 237

evaluating, 104–109

injuries to, 170–173

Periosteum, definition of, 236

Peripheral odontogenic fibroma (*See* Fibromas)

Periradicular

lesions, 125, 130

radiolucency, 133–134

Phoenix abscess, 130

Pitting, 154

Planes

axial, 236

horizontal, 7, 18, 236

vertical, 7, 18, 236

Plaque, 91, 104, 115

sclerotic, 199

Pockets, 104, 112, 23 (*See also* Gingiva; Infrabony pockets; Periodontal)

Pontic, 193, 236

Positioning patient, 24, 137–138

devices used, 13–14, 24–26

hints, 31, 41

Positions

dorsal recumbency, 30, 36–37, 42–43

lateral oblique, 25

lateral recumbency, 24–25, 31, 38–39, 43

ventral recumbency, 30–31, 34–35, 39–41

Precursors, 69

Preeruption, 206–207, 213–214

Premolars, 69–72, 75, 81–82, 89–90, 94, 237

embedded, 95

fourth maxillary, 17, 19, 34–35, 53, 68

missing, 92–93

partial gemination of, 101

projections for, 30–31, 33–34, 38, 40–41, 43–44

retained, 94–95

rotated, 96–97, 218

supernumerary, 97–98, 113, 218–220, 226

Probing, 114

Processing

automatic, 9–11, 47–49, 64

chairside, 10–11

disposing of solutions, 14

hand, 10–11, 47–48

problems with, 46–50, 61–64

rinsing and fixing, 11–12, 48, 61–63

Projection geometry, 15–22

Projections

distomesial, 35, 137

lateral, 25, 32–34, 37, 43–44, 54

mesiodistal, 34–35, 137, 143

oblique, 25, 32, 36–37, 40, 42, 70

Protective barriers, 14

Pulp, 68, 171, 237

assessing, 126

cloistered, 124

exposed, 127, 129, 131

horn, 68–69, 235

inflamed, 124

necrosis of, 124, 133, 172, 176, 228

obliterated, 101

pulpectomy, definition of, 237

radiolucent, 136

stone, 133

Pulp chambers, 68–69, 99, 124, 127, 144, 228, 237

unfilled, 143

Pulpitis, 105, 122, 125, 127, 176, 237

irreversible, 124

Pulpotomy, 140–141, 149, 182, 237

failed, 220, 238

Q

Quadrants, capturing, 24

Quality control, 47–48

R

Radiation (*See also* Fog)

safety devices, 13–14

scattered, 237

secondary, 14, 237

sources of, 14

Radiographs

artifacts in, definition of, 232

contrast of (*See* Contrast)

density of (*See* Density)

diagnostic value of, 15–16, 65–67

fixation of (*See* Processing)

imaging techniques, 23–44

normal anatomy for, 66–90

problems with images, 45–48

table-top, 24

viewing, 12, 65

Radiolucency, 67–68, 105, 237

of bite blocks, 13–14

of defects, 68, 106, 112, 114, 119–120, 122, 125, 128–129, 134, 188, 221, 229–230

false, 62

generalized, 186

halo, 168

multilocular, 187

Radiopacity, 66–68

definition of, 237

line, 105, 118, 223, 229

of metal mouth gags, 13–14

Rarefaction (*See* Bone; Periapical)

Recapitulation, 145–147

Recession, bony, 176–177, 237

advanced, 110–115

angular, 113–116

at furcation, 110

horizontal, 111–113

interproximal, 110

vertical, 110

Rectifier, 7, 238

full-wave versus half-wave, 7, 237

Reduction, 175

Reference points, 20–21

using distal teeth as, 29

Regression, bony, 106, 110

Reobturation, 145–147

Replantations, 172–173, 178

Resin, composite, 237

Resorptions, 69, 72, 93, 125, 132–133, 160–164, 166–168, 238

apical, 128, 132, 141–142, 218, 224–225
bony, 104
crestal, 156–157, 161–168, 224–225, 229
external, 114, 127–128, 151, 218, 224–225
at furcation, 104–105
horizontal, 104–105, 237
internal, 128, 135, 225, 237
intraosseous, 156–157, 160–161, 163–164, 166–168, 224
invasive, 155
of osteoclasts, definition of, 236
replacement, 223
of root, 119–120, 122, 146–147, 150–151, 225
supraosseous, 156–157, 160, 164–168, 225
vertical, 104–105, 238
Resorptive lesions, 153, 155, 157
false, 62, 225
Restorations, 157, 168–169
composite, 168
fillings, definition of, 234
lost, 146–147
pontic, 193, 236
resin, composite, 237
retrograde, 148–149
Retained deciduous teeth, 91, 93–95, 140–141, 223
Ridge, crestal, 238
Root canals, 68–69, 102, 237
burr for, 232
diameter of, 125, 131, 133–134, 228
filling (*See* Fillings)
Root fractures, 127–128, 228–229
healing, 127
horizontal, 128
oblique, 128
vertical, 128
Root fragments, 119, 135, 183, 187, 194–195, 201–202, 222–224, 229–230
callus, 232
Roots, 68, 238 (*See also* Radiolucency)
absorption of, 119
apex of, 16, 229
apical, 222
contour, irregular, 119
distal, 68, 137, 218

exposed, 108
mesial, 16–17, 19, 35, 68, 100–101
mesiobuccal, 17, 35, 137, 148–149, 218
mesiopalatal, 17, 137, 148–149
multiple, 68–69
palatal, 35, 145, 205, 218, 220
perforation of, 220
Rostroversion, 92–93, 96–97

S

Salvage technique, 149
Sclerosis, 66, 114, 122, 126, 199, 238
alveolar, 199, 202
bony, 114, 133, 186, 202
margin of, 121
of root fragments, 230
of the socket, 185
Screens (*See* Intensifying screens)
Self-rectification, 7
Sequestration, 120, 155
involucrum, 236
sequestrum, 120, 196, 238
Sialoliths, 222
Silhouette sign, 66–67, 238
SLOB rule, 20–22
Spatial orientation (*See* Film)
Spiculated periosteal reaction, 184, 238
Spiking, 198
Squamous cell carcinoma (*See* Carcinomas)
Static electricity, 48, 56, 218
Step-wedge test, 48–49
Subluxation, 171, 239
Summation effects, 67, 238
Sunburst periosteal reaction, 184, 191
Superimposition, 15–16, 19, 35, 41, 45, 53, 137, 221, 224–226
Supernumerary teeth, 97–99, 238
Surgery, evaluation after, 140–149
endodontic complications, 141–142, 206–207
frenectomy, 235
hemisected teeth, 149
pulpotomy follow-up, 140–141
retrograde therapy, 148–149
successful obturation, 143–144
unsuccessful obturation, 145–147

Swelling
facial, 205, 209, 222
at fourth premolar, 213
gingival, 211, 215, 224
mandibular, 208
maxillary, 209
Symmetry, bilateral, 65–66, 92–93
Symphysis, 44
bruising, 43
mandibular, 68

T

Technical errors, correcting, 45–64
Temporomandibular joint (TMJ), 180–181
Three-dimensional information (*See* Tube-shift technique)
Three-rooted teeth, 16–17, 19
Timing (*See* Exposure times)
Toothlets (*See* Denticles)
Trabeculation, bone, 66, 72, 106, 126, 202
Traumatic injury, 170–183, 224
radiographic interpretation of, 177
Tricks of the trade, 28
for positioning, 31, 41
Troubleshooting, 45–64
Tubehead, 7, 8, 17, 18, 20–23, 29, 31–43, 65, 66
definition of, 238
polarity of, 7
Tubercules, 68
Tube-shift technique, 19–22
Tubules, 128
Tumors (*See also* Carcinomas; Fibromas; Fibrosarcomas; Neoplasias; Osteosarcomas)
benign, 184
mandibular, 188
oral, 190
Twinning (*See* Gemination)

V

Ventral recumbency (*See* Positions)
Veterinary Dental Film System, 10, 12
Vibrissae follicles, 221, 238
Viewing frame, 12, 65–66
Views

caudal, 111–112, 116
distomesial, 137
dorsoventral, 54–55, 68
mesiodistal, 137, 143
rostral, 71, 111–112, 115
ventrodorsal, 54, 189, 193–194

White line (*See* Radiopacity, line)
Wiring teeth (*See* Cerclage wires)
Wry bite, 103

X-ray
   cone, 232
   definition of, 238
   positioning, 15, 29–44, 55
X-ray machines, 7, 15
   anode, 231
   cathode, 232
   control panel, 9
   dental, 7–8, 27–28
   head (*See* Tubehead)
   safety equipment, 13–14
   settings used, 24
   standard, 7, 27–28

Zygomatic arch, 21–22, 41, 68,
   158–159
   superimposition of, 55